Process Dynamics

ISBN 0-13-206889-3

90000

PROCESS DYNAMICS
Modeling, Analysis, and Simulation

B. Wayne Bequette
Rensselaer Polytechnic Institute

To join a Prentice Hall PTR Internet mailing list, point to:
http://www.prenhall.com/mail_lists/

Prentice Hall PTR
Upper Saddle River, New Jersey 07458

Library of Congress Cataloging-in-Publication Data

Bequette, B. Wayne.
    Process dynamics : modeling, analysis, and simulation / B. Wayne
    Bequette.
       p. cm.
    Includes bibliographical references and index.
    ISBN 0–13–206889–3
    1. Chemical processes. I. Title.
  TP155.7.B45   1998
  660′.284′01185—dc21   97–36053
                                          CIP

Acquisitions editor: Bernard M. Goodwin
Cover design director: Jerry Votta
Manufacturing manager: Alexis R. Heydt
Marketing manager: Miles Williams
Compositor/Production services: Pine Tree Composition, Inc.

 © 1998 by Prentice Hall PTR
Prentice-Hall, Inc.
A Simon & Schuster Company
Upper Saddle River, New Jersey 07458

Prentice Hall books are widely-used by corporations and
government agencies for training, marketing, and resale.

The publisher offers discounts on this book when ordered
in bulk quantities. For more information contact:

       Corporate Sales Department
       Phone: 800-382-3419
       Fax: 201-236-7141
       E-mail: corpsales@prenhall.com

       Or write:

       Prentice Hall PTR
       Corp. Sales Dept.
       One Lake Street
       Upper Saddle River, New Jersey 07458

Printed in the United States of America
10 9 8 7 6 5 4 3 2 1

ISBN: 0-13-206889-3

Prentice-Hall International (UK) Limited, *London*
Prentice-Hall of Australia Pty. Limited, *Sydney*
Prentice-Hall Canada Inc., *Toronto*
Prentice-Hall Hispanoamericana, S.A., *Mexico*
Prentice-Hall of India Private Limited, *New Delhi*
Prentice-Hall of Japan, Inc., *Tokyo*
Simon & Schuster Asia Pte. Ltd., *Singapore*
Editora Prentice-Hall do Brasil, Ltda., *Rio de Janeiro*

To Pat and Brendan

CONTENTS

Contents xiii

PREFACE

An understanding of the dynamic behavior of chemical processes is important from both process design and process control perspectives. It is easy to design a chemical process, based on steady-state considerations, which is practically uncontrollable when the process dynamics are considered. The current status of computational hardware and software has made it easy to interactively simulate the dynamic behavior of chemical processes.

It is common for process dynamics to be included as the introductory portion of a process control textbook, however, there are a number of limitations to this approach. Since the emphasis of most of the textbooks is on process control, there is too little space to give adequate depth to modeling, analysis, and simulation of dynamic systems. The focus tends to be on transfer function-based models that are used for control system design. The prime motivation for my textbook is then to provide a more comprehensive treatment of process dynamics, including modeling, analysis, and simulation. This textbook evolved from notes developed for a course on dynamic systems that I have been teaching at Rensselaer since 1991. We have been fortunate to have a two-semester sequence in dynamics and control, allowing more depth to the coverage of each topic.

Topics covered here that are not covered in a traditional text include nonlinear dynamics and the use of MATLAB for numerical analysis and simulation. Also, a significant portion of the text consists of review and learning modules. Each learning module provides model development, steady-state solutions, nonlinear dynamic results, linearization, state space and transfer function analysis and simulation. The motivation for this approach is to allow the student to "tie-together" all of the concepts, rather than treating them independently (and not understanding the connections between the different methods).

An important feature of this text is the use of MATLAB software. A set of m-files used in many of the examples and in the learning modules is available via the world wide web at the following locations:

http:/www/rpi.edu/-bequeb/Process_Dynamics
http:/www.mathworks.com/education/thirdparty.html

Additional learning modules will also be available at the RPI location.

A few acknowledgments are in order. A special thanks to Professor Jim Turpin at the University of Arkansas, who taught me the introductory course in process dynamics and control. His love of teaching should be an inspiration to us all. Many thanks to one of my graduate students, Lou Russo, who not only made a number of suggestions to improve the text, but also sparked an interest in many of the undergraduates that have taken the course. The task of developing a solutions manual has been carried out by Venkatesh Natarajan, Brian Aufderheide, Ramesh Rao, Vinay Prasad, and Kevin Schott.

Preliminary drafts of many chapters were developed over cappuccinos at the Daily Grind in Albany and Troy. Bass Ale served at the El Dorado in Troy promoted discussions about teaching (and other somewhat unrelated topics) with my graduate students; the effect of the many Buffalo wings is still unclear. Final revisions to the textbook were done under the influence of cappuccinos at Cafe Avanti in Chicago (while there is a lot of effort in developing interactive classroom environments at Rensselaer, my ideal study environment looks much like a coffee shop).

Teaching and learning should be dynamic processes. I would appreciate any comments and suggestions that you have on this textbook. I will use the WWW site to provide updated examples, additional problems with solutions, and suggestions for teaching and studying process dynamics.

B. Wayne Bequette

SECTION I

PROCESS MODELING

INTRODUCTION

1

This chapter provides a motivation for process modeling and the study of dynamic chemical processes. It also provides an overview of the structure of the textbook. After studying this chapter, the student should be able to answer the following questions:

- What is a process model?
- Why develop a process model?
- What is the difference between lumped parameter and distributed parameter systems?
- What numerical package forms the basis for the examples in this text?
- What are the major objectives of this textbook?

The major sections are:

1.1   Motivation
1.2   Models
1.3   Systems
1.4   Background of the Reader
1.5   How to Use This Textbook
1.6   Courses Where This Textbook Can Be Used

1.1   MOTIVATION

Robert Reich in *The Work of Nations* has classified three broad categories of employment in the United States. In order of increasing educational requirement these categories are:

routine production services, in-person services, and *symbolic-analytic* services. To directly quote from Reich:

> "Symbolic analysts solve, identify, and broker problems by manipulating *symbols*. They simplify reality into *abstract* images that can be rearranged, juggled, experimented with, communicated to other specialists, and then, eventually, transformed back into reality. The manipulations are done with analytic tools, sharpened by experience. The tools may be *mathematical algorithms,* legal arguments, financial gimmicks, *scientific principles,* psychological insights about how to persuade or to amuse, systems of induction or deduction, or any other set of techniques for doing conceptual puzzles." (italics added for emphasis)

Engineers (and particularly process engineers) are *symbolic analysts.* Process engineers use fundamental scientific principles as a basis for mathematical models that characterize the behavior of a chemical process. Symbols are used to represent physical variables, such as pressure, temperature or concentration. Input information is specified and numerical algorithms are used to solve the models (simulating a physical system). Process engineers analyze the results of these simulations to make decisions or recommendations regarding the design or operation of a process.

Most chemical engineers work with chemical manufacturing processes in one way or another. Often they are process engineers responsible for technical troubleshooting in the day-to-day operations of a particular chemical process. Some are responsible for designing feedback control systems so that process variables (such as temperature or pressure) can be maintained at desired values. Others may be responsible for redesigning a chemical process to provide more profitability. All of these responsibilities require an understanding of the time-dependent (dynamic) behavior of chemical processes.

1.2 MODELS

The primary objective of this textbook is to assist you in developing an understanding of the dynamic behavior of chemical processes. A requirement for assessing the dynamic behavior is a time-dependent mathematical model of the chemical process under consideration. Before proceeding, it is worth consulting with two different dictionaries for a definition of *model.*

Dictionary Definitions: *Model*

Model is derived from the Latin *modus,* which means *a measure.* Used as a noun, it means "a small representation of a planned or existing object" (*Webster's New World Dictionary*).

> "A mathematical or physical system, obeying certain specified conditions, whose behavior is used to understand a physical, biological, or social system to which it is analogous in some way" (*McGraw-Hill Dictionary of Scientific and Technical Terms*).

Notice that both definitions stress that a model is a representation of a system or object. In this textbook, when we use the term model, we will be referring to a mathematical model. We prefer to use the following definition for model (more specifically, a process model).

Working Definition: *Process Model*

A process model is a set of equations (including the necessary input data to solve the equations) that allows us to predict the behavior of a chemical process system.

The emphasis in this text is on the development and use of *fundamental* or first-principles models. By fundamental, we mean models that are based on known physical-chemical relationships. This includes the conservation of mass and conservation of energy,[1] as well as reaction kinetics, transport phenomena, and thermodynamic (phase equilibrium, etc.) relationships.

Another common model is the *empirical* model. An empirical model might be used if the process is too complex for a fundamental model (either in the formulation of the model, or the numerical solution of the model), or if the empirical model has satisfactory predictive capability. An example of an empirical model is a simple least squares fit of an equation to experimental data.

Generally, we would prefer to use models based on fundamental knowledge of chemical-physical relationships. Fundamental models will generally be accurate over a much larger range of conditions than empirical models. Empirical models may be useful for "interpolation" but are generally not useful for "extrapolation"; that is, an empirical model will only be useful over the range of conditions used for the "fit" of the data.

It should be noted that it is rare for a single process model to exist. A model is only an approximate representation of an actual process. The complexity of a process model will depend on the final use of the model. If only an approximate answer is needed, then a simplified model can often be used.

1.2.1   How Models Are Used

As we have noted, given a set of input data, a model is used to predict the output "response." A model can be used to solve the following types of problems:

- Marketing: If the price of a product is increased, how much will the demand decrease?
- Allocation: If we have several sources for raw materials, and several manufacturing

---

[1]Of course, the real conservation law is that of mass-energy, but we will neglect the interchange of mass and energy due to nuclear reactions.

plants, how do we distribute the raw materials among the plants, and decide what products each plant produces?

- Synthesis: What process (sequence of reactors, separation devices, etc.) can be used to manufacture a product?

- Design: What type and size of equipment is necessary to produce a product?

- Operation: What operating conditions will maximize the yield of a product?

- Control: How can a process input be manipulated to maintain a measured process output at a desired value?

- Safety: If an equipment failure occurs, what will be the impact on the operating personnel and other process equipment?

- Environmental: How long will it take to "biodegrade" soil contaminated with hazardous waste?

Many of the models cited above are based on a steady-state analysis. This book will extend the steady-state material and energy balance concepts, generally presented in an introductory textbook on chemical engineering principles, to dynamic systems (systems where the variables change with time). As an example of the increasing importance of knowledge of dynamic behavior, consider process design. In the past, chemical process design was based solely on steady-state analysis. A problem with performing only a steady-state design is that it is possible to design a process with desirable steady-state characteristics (minimal energy consumption, etc.) but which is dynamically inoperable. Hence, it is important to consider the dynamic operability characteristics of a process during the design phase. Also, batch processes that are commonly used in the pharmaceutical or specialty chemicals industries are inherently dynamic and cannot be simulated with steady-state models.

In the previous discussion we have characterized models as steady-state or dynamic. Another characterization is in terms of *lumped parameter systems* or *distributed parameter systems*. A lumped parameter system assumes that a variable of interest (temperature, for example) changes only with one independent variable (time, for example, but not space). A typical example of a lumped parameter system is a perfectly mixed (stirred) tank, where the temperature is uniform throughout the tank. A distributed parameter system has more than one independent variable; for example, temperature may vary with both spatial position and time.

EXAMPLE 1.1   A Lumped Parameter System

Consider a perfectly insulated, well-stirred tank where a hot liquid stream at 60°C is mixed with a cold liquid stream at 10°C (Figure 1.1). The well-mixed assumption means that the fluid temperature in the tank is uniform and equal to the temperature at the exit from the tank. This is an example of a lumped parameter system, since the temperature does not vary with spatial position.

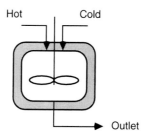

Hot   Cold

Outlet

FIGURE 1.1   Stirred tank.

Consider now the steady-state behavior of this process. If the only stream was the hot fluid, then the outlet temperature would be equal to the hot fluid temperature if the tank were perfectly insulated. Similarly, if the only stream was the cold stream, then the outlet temperature would be equal to the cold fluid temperature. A combination of the two streams yields an outlet temperature that is intermediate between the cold and hot temperatures, as shown in Figure 1.2.

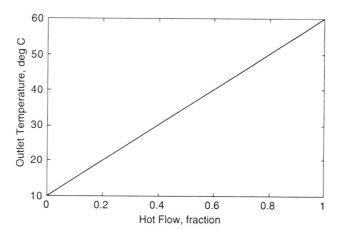

FIGURE 1.2   Relationship between hot flow and outlet temperature.

We see that there is a linear steady-state relationship between the hot flow (fraction) and the outlet temperature.

Now we consider the dynamic response to a change in the fraction of hot flow. Figure 1.3 compares outlet temperature responses for various step changes in hot flow fraction at $t = 2.5$

minutes. We see that the changes are symmetric, with the same speed of response. We will find later that these responses are indicative of a linear system.

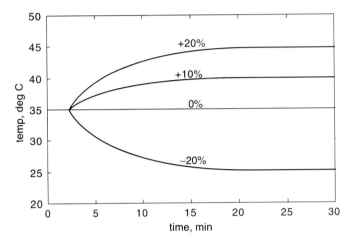

FIGURE 1.3   Response of temperature to various changes in the hot flow fraction.

EXAMPLE 1.2   A Distributed Parameter System

A simplified representation of a counterflow heat exchanger is shown in Figure 1.4. A cold water stream flows through one side of the exchanger and is heated by energy transfered from a condensing steam stream. This is a distributed parameter system because the temperature of the water stream can change with time *and* position.

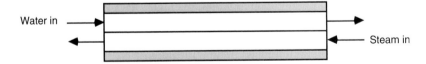

FIGURE 1.4   Counterflow heat exchanger.

The steady-state temperature profile (water temperature as a function of position) is shown in Figure 1.5. Notice that rate of change of the water temperature with respect to distance decreases as the water temperature approaches the steam temperature (100°C). This is because the temperature gradient for heat transfer decreases as the water temperature increases. The outlet water temperature as a function of inlet water temperature is shown in Figure 1.6.

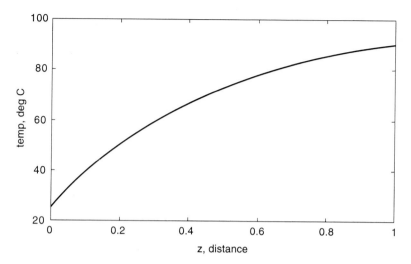

FIGURE 1.5   Water temperature as a function of position.

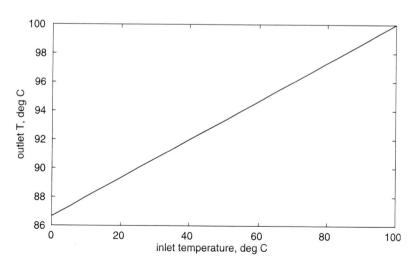

FIGURE 1.6   Outlet water temperature as a function of inlet water temperature.

Mathematical models consist of the following types of equations (including combinations)

- Algebraic equations
- Ordinary differential equations
- Partial differential equations

The emphasis in this textbook is on developing models that consist of ordinary differential equations. These equations generally result from macroscopic balances around processes, with an assumption of a perfectly mixed system. To find the steady-state solution of a set of ordinary differential equations, we must solve a set of algebraic equations. Partial differential equation models result from microscopic balances and are not covered in this textbook. One of the main techniques for solving partial differential equations is based on converting a partial differential equation to a set of ordinary differential equations. Techniques developed in this textbook can then be used to solve these problems.

1.3 SYSTEMS

We have been using the term *system* very loosely. Consider the following definition.

Definition: *System*

A combination of several pieces of equipment integrated to perform a specific function; thus a fire (artillery) control system may include a tracking radar, computer, and gun (*McGraw-Hill Dictionary of Scientific and Technical Terms*).

The example in the definition presented is of interest to electrical, aeronautical, and military engineers. For our purposes, a system will be composed of chemical unit operations, such as chemical reactors, heat exchangers, and separation devices, which are used to produce a chemical product. Indeed, we will often consider a single unit operation to be a system composed of inputs, states (to be defined later) and outputs. A series of modules in Section V of this textbook cover the behavior of a number of specific unit operations.

1.3.1 Simulation

One of the goals of this textbook is to develop numerical analysis techniques that allow us to "simulate" the behavior of a chemical process. Typically, steady-state simulation of a lumped parameter system involves the solution of algebraic equations, while dynamic simulation involves the solution of ordinary differential equations. We must be careful when using computer simulation. First of all, we must be able to say, Do the results of this simulation make sense? Common sense and "back of the envelope" calculations will tell us if the numerical results are in the ballpark.

1.3.2   Linear Systems Analysis

The mathematical tools that are used to study linear dynamic systems problems are known as *linear systems analysis* techniques. Traditionally, systems analysis techniques have been based on linear systems theory. Two basic approaches are typically used: (i) Laplace transforms are used to analyze the behavior of a single, linear, *n*th order ordinary differential equation, and (ii) state space techniques (based on the linear algebra techniques of eigenvalue and eigenvector analysis) are used to analyze the behavior of multiple first-order linear ordinary differential equations. If a system of ordinary differential equations is nonlinear, they can be linearized at a desired steady-state operating point.

1.3.3   A Broader View of Analysis

In this textbook we use *analysis* in a broader context than linear systems analysis that may be applied to a model with a specific value for the parameters. Analysis means seeking a deeper understanding of a process than simply performing a simulation or solving a set of equations for a particular set of parameters and input values. Often we want to understand how the response of system variable (temperature, for example) changes when a parameter (e.g., heat transfer coefficient) or input (flowrate or inlet temperature) changes. Rather than trying to obtain the understanding of the possible types of behavior by merely running many simulations (varying parameters, etc.), we must decide which parameters (or inputs or initial conditions) are likely to vary, and use analysis techniques to determine if a qualitative change of behavior (number of solutions or stability of a solution) can occur.

   This qualitative change is illustrated in Figure 1.7 below, which shows possible steady-state behavior for a jacketed chemical reactor. In Figure 1.7a there is a monotonic

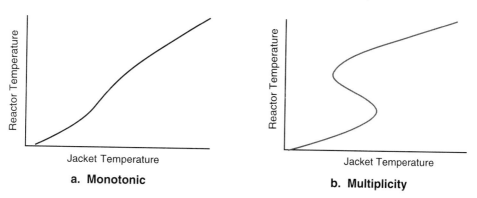

FIGURE   1.7   Two qualitatively different types of input/output behavior. Steady-state reactor temperature as a function of steady-state jacket temperature.

relationship between jacket temperature and reactor temperature, that is, as the steady-state jacket temperature increases, the steady-state reactor temperature increases. Figure 1.7b illustrates behavior known as output multiplicity, that is, there is a region of steady-state jacket temperatures where a single jacket temperature can yield three possible reactor temperatures. In Chapter 15 (and module 9) we show how to vary a reactor design parameter to change from one type of behavior to another.

Engineering problem solving can be a combination of art and science. The complexity and accuracy of a solution will depend on the information available or what is desired in the final solution. If you are simply performing a rough (back of the envelope) cost estimate for a process design, perhaps a simple steady-state material and energy balance will suffice. On the other hand, if an optimum design integrating several unit operations is required, then a more complex solution will be involved.

1.4 BACKGROUND OF THE READER

It is assumed that the reader of this textbook has a sophomore- or junior-level chemical engineering background. In addition to the standard introductory chemistry, physics, and mathematics (including differential equations) courses, the student has taken an introduction to chemical engineering (material and energy balances, reaction stoichiometry) course.

This textbook can also be used by an engineer in industry who needs to develop dynamic models to perform studies to improve a process or design controllers. Although a few years may have lapsed since the engineer took a differential equations course, the review provided in this text should be sufficient for the development and solution of models based on differential equations. Also, this textbook can serve as review material for first-year graduate student who is interested in process modeling, systems analysis, or numerical methods.

1.5 HOW TO USE THIS TEXTBOOK

The ultimate objective of this textbook is to be able to model, simulate, and (more importantly) understand the dynamic behavior of chemical processes.

1.5.1 Sections

In Section I (Chapters 1 and 2) we show how to develop dynamic models for simple chemical processes. Numerical techniques for solving algebraic and differential equations are covered in Section II (Chapters 3 and 4). Much of the textbook is based on linear system analysis techniques, which are presented in Section III (Chapters 5 through 12). Nonlinear analysis techniques are presented in Section IV (Chapters 13–17). Section V (Modules 1 through 10) consists of a number of learning modules to reinforce the concepts discussed in Sections I through IV.

1.5.2 Numerical Solutions

It is much easier to learn a new topic by "doing" rather than simply reading about it. To understand the dynamic behavior of chemical processes, one needs to be able to solve differential equations and plot response curves. We have used the MATLAB numerical analysis package to solve equations in this text. A MATLAB learning module is in the set of modules in Section V (Module 1) for readers who are not familiar with or need to be reintroduced to MATLAB. MATLAB routines are detailed within the chapter that they are used. Many of the examples have MATLAB m-files associated with them. It is recommended that the reader modify these m-files to understand the effect of parameter changes on the numerical solution.

1.5.3 Motivating Examples and Modules

This textbook contains many process examples. Often, new techniques are introduced in the examples. You are encouraged to work through each example to understand how a particular technique can be applied.

There is a limit to the complexity of an example that can be used when introducing a new technique; the examples presented in the chapters tend to be short and illustrate one or two numerical techniques. There is a set of *modules* of models, in Section V, that treats process examples in much more detail. The objective of these modules is to provide a more complete treatment of modeling and simulation of a specific process. In each process modeling module, a number of the techniques introduced in various chapters of the text are applied to the problem at hand.

1.6 COURSES WHERE THIS TEXTBOOK CAN BE USED

This textbook is based on a required course that I have taught to chemical and environmental engineering juniors at Rensselaer since 1991. This *dynamic systems* course (originally titled *lumped parameter systems*) is a prerequisite to a required course on chemical process control, normally taken in the second semester of the junior year. This textbook can be used for the first term of a two-term sequence in dynamics and control. I have not treated process control in this text because I feel that there is a need for more in-depth coverage of process dynamics than is covered in most process control textbooks.

This textbook can also be used in courses such as process modeling or numerical methods for chemical engineers. Although directed towards undergraduates, this text can also be used in a first-year graduate course on process modeling or process dynamics; in this case, much of the focus would be on Section IV (nonlinear analysis) and in-depth studies of the modules.

SUMMARY

At this point, the reader should be able to define or characterize the following

- Process model
- Lumped or distributed parameter system
- Analysis
- Simulation

FURTHER READING

The textbooks listed below are nice introductions to material and energy balances. The Russell and Denn book also provides an excellent introduction to models of dynamic systems.

Felder, R.M., & R. Rousseau. (1986). *Elementary Principles of Chemical Processes*, 2nd ed. New York: Wiley.

Himmelblau, D.M. (1996). *Basic Principles and Calculations in Chemical Engineering*, 6th ed. Upper Saddle River, NJ: Prentice-Hall.

Russell, T.R.F., & M.M. Denn. (1971). *Introduction to Chemical Engineering Analysis*, New York: Wiley.

A comprehensive tutorial and reference for MATLAB is provided by Hanselman and Littlefield. The books by Etter provide many excellent examples using MATLAB to solve engineering problems.

Hanselman, D., & B. Littlefield. (1996). *Mastering MATLAB*. Upper Saddle River, NJ: Prentice-Hall.

Etter, D.M. (1993). *Engineering Problem Solving with MATLAB*. Upper Saddle River, NJ: Prentice-Hall.

Etter, D.M. (1996). *Introduction to MATLAB for Engineers and Scientists*. Upper Saddle River, NJ: Prentice-Hall.

The following book by Denn is a graduate-level text that discusses the more philosophical issues involved in process modeling.

Denn, M.M. (1986). *Process Modeling*. New York: Longman.

An undergraduate control textbook with significant modeling and simulation is by Luyben. Numerous FORTRAN examples are presented.

Luyben, W.L. (1990). *Process Modeling, Simulation and Control for Chemical Engineers*, 2nd ed. New York: McGraw-Hill.

A number of control textbooks contain a limited amount of modeling. Examples include:

Seborg, D.E., T.F. Edgar, & D.A. Mellichamp. (1989). *Process Dynamics and Control*. New York: Wiley.

Stephanopoulos, G. (1984). *Chemical Process Control: An Introduction to Theory and Practice*. Englewood Cliffs, NJ: Prentice-Hall.

The following book by Rameriz is more of an advanced undergraduate/first-year graduate student text on numerical methods to solve chemical engineering problems. The emphasis is on FORTRAN subroutines to be used with the IMSL numerical package.

Rameriz, W.F. (1989). *Computational Methods for Process Simulation*. Boston: Butterworths.

Issues in process modeling are discussed by Himmelblau in Chapter 3 of the following book:

Bisio, A., & R.L. Kabel. (1985). *Scaleup of Chemical Processes*. New York: Wiley.

The following book by Reich provides an excellent perspective on the global economy and role played by U.S. workers

Reich, R.B. (1991). *The Work of Nations*. New York: Vintage Books.

STUDENT EXERCISES

1. Review the matrix operations module (Section V).
2. Work through the MATLAB module (Section V).
3. Consider example 2. If there is a sudden increase in steam pressure (and therefore, temperature) sketch the expected cold-side temperature profiles at 0, 25, 50, 75, and 100% of the distance through the heat exchanger.

PROCESS MODELING   2

In this chapter, a methodology for developing dynamic models of chemical processes is presented. After studying this chapter, the student should be able to:

- Write balance equations using the integral or instantaneous methods.
- Incorporate appropriate constitutive relationships into the equations.
- Determine the state, input and output variables, and parameters for a particular model (set of equations).
- Determine the necessary information to solve a system of dynamic equations.
- Define dimensionless variables and parameters to "scale" equations.

The major sections are:

2.1 BACKGROUND

Many reasons for developing process models were given in Chapter 1. Improving or understanding chemical process operation is a major overall objective for developing a dynamic process model. These models are often used for (i) operator training, (ii) process design, (iii) safety system analysis or design, or (iv) control system design.

Operator Training.   The people responsible for the operation of a chemical manufacturing process are known as *process operators*. A dynamic process model can be used to perform simulations to train process operators, in the same fashion that flight simulators are used to train airplane pilots. Process operators can learn the proper response to upset conditions, before having to experience them on the actual process.

Process Design.   A dynamic process model can be used to properly design chemical process equipment for a desired production rate. For example, a model of a batch chemical reactor can be used to determine the appropriate size of the reactor to produce a certain product at a desired rate.

Safety.   Dynamic process models can also be used to design safety systems. For example, they can be used to determine how long it will take after a valve fails for a system to reach a certain pressure.

Control System Design.   Feedback control systems are used to maintain process variables at desirable values. For example, a control system may measure a product temperature (an output) and adjust the steam flowrate (an input) to maintain that desired temperature. For complex systems, particularly those with many inputs and outputs, it is necessary to base the control system design on a process model. Also, before a complex control system is implemented on a process, it is normally tested by simulating the expected performance using computer simulation.

2.2 BALANCE EQUATIONS

The emphasis in an introductory material and energy balances textbook is on *steady-state* balance equations that have the following form:

$$\begin{bmatrix} \text{mass or energy} \\ \text{entering} \\ \text{a system} \end{bmatrix} - \begin{bmatrix} \text{mass or energy} \\ \text{leaving} \\ \text{a system} \end{bmatrix} = 0 \qquad (2.1)$$

Equation (2.1) is deceptively simple because there may be many ins and outs, particularly for component balances. The in and out terms would then include the generation and conversion of species by chemical reaction, respectively. In this text, we are interested in dynamic balances that have the form:

$$
\begin{bmatrix} \text{rate of mass or energy} \\ \text{accumulation in} \\ \text{a system} \end{bmatrix} = \begin{bmatrix} \text{rate of mass or} \\ \text{energy entering} \\ \text{a system} \end{bmatrix} - \begin{bmatrix} \text{rate of mass or} \\ \text{energy leaving} \\ \text{a system} \end{bmatrix} \quad (2.2)
$$

The rate of mass accumulation in a system has the form dM/dt where M is the total mass in the system. Similarly, the rate of energy accumulation has the form dE/dt where E is the total energy in a system. If N_i is used to represent the moles of component i in a system, then dN_i/dt represents the molar rate of accumulation of component i in the system.

When solving a problem, it is important to specify what is meant by system. In some cases the system may be microscopic in nature (a differential element, for example), while in other cases it may be macroscopic in nature (the liquid content of a mixing tank, for example). Also, when developing a dynamic model, we can take one of two general viewpoints. One viewpoint is based on an *integral* balance, while the other is based on an *instantaneous* balance. Integral balances are particularly useful when developing models for distributed parameter systems, which result in partial differential equations; the focus in this text is on ordinary differential equation-based models. Another viewpoint is the instantaneous balance where the time rate of change is written directly.

2.2.1 Integral Balances

An integral balance is developed by viewing a system at two different snapshots in time. Consider a finite time interval, Δt, and perform material balance over that time interval

$$
\begin{bmatrix} \text{mass or energy} \\ \text{inside the system} \\ \text{at } t + \Delta t \end{bmatrix} - \begin{bmatrix} \text{mass or energy} \\ \text{inside the system} \\ \text{at } t \end{bmatrix} =
$$

$$
\begin{bmatrix} \text{mass or energy} \\ \text{entering the system} \\ \text{from } t \text{ to } t + \Delta t \end{bmatrix} - \begin{bmatrix} \text{mass or energy} \\ \text{leaving the system} \\ \text{from } t \text{ to } t + \Delta t \end{bmatrix} \quad (2.3)
$$

The mean-value theorems of integral and differential calculus are then used to reduce the equations to differential equations.

For example, consider the system shown in Figure 2.1 below, where one boundary represents the mass in the system at time t, while the other boundary represents the mass in the system at $t + \Delta t$.

An integral balance on the total mass in the system is written in the form:

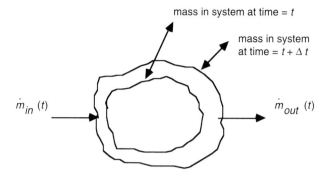

mass in system at time = t

mass in system at time = $t + \Delta t$

$\dot{m}_{in}\,(t)$

$\dot{m}_{out}\,(t)$

FIGURE 2.1 Conceptual material balance problem.

$$
\begin{bmatrix} \text{mass contained} \\ \text{in the system} \\ \text{at } t + \Delta t \end{bmatrix} - \begin{bmatrix} \text{mass contained} \\ \text{in the system} \\ \text{at } t \end{bmatrix} = \begin{bmatrix} \text{mass entering} \\ \text{the system} \\ \text{from } t \text{ to } t + \Delta t \end{bmatrix} - \begin{bmatrix} \text{mass leaving} \\ \text{the system} \\ \text{from } t \text{ to } t + \Delta t \end{bmatrix}
$$

Mathematically, this is written:

$$
M\big|_{t+\Delta t} - M\big|_t = \int_t^{t+\Delta t} \dot{m}_{in}\, dt - \int_t^{t+\Delta t} \dot{m}_{out}\, dt
$$

or

$$
M\big|_{t+\Delta t} - M\big|_t = \int_t^{t+\Delta t} (\dot{m}_{in} - \dot{m}_{out})\, dt \tag{2.4}
$$

where M represents the total mass in the system, while $\dot{m}_{in}$ and $\dot{m}_{out}$ represent the mass rates entering and leaving the system, respectively. We can write the righthand side of (2.4), using the mean value theorem of *integral* calculus, as:

$$
\int_t^{t+\Delta t} (\dot{m}_{in} - \dot{m}_{out})\, dt = (\dot{m}_{in} - \dot{m}_{out})\big|_{t+\alpha\Delta t}\Delta t
$$

where $0 < \alpha < 1$. Equation (2.4) can now be written:

$$
M\big|_{t+\Delta t} - M\big|_t = (\dot{m}_{in} - \dot{m}_{out})\big|_{t+\alpha\Delta t}\Delta t
$$

dividing by Δt,

$$
\frac{M\big|_{t+\Delta t} - M\big|_t}{\Delta t} = (\dot{m}_{in} - \dot{m}_{out})\big|_{t+\alpha\Delta t}
$$

and using the mean value theorem of *differential* calculus $(0 < \beta < 1)$ for the lefthand side,

$$
\frac{M\big|_{t+\Delta t} - M\big|_t}{\Delta t} = \frac{dM}{dt}\Big|_{t+\beta\Delta t}
$$

which yields

$$\frac{dM}{dt}\Big|_{t+\beta\Delta t} = (\dot{m}_{in} - \dot{m}_{out})\Big|_{t+\alpha\Delta t}$$

Taking the limit as Δt goes to zero, we find

$$\frac{dM}{dt} = \dot{m}_{in} - \dot{m}_{out} \tag{2.5}$$

and representing the total mass as $M = V\rho$, $\dot{m}_{in}$ as $F_{in}\rho_{in}$ and $\dot{m}_{out}$ as $F_{out}\rho$, where ρ is the mass density (mass/volume) and F is a volumetric flowrate (volume/time) we obtain the equation:

$$\frac{dV\rho}{dt} = F_{in}\rho_{in} - F_{out}\rho \tag{2.6}$$

Note that we have assumed that the system is perfectly mixed, so that the density of material leaving the system is equal to the density of material in the system.

2.2.2   Instantaneous Balances

Here we write the dynamic balance equations directly, based on an instantaneous rate-of-change:

$$\begin{bmatrix} \text{the rate of} \\ \text{accumulation of} \\ \text{mass in the system} \end{bmatrix} = \begin{bmatrix} \text{rate of} \\ \text{mass entering} \\ \text{the system} \end{bmatrix} - \begin{bmatrix} \text{rate of} \\ \text{mass leaving} \\ \text{the system} \end{bmatrix} \tag{2.7}$$

which can be written directly as,

$$\frac{dM}{dt} = \dot{m}_{in} - \dot{m}_{out} \tag{2.8}$$

or

$$\frac{dV\rho}{dt} = F_{in}\rho_{in} - F_{out}\rho \tag{2.9}$$

which is the same result obtained using an integral balance. Although the integral balance takes longer to arrive at the same result as the instantaneous balance method, the integral balance method is probably clearer when developing distributed parameter (partial differential equation-based) models. An example is shown in Section 2.6.

     Section 2.3 covers material balances and Section 2.5 covers material and energy balances. Section 2.4 discusses constitutive relationships.

2.3   MATERIAL BALANCES

The simplest modeling problems consist of material balances. In this section we use several process examples to illustrate the modeling techniques used.

EXAMPLE 2.1   Liquid Surge Tank

Surge tanks are often used to "smooth" flowrate fluctuations in liquid streams flowing between chemical processes. Consider a liquid surge tank with one inlet (flowing from process I) and one outlet stream (flowing to process II) (Figure 2.2). Assume that the density is constant. Find how the volume of the tank varies as a function of time, if the inlet and outlet flowrates vary. List the *state variables*, *parameters*, as well as the *input* and *output variables*. Give the necessary information to complete the quantitative solution to this problem.

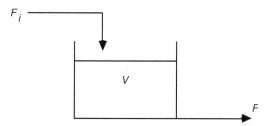

FIGURE 2.2   Liquid surge tank.

The system is the liquid in the tank, the liquid surface is the top boundary of the system. The following notation is used in the modeling equations:

F_i = inlet volumetric flowrate (volume/time)

F = outlet volumetric flowrate

V = volume of liquid in the tank

ρ = liquid density (mass/volume)

Integral Method

Consider a finite time interval, Δt. Performing a material balance over that time interval,

$$\begin{bmatrix} \text{mass of water} \\ \text{inside the tank} \\ \text{at } t + \Delta t \end{bmatrix} - \begin{bmatrix} \text{mass of water} \\ \text{inside the tank} \\ \text{at } t \end{bmatrix} =$$

$$\begin{bmatrix} \text{mass of water} \\ \text{entering tank} \\ \text{from } t \text{ to } t + \Delta t \end{bmatrix} - \begin{bmatrix} \text{mass of water} \\ \text{leaving tank} \\ \text{from } t \text{ to } t + \Delta t \end{bmatrix}$$

which we can write mathematically as:

$$V\rho|_{t+\Delta t} - V\rho|_t = \int\limits_t^{t+\Delta t} F_i\rho\, dt - \int\limits_t^{t+\Delta t} F\rho\, dt \qquad (2.10)$$

Bringing the righthand side terms under the same integral

$$V\rho|_{t+\Delta t} - V\rho|_t = \int_t^{t+\Delta t} (F_i\rho - F\rho)\, dt \tag{2.11}$$

We can use the mean value theorem of integral calculus to write the righthand side of (2.11) (where $0 \le \alpha \le 1$) as:

$$\int_t^{t+\Delta t} (F_i\rho - F\rho)\, dt = (F_i\rho - F\rho)|_{t+\alpha\Delta t}\Delta t \tag{2.12}$$

Substituting (2.12) into (2.11)

$$V\rho|_{t+\Delta t} - V\rho|_t = (F_i\rho - F\rho)|_{t+\alpha\Delta t}\Delta t \tag{2.13}$$

Dividing by Δt, we obtain

$$\frac{V\rho|_{t+\Delta t} - V\rho|_t}{\Delta t} = (F_i\rho - F\rho)|_{t+\alpha\Delta t} \tag{2.14}$$

and using the mean value theorem of differential calculus, as $\Delta t \to 0$

$$\frac{dV\rho}{dt} = F_i\rho - F\rho \tag{2.15}$$

Instantaneous Method

Here we write the balance equations based on an instantaneous rate-of-change:

$$\begin{bmatrix} \text{the rate of change of} \\ \text{mass of water in tank} \end{bmatrix} = \begin{bmatrix} \text{mass flowrate of} \\ \text{water into tank} \end{bmatrix} - \begin{bmatrix} \text{mass flowrate of} \\ \text{water out of tank} \end{bmatrix}$$

The total mass of water in the tank is $V\rho$, the rate of change is $dV\rho/dt$, and the density of the outlet stream is equal to the tank contents:

$$\frac{dV\rho}{dt} = F_i\rho - F\rho \tag{2.16}$$

which is exactly what we derived using the integral method. Given the same set of assumptions the two methods should yield the same model. You should use the approach (integral or instantaneous) that makes the most sense to you. In this text we generally use the instantaneous approach since it requires the fewest number of steps.

Notice the implicit assumption that the density of water in the tank does not depend upon position (the perfect mixing assumption). This assumption allows an ordinary differential equation (ODE) formulation. We refer to any system that can be modeled by ODEs as *lumped parameter systems*. Also notice that the outlet stream density must be equal to the density of water in the tank. This knowledge also allows us to say that the density terms in (2.16) are equal. This equation is then reduced to[1]

---

[1]It might be tempting to the reader to begin to directly write "volume balance" expressions that look similar to (2.17). We wish to make it clear that there is no such thing as a volume balance and (2.17) is only correct because of the constant density assumption. It is a good idea to always write a mass balance expression, such as (2.16), before making assumptions about the fluid density, which may lead to (2.17).

$$\frac{dV}{dt} = F_i - F \tag{2.17}$$

Equation (2.17) is a linear ordinary differential equation (ODE), which is trivial to solve if we know the inlet and outlet flowrates as a function of time, and if we know an initial condition for the volume in the tank. In equation (2.17) we refer to V as a *state variable*, and F_i and F as *input variables* (even though F is an outlet stream flowrate). If density remained in the equation, we would refer to it as a *parameter*.

In order to solve this problem we must specify the inputs $F_i(t)$ and $F(t)$ and the initial condition $V(0)$.

Example 2.1 provides an introduction to the notion of states, inputs, and parameters. This example illustrates how an overall material balance is used to find how the volume of a liquid phase system changes with time. It may be desirable to have tank height, h, rather than tank volume as the state variable. If we assume a constant tank cross-sectional area, A, we can express the tank volume as $V = Ah$ and the modeling equation as

$$\frac{dh}{dt} = \frac{F_i}{A} - \frac{F}{A} \tag{2.18}$$

If we also know that the flowrate out of the tank is proportional to the square root of the height of liquid in the tank, we can use the relationship (see student exercise 21)

$$F = \beta\sqrt{h} \tag{2.19}$$

where β is a flow coefficient, to find

$$\frac{dh}{dt} = -\frac{\beta\sqrt{h}}{A} + \frac{F_i}{A} \tag{2.20}$$

For this model we refer to h as the *state* variable, inlet flowrate (F_i) as the *input* variable and β and A as *parameters*.

Notice that a single system (in this case, the liquid surge tank) can have slightly different modeling equations and variables, depending on assumptions and the objectives used when developing the model.

EXAMPLE 2.2   An Isothermal Chemical Reactor

Assume that two chemical species, A and B, are in a solvent feedstream entering a liquid-phase chemical reactor that is maintained at a constant temperature (Figure 2.3). The two species react irreversibly to form a third species, P. Find the reactor concentration of each species as a function of time.

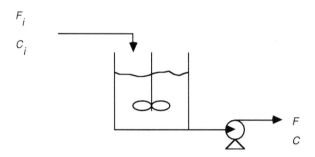

FIGURE 2.3   Isothermal chemical reactor.

Overall Material Balance

The overall mass balance is (since the tank is perfectly mixed)

$$\frac{dV\rho}{dt} = F_i \rho_i - F \rho \qquad (2.21)$$

Assumption: The liquid phase density, ρ, is not a function of concentration. The tank (and outlet) density is then equal to the inlet density, so:

$$\rho_i = \rho \qquad (2.22)$$

and we can write (2.21) as:

$$\frac{dV}{dt} = F_i - F \qquad (2.23)$$

Component Material Balances

It is convenient to work in molar units when writing component balances, particularly if chemical reactions are involved. Let C_A, C_B, and C_P represent the molar concentrations of A, B, and P (moles/volume).

Assume that the stoichiometric equation for this reaction is

$$A + 2B \ \ ---> P$$

The component material balance equations are (assuming no component P is in the feed to the reactor):

$$\frac{dVC_A}{dt} = F_i C_{Ai} - FC_A + Vr_A \qquad (2.24)$$

$$\frac{dVC_B}{dt} = F_i C_{Bi} - FC_B + Vr_B \qquad (2.25)$$

$$\frac{dVC_P}{dt} = -FC_P + Vr_P \qquad (2.26)$$

Where r_A, r_B, and r_P represent the rate of *generation* of species A, B, and P per unit volume, and C_{Ai} and C_{Bi} represent the inlet concentrations of species A and B. Assume that the rate of reaction of A per unit volume is second-order and a function of the concentration of both A and B. The reaction rate can be written

$$r_A = -kC_AC_B \qquad (2.27)$$

where k is the reaction rate constant and the minus sign indicates that A is consumed in the reaction. Each mole of A reacts with two moles of B (from the stoichiometric equation) and produces one mole of P, so the rates of generation of B and P (per unit volume) are:

$$r_B = -2\,kC_AC_B \qquad (2.28)$$

$$r_P = kC_AC_B \qquad (2.29)$$

Expanding the lefthand side of (2.24),

$$\frac{dVC_A}{dt} = V\frac{dC_A}{dt} + C_A\frac{dV}{dt} \qquad (2.30)$$

combining (2.23), (2.24), (2.27), and (2.30) we find:

$$\frac{dC_A}{dt} = \frac{F_i}{V}(C_{Ai} - C_A) - kC_AC_B \qquad (2.31)$$

Similarly, the concentrations of B and P can be written

$$\frac{dC_B}{dt} = \frac{F_i}{V}(C_{Bi} - C_B) - 2\,kC_AC_B \qquad (2.32)$$

$$\frac{dC_P}{dt} = -\frac{F_i}{V}C_P + kC_AC_B \qquad (2.33)$$

This model consists of four differential equations (2.23, 2.31, 2.32, 2.33) and, therefore, four state variables (V, C_A, C_B, and C_P). To solve these equations, we must specify the initial conditions ($V(0)$, $C_A(0)$, $C_B(0)$, and $C_P(0)$), the inputs (F_i, C_{Ai}, and C_{Bi}) as a function of time, and the parameter (k).

2.3.1 Simplifying Assumptions

The reactor model presented in Example 2.2 has four differential equations. Often other simplifying assumptions are made to reduce the number of differential equations, to make them easier to analyze and faster to solve. For example, assuming a constant volume ($dV/dt = 0$) reduces the number of equations by one. Also, it is common to feed an excess of one reactant to obtain nearly complete conversion of another reactant. If species B is maintained in a large excess, then C_B is nearly constant. The reaction rate equation can then be expressed:

$$r_A = -k\,C_AC_B \approx -k_1C_A \qquad (2.34)$$

where

$$k_1 = kC_B \tag{2.35}$$

The resulting differential equations are (since we assumed dV/dt and $dC_B/dt = 0$)

$$\frac{dC_A}{dt} = \frac{F_i}{V}(C_{Ai} - C_A) - k_1 C_A \tag{2.36}$$

$$\frac{dC_P}{dt} = \frac{F_i}{V}(C_{Pi} - C_P) + k_1 C_A \tag{2.37}$$

Notice that if we only desire to know the concentration of species A we only need to solve one differential equation, since the concentration of A is not dependent on the concentration of P.

EXAMPLE 2.3   Gas Surge Drum

Surge drums are often used as intermediate storage capacity for gas streams that are transferred between chemical process units. Consider a drum depicted in Figure 2.4, where q_i is the inlet molar flowrate and q is the outlet molar flowrate. Here we develop a model that describes how the pressure in the tank varies with time.

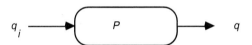

FIGURE 2.4   Gas surge drum.

Let V = volume of the drum and $\hat{V}$ = molar volume of the gas (volume/mole). The total amount of gas (moles) in the tank is then $V/\hat{V}$.

Assumption: The pressure-volume relationship is characterized by the ideal gas law, so

$$P\hat{V} = RT \tag{2.38}$$

where P is pressure, T is temperature (absolute scale), and R is the ideal gas constant. Equation (2.38) can be written

$$\frac{1}{\hat{V}} = \frac{P}{RT} \tag{2.39}$$

and, therefore, the total amount of gas in the tank is

$$\frac{V}{\hat{V}} = \frac{PV}{RT} = \text{total amount (moles) of gas in the tank} \tag{2.40}$$

the rate of accumlation of gas is then $d(PV/RT)/dt$. Assume that T is constant; since V and R are also constant, then the molar rate of accumulation of gas in the tank is:

$$\frac{V}{RT}\frac{dP}{dt} = q_i - q \tag{2.41}$$

where q_i is the molar rate of gas entering the drum and q is the molar rate of gas leaving the drum. Equation (2.41) can be written

$$\frac{dP}{dt} = \frac{RT}{V}(q_i - q) \tag{2.42}$$

To solve this equation for the state variable P, we must know the *inputs* q_i and q, the parameters R, T, and V, and the initial condition $P(0)$. Once again, although q is the molar rate *out* of the drum, we consider it an input in terms of solving the model.

2.4 CONSTITUTIVE RELATIONSHIPS

Examples 2.2 and 2.3 required more than simple material balances to define the modeling equations. These required relationships are known as *constitutive* equations; several examples of constitutive equations are shown in this section.

2.4.1 Gas Law

Process systems containing a gas will normally need a gas-law expression in the model. The ideal gas law is commonly used to relate molar volume, pressure, and temperature:

$$P\hat{V} = RT \tag{2.43}$$

The van der Waal's $P\hat{V}T$ relationship contains two parameters (a and b) that are system-specific:

$$\left(P + \frac{a}{\hat{V}^2}\right)(\hat{V} - b) = RT \tag{2.44}$$

For other gas laws, see a thermodynamics text such as Smith, Van Ness, and Abbott (1996).

2.4.2 Chemical Reactions

The rate of reaction per unit volume (mol/volume*time) is usually a function of the concentration of the reacting species. For example, consider the reaction $A + 2B \longrightarrow C + 3D$. If the rate of the reaction of A is first-order in both A and B, we use the following expression:

$$r_A = -k\,C_A C_B \tag{2.45}$$

where

$\quad r_A$ is the rate of reaction of A (mol A/volume*time)
$\quad k$ is the reaction rate constant (constant for a given temperature)
$\quad C_A$ is the concentration of A (mol A/volume)
$\quad C_B$ is the concentration of B (mol B/volume)

Reaction rates are normally expressed in terms of generation of a species. The minus sign indicates that A is consumed in the reaction above. It is good practice to associate the units with all parameters in a model. For consistency in the units for r_A, we find that k has units of (vol/mol $B*$ time). Notice that 2 mols of B react for each mol of A. Then we can write

$$r_B = 2r_A \quad = -2k\, C_A C_B$$
$$r_C = -r_A \quad = k\, C_A C_B$$
$$r_D = -3r_A = 3\,k\, C_A C_B$$

Usually, the reaction rate coefficient is a function of temperature. The most commonly used representation is the Arrhenius rate law

$$k(T) = A \exp(-E/RT) \qquad (2.46)$$

where

   $k(T)$ = reaction rate constant, as a function of temperature
   A = frequency factor or preexponential factor (same units as k)
   E = activation energy (cal/gmol)
   R = ideal gas constant (1.987 cal/gmol K, or another set of consistent units)
   T = absolute temperature (deg K or deg R)

The frequency factor and activation energy can be estimated data of the reaction constant as a function of reaction temperature. Taking the natural log of the Arrhenius rate law, we find:

$$\ln k = \ln A - \frac{E}{R}\left(\frac{1}{T}\right) \qquad (2.47)$$

and we see that A and E can be found from the slope and intercept of a plot of ($\ln k$) versus ($1/T$).

2.4.3   Equilibrium Relationships

The relationship between the liquid and vapor phase compositions of component i, when the phases are in equilibrium, can be represented by:

$$y_i = K_i\, x_i \qquad (2.48)$$

where

   x_i = liquid phase mole fraction of component i
   y_i = vapor phase mole fraction of component i
   K_i = vapor/liquid equilibrium constant for component i

   The equilibrium constant is a function of composition and temperature. Often, we will see a *constant relative volatility* assumption made to simplify vapor/liquid equilibrium models.

In a binary system, the relationship between the vapor and liquid phases for the light component often used is:

$$y = \frac{\alpha x}{1 + (\alpha - 1)x} \tag{2.49}$$

x = liquid phase mole fraction of light component
y = vapor phase mole fraction of light component
α = relative volatility ($\alpha > 1$)

2.4.4   Heat Transfer

The rate of heat transfer through a vessel wall separating two fluids (a jacketed reactor, for example) can be described by

$$Q = UA\Delta T \tag{2.50}$$

where

Q = rate of heat transfered from the hot fluid to the cold fluid
U = overall heat transfer coefficient
A = area for heat transfer
ΔT = difference between hot and cold fluid temperatures

The heat transfer coefficient is often estimated from experimental data. At the design stage it can be estimated from correlations; it is a function of fluid properties and velocities.

2.4.5   Flow-through Valves

The flow-through valves are often described by the following relationship:

$$F = C_v f(x) \sqrt{\frac{\Delta P_v}{\text{s.g.}}} \tag{2.51}$$

where

F = volumetric flowrate
C_v = valve coefficient
x = fraction of valve opening
ΔP_v = pressure drop across the valve
s.g. = specific gravity of the fluid
$f(x)$ = the flow characteristic (varies from 0 to 1, as a function of x)

Three common valve characteristics are (i) linear, (ii) equal-percentage, and (iii) quick-opening.

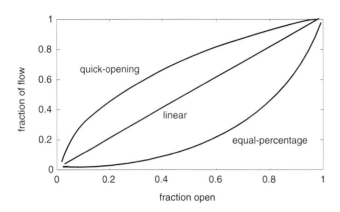

FIGURE 2.5 Flow characteristics of control valves. $\alpha = 50$ for equal-percentage valve.

For a linear valve

$$f(x) = x$$

For an equal-percentage valve

$$f(x) = \alpha^{x-1}$$

For a quick-opening valve

$$f(x) = \sqrt{x}$$

The three characteristics are compared in Figure 2.5.

Notice that for the quick-opening valve, the sensitivity of flow to valve position (fraction open) is high at low openings and low at high openings; the opposite is true for an equal-percentage valve. The sensitivity of a linear valve does not change as a function of valve position. The equal-percentage valve is commonly used in chemical processes, because of desirable characteristics when installed in piping systems where a significant piping pressure drop occurs at high flowrates. Knowledge of these characteristics will be important when developing feedback control systems.

2.5 MATERIAL AND ENERGY BALANCES

Section 2.3 covered models that consist of material balances only. These are useful if thermal effects are not important, where system properties, reaction rates, and so on do not depend on temperature, or if the system is truly isothermal (constant temperature). Many chemical processes have important thermal effects, so it is necessary to develop material and energy balance models. One key is that a basis must always be selected when evaluating an intensive property such as enthalpy.

2.5.1   Review of Thermodynamics

Developing correct energy balance equations is not trivial and the chemical engineering literature contains many incorrect derivations. Chapter 5 of the book by Denn (1986) points out numerous examples where incorrect energy balances were used to develop process models.

The total energy (TE) of a system consists of internal (U), kinetic (KE) and potential energy (PE):

$$TE = U + KE + PE$$

where the kinetic and potential energy terms are:

$$KE = \frac{1}{2} m v^2$$

$$PE = mgh$$

Often we will use energy/mole or energy/mass and write the following

$$\hat{TE} = \hat{U} + \hat{KE} + \hat{PE}$$

$$\overline{TE} = \overline{U} + \overline{KE} + \overline{PE}$$

where $^\wedge$ and $^-$ represent per mole and per mass, respectively. The kinetic and potential energy terms, on a mass basis, are

$$\overline{KE} = \frac{1}{2} v^2$$

$$\overline{PE} = gh$$

For most chemical processes where there are thermal effects, we will neglect the kinetic and potential energy terms because their contribution is generally at least two orders of magnitude less than that of the internal energy term.

When dealing with flowing systems, we will usually work with enthalpy. Total enthalpy is defined as:

$$H = U + pV$$

while the enthalpy/mole is

$$\hat{H} = \hat{U} + p\hat{V}$$

and the enthalpy/mass is (since $\rho = 1/(\overline{V})$

$$\overline{H} = \overline{U} + p\overline{V} = \overline{U} + \frac{p}{\rho}$$

we will make use of these relationships in the following example.

EXAMPLE 2.4   Stirred Tank Heater

Consider a perfectly mixed stirred-tank heater, with a single feed stream and a single product stream, as shown in Figure 2.6. Assuming that the flowrate and temperature of the inlet stream can vary, that the tank is perfectly insulated, and that the rate of heat added per unit time (Q) can vary, develop a model to find the tank temperature as a function of time. State your assumptions.

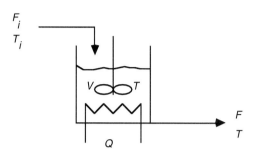

FIGURE 2.6   Stirred tank heater.

Material Balance

$$\text{accumulation} = \text{in} - \text{out}$$

$$\frac{dV\rho}{dt} = F_i\rho_i - F\rho \tag{2.52}$$

Energy Balance

$$\text{accumulation} = \text{in by flow} - \text{out by flow} + \text{in by heat transfer} + \text{work done on system}$$

$$\frac{dTE}{dt} = F_i\rho_i\overline{TE}_i - F\rho\overline{TE} + Q + W_T$$

Here we neglect the kinetic and potential energy:

$$\frac{dU}{dt} = F_i\rho_i\overline{U}_i - F\rho\overline{U} + Q + W_T \tag{2.53}$$

We write the total work done on the system as a combination of the shaft work and the energy added to the system to get the fluid into the tank and the energy that the system performs on the surroundings to force the fluid out.

$$W_T = W_s + F_ip_i - F_p \tag{2.54}$$

This allows us to write (2.53) as:

$$\frac{dU}{dt} = F_i\rho_i\left(\overline{U}_i + \frac{p_i}{\rho_i}\right) - F\rho\left(\overline{U} + \frac{p}{\rho}\right) + Q + W_s \tag{2.55}$$

and since $H = U + pV$, we can rewrite (2.55) as

$$\frac{dH}{dt} - \frac{dpV}{dt} = F_i \rho_i \overline{H}_i - F\rho \overline{H} + Q + W_s \tag{2.56}$$

Since, $dpV/dt = V\, dp/dt + p\, dV/dt$, if the *volume is constant* and the mean pressure change can be neglected (a good assumption for liquids), we can write

$$\boxed{\frac{dH}{dt} = F_i \rho_i \overline{H}_i - F\rho \overline{H} + Q + W_s} \tag{2.57}$$

We must remember the assumptions that went into the development of (2.57):

- The kinetic and potential energy effects were neglected.
- The change in pV term was neglected. This is a good assumption for a liquid system, provided Δp is not too large and constant volume is assumed.

The total enthalpy term is:

$$H = V\rho \overline{H}$$

and assuming no phase change, we select an arbitrary reference temperature (T_{ref}) for enthalpy:

$$\overline{H}(T) = \int_{T_{ref}}^{T} c_p \, dT$$

Often we assume that the heat capacity is constant, or calcuated at an average temperature, so

$$\overline{H} = c_p (T - T_{ref}) \tag{2.58}$$

$$\overline{H}_i = c_p (T_i - T_{ref}) \tag{2.59}$$

We now write the energy balance (2.57) in the following fashion:

$$\frac{d(V\rho c_p (T - T_{ref}))}{dt} = F_i \rho_i c_p (T_i - T_{ref}) + Q - F\rho c_p (T - T_{ref}) + W_s \tag{2.60}$$

using the *assumptions* of constant density and volume (so $F_i = F$, from (2.52)), we find

$$V\rho c_p \frac{d(T - T_{ref})}{dt} = F\rho c_p [(T_i - T_{ref}) - (T - T_{ref})] + Q + W_s \tag{2.61}$$

or

$$\frac{d(T - T_{ref})}{dt} = \frac{F}{V}(T_i - T) + \frac{Q}{V\rho c_p} + \frac{W_s}{V e c_p} \tag{2.62}$$

but T_{ref} is a constant, so $d(T - T_{ref})/dt = dT/dt$. Also, neglecting W_s we can write

$$\frac{dT}{dt} = \frac{F}{V}(T_i - T) + \frac{Q}{V\rho c_p} \tag{2.63}$$

In order to solve this problem, we must specify the parameters V, ρ, c_p, the inputs F, Q, and T_i (as a function of time), and the initial condition $T(0)$.

2.6 DISTRIBUTED PARAMETER SYSTEMS

In this section we show how the balance equations can be used to develop a model for a distributed parameter system, that is, a system where the state variables change with respect to position and time.

Consider a tubular reactor where a chemical reaction changes the concentration of the fluid as it moves down the tube. Here we use a volume element ΔV and a time element Δt. The total moles of species A contained in the element ΔV is written $(\Delta V)C_A$. The amount of species A entering the volume is $FC_A|_V$ and the amount of species leaving the volume is $FC_A|_{V+\Delta V}$. The rate of A leaving by reaction (assuming a first-order reaction) is $(-k\,C_A)\Delta V$.

The balance equation is then:

$$(\Delta V)C_A|_{t+\Delta t} - (\Delta V)C_A|_t = \int_t^{t+\Delta t} [FC_A|_V - FC_A|_{V+\Delta V} - kC_A\Delta V]\,dt$$

Using the mean value theorem of integral calculus and dividing by Δt, we find:

$$\frac{\Delta V[C_A|_{t+\Delta t} - C_A|_t]}{\Delta t} = FC_A|_V - FC_A|_{V+\Delta V} - kC_A\Delta V \qquad (2.64)$$

Dividing by ΔV and letting Δt and ΔV go to zero, we find:

$$\frac{\partial C_A}{\partial t} = -\frac{\partial FC_A}{\partial V} - kC_A \qquad (2.65)$$

Normally, a tube with constant cross-sectional area is used, so $dV = A\,dz$ and $F = Av_z$, where v_z is the velocity in the z-direction. Then the equation can be written:

$$\frac{\partial C_A}{\partial t} = -\frac{\partial v_z C_A}{\partial z} - kC_A \qquad (2.66)$$

Similarly, the overall material balance can be found as:

$$\frac{\partial \rho}{\partial t} = -\frac{\partial v_z \rho}{\partial z} \qquad (2.67)$$

If the fluid is at a constant density (good assumption for a liquid), then we can write the species balance as

$$\frac{\partial C_A}{\partial t} = -v_z \frac{\partial C_A}{\partial z} - kC_A \qquad (2.68)$$

To solve this problem, we must know the initial condition (concentration as a function of distance at the initial time) and one boundary condition. For example, the following boundary and initial conditions

$$C_A(z,\ t = 0) = C_{A0}(z)$$
$$C_A(0,\ t) = C_{Ain}(t) \qquad (2.69)$$

indicate that the concentration of A initially is known as a function of distance down the reactor, and that the inlet concentration as a function of time must be specified.

In deriving the tubular reactor equations we assumed that species A left a volume element only by convection (bulk flow). In addition, the molecules can leave by virtue of a concentration gradient. For example, the amount entering at V is

$$\left(FC_A + AD_{AZ} \frac{dCA}{dz} \right) \bigg| V \tag{2.70}$$

where is the diffusion coefficient. The reader should be able to derive the following reaction-diffusion equation (see exercise 19).

$$\frac{\partial C_A}{\partial t} = -v_z \frac{\partial C_A}{\partial z} + D_{AZ} \frac{\partial^2 C_A}{\partial z^2} - kC_A \tag{2.71}$$

Since this is a second-order PDE, the initial condition (C_A as a function of z) and two boundary conditions must be specified.

Partial differential equation (PDEs) models are much more difficult to solve than ordinary differential equations. Generally, PDEs are converted to ODEs by discretizing in the spatial dimension, then techniques for the solution of ODEs can be used. The focus of this text is on ODEs; with a grasp of the solution of ODEs, one can then begin to develop solutions to PDEs.

2.7 DIMENSIONLESS MODELS

Models typically contain a large number of parameters and variables that may differ in value by several orders of magnitude. It is often desirable, at least for analysis purposes, to develop models composed of dimensionless parameters and variables. To illustrate the approach, consider a constant volume, isothermal CSTR modeled by a simple first-order reaction:

$$\frac{dC_A}{dt} = \frac{F}{V} (C_{Af} - C_A) - kC_A$$

It seems natural to work with a scaled concentration. Defining

$$x = C/C_{Af0}$$

where C_{Af0} is the nominal (steady-state) feed concentration of A, we find

$$\frac{dx}{dt} = \frac{F}{V} x_f - \left(\frac{F}{V} + k \right) x$$

where $x_f = C_{Af}/C_{Af0}$. It is also natural to choose a scaled time, $\tau = t/t^*$, where t^* is a scaling parameter to be determined. We can use the relationship $dt = t^* d\tau$ to write:

$$\frac{dx}{t*d\tau} = \frac{F}{V} x_f - \left(\frac{F}{V} + k\right) x$$

A natural choice for $t*$ appears to be V/F (known as the residence time), so

$$\frac{dx}{d\tau} = x_f - \left(1 + \frac{Vk}{F}\right) x$$

The term Vk/F is dimensionless and known as a Damkholer number in the reaction engineering literature. Assuming that the feed concentration is constant, $x_f = 1$, and letting $\alpha = Vk/F$, we can write:

$$\frac{dx}{d\tau} = 1 - x + \alpha x$$

which indicates that a single parameter, α, can be used to characterize the behavior of all first-order, isothermal chemical reactions. Similar results are obtained if the dimensionless state is chosen to be conversion

$$x = (C_A - C_{Af0})/C_{Af0}$$

2.8   EXPLICIT SOLUTIONS TO DYNAMIC MODELS

Explicit solutions to nonlinear differential equations can rarely be obtained. The most common case where an analytical solution can be obtained is when a single differential equation has variables that are separable. This is a very limited class of problems. A main objective of this textbook is to present a number of techniques (analytical and numerical) to solve more general problems, particularly involving many simultaneous equations. In this section we provide an example of problems where the variables are separable.

EXAMPLE 2.5   Nonlinear Tank Height

Consider a tank height problem where the outlet flow is a nonlinear function of tank height:

$$\frac{dh}{dt} = \frac{F_i}{A} - \frac{\beta}{A} \sqrt{h}$$

Here there is not an analytical solution because of the nonlinear height relationship and the forcing function. To illustrate a problem with an analytical solution, we will assume that there is no inlet flow to the tank:

$$\frac{dh}{dt} = -\frac{\beta}{A} \sqrt{h}$$

we can see that the variables are separable, so

$$\frac{dh}{\sqrt{h}} = -\frac{\beta}{A} dt$$

$$\int_{h_o}^{h} \frac{dh}{\sqrt{h}} = -\int_{t_o}^{t} \frac{\beta}{A}\, dt$$

which has the solution

$$2\sqrt{h} - 2\sqrt{h_o} = -\frac{\beta}{A}(t - t_o)$$

or
$$\sqrt{h} = \sqrt{h_o} - \frac{\beta}{2A}(t - t_o)$$

letting $t_o = 0$, and squaring both sides, we obtain the solution

$$h(t) = \left[\sqrt{h_o} - \frac{\beta}{2A}t\right]^2$$

This analytical solution can be used, for example, to determine the time that it will take for the tank height to reach a certain level.

2.9   GENERAL FORM OF DYNAMIC MODELS

The dynamic models derived in this chapter consist of a set of first-order (meaning only first derivatives with respect to time), nonlinear, explicit, initial value ordinary differential equations. A representation of a set of first-order differential equations is

$$
\begin{aligned}
\dot{x}_1 &= f_1(x_1,...,x_n,u_1,...,u_m,p_1,...,p_r)\\
\dot{x}_2 &= f_2(x_1,...,x_n,u_1,...,u_m,p_1,...,p_r)\\
&\;\;\vdots\\
\dot{x}_n &= f_n(x_1,...,x_n,u_1,...,u_m,p_1,...,p_r)
\end{aligned}
\tag{2.72}
$$

where x_i is a state variable, u_i is an input variable and p_i is a parameter. The notation $\dot{x}_i$ is used to represent dx_i/dt. Notice that there are n equations, n state variables, m inputs, and r parameters.

2.9.1   State Variables

A state variable is a variable that arises naturally in the accumulation term of a dynamic material or energy balance. A state variable is a measurable (at least conceptually) quantity that indicates the state of a system. For example, temperature is the common state variable that arises from a dynamic energy balance. Concentration is a state variable that arises when dynamic component balances are written.

2.9.2   Input Variables

An input variable is a variable that normally must be specified before a problem can be solved or a process can be operated. Inputs are normally specified by an engineer based on knowledge of the process being considered. Input variables typically include flowrates of streams entering or leaving a process (notice that the flowrate of an outlet stream might be considered an input variable!). Compositions or temperatures of streams entering a process are also typical input variables. Input variables are often manipulated (by process controllers) in order to achieve desired performance.

2.9.3   Parameters

A parameter is typically a physical or chemical property value that must be specified or known to mathematically solve a problem. Parameters are often fixed by nature, that is, the reaction chemistry, molecular structure, existing vessel configuration, or operation. Examples include density, viscosity, thermal conductivity, heat transfer coefficient, and mass-transfer coefficient. When designing a process, a parameter might be "adjusted" to achieve some desired performance. For example, reactor volume may be an important design parameter.

2.9.4   Vector Notation

The set of differential equations shown as (2.72) above can be written more compactly in vector form.

$$\dot{\mathbf{x}} = \mathbf{f}(\mathbf{x}, \mathbf{u}, \mathbf{p}) \tag{2.73}$$

where

$\mathbf{x}$ = vector of n state variables
$\mathbf{u}$ = vector of m input variables
$\mathbf{p}$ = vector of r parameters

Notice that the dynamic models (2.73) can also be used to solve steady-state problems, since

$$\dot{\mathbf{x}} = \mathbf{0} \tag{2.74}$$

that is,

$$\mathbf{f}(\mathbf{x},\mathbf{u},\mathbf{p}) = \mathbf{0} \tag{2.75}$$

for steady-state processes. Numerical techniques (such as Newton's method) to solve algebraic equations (2.75) will be presented in Chapter 3.

The steady-state state variables from the solution of (2.75) are often used as the initial conditions for (2.73). Frequently, an input will be changed from its steady-state value, and (2.73) will be solved to understand the transient behavior of the system. The numerical solution of ordinary differential equations will be presented in Chapter 4. In the example below we show Example 2.2 (chemical reactor) in state variable form.

EXAMPLE 2.6   State Variable Form for Example 2.2

Consider the modeling equations for Example 2.2 (chemical reactor)

$$\frac{dV}{dt} = F_i - F \tag{2.23}$$

$$\frac{dC_A}{dt} = \frac{F_i}{V}(C_{Ai} - C_A) - kC_AC_B \tag{2.31}$$

$$\frac{dC_B}{dt} = \frac{F_i}{V}(C_{Bi} - C_B) - 2\,kC_AC_B \tag{2.32}$$

$$\frac{dC_P}{dt} = -\frac{F_i}{V}C_P + kC_AC_B \tag{2.33}$$

There are four states (V, C_A, C_B, and C_P), four inputs (F_i, F, C_{Ai}, C_{Bi}), and a single parameter (k). Notice that although F is the outlet flowrate, it is considered an input to the model, because it must be specified in order to solve the equations.

$$
\begin{bmatrix} \dot{V} \\[4pt] \dot{C}_A \\[4pt] \dot{C}_B \\[4pt] \dot{C}_P \end{bmatrix}
=
\begin{bmatrix}
F_i - F \\[6pt]
\dfrac{F_i}{V}(C_{Ai} - C_A) - kC_AC_B \\[6pt]
\dfrac{F_i}{V}(C_{Bi} - C_B) - 2\,kC_AC_B \\[6pt]
-\dfrac{F_i}{V}C_P + kC_AC_B
\end{bmatrix}
$$

or

$$
\begin{bmatrix} \dot{x}_1 \\[4pt] \dot{x}_2 \\[4pt] \dot{x}_3 \\[4pt] \dot{x}_4 \end{bmatrix}
=
\begin{bmatrix}
u_1 - u_2 \\[6pt]
\dfrac{u_1}{x_1}(u_3 - x_2) - p_1 x_2 x_3 \\[6pt]
\dfrac{u_1}{x_1}(u_4 - x_3) - 2\,p_1 x_2 x_3 \\[6pt]
-\dfrac{u_1}{x_1}x_4 + p_1 x_2 x_3
\end{bmatrix}
=
\begin{bmatrix}
f_1(x,u,p) \\[4pt]
f_2(x,u,p) \\[4pt]
f_3(x,u,p) \\[4pt]
f_4(x,u,p)
\end{bmatrix}
$$

SUMMARY

A number of material and energy balance examples have been presented in this chapter. The classic assumption of a perfectly stirred tank was generally used so that all models (except Section 2.6) were lumped-parameter systems. Future chapters develop the analytical and numerical techniques to analyze and simulate these models.

The student should now understand:

- that dynamic models of lumped parameter systems yield ordinary differential equations.
- that steady-state models of lumped parameter systems yield algebraic equations.
- The notion of a state, input, output, parameter.

A plethora of models are presented in modules in the final section of the textbook. More specifically, the following modules are of interest:

Module 5. Heated Mixing Tank
Module 6. Linear Equilibrium Stage Models (Absorption)
Module 7. Isothermal Continuous Stirred Tank Reactors
Module 8. Biochemical Reactor Models
Module 9. Diabatic Reactor Models
Module 10. Nonlinear Equilibrium Stage Models (Distillation)

Each of these modules covers model development and presents examples for analytical and numerical calculations.

FURTHER READING

A nice introduction to chemical engineering calculations is provided by:

Felder, R.M., & R. Rousseau. (1986). *Elementary Principles of Chemical Processes*, 2nd ed. New York: Wiley.

Excellent discussions of the issues involved in modeling a mixing tank, incorporating density effects, and energy balances is provided in the following two books:

Denn, M.M. (1986). *Process Modeling*. New York: Longman.
Russell, T.R.F., & M.M. Denn. (1971). *Introduction to Chemical Engineering Analysis*. New York: Wiley.

An introduction to chemical reaction engineering is:

Fogler, H.S. (1992). *Elements of Chemical Reaction Engineering*, 2nd ed. Englewood Cliffs, NJ: Prentice-Hall.

An excellent textbook for an introduction to chemical engineering thermodynamics is:

Smith, J.M. , H.C. Van Ness, & M.M. Abbott. (1996). *Chemical Engineering Thermodynamics*, 5th ed. New York: McGraw-Hill.

The following paper provides an advanced treatment of dimensionless variables and parameters:

Aris, R. (1993). Ends and beginnings in the mathematical modelling of chemical engineering systems. *Chemical Engineering Science*, 48(14), 2507–2517.

The relationships for mass and heat transport are shown in textbooks on transport phenomena. The chemical engineer's bible is

Bird, R.B., W.E. Stewart, & E. Lightfoot. (1960). *Transport Phenomena*. New York: Wiley.

The predator-prey model in student exercise 16 is also known as the Lotka-Volterra equations, after the researchers that developed them in the late 1920s. A presentation of the equations is in the following text:

Bailey, J.E., & D.F. Ollis. (1986). *Biochemical Engineering Fundamentals*, 2nd ed. New York: McGraw-Hill.

STUDENT EXERCISES

1. In Example 2.1 it was assumed that the input and output flowrates could be independently varied. Consider a situation in which the outlet flowrate is a function of the height of liquid in the tank. Write the modeling equation for tank height assuming two different constitutive relationships: (i) $F = \beta h$, or (ii) $F = \beta \sqrt{h}$, where β is known as a flow coefficient. You will often see these relationships expressed as $F = h/R$ or $F = \sqrt{h}/R$, where R is a flow resistance. List the state variables, parameters, as well as the input and output variables. Give the necessary information to complete the quantitative solution to this problem. If the flowrate has units of liters/min and the tank height has units of meters, find the units of the flow coefficients and flow resistances for (i) and (ii).

2. Consider a conical water tank shown below. Write the dynamic material balance equation if the flowrate out of the tank is a function of the square root of height of water in the tank ($F_o = \beta \sqrt{h}$). List state variables, input variables and parameters. (Hint: Use height as a state variable.)

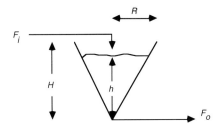

3. Extend the model developed in Example 2.2 (isothermal reaction) to handle the following stoichiometric equation: $A + B \longrightarrow 2P$. Assume that the volume is constant, but the change in concentration of component B cannot be neglected.

4. Extend the model developed in Example 2.2 (isothermal with first-order kinetics) to handle multiple reactions (assume a constant volume reactor).

$$A + B \; - - > \; 2P \quad \text{(reaction 1)}$$
$$2A + P \; - - > \; Q \quad \text{(reaction 2)}$$

Assume that no P is fed to the reactor. Assume that the reaction rate (generation) of A per unit volume for reaction 1 is characterized by expression

$$r_A = -k_1 \, C_A \, C_B$$

where the minus sign indicates that A is consumed in reaction 1. Assume that the reaction rate (generation) of A per unit volume for reaction 2 is characterized by the expression

$$r_A = -k_2 \, C_A \, C_P$$

If the concentrations are expressed in gmol/liter and the volume in liters, what are the units of the reaction rate constants?

If it is desirable to know the concentration of component Q, how many equations must be solved? If our concern is only with P, how many equations must be solved? Explain.

5. Model a mixing tank with two feedstreams, as shown below. Assume that there are two components, A and B. C represents the concentration of A. (C_1 is the mass concentration of A in stream 1 and C_2 is the mass concentration of A in stream 2). Model the following cases:

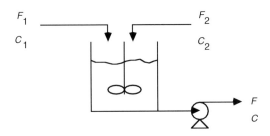

    a. Constant volume, constant density.

    b. Constant volume, density varies linearly with concentration.

    c. Variable volume, density varies linearly with concentration.

6. Consider two tanks in series where the flow out of the first tank enters the second tank. Our objective is to develop a model to describe how the height of liquid in tank 2 changes with time, given the input flowrate $F_o(t)$. Assume that the flow out of each tank is a linear function of the height of liquid in the tank ($F_1 = \beta_1 h_1$ and $F_2 = \beta_2 h_2$) and each tank has a constant cross-sectional area.

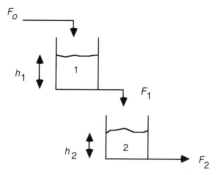

A material balance around the first tank yields (assuming constant density and $F_1 = \beta_1 h_1$)

7. Two liquid surge tanks (with constant cross-sectional area) are placed in series. Write the modeling equations for the height of liquid in the tanks assuming that the flowrate from the first tank is a function of the difference in levels of the tanks and the flowrate from the second tank is a function of the level in the second tank. Consider two cases: (i) the function is linear and (ii) the function is a square root relationship. State all other assumptions.

8. A gas surge drum has two components (hydrogen and methane) in the feedstream. Let y_i and y represent the mole fraction of methane in the feedstream and drum, respectively. Find dP/dt and dy/dt if the inlet and outlet flowrates can vary. Also assume that the inlet concentration can vary. Assume the ideal gas law for the effect of pressure and composition on density.

9. Consider a liquid surge drum that is a sphere. Develop the modeling equation using liquid height as a state variable, assuming variable inlet and outlet flows.

10. A car tire has a slow leak. The flowrate of air out of the tire is proportional to the pressure of air in the tire (we are using gauge pressure). The initial pressure is 30 psig, and after five days the pressure is down to 20 psig. How long will it take to reach 10 psig?

11. A car tire has a slow leak. The flowrate of air out of the tire is proportional to the square root of the pressure of air in the tire (we are using gauge pressure). The initial pressure is 30 psig, and after 5 days the pressure is down to 20 psig. How long will it take to reach 10 psig? Compare your results with problem 10.

12. A small room (10 ft × 10 ft × 10 ft) is perfectly sealed and contains air at 1 atm pressure (absolute). There is a large gas cylinder (100 ft^3) inside the room that contains helium with an initial pressure of 5 atm (absolute). Assume that the cylinder valve is opened (at $t = 0$) and the molar flowrate of gas leaving the cylinder is proportional to the difference in pressure between the cylinder and the room. Assume that room air does not diffuse into the cylinder.

    Write the differential equations that (if solved) would allow you to find how the cylinder pressure, the room pressure and the room mole fraction of helium change with time. State all assumptions and show all of your work.

13. A balloon expands or contracts in volume so that the pressure inside the balloon is approximately the atmospheric pressure.

    a. Develop the mathematical model (write the differential equation) for the volume of a balloon that has a slow leak. Let V represent the volume of the balloon and q represent the molar flowrate of air leaking from the balloon. State all assumptions. List state variables, inputs, and parameters.

    b. The following experimental data have been obtained for a leaking balloon.

t (minutes)	r (cm)
0	10
5	7.5

    Predict when the radius of the balloon will reach 5 cm using two different assumptions for the molar rate of air leaving the balloon:

    (i) The molar rate is constant.

    (ii) The molar rate is proportional to the surface area of the balloon.

    Reminder: The volume of a sphere is 4/3 πr^3 and the area of a sphere is $4\pi r^2$.

14. Often liquid surge tanks (particularly those containing hydrocarbons) will have a gas "blanket" of nitrogen or carbon dioxide to prevent the accumulation of explosive vapors above the liquid, as depicted below.

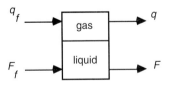

    Develop the modeling equations with gas pressure and liquid volume as the state variables. Let q_f and q represent the inlet and outlet gas molar flowrates, F_f and F the liquid volumetric flowrates, V the constant (total) volume, V_1 the liquid volume, and P the gas pressure. Assume the ideal gas law. Show that the modeling equations are:

$$\frac{dV_1}{dt} = F_f - F$$

$$\frac{dP}{dt} = \frac{P}{V - V_1}(F_f - F) + \frac{RT}{V - V_1}(q_f - q)$$

and state any other assumptions.

15. Most chemical process plants have a natural gas header that circulates through the process plant. A simplified version of such a header is shown below.

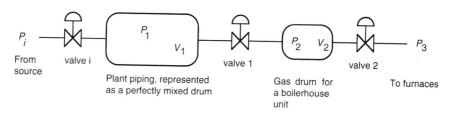

Here, the natural gas enters the process plant from a source (the natural gas pipeline) through a control valve. It flows through the plant piping, which we have represented as a perfectly mixed drum for simplicity. Another valve connects the plant piping to the gas drum for a boilerhouse unit. Gas passes through another valve to the boilerhouse furnaces.

Write modeling equations assuming that the pressures in drums 1 and 2 are the state variables. Let the input variables be h_1 (valve position 1), h_2 (valve position 2), and P_i (source pressure).

16. The Lotka-Volterra equations were developed to model the behavior of predator-prey systems, making certain assumptions about the birth and death rates of each species. Consider a system composed of sheep (prey) and coyotes (predator). In the following Lotka-Volterra equations x_1 represents the number of sheep and x_2 the number of coyotes in the system.

$$\frac{dx_1}{dt} = \alpha x_1 - \gamma x_1 x_2$$

$$\frac{dx_2}{dt} = \varepsilon \gamma x_1 x_2 - \beta x_2$$

Discuss the meaning of the parameters α, β, γ, ε and the assumptions made in the model.

17. Consider a perfectly mixed stirred-tank heater, with a single liquid feed stream and a single liquid product stream, as shown below.

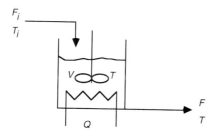

Develop the material and energy balance equations that describe this process. F_i is the volumetric flowrate into the tank, F is the volumetric flowrate out of the tank, T_i is the temperature of the fluid entering the tank, T is the temperature of the fluid in the tank, h is the height of liquid in the tank, and Q is the rate of energy added to the tank. State assumptions (such as constant density, etc.).

Assume that the volume can vary with time and that F is proportional to $\sqrt{h}$. How many differential equations does it take to model this system? What are the state variables? What are the parameters? What are the inputs? List the information necessary to solve this problem.

18. Consider a gas surge drum with variable inlet and outlet molar flowrates, q_f and q, respectively. Assume that heat is being added to the tank at a rate, Q. Write the modeling equations that describe how the temperature, T, and pressure, P, vary with time. Do not neglect the pV term in the energy balance.

19. Derive the reaction-diffusion equation

$$\frac{\partial C_A}{\partial t} = -v_z \frac{\partial C_A}{\partial z} + D_{AZ} \frac{\partial^2 C_A}{\partial z^2} - kC_A$$

using the same method to derive the tubular reactor model in Section 2.6. Assume that a chemical species enters a volume element via convection (bulk flow) and a concentration gradient (diffusion):

$$\left(FC_A - AD_{AZ} \frac{dC_A}{dz} \right)\bigg|_V$$

leaves by convection and a concentration gradient:

$$\left(FC_A - AD_{AZ} \frac{dC_A}{dz} \right)\bigg|_{V+\Delta V}$$

and also leaves by a first-order reaction.

20. Consider the nonlinear tank height model

$$\frac{dh}{dt} = \frac{F_i}{A} - \frac{\beta}{A} \sqrt{h}$$

and define the dimensionless variables $u = F_i/F_s$ and $x = h/h_s$. Where F_s and h_s are the steady-state flowrate and height, respectively ($F_s = \beta\sqrt{h_s}$). Define the dimensionless time, τ, that will yield the following dimensionless equation:

$$\frac{dx}{d\tau} = -\sqrt{x} + u$$

21. Derive the constitutive relationship $F = \beta\sqrt{h}$ by considering a steady-state energy balance around a tank with a constant flowrate. Use $P = P_o + \rho gh$ for the pressure at the bottom of the tank, where P_o is the atmospheric pressure (pressure at the top surface), h is the height of liquid in the tank, ρ is the density of fluid. Assume that

the cross-sectional area at the surface is much larger than the cross-sectional area of the exit pipe.

22. Consider the isothermal CSTR model with first-order kinetics:

$$\frac{dC_A}{dt} = \frac{F}{V}(C_{Af} - C_A) - k\,C_A$$

Use $\tau = kt$ as the dimensionless time. Develop the dimensionless equation for two cases: (i) $x = C_A/C_{Af}$ and (ii) $x = 1 - C_A/C_{Af}$. Compare and contrast the resulting equations with the example in Section 2.7.

23. Semibatch reactors are operated as a cross between batch and continuous reactors. A semibatch reactor is initially charged with a volume of material, and a continous feed of reactant is started. There is, however, no outlet stream. Develop the modeling equations for a single first-order reaction. The state variables should be volume and concentration of reactant A.

24. Pharmacokinetics is the study of how drugs infused to the body are distributed to other parts of the body. The concept of a compartmental model is often used, where it is assumed that the drug is injected into compartment 1. Some of the drug is eliminated (reacted) in compartment 1, and some of it diffuses into compartment 2 (the rest accumulates in compartment 1). Similarly, some of the drug that diffuses into compartment 2 diffuses back into compartment 1, while some is eliminated by reaction and the rest accumulates in compartment 2. Assume that the rates of diffusion and reaction are directly proportional to the concentration of drug in the compartment of interest. Show that the following balance equations arise, and discuss the meaning of each parameter (k_{ij}, units of $\min^{-1}$)

$$\frac{dx_1}{dt} = -(k_{10} + k_{12})x_1 + k_{21}\,x_2 + u$$

$$\frac{dx_2}{dt} = k_{12}x_1 - (k_{20} + k_{21})\,x_2$$

where x_1 and x_2 = drug concentrations in compartments 1 and 2 (μg/ml), and u = rate of drug input to compartment 1 (scaled by the volume of compartment 1, μg/ml min).

SECTION II

NUMERICAL TECHNIQUES

ALGEBRAIC EQUATIONS

3

The purpose of this chapter is to introduce methods to solve systems of algebraic equations. After studying this module, the student should be able to:

- Solve systems of linear algebraic equations.
- Solve nonlinear functions of one variable graphically and numerically.
- Use the MATLAB function `fzero` to solve a single algebraic equation.
- Discuss the stability of iterative techniques.
- Use the MATLAB function `fsolve` to solve sets of nonlinear algebraic equations.

The major sections in this chapter are:

3.1 Introduction

3.2 General Form for a Linear System of Equations

3.3 Nonlinear Functions of a Single Variable

3.4 MATLAB Routines for Solving Functions of a Single Variable

3.5 Multivariable Systems

3.6 MATLAB Routines for Systems of Nonlinear Algebraic Equations

3.1 INTRODUCTION

In Chapter 2 we discussed how to develop a model that consists of a set of ordinary differential equations. To solve these problems we need to know the initial conditions and how the inputs and parameters change with time. Often the initial conditions will be the steady-state values of the process variables. To obtain a steady-state solution of a system of differential equations requires the solution of a set of algebraic equations. The purpose of this chapter is to review techniques to solve algebraic equations.

Consider a set of n equations in n unknowns. The representation is

$$f_1(x_1, x_2, \ldots x_n) = 0$$
$$f_2(x_1, x_2, \ldots x_n) = 0$$
$$\cdot \qquad \cdot$$
$$\cdot$$
$$\cdot$$
$$f_n(x_1, x_2, \ldots x_n) = 0$$

(3.1)

The objective is to solve for the set of variables, x_i, that force all of the functions, f_i, to zero. A solution is called a *fixed point* or an *equilibrium point*.

Vector notation is used for a compact representation

$$\mathbf{f(x)} = \mathbf{0}$$

(3.2)

where $\mathbf{f(x)}$ is a vector valued function. Notice that these can be the same functional relationships that were developed as a set of differential equations, with $\dot{\mathbf{x}} = \mathbf{f(x)} = \mathbf{0}$. The solution $\mathbf{x}$ is then a steady-state solution to the system of differential equations.

Before we cover techniques for systems of nonlinear equations, it is instructive to review systems of linear equations.

3.2 GENERAL FORM FOR A LINEAR SYSTEM OF EQUATIONS

Consider a linear system with n equations and n unknowns. The first equation is

$$a_{11}x_1 + a_{12}x_2 + a_{13}x_3 + \ldots + a_{1n}x_n = b_1$$

where the a's and b's are known constant parameters, and the x's are the unknowns. The second equation is:

$$a_{21}x_1 + a_{22}x_2 + a_{23}x_3 + \ldots + a_{2n}x_n = b_2$$

while the nth equation is:

$$a_{n1}x_1 + a_{n2}x_2 + a_{n3}x_3 + \ldots + a_{nn}x_n = b_3$$

The coefficient, a_{ij}, relates the *j*th dependent variable to the *i*th equation.

$$\begin{bmatrix} a_{11} & a_{12} & a_{13} & \cdot & \cdot & a_{1n} \\ a_{21} & a_{22} & a_{23} & \cdot & \cdot & a_{2n} \\ \cdot & \cdot & \cdot & \cdot & \cdot & \cdot \\ a_{n1} & a_{n2} & a_{n3} & \cdot & \cdot & a_{nn} \end{bmatrix} \begin{bmatrix} x_1 \\ x_2 \\ x_3 \\ \cdot \\ \cdot \\ x_n \end{bmatrix} = \begin{bmatrix} b_1 \\ b_2 \\ b_3 \\ \cdot \\ \cdot \\ b_n \end{bmatrix}$$

(3.3)

Or, using compact matrix notation:

$$\mathbf{Ax} = \mathbf{b}$$

(3.3a)

The goal is to solve for the unknowns, $\mathbf{x}$. Notice that (3.3a) is the same form as (3.2), with

$$\mathbf{f(x)} = \mathbf{Ax} - \mathbf{b}$$

Premultiplying each side of (3a) by $\mathbf{A}^{-1}$ we find:

$$\mathbf{A}^{-1}\mathbf{A}\,\mathbf{x} = \mathbf{b}\mathbf{u}$$

and since, $\mathbf{A}^{-1}\mathbf{A} = \mathbf{I}$ (3.4)

$$\mathbf{x} = \mathbf{A}^{-1}\mathbf{b}$$

provided that the inverse of $\mathbf{A}$ exists. If the rank of $\mathbf{A}$ is less than n, then $\mathbf{A}$ is singular and the matrix inverse does not exist. If the condition number of $\mathbf{A}$ is very high, then the solution may be sensitive to model error. The concepts of rank and condition number are reviewed in Module 2 in Section V.

Equation (3.4) is used for conceptual purposes to represent the solution of the set of linear equations. In practice the solution is not obtained by finding the matrix inverse. Rather, equation (3.3a) is directly solved using a numerical technique such as Gaussian elimination or LU decomposition. Since the codes to implement these techniques are readily available in any numerical library, we do not review them here.

The next example illustrates how MATLAB can be used to solve a system of linear equations.

EXAMPLE 3.1   Linear Absorption Model, Solved Using MATLAB

Consider a 5-stage absorption column (presented in Module 6 in Section V) that has a model of the following form ($\mathbf{x}$ is a vector of stage liquid-phase compositions and $\mathbf{u}$ is a vector of column feed compositions):

$$0 = \mathbf{A}\,\mathbf{x} + \mathbf{B}\,\mathbf{u}$$

or

$$\mathbf{A}\,\mathbf{x} = -\,\mathbf{B}\,\mathbf{u}$$

the solution for $\mathbf{x}$ is $\mathbf{x} = -\mathbf{A}^{-1}\mathbf{B}\,\mathbf{u}$

The values of $\mathbf{A}$, $\mathbf{B}$, and $\mathbf{u}$ are:

```
a  =
     -0.3250    0.1250         0         0         0
      0.2000   -0.3250    0.1250         0         0
           0    0.2000   -0.3250    0.1250         0
           0         0    0.2000   -0.3250    0.1250
           0         0         0    0.2000   -0.3250
b  =
      0.2000         0
           0         0
           0         0
           0         0
           0    0.2500
u  =
           0
      0.1000
```

The following MATLAB command can be used to solve for **x**

```
» x = -inv(a)*b*u

x =
     0.0076
     0.0198
     0.0392
     0.0704
     0.1202
```

Use of the MATLAB left-division operator (\) yields the same result more efficiently (faster computation time), using the LU decomposition technique:

```
» x = -a\(b*u)
```

3.3 NONLINEAR FUNCTIONS OF A SINGLE VARIABLE

Functions of a single variable can be solved graphically by plotting $f(x)$ for many values of x and finding the values of x where $f(x) = 0$. This approach is shown in Figure 3.1. An interesting and challenging characteristic of nonlinear algebraic equations is the potential for multiple solutions, as shown in Figure 3.1b. In fact, for a single nonlinear algebraic equation it is often not possible to even know (without a detailed analysis) the number of solutions that exist. The situation is easier for polynomials because we know that an nth order polynomial has n solutions. Fortunately, many chemical process problems have a single solution that makes physical sense.

Numerical techniques for solving nonlinear algebraic equations are covered in the next section. A graphical representation will be used to provide physical insight for the numerical techniques.

Numerical methods to solve nonlinear algebraic equations are also known as *iterative techniques*. A sequence of guesses to the solution are made until we are "close enough" to the actual solution. To understand the concept of "closeness" we must use the notion of convergence tolerance.

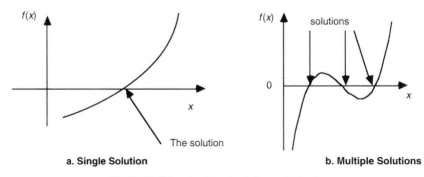

a. Single Solution   **b. Multiple Solutions**

FIGURE 3.1   Graphical solution to $f(x) = 0$.

3.3.1 Convergence Tolerance

A solution to a problem $f(x) = 0$ is considered converged at iteration k if:

$$|f(x_k)| \le \varepsilon$$

where ε is a *tolerance* that has been specified, $|f(x_k)|$ is used to denote the absolute value of function $f(x)$ evaluated at iteration k, and x_k is the variable value at iteration k. At times it is useful to base convergence on the variable rather than the function; for example, a solution can be considered converged at iteration k if the change in the variable is less than a certain *absolute tolerance*, ε_a:

$$|x_k - x_{k-1}| \le \varepsilon_a$$

The absolute tolerance is dependent on the scaling of the variable, so a *relative tolerance* specification, ε_r, is often used:

$$\frac{|x_k - x_{k-1}|}{|x_{k-1}|} < \varepsilon_r$$

 The relative tolerance specification is not useful if x is converging to 0. A combination of the relative and absolute tolerance specifications is often used, and can be expressed as

$$|x_k - x_{k-1}| < |x_{k-1}| \, \varepsilon_r + \varepsilon_a$$

The iterative methods that we present for solving single algebraic equations are: (i) direct substitution, (ii) interval halving, (iii) false position, and (iv) Newton's method.

3.3.2 Direct Substitution

Perhaps the simplest algorithm for a single variable nonlinear algebraic equation is known as direct substitution. We have been writing the relationship for a single equation in a single unknown as

$$f(x) = 0 \tag{3.5}$$

Using the direct substitution technique, we rewrite (3.5) in the form

$$x = g(x) \tag{3.6}$$

this means that our "guess" for x at iteration $k+1$ is based on the evaluation of $g(x)$ at iteration k (subscripts are used to denote the iteration)

$$x_{k+1} = g(x_k) \tag{3.7}$$

If formulated properly, (3.7) converges to a solution (within a desired tolerance)

$$x^* = g(x^*) \tag{3.8}$$

If not formulated properly, (3.7) may diverge or converge to physically unrealistic solutions, as shown by the following example.

EXAMPLE 3.2   A Reactor with Second-Order Kinetics

The dynamic model for an isothermal, constant volume, chemical reactor with a single second-order reaction is:

$$\frac{dC_A}{dt} = \frac{F}{V} C_{Af} - \frac{F}{V} C_A - kC_A^2$$

Find the steady-state concentration for the following inputs and parameters:

$$F/V = 1 \text{ min}^{-1}, C_{Af} = 1 \text{ gmol/liter}, k = 1 \text{ liter/(gmol min)}$$

At steady-state, $dC_A/dt = 0$, and substituting the parameter and input values, we find

$$1 - C_{As} - C_{As}^2 = 0$$

where the subscript s is used to denote the steady-state solution. For notational convenience, let $x = C_{As}$, and write the algebraic equation as

$$f(x) = -x^2 - x + 1 = 0$$

We can directly solve this equation using the quadratic formula to find $x = -1.618$ and 0.618 to be the solutions. Obviously a concentration cannot be negative, so the only physically meaningful solution is $x = 0.618$. Although we know the answer using the quadratic formula, our objective is to illustrate the behavior of the direct substitution method.

To use the direct substitution method, we can rewrite the function in two different ways: (i) $x^2 = -x + 1$ and (ii) $x = -x^2 + 1$. We will analyze (i) and leave (ii) as an exercise for the reader (see student exercise 4).

(i) Here we rewrite $f(x)$ to find the following direct substitution arrangement

$$x^2 = -x + 1$$

$$x = \sqrt{-x + 1} = g(x)$$

Or, using subscript k to indicate the kth iteration

$$x_{k+1} = \sqrt{-x_k + 1} = g(x_k)$$

For a first guess of $x_0 = 0.5$, we find the following sequence

$$x_1 = \sqrt{-0.5 + 1} \quad = 0.7071$$

$$x_2 = \sqrt{-0.7071 + 1} = 0.5412$$

$$x_3 = \sqrt{-0.5412 + 1} = 0.6774$$

$$x_4 = \sqrt{-0.6774 + 1} = 0.5680$$

This sequence slowly converges to 0.618, as shown in Figure 3.2.

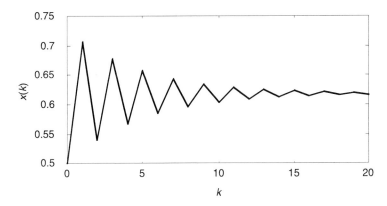

FIGURE 3.2   The iteration $x_{k+1} = \sqrt{-x_{k+1} + 1}$ with $x_0 = 0.5$. This sequence converges to 0.6180.

Notice that an initial guess of $x_0 = 0$ or 1 oscillates between 0 and 1, never converging or diverging, as shown in Figure 3.3.

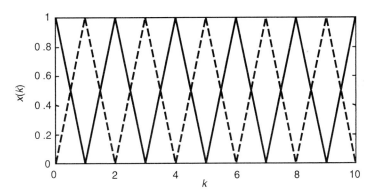

FIGURE 3.3   The iteration $x_{k+1} = \sqrt{-x + 1}$ with $x_0 = 0$ (dashed) or 1 (solid). This sequence oscillates between 0 and 1.

As noted earlier, this problem has two solutions ($x^* = -1.618$ and $x^* = 0.618$), since it is a second-order polynomial. This can be verified by plotting x versus $f(x)$ as shown in

Figure 3.4. From physical reasoning, we accept only the positive solution, since a concentration cannot be negative.

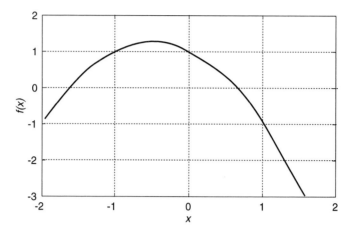

FIGURE 3.4   Plot of $f(x)$ versus x to find where $f(x) = 0$.

Example 3.2 illustrates that certain initial guesses may oscillate and never yield a solution, while other guesses may converge to a solution. It turns out that the way that a direct substitution problem is formulated may eliminate valid solutions from being reached (see student exercise 4).

These problems exist, to a certain extent, with any numerical solution technique. The potential problems appear to be worse with direct substitution; direct substitution is not generally recommended unless experience with a particular problem indicates that results are satisfactory. Direct substitution is often the easiest numerical technique to formulate for the solution of a single nonlinear algebraic equation.

If a numerical technique does not converge to a solution when the initial guess is close to the solution, we refer to the solution as unstable. The stability of iterative methods is discussed in the appendix.

3.3.3   Interval Halving (Bisection)

The interval halving technique only requires that the sign of the function value is known. The following steps are used in the interval halving technique:

1. Bracket the solution by finding two values of x, one where $f(x)$ is less than zero and another where $f(x)$ is greater than zero.
2. Evaluate the function, $f(x)$, at the midpoint of the bracket.
3. Replace the bracket limit that has the same sign as the function value at the midpoint, with the midpoint value. Check for convergence. If not converged, go back to step 2.

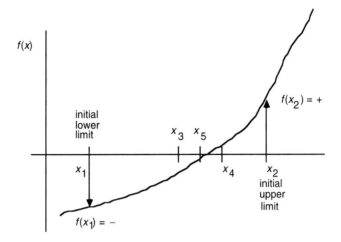

FIGURE 3.5   Illustration of the bisection technique.

An example of the interval halving approach is shown in Figure 3.5.
Notice that the solution was bracketed by

1. Finding x_1, where $f(x_1)$ is negative and x_2 where $f(x_2)$ is positive.
2. The midpoint between x_1 and x_2 was selected (x_3). The function value at x_3, $f(x_3)$, was negative, so
3. x_1 was thrown out and x_3 became the lower bracket point.
4. The midpoint between x_3 and x_2 was selected (x_4). The function value at x_4, $f(x_4)$, was positive, so
5. x_2 was thrown out and x_4 became the upper bracket point.
6. The midpoint between x_3 and x_4 was selected (x_5). The function value at x_5, $f(x_5)$, was negative, so
7. x_3 was thrown out and x_5 became the lower bracket point.

You can see that the midpoint between x_5 and x_4 will yield a positive value for $f(x_6)$, so that the x_4 point will be thrown out. You can also see that this process could go on for a very long time, depending on how close to zero you desire the solution. Engineering judgement must be used when making a convergence tolerance specification.

The advantage to interval bisection is that it is easy to understand. A disadvantage is that it is not easily extended to multivariable systems. Also, it can take a long time to reach the solution since it only uses information about the sign of the function values. The next technique is similar to interval bisection but uses the function values to determine the variable value for the next iteration.

3.3.4 False Position (Reguli Falsi)

The false position or reguli falsi technique uses the function values at two previous iterations to determine the value for the next iteration. The technique of false position consists of the following steps:

1. Select variable values x_k and x_{k+1} to bracket the solution.
2. Draw line between $f(x_k)$ and $f(x_{k+1})$ and find x_{k+2}.
3. Evaluate $f(x_{k+2})$. Replace the bracket limit that has the same sign for its function as the sign of $f(x_{k+2})$.

An example of the false position approach is shown in Figure 3.6.
   The next step would be to draw a line from $f(x_3)$ to $f(x_2)$ and find x_4. Continue until a certain tolerance is met.
   The false position method generally converges much more rapidly since it uses known function values to determine the next "guess" for the variable. We have shown graphically how each technique is used. You will have an opportunity in the student exercises to write an algorithm to implement the two techniques.

3.3.5 Newton's Method (or Newton-Raphson)

The most common method for solving nonlinear algebraic equations is known as Newton's method (or Newton-Raphson). Newton's method can be derived by performing a Taylor series expansion of $f(x)$:

$$f(x + \Delta x) = f(x) + f'(x)\Delta x + \frac{f''(x)}{2}(\Delta x)^2 + \frac{f'''(x)}{6}(\Delta x)^3 + \ldots = 0 \qquad (3.9)$$

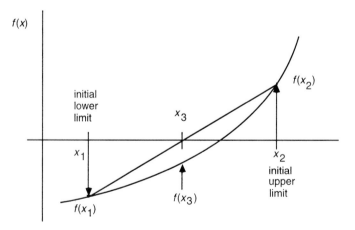

FIGURE 3.6 Illustration of the false position technique.

where

$$f'(x) = \frac{\partial f(x)}{\partial x}, f''(x) = \frac{\partial^2 f(x)}{\partial x^2}, \quad \text{and so on.}$$

Neglecting the second-order and higher derivative terms and solving for $f(x+\Delta x) = 0$, we obtain

$$\Delta x = \frac{-f(x)}{f'(x)} \tag{3.10}$$

Since this is an iterative procedure, calculate the guess for x at iteration $k+1$ as a function of the value at iteration k:

$$\text{defining} \qquad \Delta x_{k+1} = x_{k+1} - x_k \tag{3.11}$$

$$\text{from (3.10)} \quad \Delta x_{k+1} = \frac{-f(x_k)}{f'(x_k)} \tag{3.12}$$

$$\text{from (3.11)} \quad x_{k+1} - x_k = \frac{-f(x_k)}{f'(x_k)} \tag{3.13}$$

$$\boxed{x_{k+1} = x_k - \frac{f(x_k)}{f'(x_k)}} \tag{3.14}$$

Equation (3.14) is known as Newton's method for a single-variable problem.

Notice that we can obtain the following graphical representation for Newton's method (Figure 3.7).

Starting from the initial guess of x_1, we find that x_2 is the intersection of $f'(x_1)$ with the x-axis. Evaluate $f(x_2)$ and draw a line with slope $f'(x_2)$ to the x-axis to find x_3. This procedure is continued until convergence.

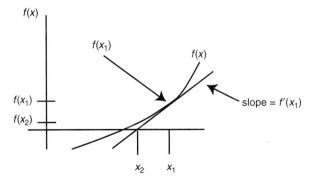

FIGURE 3.7 Illustration of Newton's method.

Advantages to Newton's method include quadratic convergence (when close to the solution) and that the method is easily extended to multivariable problems. Disadvantages include the fact that a derivative of the function is required, and that the method may not converge to a solution or may not converge to the nearest solution.

For an example of nonconvergence, consider the function shown in Figure 3.8. Here the initial guess is at a point where the derivative of function is equal to zero ($f'(x_0) = 0$), therefore there is no intersection with the x-axis to determine the next guess. We also see from (3.14) that there is no finite value for the next guess for x.

Another problem is that the solution could continuously oscillate between two values. Consider $f(x) = x^3 - x$. A plot of Newton's method for this function, with an initial guess of $x_0 = -1/\sqrt{5}$, is shown in Figure 3.9.

Notwithstanding the problems (possible division by zero or continuous oscillation) that we have shown with Newton's method, it (or some variant of Newton's) is still the most commonly used solution technique for nonlinear algebraic equations. Notice that Newton's method essentially linearizes the nonlinear model at each iteration and therefore results in successive solutions of linear models.

Notice that increasing amounts of information were needed to use the previous techniques. Interval halving required the sign of the function, reguli falsi required the value of the function, and Newton's method required the value and the derivative of the function. We also found that not all solutions to a nonlinear equation are stable when direct substitution is used. Next, we show that all solutions are stable using Newton's method.

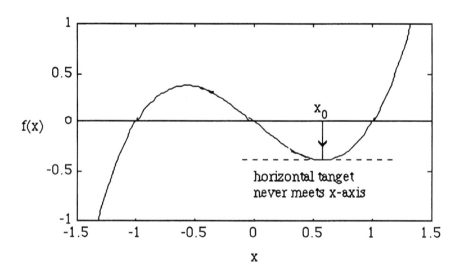

FIGURE 3.8   Problem with Newton's method when $f'(x) = 0$.

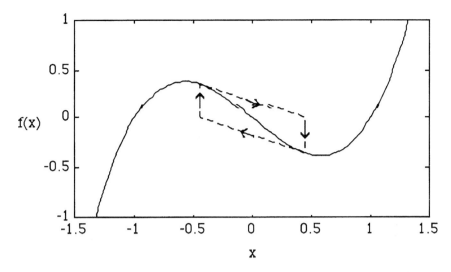

FIGURE 3.9   Oscillation of solution between two values.

3.4 MATLAB ROUTINES FOR SOLVING FUNCTIONS OF A SINGLE VARIABLE

MATLAB has two routines that can solve for the zeros of a function of a single variable. FZERO is used for a general nonlinear equation, while ROOTS can be used if the nonlinear equation is a polynomial.

3.4.1 FZERO

The first routine that we use for illustration purposes is `fzero`. `fzero` uses a combination of interval halving and false position.

In order to use `fzero`, you must first write a MATLAB m-file to generate the function that is being evaluated. Consider the function $f(x) = x^2 - 2x - 3 = 0$.

The following MATLAB m-file evaluates this function (the m-file is named `fcn1.m`):

```
function y = fcn1(x)
y = x^2 - 2*x - 3;
```

After generating the m-file `fcn1.m`, the user must provide a guess for the solution to the `fzero` routine. The following command gives an initial guess of $x = 0$.

```
y = fzero('fcn1',0)
```

MATLAB returns the answer:

$$y = -1$$

For an initial guess of $x = 2$, the user enters

$$z = \text{fzero}('\text{fcn1}',2)$$

and MATLAB returns the answer

$$z = 3$$

These results are consistent with those of Example 3.2, where we found that there were two solutions to a similar problem (we could use the quadratic formula to find them). Again, the solution obtained depends on the initial guess.

A third argument allows the user to select a relative tolerance (the default is the machine precision, eps). A fourth argument triggers a printing of the iterations.

3.4.2 ROOTS

Since the equation that we were solving was a polynomial equation, we could also use the MATLAB routine roots to find the zeros of the polynomial. Consider the polynomial function:

$$x^2 - 2x - 3 = 0$$

The user must create a vector of the coefficients of the polynomial, in descending order.

$$c = [1 \ -2 \ -3]'$$

Then the user can type the following command

$$\text{roots}(c)$$

and MATLAB returns

```
ans =
          3
         -1
```

Again, these are the two solutions that we expect.

3.5 MULTIVARIABLE SYSTEMS

In the previous sections we discussed the solution of a single algebraic equation with a single unknown variable. We covered direct substitution, bisection, reguli falsi, and Newton's method. In this section, we will discuss the reduction of a multivariable problem to a single-variable problem, as well as the multivariable Newton's method.

Consider a system of n nonlinear equations in n unknowns

$$\mathbf{f(x)} = \mathbf{0}$$

There are some special cases where $n - 1$ variables can be solved in terms of one variable—then a single variable solution technique can be used. This approach is shown in the following example.

EXAMPLE 3.3   Reducing a two-variable problem to a single-variable problem

Solve the following system of nonlinear equations.

$$f_1(x_1,x_2) = \quad x_1 - 4\,x_1^2 - \quad x_1 x_2 = 0 \tag{3.15}$$

$$f_2(x_1,x_2) = 2\,x_2 - \quad x_2^2 + 3\,x_1 x_2 = 0 \tag{3.16}$$

From (3.15) we can solve for x_2 in terms of x_1 to find:

$$x_2 = 1 - 4\,x_1 \tag{3.17}$$

Substituting (3.17) into (3.16), we find:

$$1 + 3\,x_1 - 28\,x_1^2 = 0 \tag{3.18}$$

which has the two solutions for x_1 (from the quadratic formula)

$$x_1 = 0.25 \quad \text{or} \quad x_1 = -0.1429$$

The corresponding values of x_2 (from (3.17)) are

$$x_2 = 0.0 \quad \text{and} \quad x_2 = 1.5714$$

Or, writing these solutions in vector form:

$$\textbf{solution 1 is } \mathbf{x} = \begin{bmatrix} 0.25 \\ 0 \end{bmatrix}, \text{while } \textbf{solution 2 is } \mathbf{x} = \begin{bmatrix} -0.1429 \\ 1.5714 \end{bmatrix}$$

Question: Are we certain that there are only two solutions?

Observation: We can also see by inspection that the origin (sometimes called the trivial solution), $\mathbf{x} = \begin{bmatrix} 0 \\ 0 \end{bmatrix}$, is also a solution. Another solution that is slightly less obvious is $\mathbf{x} = \begin{bmatrix} 0 \\ 2 \end{bmatrix}$, which we can see by inspection satisfies (3.15) and (3.16).

Question: How did we miss the other two solutions?

Observation: Perhaps we will find the other solutions if we solve (3.16) for x_1 in terms of x_2, then substitute this result into (3.15) to solve for x_2. When this is done, we obtain the result that

$$x_2^2 - 4\,x_2 + 4 = 0$$

and using the quadratic formula $x_2 = \dfrac{4}{2} \pm \sqrt{\dfrac{16 - 16}{2}} = 2$

which gives us the solution $\mathbf{x} = \begin{bmatrix} 0 \\ 2 \end{bmatrix}$. Notice that we are still missing the trivial solution, $\mathbf{x} = \begin{bmatrix} 0 \\ 0 \end{bmatrix}$.

The mistake that we made was back at the first step, when we solved (3.15) for x_2 in terms of x_1 to find (3.17). We must recognize that (3.15) is quadratic in x_1, therefore there are two solutions for x_1 in terms of x_2. This is more clear if we write (3.15) as

$$-4\,x_1^2 + x_1(1 - x_2) + 0 = 0$$

and solve for x_1 to find $x_1 = 0$ or $x_1 = 1/4\,(1 - x_2)$. The reader should show that substituting these values into (3.16) will lead to the four solutions:

$$\mathbf{x} = \begin{bmatrix} 0 \\ 0 \end{bmatrix}, \begin{bmatrix} 0 \\ 2 \end{bmatrix}, \begin{bmatrix} 0.25 \\ 0 \end{bmatrix}, \text{ and } \begin{bmatrix} -0.1429 \\ 1.5714 \end{bmatrix}.$$

The previous example illustrates the care that must be taken when using reduction techniques to solve several nonlinear algebraic equations.

If a problem cannot be reduced to a single variable, then a general multivariable strategy (such as that discussed in the next section) must be used.

3.5.1 Newton's Method for Multivariable Problems

Recall that we are solving the general set of equations

$$\mathbf{f(x)} = 0 \tag{3.19}$$

That is, a set of n equations in n unknowns

$$\begin{bmatrix} f_1(x_1, x_2, \ldots x_n) \\ f_2(x_1, x_2, \ldots x_n) \\ \cdot \\ \cdot \\ \cdot \\ f_n(x_1, x_2, \ldots x_n) \end{bmatrix} = \begin{bmatrix} 0 \\ 0 \\ \cdot \\ \cdot \\ \cdot \\ 0 \end{bmatrix} \tag{3.20}$$

The objective is to solve for the set of variables, x_i, that forces all of the functions, f_i, to zero. We can use a Taylor series expansion for each f_i:

$$f_i(\mathbf{x} + \mathbf{\Delta x}) = f_i(\mathbf{x}) + \sum_{j=1}^{n} \frac{\partial f_i}{\partial x_j} \Delta x_j + \text{higher order terms} \tag{3.21}$$

Neglecting the higher order terms and writing in matrix form

$$\mathbf{f(x + \Delta x)} = \mathbf{f(x)} + \mathbf{J}\,\mathbf{\Delta x} \tag{3.22}$$

where $\mathbf{J}$ is known as the *Jacobian*

$$\mathbf{J} = \begin{bmatrix} \dfrac{\partial f_1}{\partial x_1} & \dfrac{\partial f_1}{\partial x_2} & \cdot & \cdot & \dfrac{\partial f_1}{\partial x_n} \\ \cdot & \cdot & \cdot & \cdot & \cdot \\ \cdot & \cdot & \cdot & \cdot & \cdot \\ \dfrac{\partial f_n}{\partial x_1} & \dfrac{\partial f_n}{\partial x_2} & \cdot & \cdot & \dfrac{\partial f_n}{\partial x_n} \end{bmatrix} \tag{3.23}$$

Now, since we wish to solve for $\mathbf{\Delta x}$ such that $\mathbf{f(x + \Delta x)} = 0$, then (from 3.22)

$$\mathbf{f(x)} + \mathbf{J}\,\mathbf{\Delta x} = 0 \tag{3.24}$$

Solving for $\mathbf{\Delta x}$ at iteration k

$$\mathbf{\Delta x_k} = -\mathbf{J_k}^{-1}\,\mathbf{f(x_k)} \tag{3.25}$$

but $\mathbf{\Delta x}$ is simply the change in the $\mathbf{x}$ vector from the previous iteration

$$\mathbf{\Delta x_k} = \mathbf{x_{k+1}} - \mathbf{x_k} \tag{3.26}$$

Substituting (3.26) into (3.25)

$$\mathbf{x_{k+1}} = \mathbf{x_k} - \mathbf{J_k}^{-1}\,\mathbf{f(x)} \tag{3.27}$$

Remember that $\mathbf{x_k}$ is a vector of values at iteration k. Notice that for a single equation (3.27) is:

$$x_{k+1} = x_k - \frac{f(x_k)}{f'(x_k)}$$

which is the result that we obtained in Section 3.3.4.

Comment: In practice the inverse of the Jacobian is not actually used in the solution of (3.25). Actually, (3.25) is solved as a set of linear algebraic equations, using Gaussian elimination or LU decomposition.

$$\mathbf{J_k}\,\mathbf{\Delta x_k} = -\mathbf{f(x_k)}$$

where $\mathbf{J_k}$ and $\mathbf{f(x_k)}$ are known at iteration k.

EXAMPLE 3.3   Revisited. Newton's method

$$f_1(x_1,x_2) = \quad x_1 - 4\,x_1^2 - \quad x_1 x_2 = 0$$

$$f_2(x_1,x_2) = 2\,x_2 - \quad x_2^2 + 3\,x_1 x_2 = 0$$

or $$\mathbf{f(x)} = \begin{bmatrix} x_1 - 4\,x_1^2 - x_1 x_2 \\ 2\,x_2 - x_2^2 + 3\,x_1 x_2 \end{bmatrix} = \begin{bmatrix} 0 \\ 0 \end{bmatrix}$$

The Jacobian elements are

$$\mathbf{J}_{11} = \frac{\partial f_1}{\partial x_1} = 1 - 8 x_1 - x_2 \qquad \mathbf{J}_{12} = \frac{\partial f_1}{\partial x_2} = -x_1$$

$$\mathbf{J}_{21} = \frac{\partial f_2}{\partial x_1} = 3\,x_2 \qquad \mathbf{J}_{22} = \frac{\partial f_2}{\partial x_2} = 2 - 2\,x_2 + 3\,x_1$$

so the Jacobian is written

$$\mathbf{J} = \begin{bmatrix} 1 - 8x_1 - x_2 & -x_1 \\ 3\,x_2 & 2 - 2\,x_2 + 3\,x_1 \end{bmatrix}$$

Consider an initial guess of $x_1 = -1$ and $x_2 = -1$. Let $\mathbf{x}(0)$ represent the vector for this initial guess

$$\mathbf{x}(0) = \begin{bmatrix} x_1(0) \\ x_2(0) \end{bmatrix} = \begin{bmatrix} -1 \\ -1 \end{bmatrix}$$

The value of the Jacobian at this initial guess is

$$\mathbf{J}(\mathbf{x}(0)) = \begin{bmatrix} 10 & 1 \\ -3 & 1 \end{bmatrix}$$

The inverse of the Jacobian is

$$\mathbf{J}^{-1}(\mathbf{x}(0)) = \begin{bmatrix} \dfrac{1}{13} & \dfrac{-1}{13} \\ \dfrac{3}{13} & \dfrac{10}{13} \end{bmatrix}$$

The value of the function vector for the initial guess is

$$\mathbf{f}(\mathbf{x}(0)) = \begin{bmatrix} -1 - 4 - 1 \\ 2(-1) - 1 + 3 \end{bmatrix} = \begin{bmatrix} -6 \\ 0 \end{bmatrix}$$

and the guess for $\mathbf{x}(1)$, where $\mathbf{x}(1)$ represents the vector at iteration 1, is

$$\mathbf{x}(1) = \mathbf{x}(0) - \mathbf{J}^{-1}(\mathbf{x}(0))\,\mathbf{f}(\mathbf{x}(0))$$

$$\mathbf{x}(1) = \begin{bmatrix} -1 \\ -1 \end{bmatrix} - \begin{bmatrix} \dfrac{1}{13} & \dfrac{-1}{13} \\ \dfrac{3}{13} & \dfrac{10}{13} \end{bmatrix} \begin{bmatrix} -6 \\ 0 \end{bmatrix}$$

$$\mathbf{x}(1) = \begin{bmatrix} -0.5385 \\ 0.3846 \end{bmatrix}$$

Continuing with iterations 2 through 7 we find the following results

Iteration	$\mathbf{x}_1$	$\mathbf{x}_2$
0	−1	−1
1	−0.5385	0.3846
2	−0.3104	1.0688
3	−0.2016	1.3952
4	−0.1561	1.5317
5	−0.1439	1.5683
6	−0.1429	1.5714
7	−0.1429	1.5714

The sequence of iterations for an initial guess of $x_1 = 1$ and $x_2 = 1$ is

Iteration	x_1	x_2
0	1	1
1	0.6190	0.0476
2	0.3893	0.0081
3	0.2870	0.0008
4	0.2542	0.0000
5	0.2501	0.0000
6	0.2500	0.0000

Notice that a guess of $x = $ `[-1 -1]`' converged to $x = $ `[-0.1429  1.5714]`' after six iterations, and a guess of $x = $ `[ 1  1]`' converged to $x = $ `[ 0.2500  0.0000]`' after six iterations. Other initial guesses may lead to the other two known solutions that were determined analytically.

The previous example illustrates that, for systems that have multiple solutions, the solution obtained depends on the initial guess.

3.5.2   Quasi-Newton Methods

Most computer codes actually implement some variant of Newton's method; these are referred to as quasi-Newton methods. Remember that Newton's method is guaranteed to converge only if the system is nonsingular and we are "close" to the solution.

DAMPING FACTOR

Often it is desirable to "dampen" the change in the guess for x_{k+1}, to make Newton's method more stable. Applying a damping factor, α, to (3.27), we write

$$\mathbf{x}_{k+1} = \mathbf{x}_k - \alpha\, \mathbf{J}_k^{-1}\, \mathbf{f}(\mathbf{x}_k) \qquad (3.28)$$

where α is chosen so that $\| \mathbf{f}(\mathbf{x}_{k+1}) \| < \| \mathbf{f}(\mathbf{x}_k) \|$ and $0 < \alpha \le 1$ (we use the $\| \mathbf{f}(\mathbf{x}_k) \|$ notation to represent the norm of the vector $\mathbf{f}(\mathbf{x}_k)$). Often α is selected to minimize $\| \mathbf{f}(\mathbf{x}_{k+1}) \|$ using a search technique, that is, α is adjusted until $\| \mathbf{f}(\mathbf{x}_{k+1}) \|$ is minimized.

It should be noted that the single-variable equivalent to (3.28) is

$$x_{k+1} = x_k - \alpha\, \frac{f(x_k)}{f'(x_k)} \qquad (3.29)$$

HANDLING SINGULAR (OR ILL-CONDITIONED) JACOBIAN MATRICES

Notice that the Newton method with or without the damping factor requires the inverse of the Jacobian matrix (or the solution of a set of linear algebraic equations) to determine the value for the next iteration. If the Jacobian is singular, it cannot be inverted. One method that avoids this problem is known as the Levenberg-Marquardt method:

$$\mathbf{x_{k+1}} = \mathbf{x_k} - (\mathbf{J_k^T J_k} + \beta \mathbf{I})^{-1} \mathbf{J_k^T} \mathbf{f(x_k)} \tag{3.30}$$

where T represents the matrix transpose and β is an adjustable parameter used to avoid a singularity. The single variable equivalent to (3.30) is

$$x_{k+1} = x_k - \frac{f'(x_k)f(x_k)}{(f'(x_k))^2 + \beta} \tag{3.31}$$

Notice that if $\beta = 0$, the standard Newton algorithm results.

WHEN ANALYTICAL JACOBIAN MATRICES ARE NOT AVAILABLE

If an analytical Jacobian is not available, a numerical approximation to the Jacobian must be used by the quasi-Newton technique. A backward differences approximation for the Jacobian is

$$\mathbf{J_{ij}}(\mathbf{x}(k)) = \frac{\partial f_i(\mathbf{x}(k))}{\partial x_j(k)} \approx \frac{f_i(\mathbf{x}(k) + \delta x_j(k)) - f_i(\mathbf{x}(k))}{\delta x_j(k)} \tag{3.32}$$

where $\delta x_j(k)$ is a small perturbation in variable x_j at iteration k. A problem with this approach is that an n-variable problem requires $n + 1$ evaluations of the function vector at each iteration. There are other techniques that rely on infrequent function evaluations to update the Jacobian matrix.

3.6   MATLAB ROUTINES FOR SYSTEMS OF NONLINEAR ALGEBRAIC EQUATIONS

The MATLAB routine `fsolve` is used to solve sets of nonlinear algebraic equations, using a quasi-Newton method. The user must supply a routine to evaluate the function vector. It is optional to write a routine to evaluate the gradient of the function vector. As another option, the user can select the Levenburg-Marquardt method.

EXAMPLE 3.3   Reconsidered. Using MATLAB

The m-file used to implement Example 3.3 using `fsolve` is:

```
function f = nle(x)
f(1)= x(1)-4*x(1)*x(1)-x(1)*x(2);
f(2)= 2*x(2)-x(2)*x(2)+3*x(1)*x(2);
```

which is placed in an m-file called `nle.m`

The initial guess is entered

```
x0 = [1  1]';
```

and we obtain the solution by entering

$$x = \text{fsolve('nle',x0)}$$

which gives us the expected results

$$x = [0.2500 \quad 0.0000]'$$

Computationally faster results will be obtained if the analytical Jacobian is used.

$$\mathbf{J} = \begin{bmatrix} 1 - 8x_1 - x_2 & -x_1 \\ 3\,x_2 & 2 - 2\,x_2 + 3\,x_1 \end{bmatrix}$$

The following function file generates the analytical Jacobian for this problem.

```
function gf = gradnle(x)
gf(1,1)=1-8*x(1)-x(2);
gf(1,2)=-x(1);
gf(2,1)=3*x(2);
gf(2,2)=2-2*x(2)+3*x(1);
```

which we place in an m-file called `gradnle.m`. We can then solve this problem by entering

```
x0 = [1  1]';
options(5)=0;
x = fsolve('nle',x0,options,'gradnle')
```

The options vector can be used to select the Levenberg-Marquardt method by setting

```
options(5)=1;
```

SUMMARY

In this chapter we have presented a number of techniques to solve nonlinear algebraic equations. Each technique had a number of advantages and disadvantages—the approach that you use may depend on the problem at hand. If you have the option, it is a good idea to plot the function, $f(x)$, to see if the results agree with your numerical solution. This option may not be available with multivariable problems.

Notice that if a particular problem has multiple solutions, the actual solution obtained will depend on the initial guess. There are many actual chemical processes that have "multiple solutions," that is, there are several possible "steady-states" that the process may operate at. In practice, the steady-state obtained will depend on dynamic considerations, such as the way that the process is started up. These issues will be addressed when we discuss differential equation-based models.

Much research is being done in applied mathematics to develop numerical techniques that yield every one of the multiple solutions, without having to make many initial guesses. Among these is "homotopy continuation." These techniques are not well-developed and are beyond the scope of this text. For now, the student must be willing to try a number of techniques, with a number of initial guesses, to find all of the solutions to a problem. It is recommended that you generally use commerically available routines for solving these problems.

The numerical techniques covered were

- Direct substitution
- Interval bisection
- Reguli falsi
- Newton's method

Newton's method (and some variants) is the most commonly used multivariable technique. The reader should be able to understand the notions of

- Jacobian
- Fixed or equilibrium points
- Convergence and stability
- Tolerance
- Iteration

The MATLAB routines that were introduced in this chapter are

`fsolve:` Solves a system of nonlinear algebraic equations, using quasi-Newton and Levenberg-Marquardt based algorithms
`fzero:` Solves for a single equation
`roots:` Solves for the roots of a polynomial equation

A number of other numerical routines are availiable through the MATLAB NAG Toolbox.

FURTHER READING

More detailed treatments of numerical methods given in the textbooks by Davis, Finlayson, and Riggs:

Davis, M.E. (1984). *Numerical Methods and Modeling for Chemical Engineers.* New York: Wiley.

Finlayson, B.A. (1980). *Nonlinear Analysis in Chemical Engineering.* New York: McGraw-Hill.

Riggs, J.B. (1994). *Numerical Techniques for Chemical Engineers,* 2nd ed. Lubbock: Texas Tech University Press.

Example numerical techniques are also presented by Felder and Rousseau:

> Felder, R.M., & R.W. Rousseau. (1986). *Elementary Principles of Chemical Processes,* 2nd ed. New York: Wiley.

The following book is more of an advanced undergraduate/first-year graduate student text on numerical methods to solve chemical engineering problems. The emphasis is on FORTRAN subroutines to be used with the IMSL (FORTRAN-based) package.

> Rameriz, W.F. (1989). *Computational Methods for Process Simulation.* Boston: Butterworths.

STUDENT EXERCISES

Single Variable Methods

1. Real gases do not normally behave as ideal gases except at low pressure or high temperature. A number of equations of state have been developed to account for the non-idealities (van der Waal's, Redlich-Kwong, Peng-Robinson, etc.). Consider the van der Waal's $P\hat{V}T$ relationship

$$\left(P + \frac{a}{\hat{V}^2}\right)(\hat{V} - b) = RT$$

   where P is pressure (absolute units), R is the ideal gas constant, T is temperature (absolute units), $\hat{V}$ is the molar volume and a and b are van der Waal's constants. The van der Waal's constants are often calculated from the critical conditions for a particular gas.

   Assume that a, b, and R are given. We find that if P and T are given, there is an iterative solution required for $\hat{V}$.

   a. How many solutions for $\hat{V}$ are there? Why? (Hint: Expand the PVT relationship to form a polynomial.)

   b. Recall that direct substitution has the form $\hat{V}_{k+1} = g\hat{V}_k$). Write three different direct substitution formulations for this problem (call these I, II, and III).

   c. What would be your first guess for $\hat{V}$ in this problem? Why?

   d. Write the MATLAB m-files to solve for $\hat{V}$ using direct substitution, for each of the three formulations developed in **b.**

   Consider the following system: air at 50 atm and −100°C. The van der Waal's constants are (Felder and Rousseau, p. 201) $a = 1.33$ atm liter^2/gmol^2, $b = 0.0366$ liter/gmol, and $R = 0.08206$ liter atm/gmol K. Solve the following problems numerically.

   e. Plot $\hat{V}$ as a function of iteration number for each of the direct substitution methods in **b.** Use 10 to 20 iterations. Also use the first guess that you calculated in **c.** Discuss the stability of each solution (think about the stability theorem).

f. Rearrange (1) to the form of $f(\hat{V}) = 0$ and plot $f(\hat{V})$ as a function of $\hat{V}$. How many solutions are there? Does this agree with your solution for **a**? Why or why not?

g. Use the MATLAB function `roots` to solve for the roots of the polynomial developed in **a**.

h. Write and use an m-file to solve for the equation in a using Newton's method. The numerical solution is considered converged when $\left| \dfrac{\hat{V}_k - \hat{V}_{k-1}}{\hat{V}_{k-1}} \right| < \varepsilon$. Let $\varepsilon = 0.0001$.

2. Consider the van der Waals relationship for a gas without the volume correction term ($b = 0$)

$$\left(P + \frac{a}{\hat{V}2} \right) \hat{V} = RT$$

a. How many solutions for $\hat{V}$ are there? Show the solutions analytically.

b. Which solution is correct?

3. A process furnace is heating 150 lbmol/hr of vapor-phase ammonia. The rate of heat addition to the furnace is 1.0×10^6 Btu/hr. The ammonia feedstream temperature is 550°R. Use Newton's method to find the temperature of the ammonia leaving the furnace. Assume ideal gas and use the following equation for heat capacity at constant pressure:

$$C_p = a + bT + cT^{-2}$$

$$a = 7.11 \; \frac{\text{Btu}}{\text{lbmol °R}} \quad b = 3.33 \times 10^{-3} \; \frac{\text{Btu}}{\text{lbmol °R}^2} \quad c = -1.20 \times 10^5 \; \frac{\text{Btu °R}}{\text{lbmol}}$$

and remember that $Q = \dot{n} \displaystyle\int_{T_{in}}^{T_{out}} C_p \, dT$

where $\dot{n}$ is the molar flowrate of gas and Q is the rate of heat addition to the gas per unit time. How did you determine a good first guess to use?

4. Consider Example 3.2 , $f(x) = -x^2 - x + 1 = 0$, with the direct substitution method formulated as $x = -x^2 + 1 = g(x)$, so that the iteration sequence is

$$x_{k+1} = g(x_k) = -x_k^2 + 1$$

Try several different initial conditions and show whether these converge, diverge, or oscillate between values. Discuss the stability of the two solutions $x^* = 0.618$ and $x^* = -1.618$, based on an analysis of $g'(x^*)$.

5. Show why the graphical Newton's method is equivalent to $x_{k+1} = x_k - f(x_k)/f'(x_k)$.

6. Develop an algorithm (sequence of steps) to solve an algebraic equation using interval halving (bisection).

7. Develop an algorithm (sequence of steps) to solve an algebraic equation using reguli falsi (false position). Compare and contrast this algorithm with Newton's method.

8. A component material balance around a chemical reactor yields the following steady-state equation

$$0 = \frac{F}{V} C_{in} - \frac{F}{V} C - kC^3$$

where $\frac{F}{V} = 0.1 \text{ min}^{-1}$, $C_{in} = 1.0 \frac{\text{lbmol}}{\text{ft}^3}$ and $k = 0.05 \frac{\text{ft}^6}{\text{lbmol}^2 \text{ min}}$

a. How many steady-state solutions are there?
b. Write two different direct substitution methods and assess the convergence of each.
c. Perform two iterations of Newton's method using an initial guess of $C = 1.0$.

9. Consider the following direct substitution problem, which results from an energy balance problem

$$T_{k+1} = \left[\frac{15.04}{0.716 - 4.257 \times 10^{-6} T_k} \right]^2$$

Will this method converge to the solution of $T = 443.571$? Why or why not?

10. Consider the following steady-state model for an exothermic zero-order reaction in a continuous stirred-tank reactor. The variable x is the dimensionless reactor temperature.

$$f(x) = 0.42204 \exp \frac{x}{1 + \frac{x}{20}} - 1.3x = 0$$

There are two solutions to this equation, as illustrated in the figure below. The solutions are $x = 0.55946$ and 2.00000.

a. Formulate a direct substitution solution to this problem. Will your direct substitution technique converge to 0.55946? Will your direct substitution technique converge to $x = 2$? Use a rigorous mathematical argument in each case.
b. For an initial guess of $x = 1.35$, would you expect Newton's method to have any trouble with this problem? Explain.
c. What is the next guess from the reguli falsi technique if your first guess was $x = 1$ and your second guess was $x = 2.5$ (show this mathematically)?
d. Write a function routine to solve this problem using the MATLAB function `fzero`. Also show the commands that you would give in the MATLAB command window to run `fzero`.

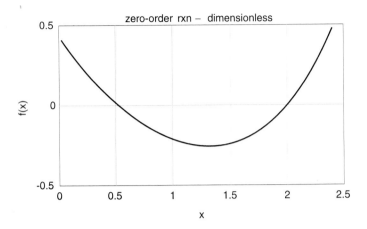

Plot of x versus $f(x)$ for problem 10.

11. A component material balance around a chemical reactor yields the following steady-state equation

$$0 = \frac{F}{V} C_{in} - \frac{F}{V} C - kC^{2.5}$$

where $\dfrac{F}{V} = 0.1 \text{ min}^{-1}$, $C_{in} = 1.0 \dfrac{\text{lbmol}}{\text{ft}^3}$ and $k = 0.05 \dfrac{\text{ft}^{4.5}}{\text{lbmol}^{1.5} \text{ min}}$

   a. Write two different direct substitution methods and assess the convergence of each to the steady-state concentration of 0.75354.

   b. Perform two iterations of Newton's method using an initial guess of $C = 1.0$.

12. Consider the dimensionless equations for an exothermic CSTR (continuous stirred tank reactor) shown in a module in the final section of the textbook.

$$f_1(x_1, x_2) = -\phi x_1 \kappa(x_2) + (1 - x_1) \quad = 0$$
$$f_2(x_1, x_2) = \beta\phi x_1 \kappa(x_2) + (1 + \delta)x_2 = 0$$

where $\kappa(x_2) = \exp\left(\dfrac{x_2}{1 + x_2/\gamma}\right)$

Use $f_1(x_1, x_2)$ to solve for x_1 in terms of x_2 and substitute into $f_2(x_1, x_2)$ to obtain the single equation

$$f(x_2) = \beta\phi \frac{1}{(\phi\kappa(x_2) + 1)} \kappa(x_2) - (1 + \delta)x_2 = 0$$

Consider the parameter values $\beta = 8$, $\phi = 0.072$, $\gamma = 20$, $\delta = 0.3$. Determine the number of solutions to this problem (graphically). Find the x_1 and x_2 values for each solution, using the single variable Newton's method.

Multivariable Methods

13. Use the multivariable Newton's method to solve the following problem. Solve this as a two-variable problem. Do not reduce it to a single-variable problem via substitution.

$$f_1(\mathbf{x}) = 2x_1 - x_2 - 5 = 0$$

$$f_2(\mathbf{x}) = -x_1 - x_2 + 4 = 0$$

How many iterations does it take to converge to the solution? Explain this result conceptually.

14. A simple bioreactor model (assuming steady-state operation) is

$$0 = \left(\frac{\mu_{max} x_2}{k_m + x_2 + k_1 x_2^2} - D\right) x_1$$

$$0 = (s_f - x_2) D - \left(\frac{x_1}{c}\right)\left(\frac{\mu_{max} x_2}{k_m + x_2 + k_1 x_2^2}\right)$$

where μ_{max} = 0.53
$\quad k_m$ = 0.12
$\quad k_1$ = 0.4545
$\quad c$ = 0.4
$\quad s_f$ = 4.0

x_1 is the biomass concentration (mass of cells) and x_2 is the substrate concentration (food source for the cells).

Find the steady-state values for x_1 and x_2 if $D = 0.3$ (There are three solutions).
a. Use fsolve and several initial guesses for the solution vector.
b. Perform a detailed analysis by hand (hint: the trivial solution $x_1 = 0$ and $x_2 = s_f$ should be easy to show.)

15. For the dimensionless CSTR problem (module 9 in Section V), use the MATLAB routine fsolve to find the solutions. Show the initial guesses and the solutions that fsolve converges to. Show your function routine as well as the calls to MATLAB.

$$f_1(x_1,x_2) = -\phi x_1 \kappa(x_2) + (1 - x_1) = 0$$
$$f_2(x_1,x_2) = \beta\phi x_1 \kappa(x_2) - (1 + \delta)x_2 = 0$$

where $\kappa(x_2) = \exp\left(\dfrac{x_2}{1 + x_2/\gamma}\right)$

and the following parameters are used

$$\beta = \quad 8 \quad \phi = 0.072$$

$$\gamma = \quad 20 \quad \delta = 0.3$$

APPENDIX

Stability of Numerial Solutions—Single Equations

If the iterates (x_k) from a numerical algorithm converge to a solution, we refer to that solution as being stable.

Definition 3.1

Let x^* represent the solution (fixed point) of $x^* = g(x^*)$, or $g(x^*) - x^* = 0$.

Theorem 3.1

x^* is a stable solution of $x^* = g(x^*)$, if $\left|\frac{\partial g}{\partial x}\right| < 1$ evaluated at x^*. x^* is an unstable solution of $x^* = g(x^*)$, if $\left|\frac{\partial g}{\partial x}\right| > 1$ evaluated at x^*.

If $\left|\frac{\partial g}{\partial x}\right| = 1$, then no conclusions can be drawn. For simplicity in notation, we generally use g' to represent $\left|\frac{\partial g}{\partial x}\right|$.

We continue with Example 3.2 to illustrate the numerical stability of direct substitution.

EXAMPLE 3.2 Continued.   Stability of direct substitution

Consider the **Case 1** formulation,

$$x = g(x) = \sqrt{-x + 1}$$

which has the derivative

$$\frac{\partial g}{\partial x} = g' = \frac{-1}{2\sqrt{-x + 1}}.$$

If $\left|\frac{-1}{2\sqrt{-x^* + 1}}\right| < 1$, then x^* is a stable solution.

1. For $x^* = 0.618$, we find that $g'(x^*) = \frac{-1}{2\sqrt{-0.618 + 1}} = 0.8090 < 1$. So $x^* = 0.618$ is a stable solution, that is, an initial "guess" for x_0 close to 0.618 will converge to $x^* = 0.618$.

2. For $x^* = -1.618$, we find $|g'(x^*)| = \frac{-1}{2\sqrt{1.618 + 1}} = 0.3090 < 1$. So $x^* = -1.618$ should be a stable solution, that is, an initial "guess" for x_0 close to -1.618 should converge to $x^* = -1.618$.

Question: Why did we find previously that an initial guess close to −1.618 did not converge?

Observation: We must realize that the square root of a positive number has two values. For example, the square root of 1 can either be +1 or −1 (after all $(-1)^2 = 1!$).

You may be questioning the utility of a stability test for the direct substitution method, which requires that the solution be known to apply the test. Our purpose is mainly to show that the direct substitution method can be unstable. Newton's method guarantees stable solutions, if the "guess" is close to the solution.

Stability of Newton's Method for Single Equations

Here we use Theorem 3.1 to show that Newton's method is stable. We see that Newton's method can be written in the form of

$$x_{k+1} = g(x_k)$$

where

$$g(x_k) = x_k - \frac{f(x_k)}{f'(x_k)}$$

Then we can find that the derivative, $g'(x_k)$ is

$$g'(x_k) = 1 - \frac{f'(x_k)}{f'(x_k)} + \frac{f(x_k)\,f''(x_k)}{[f'(x_k)]'^2}$$

or

$$g'(x_k) = \frac{f(x_k)\,f''(x_k)}{[f'(x_k)]^2}$$

At the solution, x^*, we see that

$$g'(x^*) = \frac{f(x^*)\,f''(x^*)}{[f'(x^*)]^2}$$

And since $f(x^*) = 0$, we find that the stability constraint is satisfied

$$g'(x^*) = 0$$

as long as $f'(x^*) \neq 0$. This shows that Newton's method will converge to the solution, provided an initial guess close to the solution.

NUMERICAL INTEGRATION

4

Most chemical process models are nonlinear and rarely have analytical solutions. This chapter introduces numerical solution techniques for the integration of initial value ordinary differential equations. After studying this material, the student should be able to:

- Understand the difference between explicit and implicit Euler integration.
- Write MATLAB code to implement fixed step size Euler and Runge-Kutta techniques.
- Use the MATLAB ode45 integration routine.

The major sections of this chapter are:

4.1 Background

4.2 Euler Integration

4.3 Runge-Kutta Integration

4.4 MATLAB Integration Routines

4.1 BACKGROUND

Thus far we have developed modeling equations (Chapter 2) and solved for the steady states (Chapter 3). One purpose of developing dynamic models is to be able to perform "what if" types of studies. For example, you may wish to determine how long a gas storage tank will take to reach a certain pressure if the outlet valve is closed. This requires integrating the differential equations from given initial conditions. If the dynamic equations are linear, then we can generally obtain analytical solutions; these techniques will be presented in Chapters 6 and 8 through 10. Even when systems are linear, we may wish to use numerical methods rather than analytical solutions.

At this point it is worth reviewing the difference between linear and nonlinear differential equations. An example of a linear ordinary differential equation is:

$$\frac{dx}{dt} = -x$$

since the rate of change of the dependent variable is a linear function of the dependent variable. An example of a nonlinear differential equation is:

$$\frac{dx}{dt} = -x^2$$

since the rate of change of the dependent variable is a nonlinear function of the dependent variable. Although this particulary nonlinear equation has an analytical solution, this will not normally be the case, particularly for sets of nonlinear equations. Notice that the following equation is linear:

$$\frac{dx}{dt} = -e^{-t}x$$

since the only nonlinearity is in the independent variable (t).

The purpose of this chapter is to introduce you to numerical techniques for integrating initial value ordinary differential equations. The first numerical integration technique that we will present is the *Euler* integration. In the next section we discuss two algorithms, the *explicit* and the *implicit* Euler methods.

4.2 EULER INTEGRATION

Consider a single variable ODE with the form

$$\frac{dx}{dt} = \dot{x} = f(x) \tag{4.1}$$

We consider two different approximations to the derivative. In Section 4.2.1 we consider a forward difference approximation, which leads to the explict Euler method. In Section 4.2.2 we consider a backwards differences approximation, which leads to the implicit Euler method.

4.2.1 Explicit Euler

If we use a forward difference approximation for the time derivative of (4.1), we find

$$\frac{dx}{dt} \approx \frac{x(k+1) - x(k)}{t(k+1) - t(k)} \tag{4.2}$$

where k represents the kth discrete time step of the integration. Now, assume that $f(x)$ is evaluated at $x(k)$. We will refer to this function as $f(x(k))$, and can write (4.1) and (4.2) as

$$\frac{x(k+1) - x(k)}{t(k+1) - t(k)} = f(x(k)) \tag{4.3}$$

Normally we will use a fixed increment of time, that is, $t(k+1) - t(k) = \Delta t$, where Δt is the *integration step size*. Then we write (4.3) as

$$\frac{x(k+1) - x(k)}{\Delta t} = f(x(k)) \tag{4.3}$$

Solving for $x(k+1)$

$$x(k+1) = x(k) + \Delta t\, f(x(k)) \quad \textit{Explicit Euler} \tag{4.4}$$

We can view (4.4) as a prediction of x at $k+1$ based on the value of x at k and the slope at k, as shown in Figure 4.1.

Equation (4.4) is the expression for the *explicit Euler* method for a single variable. The general statement for a multivariable problem is

$$\mathbf{x}(k+1) = \mathbf{x}(k) + \Delta t\, \mathbf{f}(\mathbf{x}(k)) \tag{4.5}$$

Where $\mathbf{x}(k)$ is a vector of state variable values at time step k and $\mathbf{f}(\mathbf{x}(k))$ is a vector of functions evaluated at step k. Equations (4.4) and (4.5) are *explicit* because the state variable value at time step $k+1$ is only a function of the variable values at step k. This method is straightforward and easy to program on either a handheld calculator or a computer. A major disadvantage is that a small step size must be used for accuracy. However, if too small of a step size is used, then numerical truncation problems may result. Explicit Euler is not often used in practice, but is covered here for illustrative purposes.

4.2.2   Implicit Euler

This method uses a backwards difference approximation for the derivative in (4.1). The function (or vector of functions) is evaluated at time step $k+1$ rather than time step k:

$$\frac{x(k+1) - x(k)}{\Delta t} = f(x(k+1)) \tag{4.6}$$

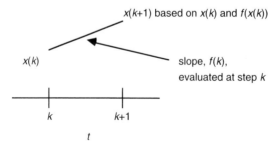

FIGURE 4.1   Pictorial representation of the explicit Euler method.

which can be written

$$x(k + 1) = x(k) + \Delta t f(x(k + 1)) \quad \textit{Implicit Euler} \tag{4.7}$$

Equation (4.7) is *implicit* because the value $x(k + 1)$ must be known in order to solve for $x(k + 1)$. What this generally requires is a nonlinear algebraic solution technique, such as Newton's method. For a linear system, equation (4.7) can be explicitly solved to obtain the form

$$x(k + 1) = g(x(k))$$

The following section compares the explicit and implicit methods for a single linear ordinary differential equation. In particular, we compare how the integration step size (Δt) affects the stability of each method.

4.2.3 Numerical Stability of Explicit and Implict Euler Methods

Consider the tank height problem covered in Example 2.1. There was no inflow, and the outlet flowrate was assumed to be linearly related to height, which gave us the following equation:

$$\frac{dx}{dt} = -\frac{1}{\tau}x = \lambda x = f(x) \tag{4.8}$$

where $\tau = A/\beta$, $\lambda = -1/\tau$ and the state variable x is the tank height. Since the variables are separable, the reader should show that the *analytical solution* is:

$$x(t) = x(0)e^{-t/\tau} = x(0)e^{\lambda t} \tag{4.9}$$

Next, we compare this analytical solution with the explicit and implicit Euler solutions.

EXPLICIT EULER

The function value at step k is:

$$f(x(k)) = -\frac{1}{\tau}x(k)$$

and the state variable value at the next time step is:

$$x(k + 1) = x(k) + -\frac{\Delta t}{\tau}x(k) = \left(1 - \frac{\Delta t}{\tau}\right)x(k) \tag{4.10}$$

IMPLICIT EULER

The implicit Euler method evaluates the function at $k + 1$ rather than k:

$$f(x(k + 1)) = -\frac{1}{\tau}x(k + 1)$$

and the state variable at step $k + 1$ is:

$$x(k + 1) = x(k) + -\frac{\Delta t}{\tau} x(k + 1) \qquad (4.11)$$

Notice that, since this is a linear problem, a nonlinear algebraic equation solver is not needed for (4.11). We can rewrite (4.11) as

$$x(k + 1) = \frac{1}{1 + \dfrac{\Delta t}{\tau}} x(k) \qquad (4.12)$$

In the next section we compare the numerical stability of the explicit and implicit Euler methods.

NUMERICAL STABILITY

The *explicit* Euler solution, written in terms of the initial condition, is (from (4.10)):

$$x(k + 1) = \left(1 - \frac{\Delta t}{\tau}\right)^{k+1} x(0)$$

which will be stable if $\left| 1 - \Delta t / \tau \right| < 1$. This is the same result if we use the representation:

$$x(k + 1) = g(x(k)) = \left(1 - \frac{\Delta t}{\tau}\right) x(k)$$

and the stability requirement that $\left| g' \right| < 1$. The explicit Euler method is then stable if:

$$-1 < 1 - \frac{\Delta t}{\tau} < 1$$

and will oscillate for:

$$-1 < 1 - \frac{\Delta t}{\tau} < 0$$

These criteria lead to the explicit Euler stability condition of:

$$0 < \Delta t < 2\,\tau$$

while the solution will have a stable, oscillatory solution for:

$$\tau < \Delta t < 2\,\tau$$

and a stable, monotonic solution for:

$$0 < \Delta t < \tau$$

The *implicit* Euler solution, written in terms of the initial condition, is (from (4.12)):

$$x(k + 1) = \left(\frac{1}{1 + \dfrac{\Delta t}{\tau}}\right)^{k+1} x(0)$$

which will be stable if $|1/1+\Delta t,\tau| < 1$. This is the same result if we use the representation

$$x(k + 1) = g(x(k)) = \left(\cfrac{1}{1 + \cfrac{\Delta t}{\tau}} \right) x(k)$$

and the stability requirement that $|g' \neq <1$. Notice that the implicit Euler method is stable for any value of Δt (as long as the sign of Δt is correct) and will not oscillate.

EXAMPLE 4.1   Numerical Comparison of Explicit and Implicit Euler

Let $x(0) = 4$, $\tau = 5$, and $\Delta t = 1$. Table 4.1 compares the exact solution (4.9) with the explicit Euler (4.10) and implicit Euler (4.12) methods.

TABLE 4.1   Linear First Order Example ($\tau = 5$, $\Delta t = 1$)

t	x, exact	x, Explicit Euler	error	x, Implicit Euler	error
0	4.0000	4.0000		4.0000	
1	3.2749	3.2000	−2.3%	3.3333	1.8%
2	2.6813	2.5600	−4.5%	2.7778	3.6%
3	2.1952	2.0480	−6.7%	2.3148	5.4%
4	1.7979	1.6384	−8.9%	1.9290	7.3%
5	1.4715	1.3107	−10.9%	1.6075	9.2%

The results shown in Table 4.1 are illustrated graphically by the curves in Figure 4.2.

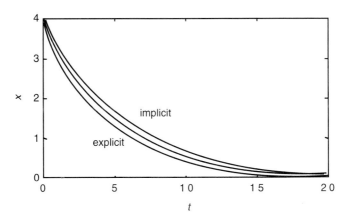

FIGURE 4.2   Comparison of the exact solution with the explicit and implicit Euler for $\Delta t = 1$.

Larger Integration Step Size.   We have seen the well-behaved response for $\Delta t = 1$ (which is $\tau/5$). Consider now a larger Δt. We can see from (4.10) that the explicit Euler method predicts $x = 0$ for all time after time 0, if $\Delta t = \tau$. Indeed, the explicit Euler solution is oscillatory for $\Delta t > \tau$, for this process. For example, let $\Delta t = 6$ for this problem. The results are shown in Table 4.2. These results are illustrated more graphically by the curves plotted in Figure 4.3. The implicit Euler technique has monotonic behavior and more closely approximates the exact solution. We see that the implicit Euler method can tolerate a larger integration step size than the explicit Euler technique.

TABLE 4.2   Linear First Order Example ($\tau = 5$, $\Delta t = 6$)

t	x, exact	x, Explicit Euler	x, Implicit Euler
0	4.0000	4.0000	4.0000
6	1.2048	−0.8000	1.8182
12	0.3629	0.1600	0.8264
18	0.1093	−0.0320	0.3757
24	0.0329	0.0064	0.1708

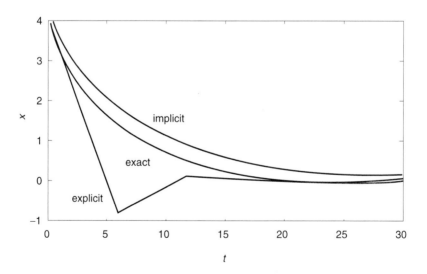

FIGURE 4.3   Comparison of exact solution with implicit and explicit Euler for $\Delta t = 6$.

We have seen that there is a limit to how large a step size can be tolerated by the explicit Euler method before it goes unstable, while the Implicit Euler method remains stable for any step size. This is true for a simple linear ordinary differential equation. The following example illustrates an important issue when solving nonlinear equations. That is, the implicit Euler method requires an iterative solution at each time step.

EXAMPLE 4.2   Explicit and Implicit Euler—Nonlinear System

Consider a surge tank with an outlet flowrate that depends on the square root of the height of liquid in the tank. When there is no inlet flow, the model has the following form

$$\frac{dx}{dt} = -a\sqrt{x} \qquad (4.13a)$$

where x is the tank height and $a = \beta/A$ (= flow coefficient/cross-sectional area).

Analytical solution (exact).   The analytical solution is

$$x(t) = [\sqrt{x(0)} - at/2]^2 \qquad (4.13b)$$

Next, we compare this solution with the explicit and implicit Euler solutions.

Explicit Euler.   The explicit Euler method yields the following equation

$$x(k + 1) = x(k) - \Delta t\, a\sqrt{x(k)}) \qquad (4.14)$$

For the numerical example of $a = 0.8$, curves for 3 different Δts are shown in Figure 4.4.

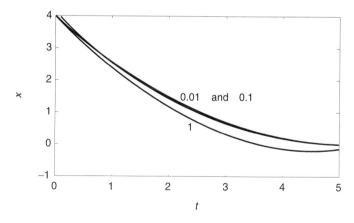

FIGURE 4.4   Explicit Euler solution. $\Delta t = 0.01$, 0.1, and 1.0. The 0.01 and 0.1 step sizes yield virtually identical results, while there is a significant error in a step size of 1.0.

Implicit Euler.   The implicit Euler method yields the following equation

$$x(k + 1) = x(k) - \Delta t\, a\sqrt{x(k + 1)} \qquad (4.15)$$

Recall that when we were dealing with a linear equation we were able to rewrite the implicit Euler equation to yield an explicit calculation of the state variable at the next time step. Here we see that this is impossible for a nonlinear equation. Rewriting (4.15),

$$x(k + 1) + \Delta t\, a\sqrt{x(k + 1)} - x(k) = 0 \qquad (4.16)$$

we see that (4.16) requires an iterative solution. That is, at time step $k + 1$ we must use a numerical method that will solve a nonlinear algebraic equation. We know from Chapter 3 that a number of techniques, including Newton's method, can be used. To illustrate clearly one approach that we can take, let y represent the value of x at step $k + 1$ for which we desire to find the solution. We can rewrite (4.16) as

$$y + b\sqrt{y} - c = 0 \tag{4.17}$$

where $y = x(k + 1)$, $b = \Delta t\, a$, and $c = x(k)$. If Newton's method is used, we can write:

$$g(y(i)) = y(i) + b\sqrt{y(i)} - c \tag{4.18}$$

where (i) is an index to indicate the ith iteration of Newton's method.

$$y(i + 1) = y(i) - \frac{g(y(i))}{g'(y(i))} \tag{4.19}$$

where

$$g'(y(i)) = 1 + \frac{b}{2\sqrt{y(i)}} \tag{4.20}$$

We can write (4.19) as (from (4.18), (4.19), and (4.20))

$$y(i + 1) = y(i) - \left[\frac{y(i)^{1.5} + b\, y(i) - c\sqrt{y(i)}}{\sqrt{y(i)} + \dfrac{b}{2}} \right] \tag{4.21}$$

Equation (4.21) is then iterated to convergence.

Summarizing, at step $k+1$ of an integration, we must iteratively solve for the value of x at $k+1$. That is, (4.21) is iteratively solved to convergence, in order to find $x(k + 1)$ in (4.16).

COMMENT ON IMPLICIT INTEGRATION TECHNIQUES

We have seen that the implicit Euler method is more accurate and stable than the explicit Euler method. We also noted that, for nonlinear systems, a nonlinear equation must be iteratively solved at each time step. The implicit Euler method allows a much larger time step, but some of the computational savings much be sacrificed in the interative solution of the nonlinear algebraic equation at each time step. There are a number of more advanced implicit methods that are used in a number of commercially available integration codes. In this text we emphasize *explicit* techniques, which are used by the MATLAB routines `ode23` and `ode45`. SIMULINK has choices of some implicit integration routines.

An explicit technique that is more accurate than the explicit Euler technique is known as the Runge-Kutta method and is shown in the next section.

4.3 RUNGE-KUTTA INTEGRATION

This technique is an extension of the Euler method. In the Euler method, the derivative at time step k was used to predict the solution at step $k+1$. *Runge-Kutta* methods use the Euler technique to predict the x value at intermediate steps, then use averages of the slopes at intermediate steps for the full prediction from the beginning of the time step.

4.3.1 Second-Order Runge-Kutta

The first Runge-Kutta approach that we discuss is the second-order Runge-Kutta method, which is also known as the midpoint Euler method, for reasons that will become clear. The Euler technique is first used to predict the state value at $\Delta t/2$. The derivative is evaluated at this midpoint, and used to predict the value of x at the end of the step, Δt, as shown in Figure 4.5.

Let m_1 represent the slope at the initial point and m_2 represent the slope (dx/dt) at the midpoint:

$$m_1 = f(t(k), x()) \tag{4.22}$$

$$m_2 = f\left(t(k) + \frac{\Delta t}{2}, x(k) + \frac{\Delta t}{2} m_1\right) \tag{4.23}$$

$$x(k + 1) = x(k) + m_2 \Delta t \tag{4.24}$$

For autonomous systems, the derivative functions are not explicitly functions of time, so (4.22) and (4.23) can be written:

$$m_1 = f(x(k)) \tag{4.22a}$$

$$m_2 = f\left(x(k) + \frac{\Delta t}{2} m_1\right) \tag{4.23a}$$

or

$$m_2 = f\left(x(k) + \frac{\Delta t}{2} f(x(k))\right) \tag{4.23b}$$

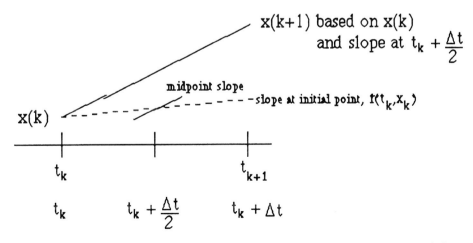

FIGURE 4.5 Pictorial representation of second-order Runge-Kutta (midpoint Euler) technique.

Equation (4.24) can now be written:

$$x(k+1) = x(k) + \Delta t f\left(f\left(x(k) + \frac{\Delta t}{2}f(x(k))\right)\right) \tag{4.24a}$$

which is of the form:

$$x(k+1) = g(x(k))$$

It should be noted that the second-order Runge-Kutta is accurate to the order of Δt^2, while the explicit Euler is accurate to the order of Δt.

EXAMPLE 4.3   Second-order Runge-Kutta (Midpoint Euler) Technique

Consider again the first-order process:

$$\frac{dx}{dt} = f(x) = -\frac{1}{\tau}x \tag{4.8}$$

$$m_1 = f(x(k)) = \left(-\frac{1}{\tau}x(k)\right) = \frac{-1}{\tau}x(k) \tag{4.25}$$

$$m_2 = f\left(x(k) + \frac{1}{2}m_1\Delta t\right) = \frac{-1}{\tau}\left[\left(x(k) - \frac{\Delta t}{2\tau}x(k)\right)\right] \tag{4.26}$$

For $\Delta t = 1$, $\tau = 5$, and $x(0) = 4.0$:

From (4.25) $m_1 = \frac{-1}{5}x(0) = \frac{-4}{5}$

From (4.26) $m_2 = \frac{-1}{5}\left(1 - \frac{1}{10}\right)x(0) = \frac{-1}{5}(1-0.1)(4) = -\frac{3.6}{5}$

From (4.24) $x(1) = x(0) + m_2\Delta t$

$$x(1) = 4.0 - 0.72 = 3.2800$$

Compare this with the analytical result of $4.0\,e^{-1/5} = 3.2749$

Notice that the error is 0.16%. Contrast this with an error of −2.3% from Table 4.1 for the Euler method. For the same step size, Runge-Kutta techniques are more accurate than the standard Explicit Euler technique.

Thus far, we have used single variable examples. The next example is for a two-state variable system.

EXAMPLE 4.4   Two-state Variable System, Second-order Runge-Kutta Method

Consider two interacting tanks in series, shown in Figure 4.6, with outlet flowrates that are a function of the square root of tank height. Notice that the flow from tank 1 is a function of $\sqrt{h_1 - h_2}$, while the flowrate out of tank 2 is a function of $\sqrt{h_2}$.

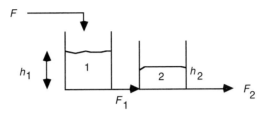

FIGURE 4.6   Interacting tanks.

The following modeling equations describe this system

$$\begin{bmatrix} \dfrac{dh_1}{dt} \\[2mm] \dfrac{dh_2}{dt} \end{bmatrix} = \begin{bmatrix} f_1(h_1,h_2,F) \\[1mm] f_2(h_1,h_2,F) \end{bmatrix} = \begin{bmatrix} \dfrac{F}{A_1} - \dfrac{\beta_1}{A_1}\sqrt{h_1 - h_2} \\[3mm] \dfrac{\beta_1}{A_2}\sqrt{h_1 - h_2} - \dfrac{\beta_2}{A_2}\sqrt{h_2} \end{bmatrix} \qquad (4.27)$$

For the following parameter values:

$$\beta_1 = 2.5\ \frac{\text{ft}^{2.5}}{\text{min}} \quad \beta_2 = \frac{5}{\sqrt{6}}\ \frac{\text{ft}^{2.5}}{\text{min}} \quad A_1 = 5\ \text{ft}^2 \ A_2 = 10\ \text{ft}^2$$

and the input: $F = 5$ ft³/min
the steady-state height values are:

$$h_{1s} = 10 \quad h_{2s} = 6$$

Numerically, we can write (4.27) as:

$$\begin{bmatrix} \dfrac{dh_1}{dt} \\[2mm] \dfrac{dh_2}{dt} \end{bmatrix} = \begin{bmatrix} f_1(h_1,h_2) \\[1mm] f_2(h_1,h_2) \end{bmatrix} = \begin{bmatrix} 1 - 0.5\sqrt{h_1 - h_2} \\[3mm] 0.25\sqrt{h_1 - h_2} - \dfrac{1}{2\sqrt{6}}\sqrt{h_2} \end{bmatrix} \qquad (4.28)$$

Since this system is autonomous (no explicit dependence on time), we can leave t out of the arguments:

$$m_1 = f(h(k)) = \begin{bmatrix} f_1(h_1(k), h_2(k)) \\[1mm] f_2(h_1(k), h_2(k)) \end{bmatrix}$$

$$m_2 = \begin{bmatrix} f_1(h_1(k) + \dfrac{\Delta t}{2} m_{11}, h_2(k) + \dfrac{\Delta t}{2} m_{21}) \\[4mm] f_2(h_1(k) + \dfrac{\Delta t}{2} m_{11}, h_2(k) + \dfrac{\Delta t}{2} m_{21}) \end{bmatrix}$$

$$h(k + 1) = \begin{bmatrix} h_1(k + 1) \\ h_2(k + 1) \end{bmatrix} = \begin{bmatrix} h_1(k) \\ h_2(k) \end{bmatrix} + m_2 \Delta t$$

Let the initial conditions be $h_1(0) = 12$ ft and $h_2(0) = 7$ ft. Also, let $\Delta t = 0.2$ minutes. For $k = 0$, we find

$$m_1 = \begin{bmatrix} 1 - 0.5 \sqrt{12 - 7} \\ 0.25 \sqrt{12 - 7} - \dfrac{1}{2\sqrt{6}} \sqrt{7} \end{bmatrix} = \begin{bmatrix} -0.118034 \\ 0.018955 \end{bmatrix}$$

so

$$\begin{bmatrix} h_1(0) \\ h_2(0) \end{bmatrix} + \frac{\Delta t}{2} m_1 = \begin{bmatrix} 12 + \dfrac{0.2}{2}(-0.118034) \\ 7 + \dfrac{0.2}{2}(0.01\text{?}\,\text{?}5) \end{bmatrix} = \begin{bmatrix} 11.988197 \\ 7.001896 \end{bmatrix}$$

$$h_1(0) + \frac{\Delta t}{2} m.\qquad) = 11.988197 \text{ ft}$$

$$)1896 \text{ ft}$$

$$197, 7.001896)$$

$$.001896)$$

$$\text{...}i16$$

$$h_1(\text{...})\qquad \text{...}\text{?}01\ (0.2) = 11.976700 \text{ ft}$$

$$h_2(1)\qquad - 7 + 0.018116\ (0.2) = 7.003623 \text{ ft}$$

and we can continue ... ...he next time step, $k = 1$. A plot of the response of this system is shown in Figure 4.7. The response of h_2 actually increases slightly before decreasing—this is missed because of the scaling.

Notice that when h_1 is greater than h_2, the flow is from tank 1 to tank 2; while when h_1 is less than h_2, the flow is from tank 2 to tank 1 (although this cannot occur at steady-state). Since we have assigned a positive value to F_1 when the flow is from tank 1 to tank 2, then a negative value of F_1 indicates the opposite flow. Care must be taken when solving this problem numerically, so that the square root of a negative number is not taken. For this purpose, the sign function is used

$$F_1 = \beta_1 \, \text{sign}(h_1 - h_2) \sqrt{|h_1 - h_2|}$$

where sign $(h_1 - h_2) = 1$ if $h_1 > h_2$ and sign $(h_1 - h_2) = -1$ if $h_2 > h_1$.

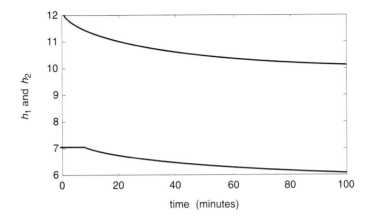

FIGURE 4.7   Transient response of the interacting tank example, using second-order Runge-Kutta.

The idea behind the second-order Runge-Kutta can be extended to higher-orders. The most commonly used method is the *fourth-order Runge-Kutta* method as outlined in the next section.

4.3.2   Fourth-Order Runge-Kutta

Using this method, the approximations are more accurate than explicit Euler or second-order Runge-Kutta. The idea is to use the initial slope (m_1) to generate a first guess for the state variable at the midpoint of the integration interval. This first guess is then used to find the slope at the midpoint (m_2). A "corrected" midpoint slope (m_3) is then found by using m_2. A final slope (m_4) is found at the end of a step using m_3. A weighted average of these slopes is then used for the integration. The algorithm is

$$m_1 = f(t(k), x(k)) \tag{4.29}$$

$$m_2 = f\left(t(k)) + \frac{1}{2}\Delta t, x(k) + \frac{1}{2}m_1\Delta t\right) \tag{4.30}$$

$$m_3 = f\left(t(k) + \frac{1}{2}\Delta t, x(k) + \frac{1}{2}m_2\Delta t\right) \tag{4.31}$$

$$m_4 = f(t(k) + \Delta t, x(k) + m_3\Delta t) \tag{4.32}$$

$$x(k + 1) = x(k) + \left[\frac{m_1}{6} + \frac{m_2}{3} + \frac{m_3}{3} + \frac{m_4}{6}\right] \Delta t \qquad (4.33)$$

To become more familiar with integration techniques, you should solve some of your initial problems using the explicit Euler method. Make certain that your step size is small enough so that the errors do not build up too rapidly. As you find a need for more accuracy, you should then use the fourth-order Runge-Kutta method. In Section 4.4 we introduce the MATLAB routines that are available for numerical integration.

SELECTION OF INTEGRATION STEP SIZE

Generally, integration step size must be "small" for Euler, can be larger for second-order Runge-Kutta (as far as accuracy is concerned), and can be still larger for fourth-order Runge-Kutta. Particularly for Euler, step sizes that are too large can be unstable or inaccurate. Step sizes that are too small may waste computer time or have numerical truncation errors since the state variables may not change much from step to step. If the student uses a fixed step size, then it is generally a good idea to try larger and smaller step sizes to see if the results change significantly. Generally, you will want to use as large a step size as possible. A particular challenge is from "stiff" systems (time constants that span a wide range), where a commerical code specifically for stiff systems should be used. One well-known implicit method for stiff systems is Gear's method, which is available in SIMULINK. Implicit methods will only work well for systems that are continous. If discontinuous systems are simulated (for example, step changes at certain times), then Runge-Kutta methods should be used.

Most commerical integration packages use a variable step size. The integration step size is automatically chosen and varied from step-to-step to assure accuracy while minimizing computation time. The integration routines in MATLAB use a variable integration step size.

4.4 MATLAB INTEGRATION ROUTINES

The primary purpose of the previous sections in this chapter was to review simple numerical techniques for integrating initial value ordinary differential equations. We have illustrated the techniques with some simple numerical examples, implemented as m-files in MATLAB. In practice, we do not recommend that you write your own integration routines. You will spend much time debugging these routines and they will generally not be as powerful as existing academic or commerical integration routines. Your goal should be to provide the correct formulation of the model, specifying the correct initial conditions and parameters. You should generally use a well-documented, commerical or public domain integration code to implement your simulation.

MATLAB has several routines for numerical integration; two are ode23 and ode45. ode23 uses second-order and ode45 uses fourth-order Runge-Kutta integration. Both routines use a *variable integration step size* (Δt is not constant). The integration step size is adjusted by the routine to provide the necessary accuracy, without taking too much computation time.

4.4.1   `ode23` and `ode45`

To use `ode23` or `ode45`, the reader must first generate an m-file to evaluate the state variable derivatives. Then the student gives the command:

$$[\texttt{t,x}]=\texttt{ode45('xprime',[t0,tf]x0)}$$

where

`xprime`	is a string variable containing the name of the m-file for the derivatives
`t0`	is the initial time
`tf`	is the final time
`x0`	is the initial condition vector for the state variables (usually a column vector)

The arrays that are returned are

`t`	a (column) vector of time
`x`	an array of state variables as a function of time (column 1 is state 1, etc.)

For example, if the time vector has 50 elements, and there are three state variables, then the state variable vector has the 50 rows and three columns. After the integration is performed, if the student wishes to plot all three variables as a function of time, she/he simply types

$$\texttt{plot(t,x)}$$

If you only want to plot the second state variable, then the command `plot(t,x(:,2))` is given.

EXAMPLE 4.4   Revisited Solution Using MATLAB Routine `ode45`

First, the following file titled `twotnk.m` was generated:

```
function hdot = twotnk(t,h);

const=(1/(2*sqrt(6)));

hdot(1) = 1-0.5*sqrt(h(1)-h(2));

   hdot(2) = 0.25*sqrt(h(1)-h(2))-const*sqrt(h(2));
```

Then, the following command is entered in the MATLAB command window:

```
[t,h]=ode45('twotnk',[0100],[12 7]');
```

Notice that we are generating two arrays, t and h, and using `ode45`. The function file is named `twotnk.m`. The initial time is $t0 = 0$ and the final time is $tf = 100$. The initial condition is $h0 = [12\ 7]'$. At the MATLAB prompt (») the following commands were given:

$$plot(t,h(:,1))$$

$$plot(t,h(:,2))$$

The transient responses are shown in Figure 4.8.

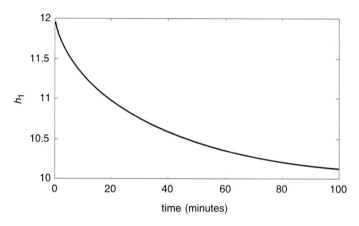

a. Height of Tank 1

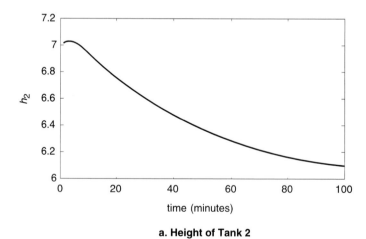

a. Height of Tank 2

FIGURE 4.8   Transient response curves for interacting tank example.

Notice the tremendous reduction in effort when compared with generating your own Runge-Kutta code.

Often it is desirable to know the state variable values at a particular time or at fixed time steps. A variable step size algorithm yields variable values that are not at a fixed step sizes. One has two options. If the variable step size is smaller than that of the variable step, then we could reduce the step size. The major disadvantage is that computation time will increase.

The best option (and that recommended by MATLAB) is to use a spline fit to interpolate or extrapolate the values to desired points. The routine used is `interp1`.

SUMMARY

It is important for the student to understand the Euler, as well as the second and fourth order Runge-Kutta integration techniques. When using your own fixed step size integration code, be careful with the selection of Δt. In practice, it is preferable to use a commercial integration code, which automatically selects the integration step size.

The MATLAB routines used were

ODE23:   Variable step size, second-order Runge-Kutta
ODE45:   Variable step size, fourth-order Runge-Kutta

FURTHER READING

A nice treatment of numerical integration is provided by:

Parker, T.S., & L.O. Chua. (1989). *Practical Numerical Algorithms for Chaotic Systems*. New York: Springer-Verlag.

A treatment of integration techniques with chemical engineering applications is presented by Davis.

Davis, M. E. (1984). *Numerical Methods and Modeling for Chemical Engineers*. New York: Wiley.

The following book is more of an advanced undergraduate/first-year graduate student text on numerical methods to solve chemical engineering problems. The emphasis is on FORTRAN subroutines to be used with the IMSL (FORTRAN-based) package.

Rameriz, W.F. (1989). *Computational Methods for Process Simulation*. Boston: Butterworths.

STUDENT EXERCISES

1. Consider the scaled predator-prey equations.

$$\frac{dy_1}{dt} = \alpha(1 - y_2)\, y_1$$

$$\frac{dy_2}{dt} = -\beta(1 - y_1)y_2$$

The parameters are $\alpha = \beta = 1.0$ and the initial conditions are $y_1(0) = 1.5$ and $y_2(0) = 0.75$. The time unit is days.

   a. Solve these equtions using explicit Euler integration. Compare various integration step sizes. What Δt do you recommend? In addition to transient responses (t versus y_1 and y_2), also plot "phase-plane" plots (y_1 versus y_2).

   b. Solve these equations using the MATLAB integration routine ode45. Compare the transient response curves with the Euler results.

   c. How do the initial conditions effect the response of y_1 and y_2? Please elaborate.

2. Consider a CSTR with a second-order reaction. Assume that the rate of reaction (per unit volume) is proportional to the square of the concentration of the reacting component. Assuming constant volume and constant density, show that the modeling equation is:

$$\frac{dC}{dt} = \frac{F}{V}C_i - \frac{F}{V}C - k_2\,C^2$$

Use the following parameters:

$$\frac{V}{F} = 5 \text{ min} \quad k_2 = 0.32 \text{ ft}^3 \text{ lbmol}^{-1} \text{ min}^{-1}$$

and a steady-state inlet concentration of

$$C_{is} = 1.25 \text{ lbmol ft}^{-3}$$

Calculate the steady-state concentration of $C_s = 0.625$ lbmol ft^{-3}.

   Assume that a step change in the inlet concentration occurs at $t = 0$. That is, C_i changes from 1.25 lbmol ft^{-3} to 1.75 lbmol ft^{-3} at $t = 0$ minutes. Use ode45 to simulate how the outlet concentration changes as a function of time.

3. Analyze the stability of the fourth order Runge-Kutta method for the classical first-order process.

$$\frac{dx}{dt} = -x$$

What is the largest integration step size before the numerical solution becomes unstable?

4. A gas surge drum has two components (hydrogen and methane) in the feedstream. Let y_i and y represent the mole fraction of methane in the feedstream and drum, respectively. Find dP/dt and dy/dt if the inlet and outlet flowrates can vary. Also assume that the inlet concentration can vary. Assume the ideal gas law for the effect of pressure and composition on density.

Assume that the gas drum volume is 100 liters. The temperature of the drum is 31.5 deg C (304.65 K).

At steady-state the drum pressure is 5 atm, the molar flowrate in and out is 2 gmol/min and the concentration is 25% methane, 75% hydrogen.

Use Euler integration and ode45 to solve the following problems. Discuss the effect of integration step size when using Euler integration. In all cases, you are initially at steady-state.

  a. Assume that the molar flowrates remain constant, but the inlet methane concentration is changed to 50%. Find how pressure and composition change with time.

  b. Assume that the molar flowrate out of the drum is proportional to the difference in pressure between the drum and the outlet header, which is at 2 atm pressure. Perform a step change in inlet concentration to 50% methane, simultaneously with a step change in inlet flowrate to 3 gmol/min.

  c. Assume that the MASS flowrate out of the drum is proportional to the square root of the difference in pressure between the drum and the outlet header (which is at 2 atm pressure). Again, perform a step change in inlet concentration to 50% methane, simultaneously with a step change in inlet flowrate to 3 gmol/min.

  d. Assume that the MASS flowrates in and out are proportional to the square root of the pressure drops. Assume that the steady-state inlet gas header is at 5 atm. Perform a step change in inlet concentration to 50% methane, simultaneously with a step change in inlet pressure to 6 atm.

5. Pharmacokinetics is the study of how drugs infused to the body are distributed to other parts of the body. The concept of a compartmental model is often used, where it is assumed that the drug is injected into compartment 1. Some of the drug is eliminated (reacted) in compartment 1, and some of it diffuses into compartment 2 (the rest accumulates in compartment 1). Similarly, some of the drug that diffuses into compartment 2 diffuses back into compartment 1, while some is eliminated by reaction and the rest accumulates in compartment 2. The rates of diffusion and reaction are directly proportional to the concentration of drug in the compartment of interest. The following balance equations describe the rate of change of drug concentration in each compartment.

$$\frac{dx_1}{dt} = -(k_{10} + k_{12}) x_1 + k_{21} x_2 + u$$

$$\frac{dx_2}{dt} = k_{12} x_1 - (k_{20} + k_{21}) x_2$$

where x_1 and x_2 = drug concentrations in compartments 1 and 2 (μg/kg patient weight), and u = rate of drug input to compartment 1 (scaled by the patient weight, μg/kg min).

Experimental studies (of the response of the compartment 1 concentration to various drug infusions) have led to the following parameter values:

$$(k_{10} + k_{12}) = 0.26 \text{ min}^{-1}$$
$$(k_{20} + k_{21}) = 0.094 \text{ min}^{-1}$$
$$k_{12}k_{21} = 0.015 \text{ min}^{-1}$$

for the drug atracurium, which is a muscle relaxant. Notice that the parameters have not been independently determined. Show (through numerical simulation) that all of the following values lead to the same results for the behavior of x_1, while the results for x_2 are different. Let the initial concentration be 0 for each compartment, and assume a constant drug infusion rate of 5.2 μg/kg min.

   a. $k_{12} = k_{21}$
   b. $k_{12} = 2\,k_{21}$
   c. $k_{12} = 0.5\,k_{21}$

Discuss how the concentration of compartment 2 (if measurable) could be used to determine the actual values of k_{12} and k_{21}.

Use the MATLAB function `ode45` for your simulations.

6. A stream contains a waste chemical, W, with a concentration of 1 mol/liter. To meet EPA and state standards, at least 90% of the chemical must be removed by reaction. The chemical decomposes by a second-order reaction with a rate constant of 1.5 liter/(mol hr). The stream flowrate is 100 liter/hr and two available reactors (400 and 2000 liters) have been placed in series (the smaller reactor is placed before the larger one).

  a. Write the modeling equations for the concentration of the waste chemical. Assume constant volume and constant density. Let

      C_{w1} = concentration in reactor 1, mol/liter
      C_{w2} = concentration in reactor 2, mol/liter
      F = volumetric flowrate, liter/hr
      V_1 = liquid volume in reactor 1, liters
      V_2 = liquid volume in reactor 2, liters
      k = second-order rate constant, liter/(mol hr)

  b. Show that the steady-state concentrations are 0.33333 mol/liter (reactor 1) and 0.09005 mol/liter (reactor 2), so the specification is met.

   (*Hint:* You need to solve quadratic equations to obtain the concentrations.)

  c. The system is not initially at steady-state. Write a function file and use `ode45` for the following:

(i). If $C_{w1}(0) = 0.3833$ and $C_{w2}(0) = 0.09005$, find how the concentrations change with time.

(ii). If $C_{w1}(0) = 0.3333$ and $C_{w2}(0) = 0.14005$, find how the concentrations change with time.

7. Consider a batch reactor with a series reaction where component A reacts to form the desired component B *reversibly*. Component B can also react to form the undesired component C. The process objective is to maximize the yield of component B. A mathematical model is used to predict the time required to achieve the maximum yield of B.

The reaction scheme can be characterized by

$$A \begin{array}{c} k_{1f} \\ \dashrightarrow \\ \dashleftarrow \\ k_{1r} \end{array} B \xrightarrow{k_2} C$$

Here k_{1f} and k_{1r} represent the kinetic rate constants for the forward and reverse reactions for the conversion of A to B, while k_2 represents the rate constant for the conversion of B to C.

Assuming that each of the reactions is first-order, the reactor operates at constant volume, and there are no feed or product streams, the modeling equations are:

$$\frac{dC_A}{dt} = -k_{1f} C_A + k_{1r} C_B$$

$$\frac{dC_B}{dt} = k_{1f} C_A - k_{1r} C_B - k_2 C_B$$

$$\frac{dC_C}{dt} = k_2 C_B$$

where C_A, C_B, and C_C represent the concentrations (mol/volume) of components A, B, and C, respectively.

a. For $k_{1f} = 2$, $k_{1r} = 1$, and $k_2 = 1.25$ hr^{-1}, use ode45 to solve for the concentrations as a function of time. Assume an initial concentration of A of $C_{A0} = 1$ mol/liter. Then plot the concentrations as a function of time. For what time is the concentration of B maximized?

b. Usually there is some uncertainty in the rate constants. If the real value of k_2 is 1.5 hr^{-1} find how the concentrations vary with time and compare with part a.

SECTION III

LINEAR SYSTEMS ANALYSIS

LINEARIZATION OF NONLINEAR MODELS: THE STATE-SPACE FORMULATION

5

Many dynamic chemical processes are modeled by a set of nonlinear, first-order differential equations that generally arise from material and energy balances around the system. Common analysis techniques are based on linear systems theory and require a *state-space* model. Also, most control system design techniques are based on linear models. The purpose of this chapter is to provide an introduction to state-space models and linearization of nonlinear systems. After studying this chapter the reader should be able to:

- Write a linear model in state-space form.
- Linearize a nonlinear model and place in state-space form.
- Use the MATLAB `eig` function to analyze the stability of a state-space model.
- Develop the analytical solution of state-space models.
- Understand stability and transient response characterstics as a function of the eigenvalues.
- Understand the importance of initial condition "direction".
- Be able to use the MATLAB routines `step` and `initial` for simulation of state-space models.

The major sections in this chapter are:

5.1   State-Space Models
5.2   Linearization of Nonlinear Models
5.3   Geometrical Interpretation of Linearization
5.4   Solution of the Zero-Input Form
5.5   Solution of the General State-Space Form
5.6   MATLAB Routines `step` and `initial`

5.1   STATE-SPACE MODELS

Thus far in this text we have discussed dynamic models of the general form:

$$\dot{\mathbf{x}} = \mathbf{f(x,u)} \tag{5.1}$$

where $\mathbf{f(x,u)}$ is, in general, a nonlinear function vector.

A linear model is a subset of the more general modeling equation (5.1). The form of linear model that we discuss in this chapter is known as a *state-space* model. First, we show how to write state-space models for systems that are inherently linear. Then, we show how to approximate nonlinear systems with linear models.

Example 5.1 illustrates the form of a state-space model.

EXAMPLE 5.1   Noninteracting Tanks

Consider two tanks in series where the flow out of the first tank enters the second tank (Figure 5.1). Our objective is to develop a model to describe how the height of liquid in tank 2 changes with time, given the input flowrate $F_o(t)$. We assume that the flow out of each tank is a linear function of the height of liquid in the tank.

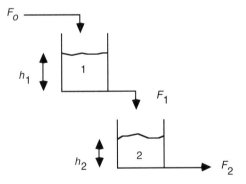

FIGURE 5.1   Noninteracting tanks.

A material balance around the first tank yields (assuming constant density and $F_1 = \beta_1 h_1$)

$$\frac{dh_1}{dt} = \frac{F_o}{A_1} - \frac{\beta_1}{A_1} h_1 \tag{5.2}$$

where A_1 is the constant cross-sectional area (parameter), β_1 is the flow coefficient (parameter), F_o is the flowrate into the tank (input), and h_1 is the tank height (state).

Writing a material balance around the second tank (since $F_2 = \beta_2 h_2$)

$$\frac{dh_2}{dt} = \frac{F_1}{A_2} - \frac{\beta_2}{A_2} h_2$$

where A_2 is the constant cross-sectional area for tank 2 (parameter), β_2 is the flow coefficient (parameter), F_1 is the flowrate into the tank, and h_2 is the tank height (state). In this case, F_1 is not an independent input variable that can be manipulated, since $F_1 = \beta_1 h_1$. We can write the previous equation as

$$\frac{dh_2}{dt} = \frac{\beta_1}{A_2} h_1 - \frac{\beta_2}{A_2} h_2 \qquad (5.3)$$

Notice that we can write (5.2) and (5.3) in the following matrix form:

$$\begin{bmatrix} \dot{h}_1 \\ \dot{h}_2 \end{bmatrix} = \begin{bmatrix} \dfrac{-\beta_1}{A_1} & 0 \\ \dfrac{\beta_1}{A_2} & \dfrac{-\beta_2}{A_2} \end{bmatrix} \begin{bmatrix} h_1 \\ h_2 \end{bmatrix} + \begin{bmatrix} \dfrac{1}{A_1} \\ 0 \end{bmatrix} F_o \qquad (5.4)$$

which has the general form:

$$\dot{\mathbf{x}} = \mathbf{A}\,\mathbf{x} + \mathbf{B}\,\mathbf{u} \qquad (5.5)$$

where:

$$\mathbf{A} = \begin{bmatrix} \dfrac{\beta_1}{A_1} & 0 \\ \dfrac{\beta_1}{A_2} & \dfrac{-\beta_2}{A_2} \end{bmatrix} \quad \text{and} \quad \mathbf{B} = \begin{bmatrix} \dfrac{1}{A_1} \\ 0 \end{bmatrix}$$

The state and input vectors are (notice that the input is a scalar):

$$\mathbf{x} = \begin{bmatrix} h_1 \\ h_2 \end{bmatrix} \quad \text{and} \quad \mathbf{u} = F_o$$

The additional equation that is normally associated with a state space model is

$$\mathbf{y} = \mathbf{C}\,\mathbf{x} + \mathbf{D}\,\mathbf{u} \qquad (5.6)$$

where $\mathbf{y}$ is a vector of output variables. Generally, output variables are variables that can be measured (at least conceptually) or are of particular interest in a simulation study. Here, we will consider the case where both tank heights are outputs. Let output 1 be the first tank height and output 2 be the second tank height

$$y_1 = h_1$$

$$y_2 = h_2$$

The matrix-vector form is:

$$\mathbf{y} = \begin{bmatrix} 1 & 0 \\ 0 & 1 \end{bmatrix} \begin{bmatrix} h_1 \\ h_2 \end{bmatrix} = \mathbf{C}\,\mathbf{x}$$

where:

$$\mathbf{C} = \begin{bmatrix} 1 & 0 \\ 0 & 1 \end{bmatrix}$$

If we also consider the input, F_o, to be the third output variable, we have the following relationship:

$$\mathbf{y} = \begin{bmatrix} 1 & 0 \\ 0 & 1 \\ 0 & 0 \end{bmatrix} \begin{bmatrix} h_1 \\ h_2 \end{bmatrix} + \begin{bmatrix} 0 \\ 0 \\ 1 \end{bmatrix} [F_o]$$

which is the form of (5.6), with

$$\mathbf{C} = \begin{bmatrix} 1 & 0 \\ 0 & 1 \\ 0 & 0 \end{bmatrix} \quad \text{and} \quad \mathbf{D} = \begin{bmatrix} 0 \\ 0 \\ 1 \end{bmatrix}$$

5.5.1   General Form of State Space Models

Example 5.1 illustrated a specific case of a state-space model. In general, a state-space model has the following form:

$$\frac{dx_1}{dt} = a_{11} x_1 + a_{12} x_2 + \ldots + a_{1n} x_n + b_{11} u_1 + \ldots + b_{1m} u_m$$

$$\frac{dx_n}{dt} = a_{n1} x_1 + a_{n2} x_2 + \ldots + a_{nn} x_n + b_{n1} u_1 + \ldots + b_{nm} u_m$$

$$y_1 = c_{11} x_1 + c_{12} x_2 + \ldots + a_{1n} x_n + d_{11} u_1 + \ldots + d_{1m} u_m$$

$$y_r = c_{r1} x_1 + c_{r2} x_2 + \ldots + c_{rn} x_n + d_{r1} u_1 + \ldots + d_{rm} u_m$$

which has n state variables (x), m input variables (u) and r output variables (y). This relationship is normally written in the matrix form:

$$\begin{bmatrix} \dot{x}_1 \\ . \\ . \\ \dot{x}_n \end{bmatrix} = \begin{bmatrix} a_{11} & a_{12} & . & a_{1n} \\ . & . & . & . \\ . & . & . & . \\ a_{n1} & a_{n2} & . & a_{nn} \end{bmatrix} \begin{bmatrix} x_1 \\ . \\ . \\ x_n \end{bmatrix} + \begin{bmatrix} b_{11} & b_{12} & . & b_{1m} \\ . & . & . & . \\ . & . & . & . \\ b_{n1} & b_{n2} & . & b_{nm} \end{bmatrix} \begin{bmatrix} u_1 \\ . \\ . \\ u_m \end{bmatrix}$$

$$\begin{bmatrix} y_1 \\ . \\ . \\ y_r \end{bmatrix} = \begin{bmatrix} c_{11} & c_{12} & . & c_{1n} \\ . & . & . & . \\ . & . & . & . \\ c_{r1} & c_{r2} & . & c_{rn} \end{bmatrix} \begin{bmatrix} x_1 \\ . \\ . \\ x_n \end{bmatrix} + \begin{bmatrix} d_{11} & d_{12} & . & d_{1m} \\ . & . & . & . \\ . & . & . & . \\ d_{r1} & d_{r2} & . & d_{rm} \end{bmatrix} \begin{bmatrix} u_1 \\ . \\ . \\ u_m \end{bmatrix}$$

which has the general (*state-space*) form:

$$\dot{x} = \mathbf{A}\,x + \mathbf{B}\,u$$
$$y = \mathbf{C}\,x + \mathbf{D}\,u \tag{5.7}$$

where the dot over a state variable indicates the derivative with respect to time. As shown in Section 5.4, the eigenvalues of the Jacobian matrix ($\mathbf{A}$) determine the stability of the system of equations and the "speed" of response.

The a_{ij} coefficient relates state variable j to the rate of change of state variable i. Similarly, the b_{ij} coefficient relates input j to the rate of change of state variable i. Also, c_{ij} relates state j to output i, while d_{ij} relates input j to output i. We can also say that the kth row of C relates all states to the kth output, while the kth column of C relates state k to all outputs.

In this section we have shown how to write modeling equations that are naturally linear in the state-space form. In the next section we show how to linearize nonlinear models and write them in the state-space form. Linear models are easier to analyze for stability and expected dynamic behavior.

5.2 LINEARIZATION OF NONLINEAR MODELS

Most chemical process models are nonlinear, but they are often linearized to perform a stability analysis. Linear models are easier to understand (than nonlinear models) and are necessary for most control system design methods.

Before we generalize our results, we will illustrate linearization for a single variable problem.

5.2.1 Single Variable Example

A general single variable nonlinear model is:

$$\frac{dx}{dt} = f(x) \tag{5.8}$$

The function of a single variable, $f(x)$, can be approximated by a truncated Taylor series approximation around the steady-state operating point (x_s):

$$f(x) = f(x_s) + \left.\frac{\partial f}{\partial x}\right|_{x_s}(x - x_s) + \frac{1}{2}\left.\frac{\partial^2 f}{\partial x^2}\right|_{x_s}(x - x_s)^2 + \text{higher order terms} \tag{5.9}$$

Neglecting the quadratic and higher order terms, we obtain:

$$f(x) \approx f(x_s) + \left.\frac{\partial f}{\partial x}\right|_{x_s}(x - x_s) \tag{5.10}$$

Note that:

$$\frac{dx_s}{dt} = f(x_s) = 0 \tag{5.11}$$

by definition of a steady-state, so:

$$\frac{dx}{dt} = f(x) \approx \left.\frac{\partial f}{\partial x}\right|_{x_s} (x - x_s) \tag{5.12}$$

where the notation $\left.\partial f/\partial x\right|_{x_s}$ is used to indicate the partial derivative of $f(x)$ with respect to x, evaluated at the steady-state. Since the derivative of a constant (x_s) is zero, we can write:

$$\frac{dx}{dt} = \frac{d(x - x_s)}{dt} \tag{5.13}$$

which leads to:

$$\frac{d(x - x_s)}{dt} \approx \left.\frac{\partial f}{\partial x}\right|_{x_s} (x - x_s) \tag{5.14}$$

The reason for using the expression above is that we are often interested in deviations in a state from a steady-state operating point. Sometimes the $'$ symbol is used to represent *deviation variables*, $x' = x - x_s$. We can see that a deviation variable represents the change or perturbation (deviation) from a steady-state value.

$$\frac{dx'}{dt} = \left.\frac{\partial f}{\partial x}\right|_{x_s} x' \tag{5.15}$$

This can be written in state-space form:

$$\frac{dx'}{dt} = a\,x' \tag{5.16}$$

where $a = \left.\partial f/\partial x\right|_{x_s}$.

We have shown how to linearize a single variable equation. Next, we consider a system with one state and one input.

5.2.2   One State Variable and One Input Variable

Similarly, consider a function with one state variable and one input variable

$$\dot{x} = \frac{dx}{dt} = f(x,u) \tag{5.17}$$

Using a Taylor Series Expansion for $f(x,u)$:

$$\dot{x} = f(x_s,u_s) + \left.\frac{\partial f}{\partial x}\right|_{x_s,u_s} (x - x_s) + \left.\frac{\partial f}{\partial u}\right|_{x_s,u_s} (u - u_s)$$

$$+ \frac{1}{2}\left.\frac{\partial^2 f}{\partial x^2}\right|_{x_s,u_s} (x - x_s)^2 + \left.\frac{\partial^2 f}{\partial x \partial u}\right|_{x_s,u_s} (x - x_s)(u - u_s) + \frac{1}{2}\left.\frac{\partial^2 f}{\partial u^2}\right|_{xs,us} (u - u_s)^2$$

$$+ \text{ higher order terms}$$

and truncating after the linear terms, we have:

$$\dot{x} \approx f(x_s, u_s) + \left.\frac{\partial f}{\partial x}\right|_{x_s, u_s} (x - x_s) + \left.\frac{\partial f}{\partial u}\right|_{x_s, u_s} (u - u_s) \tag{5.18}$$

and realizing that $f(x_s, u_s) = 0$ and $dx/dt = d(x - x_s)/dt$:

$$\frac{d(x - x_s)}{dt} \approx \left.\frac{\partial f}{\partial x}\right|_{x_s, u_s} (x - x_s) + \left.\frac{\partial f}{\partial u}\right|_{x_s, u_s} (u - u_s)$$

Using deviation variables, $x' = x - x_s$ and $u' = u - u_s$:

$$\frac{dx'}{dt} \approx \left.\frac{\partial f}{\partial x}\right|_{x_s, u_s} x' + \left.\frac{\partial f}{\partial u}\right|_{x_s, u_s} u'$$

which can be written:

$$\frac{dx'}{dt} = a\,x' + b\,u' \tag{5.19}$$

where $a = \partial f/\partial x|_{x_s}$ and $b = \partial f/\partial u|_{x_s, u_s}$

If there is a single output that is a function of the states and inputs, then:

$$y = g(x, u) \tag{5.20}$$

Again, performing a Taylor series expansion and truncating the quadratic and higher terms:

$$g(x, u) \approx g(x_s, u_s) + \left.\frac{\partial g}{\partial x}\right|_{x_s, u_s} (x - x_s) + \left.\frac{\partial g}{\partial u}\right|_{x_s, u_s} (u - u_s) \tag{5.21}$$

Since $g(x_s, u_s)$ is simply the steady-state value of the output (y_s), we can write:

$$y \approx g(x_s, u_s) + \left.\frac{\partial g}{\partial x}\right|_{x_s, u_s} (x - x_s) + \left.\frac{\partial g}{\partial u}\right|_{x_s, u_s} (u - u_s) \tag{5.22}$$

or

$$y - y_s = c\,(x - x_s) + d\,(u - u_s)$$

where $c = \partial g/\partial x|_{x_s, u_s}$ and $d = \partial g/\partial u|_{x_s, u_s}$

Using deviation notation:

$$y' = c\,x' + d\,u' \tag{5.23}$$

Example 5.2 illustrates the application of linearization to a one-input, one-state nonlinear system.

EXAMPLE 5.2   Consider a Nonlinear Tank Height Problem

$$\frac{dh}{dt} = \frac{F}{A} - \frac{\beta}{A}\sqrt{h} \qquad (5.24)$$

where h is the state variable, F is the input variable, β and A are parameters. The righthand side is:

$$f(h,F) = \frac{F}{A} - \frac{\beta}{A}\sqrt{h}$$

Using a truncated Taylor series expansion, we find:

$$f(h,F) \approx \left[\frac{F_s}{A} - \frac{\beta}{A}\sqrt{h_s}\right] + \frac{1}{A}[F - F_s] - \frac{\beta}{2A\sqrt{h_s}}[h - h_s] \qquad (5.25)$$

The first term on the righthand side is zero, because the linearization is about a steady-state point. That is,

$$\left.\frac{dh}{dt}\right|_{h_s,F_s} = \frac{F_s}{A} - \frac{\beta}{A}\sqrt{h_s} = 0$$

We can now write:

$$\frac{d(h - h_s)}{dt} \approx -\frac{\beta}{2A\sqrt{h_s}}[h - h_s] + \frac{1}{A}[F - F_s]$$

and using deviation variable notation ($h' = h - h_s$ and $u' = F - F_s$), and dropping the $\approx$

$$\frac{dh'}{dt} = -\frac{\beta}{2A\sqrt{h_s}}h' + \frac{1}{A}F'$$

For convenience (simplicity in notation) we often drop the ($'$) notation and assume that x and u are deviation variables ($x = h - h_s$, $u = F - F_s$) and write:

$$\frac{dx}{dt} = -\frac{\beta}{2A\sqrt{h_s}}x + \frac{1}{A}u \qquad (5.26)$$

which is in the state-space form

$$\frac{dx}{dt} = a\,x + b\,u \qquad (5.27)$$

5.2.3   Linearization of Multistate Models

The previous examples showed how to linearize single-state variable systems. In this section we generalize the technique for any number of states. Before we generalize the technique, it is worthwhile to consider an example system with two states, one input and one output.

EXAMPLE 5.3   Two-state System

$$\dot{x}_1 = \frac{dx_1}{dt} = f_1(x_1, x_2, u) \tag{5.28}$$

$$\dot{x}_2 = \frac{dx_2}{dt} = f_2(x_1, x_2, u) \tag{5.29}$$

$$y = g(x_1, x_2, u) \tag{5.30}$$

Performing a Taylor series expansion of the nonlinear functions, and neglecting the quadratic and higher terms:

$$f_1(x_1, x_2, u) = f_1(x_{1s}, x_{2s}, u_s) + \left.\frac{\partial f_1}{\partial x_1}\right|_{x_{1s}, x_{2s}, u_s} (x_1 - x_{1s})$$

$$+ \left.\frac{\partial f_1}{\partial x_2}\right|_{x_{1s}, x_{2s}, u_s} (x_2 - x_{2s}) + \left.\frac{\partial f_1}{\partial u}\right|_{x_{1s}, x_{2s}, u_s} (u - u_s) + \text{higher order terms}$$

$$f_2(x_1, x_2, u) = f_2(x_{1s}, x_{2s}, u_s) + \left.\frac{\partial f_2}{\partial x_1}\right|_{x_{1s}, x_{2s}, u_s} (x_1 - x_{1s})$$

$$+ \left.\frac{\partial f}{\partial x_2}\right|_{x_{1s}, x_{2s}, u_s} (x_2 - x_{2s}) + \left.\frac{\partial f_2}{\partial u}\right|_{x_{1s}, x_{2s}, u_s} (u - u_s) + \text{higher order terms}$$

$$g(x_1, x_2, u) = g(x_{1s}, x_{2s}, u_s) + \left.\frac{\partial f}{\partial x_1}\right|_{x_{1s}, x_{2s}, us} (x_1 - x_{1s})$$

$$+ \left.\frac{\partial g}{\partial x_2}\right|_{x_{1s}, x_{2s}, u_s} (x_2 - x_{2s}) + \left.\frac{\partial g}{\partial u}\right|_{x_{1s}, x_{2s}, u_s} (u - u_s) + \text{higher order terms}$$

From the linearization about the steady-state:

$$f_1(x_{1s}, x_{2s}, u_s) = f_2(x_{1s}, x_{2s}, u_s) = 0$$

and:

$$g(x_{1s}, x_{2s}, u_s) = y_s$$

Since the derivative of a constant is zero:

$$\frac{dx_1}{dt} = \frac{d(x_1 - x_{1s})}{dt} \quad \text{and} \quad \frac{dx_2}{dt} = \frac{d(x_2 - x_{2s})}{dt}$$

we can write the state-space model:

$$\begin{bmatrix} \dfrac{d(x_1 - x_{1s})}{dt} \\[2ex] \dfrac{d(x_2 - x_{2s})}{dt} \end{bmatrix} = \begin{bmatrix} \left.\dfrac{\partial f_1}{\partial x_1}\right|_{x_{1s}, x_{2s}, u_s} & \left.\dfrac{\partial f_1}{\partial x_2}\right|_{x_{1s}, x_{2s}, u_s} \\[2ex] \left.\dfrac{\partial f_2}{\partial x_1}\right|_{x_{1s}, x_{2s}, u_s} & \left.\dfrac{\partial f_2}{\partial x_2}\right|_{x_{1s}, x_{2s}, u_s} \end{bmatrix} \begin{bmatrix} x_1 - x_{1s} \\ x_2 - x_{2s} \end{bmatrix}$$

$$+ \left[\begin{array}{c} \left. \dfrac{\partial f_1}{\partial u} \right|_{x_1,x_2,u_s} \\ \left. \dfrac{\partial f_2}{\partial u} \right|_{x_1,x_2,u_s} \end{array} \right] [u - u_s] \tag{5.31}$$

$$y - y_s = \left[\left. \dfrac{\partial g}{\partial x_1} \right|_{x_1,x_2,u_s} \left. \dfrac{\partial g}{\partial x_2} \right|_{x_1,x_2,u_s} \right] \left[\begin{array}{c} x_1 - x_{1s} \\ x_2 - x_{2s} \end{array} \right]$$

$$+ \left[\left. \dfrac{\partial g}{\partial u} \right|_{x_1,x_2,u_s} \right] [u - u_s] \tag{5.32}$$

which is the form of a state-space model:

$$\dot{\mathbf{x}}' = \mathbf{A}\,\mathbf{x}' + \mathbf{B}\,\mathbf{u}'$$
$$\mathbf{y}' = \mathbf{C}\,\mathbf{x}' + \mathbf{D}\,\mathbf{u}'$$

where (′) indicates deviation variables.

5.2.4 Generalization

Now consider the general nonlinear model where $\mathbf{x}$ is a vector of n state variables, $\mathbf{u}$ is a vector of m input variables and $\mathbf{y}$ is a vector of r output variables:

$$\dot{\mathbf{x}}_1 = f_1(x_1,...,x_n,u_1,...,u_m)$$

$$\dot{\mathbf{x}}_n = f_n(x_1,...,x_n,u_1,...,u_m)$$
$$\mathbf{y}_1 = g_1(x_1,...,x_n,u_1,...,u_m)$$

$$\mathbf{y}_r = g_r(x_1,...,x_n,u_1,...,u_m)$$

In vector notation:

$$\dot{\mathbf{x}} = \mathbf{f(x,u)} \tag{5.33}$$

$$\mathbf{y} = \mathbf{g(x,u)} \tag{5.34}$$

Elements of the linearization matrices are defined in the following fashion:

$$A_{ij} = \left. \dfrac{\partial f_i}{\partial x_j} \right|_{x_s,u_s} \tag{5.35}$$

$$B_{ij} = \left. \dfrac{\partial f_i}{\partial u_j} \right|_{x_s,u_s} \tag{5.36}$$

$$C_{ij} = \frac{\partial g_i}{\partial x_j}\bigg|_{x_s,u_s} \tag{5.37}$$

$$D_{ij} = \frac{\partial g_i}{\partial u_j}\bigg|_{x_s,u_s} \tag{5.38}$$

After linearization, we have the state-space form:

$$\dot{\mathbf{x}}' = \mathbf{A}\,\mathbf{x}' + \mathbf{B}\,\mathbf{u}'$$

$$\mathbf{y}' = \mathbf{C}\,\mathbf{x}' + \mathbf{D}\,\mathbf{u}'$$

Generally, the (′) notation is dropped and it is understood that the model is in deviation variable form:

$$\dot{\mathbf{x}} = \mathbf{A}\,\mathbf{x} + \mathbf{B}\,\mathbf{u}$$

$$\mathbf{y} = \mathbf{C}\,\mathbf{x} + \mathbf{D}\,\mathbf{u}$$

Usually, the measured (output) variable is not a direct function of the input variable, so it is more common to see the following state-space model:

$$\dot{\mathbf{x}} = \mathbf{A}\,\mathbf{x} + \mathbf{B}\,\mathbf{u}$$

$$\mathbf{y} = \mathbf{C}\,\mathbf{x}$$

This procedure is applied in Example 5.4.

EXAMPLE 5.4   Interacting Tanks

Consider the interacting tank height problem shown in Figure 5.2:

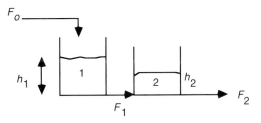

FIGURE 5.2   Interacting tanks.

Assume that the flowrate out of tanks is a nonlinear function of tank height. The flowrate out of tank one is a function of the difference in levels between tank 1 and tank 2.

$$\frac{dh_1}{dt} = f_1(h_1, h_2, F) = \frac{F}{A_1} - \frac{\beta_1}{A_1}\sqrt{h_1 - h_2}$$

$$\frac{dh_2}{dt} = f_2(h_1, h_2, F) = \frac{\beta_1}{A_2}\sqrt{h_1 - h_2} - \frac{\beta_2}{A_2}\sqrt{h_2}$$

Also, assume that only the second tank height is measured. The output, in deviation variable form is

$$y = h_2 - h_{2s}$$

Notice that there are two state variables, one input variable, and one output variable. Let

$$\mathbf{h_s} = \begin{bmatrix} h_{1s} \\ h_{2s} \end{bmatrix}$$

$$\mathbf{x} = \begin{bmatrix} x_1 \\ x_2 \end{bmatrix} = \begin{bmatrix} h_1 - h_{1s} \\ h_2 - h_{2s} \end{bmatrix}$$

$$\mathbf{u} = F - F_s$$

The elements of the **A** (Jacobian) and **B** matrices ((5.35) and (5.36)) are:

$$A_{11} = \left.\frac{\partial f_1}{\partial h_1}\right|_{\mathbf{h}_s, F_s} = -\frac{\beta_1}{2A_1\sqrt{h_{1s} - h_{2s}}}$$

$$A_{12} = \left.\frac{\partial f_1}{\partial h_2}\right|_{\mathbf{h}_s, F_s} = \frac{\beta_1}{2A_1\sqrt{h_{1s} - h_{2s}}}$$

$$A_{21} = \left.\frac{\partial f_2}{\partial h_1}\right|_{\mathbf{h}_s, F_s} = \frac{\beta_1}{2A_2\sqrt{h_{1s} - h_{2s}}}$$

$$A_{22} = \left.\frac{\partial f_2}{\partial h_2}\right|_{\mathbf{h}_s, F_s} = -\frac{\beta_1}{2A_2\sqrt{h_{1s} - h_{2s}}} - \frac{\beta_2}{2A_2\sqrt{h_{2s}}}$$

$$B_{11} = \left.\frac{\partial f_1}{\partial F}\right|_{\mathbf{h}_s, F_s} = \frac{1}{A_1}$$

$$B_{21} = \left.\frac{\partial f_2}{\partial F}\right|_{\mathbf{h}_s, F_s} = 0$$

Since only the height of the second tank is measured, $y = g(h_1, h_2, F) = h_2 - h_{2s}$ (from (5.37)):

$$C_{11} = \left.\frac{\partial g}{\partial h_1}\right|_{\mathbf{h}_s, u_s} = 0$$

$$C_{12} = \left.\frac{\partial g}{\partial h_2}\right|_{\mathbf{h}_s, u_s} = 1$$

and the state-space model is:

$$\begin{bmatrix} \dfrac{dx_1}{dt} \\[2ex] \dfrac{dx_2}{dt} \end{bmatrix} = \begin{bmatrix} -\dfrac{\beta_1}{2A_1\sqrt{h_{1s} - h_{2s}}} & \dfrac{\beta_1}{2A_1\sqrt{h_{1s} - h_{2s}}} \\[3ex] \dfrac{\beta_1}{2A_2\sqrt{h_{1s} - h_{2s}}} & -\dfrac{\beta_1}{2A_2\sqrt{h_{1s} - h_{2s}}} - \dfrac{\beta_2}{2A_2\sqrt{h_{2s}}} \end{bmatrix} \begin{bmatrix} x_1 \\ x_2 \end{bmatrix}$$

$$+ \begin{bmatrix} \dfrac{1}{A_1} \\[2ex] 0 \end{bmatrix} [u]$$

$$y = [0 \;\; 1] \begin{bmatrix} x_1 \\ x_2 \end{bmatrix}$$

where:

$$y = x_2 = h_2 - h_{2s}$$

In this section we have shown how to linearize a nonlinear process model and put it in state-space form. The states in this model are in deviation (perturbation) variable form; that is, the states are perturbations from a nominal steady-state. A state-space model provides a good approximation to the physical system when the operating point is "close" to the linearization point (nominal steady-state).

5.3 INTERPRETATION OF LINEARIZATION

In Section 5.2 we illustrated the method of linearization of models into state-space form. The objective of this section is to illustrate what is meant by linearization of a function. Consider the single tank height problem, which has the following model:

$$\frac{dh}{dt} = f(h,F) = \frac{F}{A} - \frac{\beta}{A}\sqrt{h} \tag{5.39}$$

for a system with $A = 1$ ft^2, $h_s = 5$ ft, and $\beta = 1/\sqrt{5}$ ft$^{2.5}$/min the steady-state flowrate is $F_s = 1$ ft^3/min. To focus our analysis on the meaning of the linearization with respect to the state variable, consider the case where the input is constant. Then, from (5.39) and the given parameter values:

$$f(h,F_s) = 1 - \frac{1}{\sqrt{5}}\sqrt{h} \tag{5.40}$$

performing the linearization:

$$f(h,F_s) \approx f(h_s,F_s) + \frac{\partial f}{\partial h}\bigg|_{h_s,Fs} (h - h_s)$$

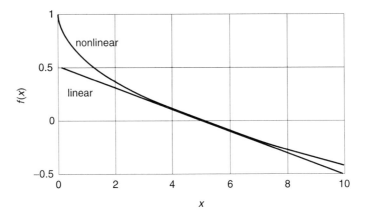

FIGURE 5.3   Basic idea of linearization. The linear approximation is exact for the steady-state value of $x = 5$.

for our parameter values

$$f(h,F_s) \approx 0 + \frac{-1}{2\sqrt{5}\ \sqrt{h_s}}\ (h - h_s)$$

or

$$f(h,F_s) \approx 0 - \frac{1}{10}\ (h - h_s) \qquad\qquad (5.41)$$

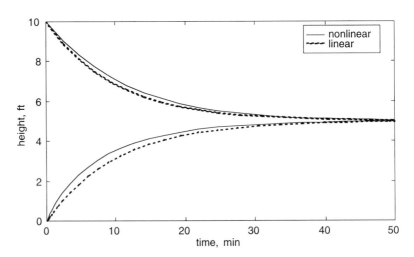

FIGURE 5.4   Comparison of linear and nonlinear responses for two different initial conditions.

We can see how good the linear approximation is by plotting both the nonlinear function (5.40) and the linear function (5.41), as shown in Figure 5.3 on p. 118. Here we have used x to represent tank height and $f(x)$ to represent the nonlinear and linear functions. Notice that the linear approximation works well between roughly 3.5 to 7 feet. Of course, the two functions are exactly equal at the steady-state value of 5 feet, which was the point at which the Taylor series expansion was performed. Realize that $f(x)$ is dx/dt, which is the rate of change of tank height. It makes sense that the rate of change is positive at a tank height less than 5 feet, because the system "seeks" to achieve a steady-state level of 5 feet. Similarly, for a tank height greater that 5 feet, the rate of change of tank height is negative, because the level "desires" to decrease to 5 feet. We can also see that the linear system will be slower than the nonlinear system, if the tank height is less that 5 feet, but will be faster if the height is greater than 5 feet, as shown in Figure 5.4.

5.4 SOLUTION OF THE ZERO-INPUT FORM

We have previously written the general state space model in the following form:

$$\dot{\mathbf{x}} = \mathbf{A}\,\mathbf{x} + \mathbf{B}\,\mathbf{u}$$

where $\mathbf{x}$ and $\mathbf{u}$ are deviation variable vectors for the states and inputs, respectively. In this section we assume that the inputs are held constant at their steady-state values, but that the states may be initially perturbed from steady-state. The "zero-input" form of the state space model is then:

$$\begin{bmatrix} \dot{x}_1 \\ \cdot \\ \cdot \\ \dot{x}_n \end{bmatrix} = \begin{bmatrix} a_{11} & a_{12} & \cdot & a_{1n} \\ \cdot & \cdot & \cdot & \cdot \\ \cdot & \cdot & \cdot & \cdot \\ a_{n1} & a_{n2} & \cdot & a_{nn} \end{bmatrix} \begin{bmatrix} x_1 \\ \cdot \\ \cdot \\ x_n \end{bmatrix}$$

or

$$\dot{\mathbf{x}} = \mathbf{A}\,\mathbf{x} \qquad (5.42)$$

This form is used to analyze the stability of a system and to understand the dynamic behavior of a system that has had its states perturbed from the steady-state values.

Recall that the single variable equation:

$$\dot{\mathbf{x}} = \mathbf{a}\,\mathbf{x}$$

has the solution:

$$x(t) = e^{at}\,x(0)$$

which is stable if $a < 0$. In a similar fashion, the solution to (5.42) is

$$x(t) = e^{At}x(0) \qquad (5.43)$$

and the solution to (5.43) is stable if all of the eigenvalues of A are less than zero. The response of (5.43) is oscillatory if the eigenvalues are complex.

There are many different ways to calculate the exponential of a matrix; in this chapter we discuss only the *similarity transform* method.

Recall that the eigenvector/eigenvalue problem is written (see Module 2 for a review):

$$\mathbf{A}\,\mathbf{V} = \mathbf{V}\,\Lambda \qquad (5.44)$$

For a $2{\times}2$ **A** matrix we have the following eigenvector matrix:

$$\mathbf{V} = [\xi_1 \;\; \xi_2] = \begin{bmatrix} v_{11} & v_{12} \\ v_{21} & v_{22} \end{bmatrix} \qquad (5.45)$$

where $\xi_1 = \begin{bmatrix} v_{11} \\ v_{21} \end{bmatrix}$ = first eigenvector (associated with λ_1)

$\xi_2 = \begin{bmatrix} v_{12} \\ v_{22} \end{bmatrix}$ = second eigenvector (associated with λ_2)

and the following eigenvalue matrix:

$$\Lambda = \begin{bmatrix} \lambda_1 & 0 \\ 0 & \lambda_2 \end{bmatrix} \qquad (5.46)$$

Multiplying (5.44) on the right side by $\mathbf{V}^{-1}$ we find:

$$\mathbf{A} = \mathbf{V}\,\Lambda\,\mathbf{V}^{-1} \qquad (5.47)$$

multiplying by the scalar t and taking the matrix exponential, we find:

$$e^{At} = \mathbf{V}\,e^{\Lambda t}\,\mathbf{V}^{-1} \qquad (5.48)$$

where

$$e^{\Lambda t} = \begin{bmatrix} e^{\lambda_1 t} & 0 \\ 0 & e^{\lambda_2 t} \end{bmatrix} \qquad (5.49)$$

and we see immediately why $\lambda_i < 0$ is required for a stable solution. The solution for $\mathbf{x}(t)$ is

$$\mathbf{x}(t) = \mathbf{V}\,e^{\Lambda t}\,\mathbf{V}^{-1}\,x(0) \qquad (5.50)$$

An interesting result is that an initial condition vector in the same direction as ξ_i has a response in the direction of ξ_i with a "speed or response" of λ_i. This is shown by the following analysis.

5.4.1 Effect of Initial Condition Direction (Use of Similarity Transform)

Recall that we are solving the following model

$$\dot{\mathbf{x}} = \mathbf{A}\,\mathbf{x} \qquad (5.42)$$

Define a new vector **z**, such that

$$\mathbf{x} = \mathbf{V}\,\mathbf{z} \qquad (5.51)$$

or

$$\mathbf{z} = \mathbf{V}^{-1}\,\mathbf{x} \tag{5.52}$$

and notice that (from (5.51)):

$$\dot{\mathbf{x}} = \mathbf{V}\,\dot{\mathbf{z}} \tag{5.53}$$

Substituting (5.53) and (5.51) into (5.42):

$$\mathbf{V}\,\dot{\mathbf{z}} = \mathbf{A}\,\mathbf{V}\,\mathbf{z} \tag{5.54}$$

or, left multiplying by $\mathbf{V}^{-1}$

$$\dot{\mathbf{z}} = \mathbf{V}^{-1}\mathbf{A}\,\mathbf{V}\,\mathbf{z} \tag{5.55}$$

But, from (5.44) $\mathbf{A}\,\mathbf{V} = \mathbf{V}\,\mathbf{\Lambda}$

so we can write:

$$\mathbf{V}^{-1}\,\mathbf{A}\,\mathbf{V} = \mathbf{\Lambda} \tag{5.56}$$

which yields (from (5.55) and (5.56)):

$$\dot{\mathbf{z}} = \mathbf{\Lambda}\,\mathbf{z} \tag{5.57}$$

But $\mathbf{\Lambda}$ is a diagonal matrix (see (5.46)), so we have:

$$\begin{bmatrix} \dot{z}_1 \\ \dot{z}_2 \end{bmatrix} = \begin{bmatrix} \lambda_1 & 0 \\ 0 & \lambda_2 \end{bmatrix} \begin{bmatrix} z_1 \\ z_2 \end{bmatrix} \tag{5.58}$$

Notice that (5.58) represents two independent equations:

$$\dot{z}_1 = \lambda_1 z_1 \tag{5.59}$$

$$\dot{z}_2 = \lambda_2 z_2 \tag{5.60}$$

which have the solutions:

$$z_1(t) = z_1(0)\,e^{\lambda_1 t} \tag{5.61}$$

$$z_2(t) = z_2(0)\,e^{\lambda_2 t} \tag{5.62}$$

and we can write:

$$\begin{bmatrix} z_1(t) \\ z_2(t) \end{bmatrix} = \begin{bmatrix} e^{\lambda_1 t} & 0 \\ 0 & e^{\lambda_2 t} \end{bmatrix} \begin{bmatrix} z_1(0) \\ z_2(0) \end{bmatrix} \tag{5.63}$$

or

$$\mathbf{z}(t) = e^{\mathbf{\Lambda} t}\,\mathbf{z}(0) \tag{5.64}$$

Notice that, if the $\mathbf{z}(0)$ vector has the form:

$$\mathbf{z}(0) = \begin{bmatrix} z_1(0) \\ 0 \end{bmatrix} \tag{5.65}$$

then:

$$\mathbf{z}(t) = \begin{bmatrix} z_1(0)e^{\lambda_1 t} \\ 0 \end{bmatrix} \tag{5.66}$$

and, if the $\mathbf{z}(0)$ vector has the form:

$$\mathbf{z}(0) = \begin{bmatrix} 0 \\ z_2(0) \end{bmatrix} \tag{5.67}$$

then,

$$\mathbf{z}(t) = \begin{bmatrix} 0 \\ z_2(0)e^{\lambda_2 t} \end{bmatrix} \tag{5.68}$$

that is, initial conditions of $\mathbf{z}(0) = \begin{bmatrix} z_1(0) \\ 0 \end{bmatrix}$ will yield a "speed of response" associated with λ_1, while initial conditions of $\mathbf{z}(0) = \begin{bmatrix} 0 \\ z_2(0) \end{bmatrix}$ will yield a "speed of response" associated with λ_2.

This means that state variable initial conditions in the "direction" of the first eigenvector will have a speed or response associated with the first eigenvalue:

$$\mathbf{x}(t) = \mathbf{V}\,\mathbf{z}(t) = \begin{bmatrix} v_{11} & v_{12} \\ v_{21} & v_{22} \end{bmatrix}\begin{bmatrix} z_1(t) \\ z_2(t) \end{bmatrix} = \begin{bmatrix} v_{11} & v_{12} \\ v_{21} & v_{22} \end{bmatrix}\begin{bmatrix} z_1(0)e^{\lambda_1 t} \\ 0 \end{bmatrix} \tag{5.69}$$

$$\mathbf{x}(t) = \begin{bmatrix} x_1(t) \\ x_2(t) \end{bmatrix} = \begin{bmatrix} v_{11}z_1(0)e^{\lambda_1 t} \\ v_{21}z_1(0)e^{\lambda_1 t} \end{bmatrix}$$

and state variable initial conditions in the "direction" of the second eigenvector will have a speed or response associated with the second eigenvalue

$$\mathbf{x}(t) = \mathbf{V}\,\mathbf{z}(t) = \begin{bmatrix} v_{11} & v_{12} \\ v_{21} & v_{22} \end{bmatrix}\begin{bmatrix} z_1(t) \\ z_2(t) \end{bmatrix} = \begin{bmatrix} v_{11} & v_{12} \\ v_{21} & v_{22} \end{bmatrix}\begin{bmatrix} 0 \\ z_2(0)e^{\lambda_2 t} \end{bmatrix} \tag{5.70}$$

$$\mathbf{x}(t) = \begin{bmatrix} x_1(t) \\ x_2(t) \end{bmatrix} = \begin{bmatrix} v_{12}z_2(0)e^{\lambda_2 t} \\ v_{22}z_2(0)e^{\lambda_2 t} \end{bmatrix}$$

Knowing the effect of the initial condition "direction" is important. If a random case study approach was taken, then we might arbitrarily select initial conditions that were "fast," while other (missed) initial conditions could cause a much slower response.

We will show two examples of the effect of initial condition: Example 5.5, where the system is stable, and Example 5.6, where the system is unstable.

EXAMPLE 5.5   A Stable System

Consider the following system of equations

$$\dot{x}_1 = -0.5\,x_1 + x_2 \tag{5.71}$$

$$\dot{x}_2 = \qquad\quad -2\,x_2 \tag{5.72}$$

Using standard state-space notation

$$\dot{x} = A\,x \tag{5.42}$$

The Jacobian matrix is

$$A = \begin{bmatrix} -0.5 & 1 \\ 0 & -2 \end{bmatrix} \tag{5.73}$$

the eigenvalues are the solution to $\det(\lambda I - A) = 0$, which yields

$$\det\left(\begin{bmatrix} \lambda + 0.5 & -1 \\ 0 & \lambda + 2 \end{bmatrix}\right) = (\lambda + 0.5)(\lambda + 2) = 0$$

so

$$\lambda_1 = -0.5 \quad \lambda_2 = -2$$

and the eigenvectors are

$$\xi_1 = \begin{bmatrix} 1 \\ 0 \end{bmatrix} \quad \xi_2 = \begin{bmatrix} -0.5547 \\ 0.8321 \end{bmatrix}$$

Note that ξ_1 is the "slow" subspace, since it corresponds to $\lambda_1 = -0.5$ and ξ_2 is the "fast" subspace, since it corresponds to $\lambda_2 = -2$.

The numerical values of (5.50) for this problem are

$$x(t) = V\,e^{\Lambda t}\,V^{-1}\,x(0)$$

$$x(t) = \begin{bmatrix} 1 & -0.5547 \\ 0 & 0.8321 \end{bmatrix}\begin{bmatrix} e^{-0.5t} & 0 \\ 0 & e^{-2t} \end{bmatrix}\begin{bmatrix} 1 & 0.6667 \\ 0 & 1.2019 \end{bmatrix} x(0) \tag{5.74}$$

If the initial condition is in the direction of ξ_1, that is

$$x(0) = \begin{bmatrix} 1 \\ 0 \end{bmatrix} \tag{5.75}$$

we find the following state solution (from (5.74) and (5.75)):

$$x(t) = \begin{bmatrix} 1e^{-0.5t} \\ 0 \end{bmatrix} \tag{5.76}$$

If the initial condition is in the direction of ξ_2, that is,

$$x(0) = \begin{bmatrix} -0.5547 \\ 0.8321 \end{bmatrix} \tag{5.77}$$

we find the following state solution (from (5.74) and (5.77)):

$$\mathbf{x}(t) = \begin{bmatrix} -0.5547 \ e^{-2t} \\ 0.8321 \ e^{-2t} \end{bmatrix} \tag{5.78}$$

Note that $\mathbf{x}(0) = \xi_1 = \begin{bmatrix} 1 \\ 0 \end{bmatrix}$ is the slow initial condition and $\mathbf{x}(0) = \xi_2 = \begin{bmatrix} -0.5547 \\ 0.8321 \end{bmatrix}$ is the fast initial condition, as shown in Figures 5.4 and 5.5. The initial conditions in the fast subspace have reached the steady-state in roughly 2.5 minutes (Figure 5.5), while the initial conditions in the slow subspace are roughly 75% complete in 2.5 minutes (Figure 5.6).

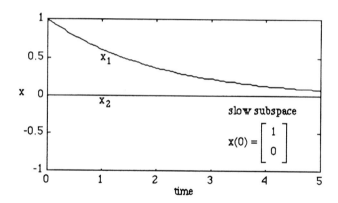

FIGURE 5.5   Transient response for initial condition in the slow subspace.

The expm (matrix exponential) function from MATLAB can be used to verify these simulations. Using $t = 0.5$ and the fast initial condition, we find

```
»a = [-0.5, 1; 0,-2];
»x = expm(a*0.5)*[-0.5547 ; 0.8321]
x =
      -0.2040
       0.3061
```

which agrees with the plot shown in Figure 5.6.

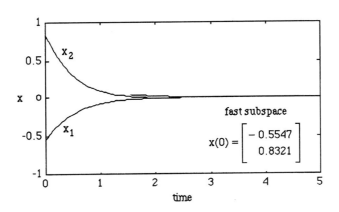

FIGURE 5.6 Transient response for initial condition in the fast subspace.

The previous example was a stable system. The next example is an unstable system.

EXAMPLE 5.6   An Unstable System (Saddle)

Consider the following system of equations:

$$\dot{x}_1 = 2\,x_1 + x_2$$
$$\dot{x}_2 = 2\,x_1 - x_2$$

The Jacobian matrix is
$$\mathbf{A} = \begin{bmatrix} 2 & 1 \\ 2 & -1 \end{bmatrix}$$

the eigenvalues are
$$\lambda_1 = -1.5616 \qquad \lambda_2 = 2.5616$$

and the eigenvectors are
$$\xi_1 = \begin{bmatrix} 0.2703 \\ -0.9628 \end{bmatrix} \qquad \xi_2 = \begin{bmatrix} 0.8719 \\ 0.4896 \end{bmatrix}$$

since $\lambda_1 < 0$, ξ_1 is a stable subspace; since $\lambda_2 > 0$, ξ_2 is an unstable subspace.
The solution for this system is:

$$\mathbf{x}(t) = \begin{bmatrix} 0.2703 & 0.8719 \\ -0.9628 & 0.4896 \end{bmatrix} \begin{bmatrix} e^{-1.5616t} & 0 \\ 0 & e^{2.5616t} \end{bmatrix} \begin{bmatrix} 0.5038 & -0.8972 \\ 0.9907 & 0.2782 \end{bmatrix} \mathbf{x}(0)$$

If the initial condition is in the direction of ξ_1, that is:

$$\mathbf{x}(0) = \begin{bmatrix} 0.2703 \\ -0.9628 \end{bmatrix}$$

we find the following state solution

$$\mathbf{x}(t) = \begin{bmatrix} 0.2703 \ e^{-1.5616t} \\ -0.9628 \ e^{-1.5616t} \end{bmatrix}$$

which is a stable solution, as shown in Figure 5.7.

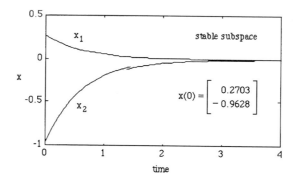

FIGURE 5.7   Initial condition in the stable subspace.

If the initial condition is in the direction of ξ_2, that is,

$$\mathbf{x}(0) = \begin{bmatrix} 0.8719 \\ 0.4896 \end{bmatrix}$$

we find the following state solution:

$$\mathbf{x}(t) = \begin{bmatrix} 0.8719 \ e^{2.5616t} \\ 0.4896 \ e^{2.5616t} \end{bmatrix}$$

which is an unstable solution, as shown in Figure 5.8.

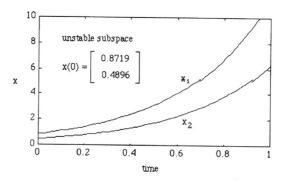

FIGURE 5.8   Initial condition in the unstable subspace.

It should be noted that if the initial condition is not *exactly* in the stable subspace, the solution will begin to diverge and become unstable. That is, if the initial condition is off by, say 10^{-10}, the response will eventually become unbounded.

5.5   SOLUTION OF THE GENERAL STATE-SPACE FORM

Now, consider the general form:

$$\begin{bmatrix} \dot{x}_1 \\ \cdot \\ \cdot \\ \dot{x}_n \end{bmatrix} = \begin{bmatrix} a_{11} & a_{12} & \cdot & a_{1n} \\ \cdot & \cdot & \cdot & \cdot \\ \cdot & \cdot & \cdot & \cdot \\ a_{n1} & a_{n2} & \cdot & a_{nn} \end{bmatrix} \begin{bmatrix} x_1 \\ \cdot \\ \cdot \\ x_n \end{bmatrix} + \begin{bmatrix} b_{11} & b_{12} & \cdot & b_{1m} \\ \cdot & \cdot & \cdot & \cdot \\ \cdot & \cdot & \cdot & \cdot \\ b_{n1} & b_{n2} & \cdot & b_{nm} \end{bmatrix} \begin{bmatrix} u_1 \\ \cdot \\ \cdot \\ u_m \end{bmatrix}$$

or

$$\dot{\mathbf{x}} = \mathbf{A}\,\mathbf{x} + \mathbf{B}\,\mathbf{u} \qquad (5.79)$$

Recall that the single variable equation:

$$\dot{x} = a\,x + b\,u \qquad (5.80)$$

has the solution:

$$x(t) = e^{at}\,x(0) + (e^{at} - 1)\frac{b}{a}\,u(0) \qquad (5.81)$$

when $u(t) = $ constant $= u(0)$.

   In a similar fashion, the solution to (5.79), for a constant input ($\mathbf{u}(t) = \mathbf{u}(0)$) from $t = 0$ to t is

$$\mathbf{x}(t) = \mathbf{P}\,\mathbf{x}(0) + \mathbf{Q}\,\mathbf{u}(0) \qquad (5.82)$$

where

$$\mathbf{P} = e^{At} \qquad (5.83)$$

and

$$\mathbf{Q} = (\mathbf{P} - \mathbf{I})\,\mathbf{A}^{-1}\mathbf{B} \qquad (5.84)$$

Equation (5.82) can be used to solve for a system where the inputs change from time step to time step by using:

$$\mathbf{x}(t + \Delta t) = \mathbf{P}\,\mathbf{x}(t) + \mathbf{Q}\,\mathbf{u}(t) \qquad (5.85)$$

More often this is written as

$$\mathbf{x}(k + 1) = \mathbf{P}\,\mathbf{x}(k) + \mathbf{Q}\,\mathbf{u}(k) \qquad (5.86)$$

where k represents the kth time step. Often a general purpose numerical integration technique (such as one presented in Chapter 4) will be used to solve (5.79).

5.6   MATLAB ROUTINES step AND initial

We show the use of step and initial by way of the following example.

EXAMPLE 5.7 A Linearized Bioreactor Model

Consider the following linearized form of a bioreactor model with substrate inhibition kinetics (see Module 8 for details):

$$\dot{x} = A\,x + B\,u$$

$$y = C\,x + D\,u$$

where:

$$A = \begin{bmatrix} 0 & 0.9056 \\ -0.7500 & -2.5640 \end{bmatrix}$$

$$B = \begin{bmatrix} -1.5302 \\ 3.8255 \end{bmatrix}$$

$$C = \begin{bmatrix} 1 & 0 \\ 0 & 1 \end{bmatrix}$$

$$D = \begin{bmatrix} 0 \\ 0 \end{bmatrix}$$

Enter the state space model:

```
» a = [0, 0.9056;-0.7500,-2.5640]

a =
         0   0.9056
   -0.7500  -2.5640
» b = [-1.5302;3.8255]

b =
   -1.5302
    3.8255
» c = [1, 0 ; 1 , 0]

c =
        1    0
        0    1
» d = [0;0]

d =
        0
        0
```

Check the stability

```
» eig(a)

ans =
    -0.3000
    -2.2640
```

The system is stable.

Assume the process is initially at steady-state. Since this model is in deviation variable form, the initial condition is the zero vector.

5.6.1   The MATLAB `step` Function

The MATLAB `step` function assumes a deviation variable form (the initial conditions are zero). The commands are:

```
» [y,x,t] = step(a,b,c,d,1);
» plot(t,y)
```

which yields the plot shown in Figure 5.9.

Notice that the `step` function automatically determined the length of the time vector. You may also provide an equal-spaced time vector and use the following command

```
[y,x] = step(a,b,c,d,1,t)
```

5.6.2   The MATLAB `initial` Function

The MATLAB `initial` function assumes a deviation variable form, with the initial conditions perturbed from zero. The commands are:

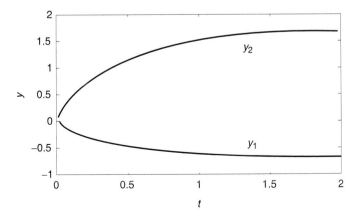

FIGURE 5.9   Plot of outputs, for a step input change.

```
» [y,x,t] = initial(a,b,c,d,1);
» plot(t,y)
```

Notice that the b and d matrices are not really used by the initial function, since it is assumed that there is no input change.

SUMMARY

In this chapter we have developed a state-space model of a chemical process that is inherently linear (e.g., the tank height example). We have also shown how to linearize models that are nonlinear. The models obtained in this fashion are based on *deviation variables,* that is, the states and inputs are *perturbations* from the steady-state operating point where the linearization is performed. The stability of a nonlinear system is determined from the eigenvalues of the Jacobian matrix in the linearized model (state-space form).

Several important concepts were presented in this chapter.

- For unforced systems (zero input), the initial condition vector will determine the "speed" of response. For stable systems (all $\lambda < 0$), the eigenvector associated with the largest magnitude λ is the fast direction, while the eigenvector associated with the smallest λ is the slow direction.
- Although it is possible for a system with both negative (stable) and positive (unstable) eigenvalues to have stable behavior if the initial condition is in the stable subspace, this is impossible in practice. Any perturbation from the stable trajectory will cause the solution to become unbounded (unstable).

The MATLAB routines that were used include:

expm: Matrix exponential

step: Step response of a state-space (or transfer function) model

State-space models can be transformed to Laplace transfer function form, which is particularly useful for control system design. Applications of Laplace transforms will be presented in Chapters 7 through 10.

Eigenvector/eigenvalue analysis will be useful in performing phase-plane analysis, which is covered in Chapter 13.

The reader should understand the following terms:

state-space

Jacobian

deviation or perturbation variable

eigenvalue

eigenvector
linearization
stability
Taylor series

FURTHER READING

Linearization is discussed briefly in most books on process control, including:

Luyben, W.L. (1990). Process Modeling, Simulation and Control for Chemical Engineers, 2nd Ed., New York: McGraw-Hill.

Marlin, T.E. (1995). Process Control. Designing Processes and Control Systems for Dynamic Performance, New York: McGraw-Hill.

Ogunnaike, B.A. and W. H. Ray (1994). Process Dynamics, Modeling and Control, New York: Oxford University Press.

Seborg, D.E., T.F. Edgar, and D.A. Mellichamp (1989). Process Dynamics and Control, New York: Wiley.

Stephanopoulos, G. (1984). Chemical Process Control: An Introduction to Theory and Practice, Englewood Cliffs, NJ: Prentice-Hall.

STUDENT EXERCISES

1. As a process development engineer you are working on a process with three continuous-stirred-tank reactors (CSTRs) in series. A constant volumetric flowrate (flowrate does not vary with time) is maintained throughout the system, however the volume in each reactor is different (but constant). Since the temperature varies from reactor to reactor (but is constant in an individual reactor) the reaction rate parameter is different for each reactor. The molar concentration of the inlet stream varies.

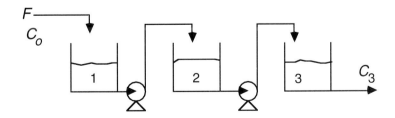

Assume that the density of the streams remains constant (independent of concentration). The reaction is a first-order (irreversible) decomposition ($A \rightarrow B$). Molar rate of decomposition of A (per unit volume) $= k\,CA$

   a. Write the 3 dynamic model equations.
   b. Write the state-space model $o(\dot{x}) = A\,x + B\,u$
   c. The values of the parameters and variables are
$$F = 1 \text{ ft}^3/\text{min} \quad C_o = 1 \text{ lbmol/ft}^3$$
$$V_1 = 10 \text{ ft}^3 \quad V_2 = \text{ ft}^3 \quad V_3 = 5 \text{ ft}^3$$
$$k_1 = 0.0333 \text{ min}^{-1} \quad k_2 = 0.2 \text{ min}^{-1} \quad k_3 = 0.55 \text{ min}^{-1}$$
   i. Find the steady-state concentrations in each reactor
   ii. Evaluate the A matrix (Jacobian) and find the eigenvalues

2. Consider a chemical reactor with bypass, as shown below. Assume that the reaction rate (per unit volume) is first-order ($r = kC_1$) and C_1 is the concentration in the reactor (the reactor is perfectly mixed). Assume that the volume in the reactor (V) and the feed flowrate (F) remain constant. The fraction of feed bypassing the reactor is $(1-a)F$ and that entering the reactor is aF. Assume that the fraction bypassing the reactor does not change. The inlet concentration (C_{in}) is the input variable and the mixed outlet stream composition (C_2) is the output variable. Write this model in state-space form (this model is inherently linear, so deviation variables are not needed).

$$o(\dot{x}) = A\,x + B\,u$$
$$y = C\,x + D\,u$$

3. Consider the following set of series and parallel reactions

$$A \xrightarrow{k_1} B \xrightarrow{k_2} C$$

$$A + A \xrightarrow[k_3]{} D$$

Material balances on components A and B yield the following two equations

$$\frac{dC_A}{dt} = \frac{F}{V}(C_{Af} - C_A) - k, C_A - k_3\, C_A^2$$
$$\frac{dC_B}{dt} = \frac{F}{V}(-C_B) + (k_1\, C_A - k_2\, C_B)$$

where
$$k_1 = \frac{5}{6} \text{ min}^{-1} \quad k_2 = \frac{5}{3} \text{ min}^{-1} \quad k_3 = \frac{1}{6} \frac{\text{liters}}{\text{mol min}}$$

$$C_{Af} = 10 \frac{\text{mol}}{\text{liter}} \quad C_{As} = 3 \frac{\text{mol}}{\text{liter}}$$

   a. Find the steady-state dilution rate (F/V) and concentration of B (show all units).
   b. Linearize and put in state-space form (find the numerical values of the A, B, and C matrices), assuming that the manipulated variable is dilution rate (F/V), and the output variable is C_B.
   c. Find the eigenvalues (show units).
   d. Find perturbations in initial conditions that are in the fastest and slowest directions.

4. A chemical reactor that has a single second-order reaction and an outlet flowrate that is a linear function of height has the following model:

$$\frac{dVC}{dt} = F_{in}C_{in} - FC - kVC^2 \tag{5.87}$$

$$\frac{dV}{dt} = F_{in} - F \tag{5.88}$$

where the outlet flowrate is linearly related to the volume of liquid in the reactor $(F = \beta V)$. The parameters, variables and their steady-state values are shown below.

F_{in} = inlet flowrate (1 liter/min)
C_{in} = inlet concentration (1 gmol/liter)
C = tank concentration (0.5 gmol/liter)
V = tank volume (1 liter)
k = reaction rate constant (2 liter/(gmol min))
β = 1 min^{-1}

Equations (5.87) and (5.88) can be written in physical state variable form as

$$\frac{dC}{dt} = \frac{F_{in}}{V}(C_{in} - C) - kC^2 \tag{5.89}$$

$$\frac{dV}{dt} = F_{in} - \beta V \tag{5.90}$$

a. List the states, outputs, inputs and parameters for the nonlinear equations (5.89) and (5.90).
b. Linearize (5.89) and (5.90) and write the state space model (find the numerical values for the A, B, and C matrices), assuming that the inlet flowrate is the input variable and that both states are output variables. Define the deviation variables for states, inputs, and outputs.

5. Find the "fast" and "slow" initial conditions for the following model

$$\dot{x}_1 = -x_1$$
$$\dot{x}_2 = -4x_2$$

6. Find the stable and unstable subspaces for the following system of equations

$$\dot{x}_1 = -x_1$$
$$\dot{x}_2 = 4x_2$$

Plot the transient responses for initial conditions in both the stable and unstable subspaces. Show that a small perturbation from the stable initial condition will lead to an unstable solution.

7. The noninteracting tank model is (see Example 5.1)

$$
\begin{bmatrix} \dot{h}_1 \\ \dot{h}_2 \end{bmatrix} = \begin{bmatrix} \dfrac{-\beta_1}{A_1} & 0 \\ \dfrac{\beta_1}{A_2} & \dfrac{-\beta_2}{A_2} \end{bmatrix} \begin{bmatrix} h_1 \\ h_2 \end{bmatrix} + \begin{bmatrix} \dfrac{1}{A_1} \\ 0 \end{bmatrix} F_o
$$

Consider a system where the steady-state flowrates are 5 ft³/min, and the following cross-sectional areas and steady-state heights:

$$
A_1 = 2 \text{ ft}^2 \quad A_2 = 10 \text{ ft}^2
$$

$$
h_1 = 2.5 \text{ ft} \quad h_2 = 5 \text{ ft}
$$

We find (from $F_1 = \beta_1 h_1$ and $F_2 = \beta_2 h_2$), then, that:

$$
\beta_1 = 2 \frac{\text{ft}^2}{\text{min}} \quad \beta_2 = 1 \frac{\text{ft}^2}{\text{min}}
$$

and the state-space model (in physical variables) becomes:

$$
\begin{bmatrix} \dot{h}_1 \\ \dot{h}_2 \end{bmatrix} = \begin{bmatrix} -1 & 0 \\ 0.2 & -0.1 \end{bmatrix} \begin{bmatrix} h_1 \\ h_2 \end{bmatrix} + \begin{bmatrix} 0.5 \\ 0 \end{bmatrix} F_o
$$

$$
\begin{bmatrix} h_1 \\ h_2 \\ F_2 \\ F_o \end{bmatrix} = \begin{bmatrix} 1 & 0 \\ 0 & 1 \\ 0 & 1 \\ 0 & 0 \end{bmatrix} \begin{bmatrix} h_1 \\ h_2 \end{bmatrix} + \begin{bmatrix} 0 \\ 0 \\ 0 \\ 1 \end{bmatrix} [F_o]
$$

a. Work in deviation variable form and find the fast and slow subspaces. Use `ini-tial` to simulate the unforced deviation variable system (input deviation remains constant at 0), from initial conditions in both the fast and slow subspaces.
b. Use the results from part a, and convert to the actual physical variables.
c. Work in physical variable form. Use `initial` to simulate the unforced deviation variable system (input remains constant), from initial conditions in both the fast and slow subspaces. Show that the results obtained are the same as those in part b.

8. As a chemical engineer in the pharmaceutical industry you are responsible for a process that uses a bacteria to produce an antibiotic. The reactor has been contaminated with a protozoan that consumes the bacteria. The predator-prey equations are used to model the system (b = bacteria (prey), p = protozoa (predator)). The time unit is days.

$$
\frac{db}{dt} = \alpha\, b - \gamma\, bp
$$

$$
\frac{dp}{dt} = \varepsilon\, \gamma\, bp - \beta\, p
$$

a. Show that the steady-state values are

$$b_s = \frac{\beta}{\varepsilon \gamma} \qquad p_s = \frac{\alpha}{\gamma}$$

b. Use the scaled variables, w and z, to find the following scaled modeling equations

$$w = \frac{b}{b_s} \qquad z = \frac{p}{p_s}$$

$$\frac{dw}{dt} = \alpha \left(1 - z\right) w$$

$$\frac{dz}{dt} = -\beta \left(1 - w\right) z$$

c. Find the eigenvalues of the Jacobian matrix for the scaled equations, evaluated at w_s and z_s. Realize that w_s and z_s are 1.0 by definition. Find the eigenvalues in terms of α and β.

d. The parameters are $\alpha = \beta = 1.0$ and the initial conditions are $w(0) = 1.5$ and $z(0) = 0.75$.

    i. Linearize and write the state-space form (let the state variables be $x_1 = w - w_s$ and $x_2 = z - z_s$). Find the initial condition vector $x_0 = \begin{bmatrix} x_1(0) \\ x_2(0) \end{bmatrix}$, to use with `initial`.

    ii. Solve the state space model from (i) using `lsim` and plot the transient response of x_1 and x_2 as a function of time (plot these curves on the same graph), simulating to at least $t = 20$.

    iii. Show a phase-plane plot, placing x_1 on the x-axis and x_2 on the y-axis.

    iv. What is the "peak-to-peak" time for the bacteria? By how much time does the protozoan "lag" the bacteria?

9. Consider the state-space model

$$\begin{bmatrix} \dot{x}_1 \\ \dot{x}_2 \end{bmatrix} = \begin{bmatrix} -1.0 & 0.0 \\ 4.0 & -5.0 \end{bmatrix} \begin{bmatrix} x_1 \\ x_2 \end{bmatrix}$$

a. Find the "fast" and "slow" initial condition directions.

10. Consider the following system of two reactors.

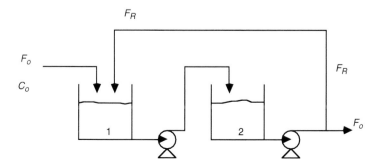

Assume a first-order decomposition of $A \rightarrow B$. Assume that all flowrates are constant (volumes are constant).

a. Write the modeling equations for concentration of A, using either the instantaneous or integral method.

b. Write these in state-space form:

$$\dot{x} = Ax + Bu$$

c. Given the following constants, calculate the steady-state concentrations:

$$F_o = 1.25 \frac{m^3}{hr} \qquad F_R = 1.75 \frac{m^3}{hr}$$

$$C_o = 1.5 \frac{kgmol}{m^3} \qquad k_1 = 0.10833 \ hr^{-1} \qquad k_2 = 0.33333 \ hr^{-1}$$

$$V_1 = 15 \ m^3 \qquad V_2 = 9 \ m^3$$

d. Find the eigenvalues of the A matrix. Discuss the stability of this system.

e. The inlet concentration, C_o, is changed from 1.5 to 1.75 at $t = 0$. Use `step` to simulate the behavior of this system.

11. A stirred tank heater is used to supply a chemical process with a fluid at a constant temperature. The heater receives fluid from an upstream process unit, which may cause the flowrate or temperature to change.

Consider the diagram of the stirred tank heater shown below, where the tank inlet stream is received from another process unit. A heat transfer fluid is circulated through a jacket to heat the fluid in the tank. Assume that no change of phase occurs in either the tank liquid or the jacket liquid.

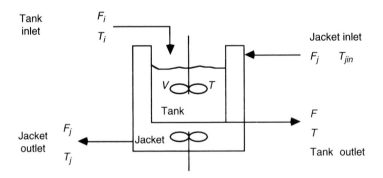

Part 1

a. Write the dynamic modeling equations to find the tank and jacket temperatures. Do not use any numerical values—leave these equations in terms of the process parameters and variables. State any additional assumptions needed to solve the problem.

Assume: Constant level.

Perfect mixing in both the tank and jacket.

The tank inlet flowrate, jacket flowrate, tank inlet temperature, and jacket inlet temperature may change.

The rate of heat transfer from the jacket to the tank is governed by the equation $Q = UA(T_j - T)$, where U is the overall heat transfer coefficient and A is the area for heat exchange.

b. State the major objective of this process.

c. What do you consider the most important measured variable?

d. What is a likely input variable variable that you would use to maintain a desired tank temperature?

Part 2

Assume that both the tank fluid and the jacket fluid are water. The steady-state values of this system variables and some parameters are:

$$F = 1\ \frac{ft^3}{min} \qquad \rho C_p = 61.3\ \frac{Btu}{°F\ ft^3} \qquad \rho_j C_{pj} = 61.3\ \frac{Btu}{°F\ ft^3}$$

$$T_i = 50°F \qquad T = 125°F \qquad V = 10\ ft^3$$

$$T_{jin} = 200°F \qquad T_j = 150°F \qquad V_j = 1\ ft^3$$

e. Find F_j and UA (show units) at steady-state.

f. Linearize the set of two nonlinear ODEs obtained in problem a, to obtain the state space form:

$$\dot{x} = A\,x + B\,u$$

$$y = C\,x$$

where

$$x = \begin{bmatrix} T - T_s \\ T_j - T_{js} \end{bmatrix} = \text{state variables}$$

$$u = \begin{bmatrix} F_j - F_{js} \\ F - F_s \\ T_i - T_{is} \\ T_{jin} - T_{jins} \end{bmatrix} = \text{input variables}$$

$$y = \begin{bmatrix} T - T_s \\ T_j - T_{js} \end{bmatrix} = \text{output variables}$$

Determine the **A, B,** and **C** matrices (symbolically and numerically)

g. Find the eigenvalues of **A.**

Part 3

h. Simulate the system of state-space equations for a step change in the jacket flowrate from $F_j = 1.5$ ft³/min to $F_j = 1.75$ ft³/min F at time = 5 minutes (work in deviation variables, but remember to convert back to physical variables before plotting). What is the final value of the states, in the physical variables (T and T_j)? Plot the response.

i. Perform some simulations with step changes on some of the other input variables. Comment on any different behavior that you may observe.

12. Consider the following model of 2-stage absorption column:

$$\frac{dw}{dt} = -\left(\frac{L + Va}{M}\right)w + \left(\frac{Va}{M}\right)z$$

$$\frac{dz}{dt} = \left(\frac{L}{M}\right)w - \left(\frac{L + Va}{M}\right)z + \frac{V}{M}z_f$$

where w and z are the liquid concentrations on stage 1 and stage 2, respectively. L and V are the liquid and vapor molar flowrates. z_f is the concentration of the vapor stream entering the column.

The steady-state input values are $L = 80$ gmol inert liquid/min and $V = 100$ gmol inert vapor/min.

The parameter values are $M = 20$ gmol inert liquid, $a = 0.5$, and $z_f = 0.1$ gmol solute/gmol inert vapor.

a. Find the steady-state values of w and z.

b. Linearize and find the state space model, assuming that L and V are the inputs.

c. Find the eigenvalues and eigenvectors of $\mathbf{A}$ (Jacobian).

d. Find the expected "slowest" and "fastest" initial conditions (perturbations from steady-state).

13. Most chemical process plants have a natural gas header that circulates through the process plant. A simplified version of such a header is shown below.

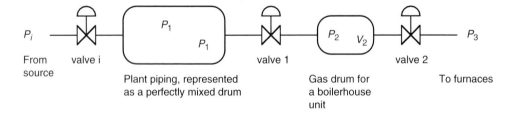

Here, the natural gas enters the process plant from a source (the natural gas pipeline) through a control valve. It flows through the plant piping, which we have represented as a perfectly mixed drum for simplicity. Another valve connects the plant piping to the gas drum for a boilerhouse unit. Gas passes through another valve to the boilerhouse furnaces.

The objective of this problem is to develop a linear model that relates changes in valve position to changes in drum pressures.

a. Write modeling equations assuming that the pressures in drums 1 and 2 are the state variables. Let the input variables be (1) valve position 1, (2) valve position 2, and (3) source pressure.

b. Solve for the steady-state conditions and write the modeling equations in linear, deviation variable form.

$$\dot{\mathbf{x}} = \mathbf{A}\,\mathbf{x} + \mathbf{B}\,\mathbf{u}$$

$$\mathbf{y} = \mathbf{C}\,\mathbf{x}$$

$$\mathbf{x} = \begin{bmatrix} P_1 - P_{1s} \\ P_2 - P_{2s} \end{bmatrix} = \text{state variables}$$

$$\mathbf{u} = \begin{bmatrix} h_1 - h_{1s} \\ h_2 - h_{2s} \\ P_i - P_{is} \end{bmatrix} = \begin{bmatrix} \text{change in valve position 1} \\ \text{change in valve position 2} \\ \text{change in source pressure} \end{bmatrix} = \text{input variables}$$

$$\mathbf{y} = \begin{bmatrix} P_1 - P_{1s} \\ P_2 - P_{2s} \end{bmatrix} = \text{output variables}$$

c. Study the effect of step changes in each input on each tank pressure.

HINTS: For simplicity, assume that the following equations can be used for the flow through the valves:

$$q_i = \alpha_i\, h_i\, (pi - p_1) = \text{flow through valve } i$$

$$q_1 = \alpha_1\, h_1\, (p_1 - p_2) = \text{flow through valve 1}$$

$$q_2 = \alpha_2\, h_2\, (p_2 - p_3) = \text{flow through valve 2}$$

where the flowrate is in lbmol/min, h is the fraction that a valve is open (varies between 0 and 1), and α is a valve coefficient.

STEADY STATE DATA:

gas flowrate = 1000 std ft^3/min

$P_{is} = 250$ psig, $P_{1s} = 50$ psig, $P_{2s} = 30$ psig, $P_{3s} = 5$ psig

assume that each valve is 1/2 open under these conditions ($h_{is} = h_{1s} = h_{2s} = 0.5$)

CONSTANTS:

$V_1 = 1135$ ft^3, $V_2 = 329$ ft^3, Temperature = 32 °F

$$R \text{ (gas constant)} = 10.73 \, \frac{\text{psia ft}^3}{\text{lbmol °R}}$$

MAGNITUDE OF STEP CHANGES:

Make separate step changes of 0.1 (10%) in the valve openings, and 10 psia in the inlet pressure. Simulate for $t = 0$ to $t = 15$ minutes.

14. A stream contains a waste chemical, W, with a concentration of 1 mol/liter. To meet EPA and state standards, at least 90% of the chemical must be removed by reaction. The chemical decomposes by a second-order reaction with a rate constant of 1.5 liter/(mol hr). The stream flowrate is 100 liter/hr and two available reactors (400 and 2000 liters) have been placed in series (the smaller reactor is placed before the larger one).
    a. Write the modeling equations for the concentration of the waste chemical. Assume constant volume and constant density. Let

    C_{w1} = concentration in reactor 1, mol/liter
    C_{w2} = concentration in reactor 2, mol/liter
    F = volumetric flowrate, liter/hr
    V_1 = liquid volume in reactor 1, liters
    V_2 = liquid volume in reactor 2, liters
    k = second-order rate constant, liter/(mol hr)

    b. Show that the steady-state concentrations are 0.33333 mol/liter (reactor 1) and 0.09005 mol/liter (reactor 2), so the specification is met.
    (*Hint:* You need to solve quadratic equations to obtain the concentrations.)
    c. Linearize at steady-state and develop the state space model (analytical), of the form:

    $$\dot{x} = A x + B u$$

    where:

    $$x = \begin{bmatrix} C_{w1} - C_{w1s} \\ C_{w2} - C_{w2s} \end{bmatrix} \quad u = \begin{bmatrix} F - F_s \\ C_{win} - C_{wins} \end{bmatrix}$$

    d. Show that the A and B matrices are:

    $$A = \begin{bmatrix} -1.25 & 0 \\ 0.05 & -0.32015 \end{bmatrix} \quad B = \begin{bmatrix} 0.0016667 & 0.25 \\ 0.0001216 & 0 \end{bmatrix}$$

    (also, show the units associated with each coefficient)
    e. i. Find the eigenvalues and eigenvectors using the MATLAB `eig` function.
       ii. Find the eigenvalues by hand, by solving det($\lambda I - A$) = 0.

f. The system is not initially at steady-state. Solve the following for the linearized model, using the MATLAB function `initial` (first, convert the physical variables to deviation variables)

   i. If $C_{w1}(0) = 0.3833$ and $C_{w2}(0) = 0.09005$, find how the concentrations change with time.

   ii. If $C_{w1}(0) = 0.3333$ and $C_{w2}(0) = 0.14005$, find how the concentrations change with time.

Relate these responses to the eigenvalues/eigenvector analysis of e. Discuss the differences in speeds of response (you should find that a perturbation in the first reactor concentration responds more rapidly and a perturbation in the second reactor concentration).

The MATLAB `initial` function needs you to create the following matrices before using it:

$$\mathbf{C} = \begin{bmatrix} 1 & 0 \\ 0 & 1 \end{bmatrix} \quad \mathbf{D} = \begin{bmatrix} 0 & 0 \\ 0 & 0 \end{bmatrix}$$

g. Solve f for the nonlinear equations, using `ode45`. Compare the linear and nonlinear variables on the same plots (make certain you convert from deviation to physical variables for the linear results).

h. Now, consider a step change in the flowrate from 100 liters/hour to 110 liters/hour. Assume the initial concentrations are the steady-state values (0.3333 and 0.09005). Compare the linear and nonlinear responses of the reactor concentrations. Is the removal specification still obtained?

i. Would better steady-state removal of W be obtained if the order of the reaction vessels was reversed? Why or why not? (Show your calculations.)

SOLVING LINEAR nTH ORDER ODE MODELS

6

The purpose of this chapter is to review methods to solve solve linear nth order ODEs. After studying this material, the student will be able to:

- Transform a linear state-space model with n states to a single nth order ordinary differential equation.
- Solve an nth order constant coefficient coefficient homogeneous ODE.
- Solve an nth order constant coefficient coefficient heterogeneous ODE.
- Solve a first-order ODE with a time-varying coefficient.
- Use the Routh stability criterion for stability analysis.

The major sections in this chapter are:

6.1 BACKGROUND

A model composed of a single, nth order linear ordinary differential equation has the following form:

$$a_n(t)\frac{d^n x}{dt^n} + a_{n-1}(t)\frac{d^{n-1}x}{dt^{n-1}} + \ldots + a_1(t)\frac{dx}{dt} + a_o(t)\,x$$
$$= b_m(t)\frac{d^m u}{dt^m} + b_{m-1}(t)\frac{d^{m-1}u}{dt^{m-1}} + \ldots + b_1(t)\frac{du}{dt} + b_o(t)\,u \tag{6.1}$$

where the state variable is x and the input variable is u. This general model is linear because the state (x) and input (u) and all of their derivatives with respect to time appear linearly. Notice that the coefficients do not have to be linear functions of time, however.

Models of the form of (6.1) do not arise naturally when chemical processes are modeled. As shown in previous chapters, dynamic chemical process models are generally sets of first-order (either linear or nonlinear ordinary differential equations. The advantage of the form of (6.1) is that there exist a number of techniques to obtain analytical solutions.

In this chapter we show how to transform sets of linear, first-order differential equations to a single nth order differential equation. We then review several techniques for solving this type of equation. For motivation, we use a batch reactor example to illustrate each of the techniques. It should be noted that there are many good mathematics texts that cover each of these techniques in more depth (see Boyce and DiPrima, 1992, for example). Our goal here is to provide a concise overview of some more useful techniques to solve dynamic chemical process problems.

EXAMPLE 6.1   Batch Chemical Reactor

Consider a batch chemical reactor, where there is no flow in or out of the vessel. The reactor is initially charged with a liquid of volume V and an initial concentration (mol/liter) of reactant A of C_{A0}.

We consider a series reaction where component A reacts to form the desired component B. Component B can further react to form the undesired component C. Each of the reactions is irreversible, so A can react to form B, but B does not react to form A.

$$A \overset{k_1}{\to} B \overset{k_2}{\to} C$$

Here k_1 represents kinetic rate constant (time^{-1}) for the conversion of A to B, while k_2 represents the rate constant for the conversion of B to C.

Since component B is the desired product, we would like to know how long to run the reaction in order to maximize the amount of B produced. If the reaction time is too long, all of B will eventually be converted to C.

Develop the Modeling Equations.   Assume that each of the reactions is first-order. Since the volume is constant ($dV/dt = 0$), and there is no flow in or out, the modeling equations are (the reader should be able to derive these, based on material balances on each component):

$$\frac{dC_A}{dt} = -k_1 C_A \tag{6.2}$$

$$\frac{dC_B}{dt} = k_1 C_A - k_2 C_B \tag{6.3}$$

$$\frac{dC_C}{dt} = k_2 C_B \tag{6.4}$$

where C_A, C_B, and C_C represent the concentrations (mol/volume) of components A, B, and C, respectively. The units for the rate constants (k_1 and k_2) are time^{-1}.

Notice that the time rate of change of component A is only a function of the concentration of A. Then equation (6.2) can be solved, since C_A and t are separable, to find

$$C_A(t) = C_{A0} e^{-k_1 t} \tag{6.5}$$

where C_{A0} is the initial condition for the concentration of A. If we define the conversion of A as $x = (C_{A0} - C_A)/C_{A0}$ and the dimensionless time $\tau = k_1 t$, (6.5) can be represented by the single curve shown in Figure 6.1 ($x = 1 - e^{-\tau}$).

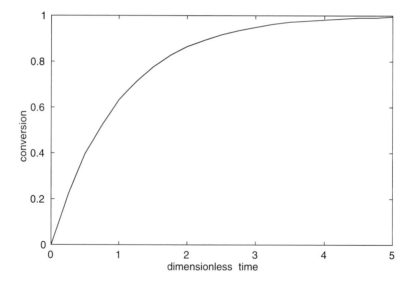

FIGURE 6.1   Conversion of A as a function of the dimensionless time.

Now we wish to find a single differential equation to solve for C_B.

Reduce to a Single Equation for CB.   Here we have two different ways to solve for C_B.

Method 1. Substitute (6.5) into (6.3) to obtain the expression

$$\frac{dC_B}{dt} + k_2 C_B = k_1 C_{A0} e^{-k_1 t} \tag{6.6}$$

Equation (6.6) is a linear, constant coefficient, *heterogeneous* differential equation. It is heterogeneous because of the "forcing function" on the righthand side. Heterogeneous equations are solved in Section 6.3.

Method 2. Here we can rewrite (6.3) to solve for C_A in terms of C_B:

$$C_A = \frac{1}{k_1}\frac{dC_B}{dt} + \frac{k_2}{k_1}C_B \tag{6.7}$$

Taking the first derivative of (6.7) with respect to time, we find:

$$\frac{dC_A}{dt} = \frac{1}{k_1}\frac{d^2C_B}{dt^2} + \frac{k_2}{k_1}\frac{dC_B}{dt} \tag{6.8}$$

Substituting (6.7) and (6.8) into (6.2), we find the second-order equation:

$$\frac{d^2C_B}{dt^2} + (k_1 + k_2)\frac{dC_B}{dt} + k_1 k_2 C_B = 0 \tag{6.9}$$

Notice that (6.9) has the form

$$a_2 \frac{d^2x}{dt^2} + a_1 \frac{dx}{dt} + a_o x = 0 \tag{6.10}$$

Equation (6.10) is known as a linear, constant coefficient, homogeneous differential equation. The term homogeneous means that there is no "forcing function" on the righthand side. In Section 6.2 we cover the solution of these equations.

6.2 SOLVING HOMOGENEOUS, LINEAR ODES WITH CONSTANT COEFFICIENTS

Homogeneous *n*th order linear differential equations have the form

$$a_n \frac{d^n x}{dt^n} + a_{n-1}\frac{d^{n-1}x}{dt^{n-1}} + \ldots + a_1 \frac{dx}{dt} + a_o x = 0 \tag{6.11}$$

To solve equation (6.11) we replace all $d^i x/dt^i$ terms by λ^i

$$a_n \lambda^n + a_{n-1}\lambda^{n-1} + \ldots + a_1 \lambda + a_o = 0 \tag{6.12}$$

Equation (6.12) is called the *characteristic equation*. The *n* roots of the characteristic equation are called *eigenvalues* (in control textbooks the roots are often called *poles*). The eigenvalues are used to solve (6.11). Two related methods are used, depending on whether the eigenvalues are distinct (all are different) or repeated (some are the same).

6.2.1　Distinct Eigenvalues

We see that (6.12) is an *n*th order polynomial that will have *n* roots, λ_i. If all of the roots are distinct (not repeated), the solution to (6.11) is

$$x(t) = c_1 e^{\lambda_1 t} + c_2 e^{\lambda_2 t} + \ldots + c_n e^{\lambda_n t} \tag{6.13}$$

where each of the constants c_1 through c_n is found from the initial conditions, $x(0)$, .., $dx^{n-1}(0)/dt^{n-1}$.

In order to find the coefficients, c_i, we must know the *initial conditions* for x and its derivatives.

EXAMPLE 6.1 Continued.　Solution for Component A

We see from equation (6.2) that the concentration of A does not depend on the values of B and C.

$$\frac{dC_A}{dt} + k_1 C_A = 0 \tag{6.14}$$

The characteristic equation is

$$\lambda + k_1 = 0 \tag{6.15}$$

and the eigenvalue is $\lambda = -k_1$. The solution is then

$$C_A(t) = c_1 e^{-k_1 t} \tag{6.16}$$

and we can solve for c_1 from initial condition $C_A(0) = C_{A0}$, to obtain

$$C_A(t) = C_{A0} e^{-k_1 t} \tag{6.17}$$

which is the same result obtained in (6.5) using separation of variables and integration.

Now, let's continue and use the general procedure to solve a second-order differential equation.

EXAMPLE 6.1 Continued.　Solution for Component B

Recall that the equation for the concentration of B is:

$$\frac{d^2 C_B}{dt^2} + (k_1 + k_2) \frac{dC_B}{dt} + k_1 k_2 C_B = 0 \tag{6.9}$$

and the characteristic equation is:

$$\lambda^2 + (k_1 + k_2)\lambda + k_1 k_2 = 0 \tag{6.18}$$

which can be written:

$$(\lambda + k_1)(\lambda + k_2) = 0 \tag{6.19}$$

So the eigenvalues are:

$$\lambda_1 = -k_1 \quad \text{and} \quad \lambda_2 = -k_2$$

and the solution can be written:

$$C_B(t) = c_1 \exp(-k_1 t) + c_2 \exp(-k_2 t) \tag{6.20}$$

We need two initial conditions, $C_B(0)$ and $dC_B(0)/dt$, to evaluate the constants, c_1 and c_2.

We assumed that there is no component B in the reactor initially, so $C_B(0) = 0$. From (6.20) we then find:

$$c_1 = -c_2 \tag{6.21}$$

From (3) we see that:

$$\frac{dC_B(0)}{dt} = k_1 C_{A0} \tag{6.22}$$

Taking the derivative of (6.20) and using (6.21) and (6.22), we find:

$$C_B(t) = \frac{k_1 C_{A0}}{k_2 - k_1} [\exp(-k_1 t) - \exp(-k_2 t)] \tag{6.23}$$

This expression can be used, for example, to solve for the amount of time that will yield the maximum amount of C_B (see student exercise 15).

Dimensionless Equation.   It should also be noted that (6.23) can be made dimensionless by defining the following variables:

$$x = C_B/C_{A0} = \text{conversion of } A \text{ to } B$$

$$\tau = k_1 t \qquad = \text{dimensionless time}$$

$$\alpha = k_2/k_1 \quad = \text{rate constant ratio}$$

to find:

$$x(\tau) = \frac{1}{\alpha - 1} [\exp(-\tau) - \exp(-\alpha\tau)]$$

which is shown in Figure 6.2, for $\alpha = 0.5$ and 2. Notice that when the first reaction is faster than the second ($\alpha = 0.5$), there is a higher concentration of B than when the first reaction is slower than the second ($\alpha = 2$). When the second reaction is faster than the first, component B reacts further to form C, before a substantial amount of B is formed.

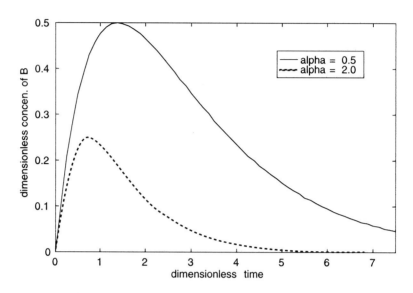

FIGURE 6.2   Concentration of B as a function of time. When the rate for the second reaction is faster than the first ($\alpha = 2$), the peak concentration of B is lower.

We notice in the previous example that (6.23) cannot be used if $k_2 = k_1$. This is a case where the eigenvalues are repeated. The procedure for repeated eigenvalues is shown next.

6.2.2   Repeated Eigenvalues

If a particular root in the solution of (6.12), λ_i, occurs r times, then the corresponding terms in the solution to (6.11) are:

$$(c_i + c_{i+1}\, t + c_{i+2}\, t^2 + \ldots + c_{i+r-1}\, t^{r-1})\, e^{\lambda_i t} \qquad (6.24)$$

EXAMPLE 6.1 Continued.   Repeated Roots

The equation for the concentration of B is when $k_2 = k_1 = k$ is:

$$\frac{d^2 C_B}{dt^2} + 2\,k\,\frac{dC_B}{dt} + k^2\,C_B = 0 \qquad (6.25)$$

and the characteristic equation is:

$$\lambda^2 + 2 k \lambda + k^2 = 0 \tag{6.26}$$

which can be factored as:

$$(\lambda + k)(\lambda + k) = 0$$

so the eigenvalues (roots) are:

$$\lambda_1 = \lambda_2 = -k$$

The solution can be written:

$$C_B(t) = (c_1 + c_2\, t) \exp(-k_t) \tag{6.27}$$

Notice that we can find c_1 from the initial condition for C_B. From (6.27) at $t = 0$,

$$c_1 = C_B(0)$$

But $C_B(0) = 0$, since there is no B initially, so:

$$C_B(t) = c_2\, t \exp(-kt) \tag{6.28}$$

The derivative of (6.28) with respect to time is:

$$\dot{C}_B(0) = c_2 e^{-kt} - c_2\, k\, t\, e^{-kt}$$

at $t = 0$,

$$c_2 = \dot{C}_B(0)$$

We also know from (6.3) that:

$$\frac{dC_B(0)}{dt} = k\, C_{A0}$$

so:

$$C_B(t) = k\, C_{A0}\, t \exp(-kt) \tag{6.29}$$

If we define the conversion of A to B as $x = C_B/C_{A0}$, and the dimensionless time as $\tau = kt$, then (6.29) can be written:

$$x(t) = \tau \exp(-\tau) \tag{6.30}$$

which is shown in Figure 6.3. The reader should be able to find the maximum value for the conversion of A to B and the reaction time required for this conversion.

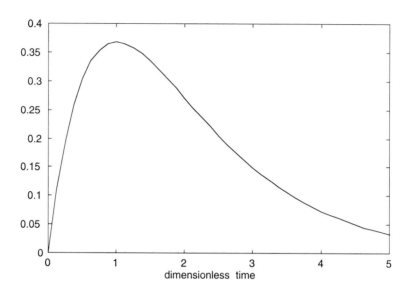

FIGURE 6.3   Conversion of A to B as a function of dimensionless time ($\tau = kt$), for the case of equal rate constants.

The previous example illustrated the solution for systems with real roots. The next example illustrates a system with complex roots.

EXAMPLE 6.2   Complex Roots

Consider the second-order equation:

$$\frac{d^2x}{dt^2} + \frac{dx}{dt} + x = 0 \tag{6.31}$$

The characteristic equation is:

$$\lambda^2 + \lambda + 1 = 0 \tag{6.32}$$

Solving for the roots using the quadratic formula, we find that the roots are complex:

$$\lambda = \frac{-1 \pm \sqrt{1 - 4}}{2}$$

$$\lambda = -\frac{1}{2} \pm \frac{\sqrt{3}}{2}j$$

where $j = \sqrt{-1}$. The solution is

$$x(t) = c_1 e\left(-\frac{1}{2} + \frac{\sqrt{3}}{2}j\right)t + c_2 e\left(-\frac{1}{2} - \frac{\sqrt{3}}{2}j\right)t \tag{6.33}$$

We can use the following Euler identities:

$$e^{j\theta} = \cos\theta + j\sin\theta \tag{6.34}$$

$$e^{-j\theta} = \cos\theta - j\sin\theta \tag{6.35}$$

and the property that $e^{x+y} = e^x e^y$ to write (6.33) as:

$$x(t) = c_1 e^{-t/2}\left[\cos\frac{\sqrt{3}}{2}t + j\sin\frac{\sqrt{3}}{2}t\right] + c_2 e^{-t/2}\left[\cos\frac{\sqrt{3}}{2}t - j\sin\frac{\sqrt{3}}{2}t\right] \tag{6.36}$$

$$x(t) = e^{-t/2}\left[c_1\cos\frac{\sqrt{3}}{2}t + c_2\cos\frac{\sqrt{3}}{2}t + c_1 j\sin\frac{\sqrt{3}}{2}t - c_2 j\sin\frac{\sqrt{3}}{2}t\right] \tag{6.37}$$

which can be written:

$$x(t) = e^{-t/2}\left[(c_1 + c_2)\cos\frac{\sqrt{3}}{2}t + (c_1 - c_2)j\sin\frac{\sqrt{3}}{2}t\right] \tag{6.38}$$

Defining $c_3 = c_1 + c_2$ and $c_4 = (c_1 - c_2)j$,

$$x(t) = e^{-t/2}\left[c_3\cos\frac{\sqrt{3}}{2}t + c_4\sin\frac{\sqrt{3}}{2}t\right] \tag{6.39}$$

Again, initial conditions for $x(0)$ and $\dot{x}(0)$ can be used to determine c_3 and c_4. The student should verify that if $x(0) = 1$ and $\dot{x}(0) = 1$, then $c_3 = 1.0$ and $c_4 = 1.5$. A plot is shown in Figure 6.4.

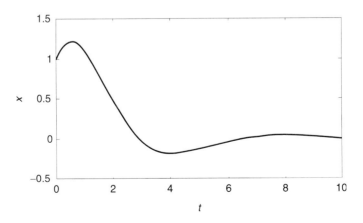

FIGURE 6.4   Plot of (6.39) with $x(0) = 1$ and $\dot{x}(0) = 1$.

Chemical process system models with complex roots include some exothermic chemical reactors. Also, models including feedback control will often have complex roots (leading to oscillatory behavior).

6.2.3 General Result for Complex Roots

We can now generalize the results of Example 6.2 for any equation that has pairs of complex roots. For each pair of complex roots, $\lambda = \lambda_r \pm j\,\lambda_i$, where λ_r and λ_i are the real and imaginary portions, the solution is:

$$x(t) = e^{\lambda_r t}[c_1 \cos \lambda_i t + c_2 \sin \lambda_i\, t] \tag{6.40}$$

In the previous example the real part of the complex roots was negative (stable). Notice that the state variables decayed to zero with time. Notice from (6.40) that there will be no decay (simply a continuous oscillation) if the real portion is 0. We can also see from (6.40) that a positive real portion of the complex root will lead to an ever growing (unstable) solution. This behavior is shown by student exercise 19.

Most chemical processes are stable; however, some exothermic chemical reactors have unstable operating points. Also, improperly tuned feedback control systems can be unstable.

Thus far in this chapter we have solved the *homogeneous* problems. Homogeneous problems result from models that are "unforced," that is, there is no input. This usually occurs when the process model is in deviation variable form, and there is no change in the input variable. They are based on a perturbation from steady-state in the state variable values.

In Section 6.3 we will solve *nonhomogeneous* problems using the method of undetermined coefficients. These types of problems arise when there are input changes to a process.

6.3 SOLVING NONHOMOGENEOUS, LINEAR ODES WITH CONSTANT COEFFICIENTS

In Section 6.2 we solved homogeneous problems with constant coefficients:

$$\frac{d^n x}{dt^n} + a_{n-1}\frac{d^{n-1}x}{dt^{n-1}} + \ldots + a_1\frac{dx}{dt} + a_o x = 0 \tag{6.41}$$

In this section we will solve nonhomogeneous problems with the following form:

$$\frac{d^n x}{dt^n} + a_{n-1}\frac{d^{n-1}x}{dt^{n-1}} + \ldots + a_1\frac{dx}{dt} + a_o x = q(t) \tag{6.42}$$

using the *method of undetermined coefficients*, which is outlined below.

Method of Undetermined Coefficients

The method of undetermined coefficients consists of the following steps:

1. Solve the *homogeneous* problem to find

$$x_H(t)$$

2. Solve for the particular solution by determining the coefficients of a trial function (see Table 6.1) that satisfy the *nonhomogeneous* equation

$$x_p(t)$$

3. Combine the two solutions for

$$x(t) = x_H(t) + x_P(t)$$

TABLE 6.1   Trial Functions for Method of Undetermined Coefficients (Boyce and DiPrima, 1992)

Forcing Function	Trial Function
A (a constant)	B (a constant)
$Ae^{\alpha t}$	$Be^{\alpha t}$
$A \cos \alpha t$ or $A \sin \alpha t$	$B_1 \cos \alpha t + B_2 \sin \alpha t$
$A\, t^n$	$B_n t^n + B_{n-1} t^{n-1} + \ldots + B_o$

We illustrate the method by use of an illustrative example.

EXAMPLE 6.1 Continued.   First Order Heterogeneous System

Notice that we can take the solution for C_A as a function of time

$$c_A(t) = C_{A0} e^{-k_1 t} \tag{6.14}$$

and substitute it into (6.3) to obtain

$$\frac{dC_B}{dt} + k_2\, C_B = k_1\, C_{A0}\, e^{-k_1 t} \tag{6.43}$$

Step 1.   The homogeneous solution to (6.43) is

$$x_H(t) = c_1\, e^{-k_2 t} \tag{6.44}$$

Step 2.   Since the forcing function is $k_1\, C_{A0}\, e^{-k_1 t}$, we use $c_2\, e^{-k_1 t}$ (Table 6.1) as our trial function for the particular solution:

$$x_P(t) = c_2\, e^{-k_1 t} \tag{6.45}$$

substituting this solution into the original equation (6.47),

$$-k_1\, c_2\, e^{-k_1 t} + k_2\, c_2\, e^{-k_1 t} = k_1\, C_{A0}\, e^{-k_1 t} \tag{6.46}$$

which we can solve for c_2:

$$c_2 = \frac{k_1 \, C_{A0}}{k_2 - k_1} \tag{6.47}$$

Step 3.   Now find the complete solution as $x(t) = x_H(t) + x_P(t)$

$$C_A(t) = c_1 \, e^{-k_2 t} + \frac{k_1 \, C_{A0}}{k_2 - k_1} \, e^{-k_1 t} \tag{6.48}$$

We can evaluate c_1 from the initial conditions, $C_{B0} = 0$:

$$c_1 = -\frac{k_1 \, C_{A0}}{k_2 - k_1}$$

and the total solution is:

$$C_B(t) = \frac{k_1 \, C_{A0}}{k_2 - k_1} \left[\exp(-k_1 t) - \exp(-k_2 t) \right]$$

which, of course, is the same result obtained previously (6.23) by solving the second-order homogeneous equation in C_B.

We have used a single first-order equation to illustrate the procedure for heterogeneous equations. The same procedure is used for higher-order equations.

6.4   EQUATIONS WITH TIME-VARYING PARAMETERS

Consider a first-order equation with the following form:

$$\frac{dx}{dt} + p(t) \, x = q(t) \tag{6.49}$$

Notice that the coefficient is time-varying and the equation is heterogeneous. One approach to solve this type of problem is to use an *integrating factor*.

Let the integrating factor be represented by $\mu(t)$

$$\mu(t) = \exp \left[\int p(t) \, dt \right] \tag{6.50}$$

Equation (6.49) is solved by multiplying each term by the integrating factor:

$$\mu(t) \frac{dx}{dt} + \mu(t) \, p(t) \, x = \mu(t) \, q(t) \tag{6.51}$$

$$\exp \left[\int p(t) \, dt \right] \frac{dx}{dt} + \exp \left[\int p(t) \, dt \right] p(t) \, x = \exp \left[\int p(t) \, dt \right] q(t) \tag{6.52}$$

Notice that the lefthand side of (6.52) is simply the expansion of:

$$\frac{d}{dt} \left[x(t) \exp \left\{ \int p(t) \, dt \right\} \right] = \exp \left[\int p(t) \, dt \right] \frac{dx}{dt} + \exp \left[\int p(t) \, dt \right] p(t) x \tag{6.53}$$

so we can write:

$$\frac{d}{dt}\left[x(t)\exp\left\{\int p(t)dt\right\}\right] = \exp\left[\int p(t)\,dt\right]q(t) \tag{6.54}$$

which is a separable equation. Separating and integrating, we find:

$$x(t)\exp\left[\int p(t)\,dt\right] = \int q(t)\exp\left[\int p(t)\,dt\right]dt + c \tag{6.55}$$

and, evaluating c using the initial conditions,

$$x(t) = \exp\left[-\int p(t)\,dt\right]\left\{x(0) + \int q(t)\exp\left[\int p(t)\,dt\right]dt\right\} \tag{6.56}$$

EXAMPLE 6.3   Semi-batch Reactor

Consider the case where the batch reactor is being filled. Assume a single, first-order reaction $(A \rightarrow B)$ and a constant volumetric flowrate into the reactor (F), with no flow out of the reactor.
    The modeling equations are:

$$\frac{dV}{dt} = F \tag{6.57}$$

$$\frac{dVC_A}{dt} = -k_1 V C_A + FC_{AF} \tag{6.58}$$

Expanding the LHS of (6.58) as:

$$\frac{dVC_A}{dt} = C_A \frac{dV}{dt} + V \frac{dC_A}{dt} \tag{6.59}$$

we find:

$$\frac{dC_A}{dt} + \left[\frac{F}{V} + k_1\right]C_A = \frac{F}{V}C_{AF} \tag{6.60}$$

If the flowrate is constant and the initial volume is 0, then:

$$V = Ft \tag{6.61}$$

and:

$$\frac{dC_A}{dt} + \left[\frac{1}{t} + k_1\right]C_A = \frac{C_{AF}}{t} \tag{6.62}$$

Let:

$$\int p(t)\,dt = \int\left(\frac{1}{t} + k_1\right)dt = \ln t + k_1 t + c_1 \tag{6.63}$$

$$\exp\left[\int p(t)\,dt\right] = \exp\left[\ln t + k_1 t + c_1\right] = \exp\left[\ln t\right]\exp\left[k_1 t\right]\exp\left[c_1\right]$$
$$= c_2\,t\exp\left[k_1 t\right] \tag{6.64}$$

Multiplying through on each side of (6.64) by $c_2 t \exp[k_1 t]$ and dividing by c_2

$$t \exp[k_1 t] \frac{dC_A}{dt} + \exp[k_1 t] [1 + k_1 t] C_A = \exp[k_1 t] C_{AF} \qquad (6.65)$$

and noting that the lefthand side is simply

$$\frac{d[t \exp[k_1 t] C_A]}{dt} \qquad (6.66)$$

we multiply by dt and integrate to find

$$\exp[k_1 t] C_A t = \frac{C_{AF}}{k_1} \{\exp[k_1 t] - 1\} \qquad (6.67)$$

multiplying by $\exp[-k_1 t]$ and dividing by t, we find the solution

$$C_A = \frac{C_{AF}}{k_1 t} \{1 - \exp[-k_1 t]\} \qquad (6.68)$$

The division by t is bothersome at $t = 0$; the reader should use L'Hospital's rule to show that the correct initial condition is obtained with this expression.

Notice that we can define a dimensionless concentration and time as

$$y = C_A / C_{AF} \quad \text{and} \quad \tau = k_1 t$$

to find

$$y(\tau) = \frac{1}{\tau} \{1 - \exp[-\tau]\} \qquad (6.69)$$

which is shown in Figure 6.5.

Notice that this solution holds while the reactor is being "fed". After the feed is stopped the model is simply $dC_A/dt = -k_1 C_A$ with appropriate intial conditions (see student exercise 20).

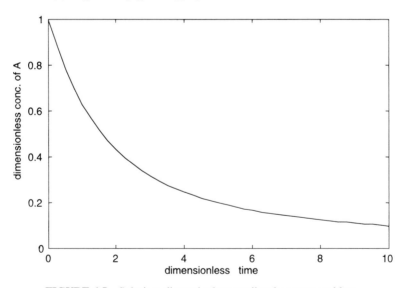

FIGURE 6.5   Solution, dimensionless semibatch reactor problem.

6.5 ROUTH STABILITY CRITERION—DETERMINING STABILITY WITHOUT CALCULATING EIGENVALUES

The stability of the characteristic equation is determined from the values of its roots (eigenvalues). This is easy for first and second order equations (and not too hard for third) since there is an analytical solution for the roots of polynomials through third order. If the polynomial is fourth order or higher, the roots must be determined numerically. There is a method for determining if any of the roots are positive (unstable) without actually calculating the roots (Routh, 1905). This method involves an analysis of the coefficients of the characteristic polynomial by setting up the *Routh Array*. The test of the coefficients in the Routh Array is called the *Routh Stability Criterion*.

The Routh Stability Criterion is based on the characteristic equation that has the following polynomial form

$$a_n \lambda^n + a_{n-1} \lambda^{n-1} + \ldots + a_1 \lambda + a_o = 0 \qquad (6.70)$$

We can arbitrarily assume that $a_n > 0$. If $a_n < 0$ then multiply (6.73) by -1. A *necessary* condition for stability is that all of the coefficients in (6.70) must be positive. If any of the coefficients are negative or zero then at least one eigenvalue (root of the characteristic equation) is positive or zero, indicating that the equation is unstable. Even if all of the coefficients are positive, we cannot state that the system is stable. What is needed is a *sufficient* condition for stability. To determine that the system is stable, we must construct the Routh array and use the Routh stability criterion, which provides necessary and sufficient conditions for stability.

Sometimes we simply wish to determine if a particular system is stable or not, without actually evaluating the eigenvalues. This is particularly true if we wish to determine values of system parameters that will cause a system to lose stability. This approach will be useful in performing a bifurcation analysis in later chapters, and in tuning control systems in chemical process control.

6.5.1 Routh Array

If all of the coeffients of the characteristic equation (6.70) are positive, the *necessary* condition for stability is satisfied. The following Routh array (Seborg, Edgar, & Mellichamp, 1989) is developed to test for the *sufficient* conditions for stability:

Row				
1	a_n	a_{n-2}	a_{n-4}	...
2	a_{n-1}	a_{n-3}	a_{n-5}	...
3	b_1	b_2	b_3	...
4	c_1	c_2	...	
.	.			
$n+1$				

where n is the order of the characteristic polynomial. Notice that the first two rows consist of the coefficients of the characteristic polynomial. The elements of the third row are calculated in the following fashion:

$$b_1 = \frac{a_{n-1}a_{n-2} - a_n a_{n-3}}{a_{n-1}} \qquad b_2 = \frac{a_{n-1}a_{n-4} - a_n a_{n-5}}{a_{n-1}}$$

and so on. Elements of the fourth and larger rows are calculated in a similar fashion:

$$c_1 = \frac{b_1 a_{n-3} - a_{n-1}b_2}{b_1} \qquad c_2 = \frac{b_1 a_{n-5} - a_{n-1}b_3}{b_1}$$

and so on.

Routh Stability Criterion

A necessary and sufficient condition for all roots of the characteristic polynomial to have negative real parts is that all of the coefficients of the polynomial are positive and all of the elements in the left column of the Routh array are positive.

EXAMPLE 6.4   Second-order Characteristic Equations

Consider the second-order ODE:

$$a_2 \frac{d^2x}{dt^2} + a_1 \frac{dx}{dt} + a_0 x = 0 \tag{6.71}$$

The characteristic polynomial is:

$$a_2 \lambda^2 + a_1 \lambda + a_0 = 0 \tag{6.72}$$

If all of the coefficients a_2, a_1, and a_0 are positive, then the necessary condition is satisfied. We can form the Routh array to test for the sufficient condition:

Row		
1	a_2	a_0
2	a_1	
3	a_0	

Since the left column consists of the polynomial coefficients, if all of the coefficients in the second order system are positive, the system is stable.

Notice that, for second-order systems, a test for positive coefficients is necessary *and* sufficient for stability.

EXAMPLE 6.5   Third-order System

The system:

$$\frac{d^3x}{dt^3} + 2\frac{d^2x}{dt^2} + 3\frac{dx}{dt} + x = 0$$

has the characteristic polynomial:

$$\lambda^3 + 2\lambda^2 + 3\lambda + 1 = 0$$

and the following Routh array:

Row		
1	1	3
2	2	1
3	5/2	
4	1	

All of the coefficients of the characteristic polynomial are positive and all of the elements in the left column of the Routh array are positive, so the system is stable.

The Routh array is particularly useful for determining how much a parameter can vary before a system loses stability. The following example illustrates such a system.

EXAMPLE 6.6   Third-order, System With a Variable Parameter

The system:

$$\frac{d^3x}{dt^3} + 2\frac{d^2x}{dt^2} + 3\frac{dx}{dt} + \mu x = 0$$

has the characteristic polynomial:

$$\lambda^3 + 2\lambda^2 + 3\lambda + \mu = 0$$

where μ is a parameter that may vary. The Routh array is:

Row		
1	1	3
2	2	μ
3	b_1	
4	c_1	

where $b_1 = 3 - \mu/2$ and $c_1 = \mu$. From the characteristic polynomial, we see that $\mu > 0$ is required. The same result holds true for the requirement of $c_1 > 0$. We notice that b_1 will be positive only if $\mu < 6$.

From these conditions, we find that the stability requirement is $0 < \mu < 6$.

For complex, high order (3 or greater), it is not uncommon for a system to have parameters that stabilize the system only over a certain range of parameter values. This is particularly true of feedback control systems.

SUMMARY

We have reviewed techniques to solve homogeneous and nonhomogeneous (heterogeneous) nth order ODEs.

- Homogeneous problems are solved using the roots of the characteristic equation, forming the solution as a sum of exponential terms. Homogeneous equations generally occur if the system is unforced, but there is an initial deviation from steady-state in the state variables.

- The *method of undetermined coefficients* is useful for solving nonhomogeneous (heterogeneous) problems. These generally occur if the system is forced by a changing input.

- The *integrating factor* method was useful for solving a first-order heterogeneous equation with a time-varying coefficient.

- The *Routh array* was used to test for the stability of a differential equation. This is useful for finding values of a parameter that cause a system to lose stability, such as in feedback control system design or bifurcation analysis.

- The type of dynamic behavior of an nth order differential equation is a function of the eigenvalues (roots of the characteristic equation). Eigenvalues that are further in the left half plane are "fast." The larger the ratio of the imaginary portion to real portion of a complex eigenvalue, the more oscillatory the response. Stability is determined by the real portion of the complex eigenvalue. If all eigenvalues have a real portion that is negative, then the system is stable. If any single eigenvalue has a real portion that is positive, then the system is unstable.

Often engineers study the dynamic behavior of processes by starting out at steady-state, then applying a changing input to the process. Although the method of undetermined coefficients can be used to solve these problems, the Laplace transform technique is used more often. The Laplace transform method is introduced in Chapter 7.

FURTHER READING

Boyce, W., & R. DiPrima. (1992). *Ordinary Differential Equations and Boundary Value Problems,* 5th ed. New York: Wiley.

Routh, E.J. (1905). *Dynamics of a System of Rigid Bodies, Part II.* London: Macmillan.

Seborg, D.E., T.F. Edgar, & D.A. Mellichamp. (1989). *Process Dynamics and Control.* New York: Wiley.

STUDENT EXERCISES

1. Consider the state-space model:

$$\begin{bmatrix} \dot{x}_1 \\ \dot{x}_2 \end{bmatrix} = \begin{bmatrix} -1 & -2 \\ 1 & 0 \end{bmatrix} \begin{bmatrix} x_1 \\ x_2 \end{bmatrix}$$

  a. Find the second order ODE in terms of x_1.

  b. Find the second order ODE in terms of x_2.

  c. For $x_1(0) = -1$ and $x_2(0) = 1$, obtain the analytical solution for $x_1(t)$ and $x_2(t)$.

  d. Use `ode45` or `initial` to solve the set of two differential equations, given the initial conditions in part c.

2. Consider the following linearized form of a bioreactor model with substrate inhibition kinetics (see Module 8 for details):

$$\dot{\mathbf{x}} = \mathbf{A}\,\mathbf{x} + \mathbf{B}\,\mathbf{u}$$
$$\mathbf{y} = \mathbf{C}\,\mathbf{x} + \mathbf{D}\,\mathbf{u}$$

where:

$$\mathbf{A} = \begin{bmatrix} 0 & 0.9056 \\ -0.7500 & -2.5640 \end{bmatrix} \quad \mathbf{B} = \begin{bmatrix} -1.5302 \\ 3.8255 \end{bmatrix}$$

$$\mathbf{C} = \begin{bmatrix} 1 & 0 \\ 0 & 1 \end{bmatrix} \quad \mathbf{D} = \begin{bmatrix} 0 \\ 0 \end{bmatrix}$$

  a. Find the second-order ODE in terms of x_1, assuming $u = 0$.

  b. Find the second-order ODE in terms of x_2, assuming $u = 0$.

  c. For $x_1(0) = -1$ and $x_2(0) = 1$, obtain the analytical solution for $x_1(t)$ and $x_2(t)$.

  d. Use `ode45` or `initial` to solve the set of two differential equations, given the initial conditions in part c.

3. Consider the state space model for a two-state system:

$$\dot{\mathbf{x}} = \mathbf{A}\,\mathbf{x}$$

$$\begin{bmatrix} \dot{x}_1 \\ \dot{x}_2 \end{bmatrix} = \begin{bmatrix} a_{11} & a_{12} \\ a_{21} & a_{22} \end{bmatrix} \begin{bmatrix} x_1 \\ x_2 \end{bmatrix}$$

a. Let $y = x_1$, and derive the following relationship:

$$\ddot{y} - [a_{11} + a_{22}]\, \dot{y} + [a_{11}a_{22} - a_{21}a_{12}]\, y = 0$$

which has the characteristic equation:

$$\lambda^2 - [a_{11} + a_{22}]\, \lambda + [a_{11}a_{22} - a_{21}a_{12}] = 0$$

Recall that the eigenvalues of the **A** matrix are calculated by:

$$\det(\lambda \mathbf{I} - \mathbf{A}) = 0$$

Show that $\det(\lambda \mathbf{I} - \mathbf{A}) = 0$ applied to this general two-state example, yields:

$$\lambda^2 - [a_{11} + a_{22}]\, \lambda + [a_{11}a_{22} - a_{21}a_{12}] = 0$$

4. Solve the following differential equation with the given initial conditions:

$$\frac{d^2y}{dx^2} + 5\frac{dy}{dx} + 6y = 0$$

$$y(0) = 0 \quad \text{and} \quad \frac{dy(0)}{dx} = 1$$

5. Find the particular solution of the differential equation:

$$\frac{d^2y}{dx^2} - 3\frac{dy}{dx} - 4y = 2\sin x$$

6. Consider the following second-order homogeneous ODE:

$$\frac{d^2x}{dt^2} - 3\frac{dx}{dt} + 3x = 0$$

a. Write the characteristic equation for this ODE.
b. Find the solution (solve for any constants), $x(t)$, if the initial conditions are $x(0) = 2.0$ and $\dot{x}(0) = 3.0$.
c. Discuss the stability of this system.

7. Consider the following first-order heterogeneous ODE:

$$3\frac{dx}{dt} + x = 2(1 - e^{-4t})$$

a. Write the characteristic equation for the homogeneous part of this ODE.
b. Find the solution to the heterogenous problem. Show all steps. The initial condition is $x(0) = 2.0$.

8. Consider the following state-space model that results from a linearization of the predator-prey equations:

$$\begin{bmatrix} \dot{x}_1 \\ \dot{x}_2 \end{bmatrix} = \begin{bmatrix} 0 & -1.0 \\ 1.0 & 0.0 \end{bmatrix} \begin{bmatrix} x_1 \\ x_2 \end{bmatrix}$$

with initial conditions $x_1(0) = 0.5$ and $x_2(0) = -0.25$.

    a. What are the eigenvalues of the **A** matrix? Use both MATLAB and your own analytical solution.

    b. Write the second-order ODE that corresponds to x_1. Use the method of Section 6.2 to solve for $x_1(t)$. Plot $x_1(t)$.

    c. Write the second-order ODE that corresponds to x_2. Use the method of Section 6.2 to solve for $x_2(t)$. Plot $x_2(t)$.

    d. Compare the results from **b** and **c** with those obtained by integrating the state-space equations using either `ode45` or `initial`.

    e. Show a phase-plane plot (x_1 versus x_2), placing x_1 on the x-axis and x_2 on the y-axis.

9. Consider a system described by the following third-order ODE:

$$\frac{d^3y}{dt^3} + 1.5\frac{d^2y}{dt^2} + \frac{dy}{dt} + 2y = 0$$

Is the system described by this equation stable? Why or why not?

10. For a general third-order polynomial:

$$a_3\lambda^3 + a_2\lambda^2 + a_1\lambda + a_0 = 0$$

show that $a_i > 0$ and $a_1a_2 - a_0a_3 > 0$ are necessary and sufficient for stability.

11. For a general fourth-order polynomial:

$$a_4\lambda^4 + a_3\lambda^3 + a_2\lambda^2 + a_1\lambda + a_0 = 0$$

show that $a_i > 0$, $a_2a_3 - a_1a_4 > 0$, and $a_1a_2a_3 - a_4a_1^2 - a_0a_3^2 > 0$ are necessary and sufficient for stability.

12. Consider the following third-order ODE:

$$\frac{d^3y}{dt^3} + 2\frac{d^2y}{dt^2} + (\alpha - 1)\frac{dy}{dt} + \alpha y = 0$$

where α is a parameter. Find the range of α that will cause this equation to be stable.

13. Consider the following second-order ODE:

$$\frac{d^2y}{dt^2} + 2\frac{dy}{dt} + 2y = 0$$

which has eigenvalues of $-1 \pm 1j$ and initial conditions $y(0) = 2$ and $\dot{y}(0) = -2$. Find $y(t)$.

14. Consider the series of two tanks, where the levels interact.

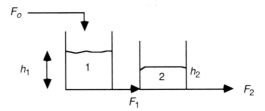

a. Assuming that the flow from the first tank is linearly proportional to the difference in the tank heights ($F_1 = \beta_1 (h_1 - h_2)$), the flowrate from tank 2 is proportional to the height in tank 2 ($F_2 = \beta_1 h_2$), and the tanks are of constant cross-sectional area (A_1 and A_2) show that the modeling equations are

$$\frac{dh_1}{dt} = \frac{F_o}{A_1} - \frac{\beta_1}{A_1}(h_1 - h_2)$$

$$\frac{dh_2}{dt} = \frac{\beta_1}{A_2}(h_1 - h_2) - \frac{\beta_2}{A_2}h_2$$

b. Reduce these two equations to a single second-order equation in h_2.
c. Assume that the steady-state flowrate is 3 ft³/min, and the steady-state tank heights for tanks 1 and 2 are 7 and 3 feet, respectively. The constant cross-sectional area is 5 ft² for each tank. The initial conditions are $h_1(0) = 6$ feet and $h_2(0) = 5$ feet. Solve for the heights of tanks 1 and 2 as a function of time. Plot the tank heights as a function of time. Discuss your results.
d. Write a MATLAB m-file and use `ode45` to integrate the two equations shown above. Show that the numerical integration agrees with your solution in part c.

15. For the batch series reaction (Example 6.1):

$$A \xrightarrow{k_1} B \xrightarrow{k_2} C$$

a. Find the reaction time that maximizes the production of B. Recall that the solution for the concentration of B is:

$$C_B(t) = \frac{k_1 C_{A0}}{k_2 - k_1}[\exp(-k_1 t) - \exp(-k_2 t)]$$

and that the maximum occurs when the condition $dC_B/dt = 0$ is satisfied.
b. For $k_1 = 1$ and $k_2 = 5$ min⁻¹, find the maximum conversion of A to B (express as C_B/C_{A0}) and the time required for this conversion.
c. In practice there is uncertainty in the rate constants. If the actual value of k_2 is 7.5 min⁻¹, and the reaction time from **b** is used, find the actual conversion of A to B.
d. Use the MATLAB routine `ode45` to integrate the three state variable equations and solve for C_A, C_B, and C_C as a function of time, for the parameter values in

b, with $C_{A0} = 1.5$ and $C_{B0} = C_{C0} = 0$ mol/liter. Make a comparative plot for the parameter values in **c**. What do you observe about the concentrations of A, B, and C?

16. For the batch series reaction with irreversible reactions ($A \rightarrow B \rightarrow C$):
    a. Find the reaction time (t_{max}) that maximizes the conversion of A to B for the case where $k_2 = k_1 = k$. Also find the value for the maximum conversion of A to B. Recall that the solution for the conversion is

$$x(t) = C_B(t)/C_{A0} = k\,t\,\exp(-kt)$$

    b. Assume that the reactor is run for the period t_{max} found in **a**. Now consider the effect of an error in the reaction rate constant of +50%. What is the actual conversion of A to B obtained at t_{max}?

17. Consider a batch reactor with a series reaction where component A reacts to form the desired component B *reversibly*. Component B can also react to form the undesired component C. The reaction scheme can be characterized by:

$$A \underset{k_{1r}}{\overset{k_{1f}}{\rightleftarrows}} B \overset{k_2}{\rightarrow} C$$

Here k_{1f} and k_{1r} represent the kinetic rate constants for the forward and reverse reactions for the conversion of A to B, while k_2 represents the rate constant for the conversion of B to C.

Assuming that each of the reactions is first-order and constant volume, the modeling equations are

$$\frac{dC_A}{dt} = -k_{1f}\,C_A + k_{1r}\,C_B$$

$$\frac{dC_B}{dt} = k_{1f}\,C_A - k_{1r}\,C_B$$

$$\frac{dC_C}{dt} = k_2\,C_B$$

where C_A, C_B, and C_C represent the concentrations (mol/volume) of components A, B, and C, respectively. Using the following definitions:

Dimensionless time,	$\tau = k_1 t$
Conversion of A,	$x_1 = (C_{A0} - C_A)/C_{A0}$
Dimensionless concentration of B,	$x_2 = C_B/C_{A0}$
Ratio of rate constants,	$\alpha = k_2/k_{1f}$
Ratio of forward and reverse rate constants,	$\beta = k_{1r}/k_{1f}$

    a. Show that the equation for the dimensionless concentration of B is

$$\frac{d^2x_2}{d\tau^2} + (\beta + \alpha + 1)\frac{dx_2}{d\tau} + \alpha x_2 = 0$$

and that the roots of the characteristic equation can never be complex or unstable (assuming that the rate constants are positive).

    b. Solve the previous equation to find x_2 as a function of τ and α and β.

    c. For $k_{1f} = 2$, $k_{1r} = 1$, and $k_2 = 1.25$ hr^{-1}, find the maximum conversion of A to B and the reaction time required for this conversion.

    d. Usually there is some uncertainty in the rate constants. If the real value of k_2 is 1.5 hr^{-1} and the reaction is run for the time found in c, what will be the actual conversion of A to B?

18. Consider the series reaction:

$$A \overset{k_1}{\to} B \overset{k_2}{\to} C \overset{k_3}{\to} D$$

The modeling equations for a constant volume batch reactor are

$$\frac{dC_A}{dt} = -k_1 C_A$$

$$\frac{dC_B}{dt} = k_1 C_A - k_2 C_B$$

$$\frac{dC_C}{dt} = k_2 C_B - k_3 C_C$$

    a. Show that the third-order ODE describing the concentration of C is:

$$\frac{d^3C_C}{dt^3} + [k_1 + k_2 + k_3]\frac{d^2C_C}{dt^2} + [k_1k_2 + k_1k_3 + k_2k_3]\frac{dC_C}{dt} + k_1k_2k_3 C_C = 0$$

    [*Hint*: Solve for C_B from the third equation and take the derivative to find dC_B/dt.]

    b. Assuming that all of the kinetic parameters are positive, show that this system is stable.

19. Consider the second-order equation:

$$\frac{d^2x}{dt^2} - \frac{dx}{dt} + x = 0$$

For initial conditions $x(0) = 1$ and $\dot{x}(0) = 1$, find the analytical solution and show that the following plot describes how x changes with time.

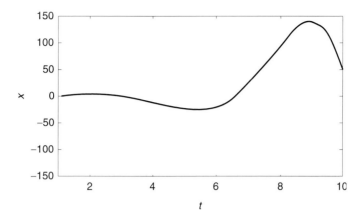

20. Consider a semibatch reactor (Example 6.3) with a first-order kinetic parameter of $k = 1$ hr^{-1}. For a flowrate of 10 liters/hour, a feed concentration of 5 mol/liter, and a feed time of 2 hours, find (and plot) how the concentration changes from 0 to 10 hours.

AN INTRODUCTION TO LAPLACE TRANSFORMS

7

After studying this chapter, the reader should be able to:

- Define the Laplace transform and apply it to several example functions.
- Use Laplace transforms to convert an nth order ODE to the Laplace domain.
- Manipulate the algebraic equations by performing a partial fraction expansion.
- "Invert" the Laplace domain functions to obtain the time domain solution.
- Use the final value theorem to compute the long-term behavior of a system.

The important sections of this chapter are:

7.1 MOTIVATION

In this chapter we introduce a mathematical tool, the *Laplace transform,* which is very useful in the analysis of linear dynamic systems. The purpose of the Laplace transform, as used in this textbook, is to convert linear differential equations into algebraic equations. Algebraic equations are much easier to manipulate than differential equations. An analogy is the use of logarithms to change the operation of multiplication into that of addition. Laplace transforms are useful for solving linear dynamic systems problems, particularly nonhomogeneous (heterogeneous) problems (i.e., where the input to the process system is changed), and are commonly used in process control system design and analysis.

7.2 DEFINITION OF THE LAPLACE TRANSFORM

Definition: *Laplace transform*

Consider the time domain function $f(t)$. The Laplace transform of $f(t)$ is represented by $L[f(t)]$ and is defined as

$$L[f(t)] \equiv F(s) = \int_0^\infty f(t)\, e^{-st}\, dt \tag{7.1}$$

This operation transforms a variable from the time domain to the s (or Laplace) domain. Note that some texts use an overbar or capital letters for the transformed variable. In this initial development we will let $f(t)$ represent the time domain function and $F(s)$ represent the Laplace domain function. Later we may be more relaxed in our notation and let $f(s)$ represent the Laplace domain function.

The Laplace transform is a linear operation, as shown below.

$$L[a_1 f_1(t) + a_2 f_2(t)] = \int_0^\infty [a_1 f_1(t) + a_2 f_2(t)]\, e^{-st}\, dt$$

$$= \int_0^\infty a_1 f_1(t)\, e^{-st}\, dt + \int_0^\infty a_2 f_2(t)\, e^{-st}\, dt$$

$$= a_1 \int_0^\infty f_1(t)\, e^{-st}\, dt + a_2 \int_0^\infty f_2(t)\, e^{-st}\, dt \tag{7.2}$$

$$L[a_1 f_1(t) + a_2 f_2(t)] = a_1 L[f_1(t)] + a_2 L[f_2(t)]$$

Equation (7.2) satisfies the definition of a linear operation.

In (7.1) we used $L[f(t)] \equiv F(s)$ to define the transform of a time domain function. If we wish to transform a Laplace domain (sometimes called the s-domain) function to the time domain, we use the notion of an inverse transform

$$L^{-1}[F(s)] = f(t)$$

Although not emphasized in this text, Laplace transforms can also be used to solve linear partial differential equations (PDEs).

7.3 EXAMPLES OF LAPLACE TRANSFORMS

In this section we develop transforms of some functions that commonly occur in the solution of linear dynamic problems. These functions are: (i) exponential function, (ii) step function, (iii) time-delay, (iv) derivatives, (v) integrals, and (vii) impulse.

7.3.1 Exponential Function

Exponential functions commonly arise in the solutions of linear, constant coefficient, ordinary differential equations:

$$f(t) = e^{-at}$$

A plot of this function is shown in Figure 7.1.

Recall that the transform is defined for $t > 0$ (we also use the identity that $e^{x+y} = e^x e^y$)

$$L[e^{-at}] = \int_0^\infty f(t)\, e^{-st}\, dt = \int_0^\infty e^{-at}\, e^{-st}\, dt = \int_0^\infty e^{-(s+a)t}\, dt$$

$$= -\frac{1}{s+a}\, [e^{-(s+a)t}]_0^\infty = -\frac{1}{s+a}\, [0-1] = \frac{1}{s+a}$$

$$L[e^{-at}] = \frac{1}{s+a}$$

Exponential

$$L^{-1}\left[\frac{1}{s+a}\right] = e^{-at}$$

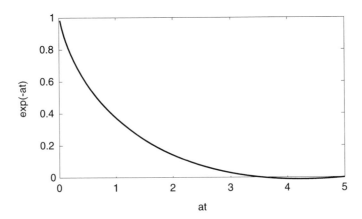

FIGURE 7.1 Exponential function.

Notice that the way we have solved for the limits of integration is only rigorously true for $a > 0$. For $a < 0$ the solution still holds for $s > -a$; we will assume that this condition is always satisfied.

7.3.2   Step Function

The step function is used to solve dynamic problems where a sudden change in an input variable occurs (a flowrate could be rapidly changed from one value to another, for example). The step function is defined as 0 before $t = 0$ and A after $t = 0$, as shown in Figure 7.2.

$$f(t) = \begin{Bmatrix} 0 \text{ for } t < 0 \\ A \text{ for } t > 0 \end{Bmatrix}$$

We must use the "more precise" definition of the Laplace transform, because of the discontinuity at $t = 0$:

$$L[f(t)] = \lim_{\substack{\varepsilon \to 0 \\ T \to \infty}} + \int_{\varepsilon}^{T} f(t)\, e^{-st}\, dt$$

Since the transform is defined for $t > 0$,

$$L[A] = \int_{0+}^{\infty} Ae^{-st}\, dt = -\frac{A}{s}\left[e^{-st}\right]_{0}^{\infty} = -\frac{A}{s}[0 - 1] = \frac{A}{s}$$

$$L[A] = \frac{A}{s} \qquad \qquad \textit{Step}$$

$$L^{-1}\left[\frac{A}{s}\right] = A$$

Notice that the same expression is used for the Laplace transform of a constant.

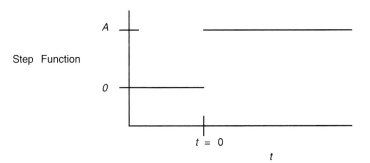

Step Function

FIGURE 7.2   Step function.

7.3.3 Time-Delay (Dead Time)

This is important for systems with transport delays (flow through pipes, etc.), or delays due to measurements. Let t_d represent the time delay. If the undelayed time domain function is $f(t)$, then the delayed function is $f(t - t_d)$, as shown in Figure 7.3.
  The Laplace transform of the delayed function is:

$$L[f(t - t_d)] = \int_0^\infty f(t - t_d) e^{-st}\, dt = \int_0^\infty f(t - t_d) e^{-s(t-t_d+t_d)}\, dt = \int_0^\infty f(t - t_d) e^{-s(t-t_d)} e^{-st_d}\, dt$$

$$= \int_0^\infty f(t - t_d)\, e^{-s(t-t_d)}\, e^{-st_d}\, d(t - t_d) = e^{-st_d} \int_0^\infty f(t - t_d)\, e^{-s(t-t_d)}\, d(t - t_d)$$

$$= e^{-st_d} \int_0^\infty f(t')\, e^{-st'}\, dt' = e^{-st_d} F(s)$$

Notice that the lower limit of integration did not change with the change of variable, because the function $f(t)$ is defined as $f(t) = 0$ for $t < 0$.

$$\boxed{L[f(t - t_d)] = e^{-st_d} F(s) \qquad \qquad \textit{Time-Delay}}$$

The transform of a delayed function is simply e^{-st_d} times the transform of the undelayed function.

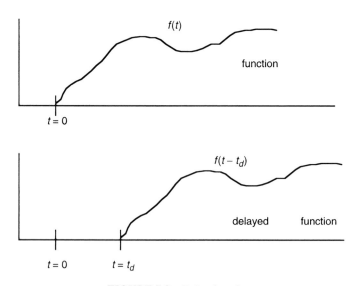

FIGURE 7.3 Delay function.

7.3.4   Derivatives

This will be important in transforming the derivative (accumulation) term in a dynamic equation to the Laplace domain.

$$L\left[\frac{df(t)}{dt}\right] = \int_0^\infty \frac{df(t)}{dt} e^{-st} \, dt$$

Using integration by parts ($\int u\,dv = uv - \int v\,du$)

Let $\qquad u = e^{-st}$ and $v = f(t)$

$$L\left[\frac{df(t)}{dt}\right] = \int_0^\infty \frac{df(t)}{dt} e^{-st} \, dt = \left[e^{-st} f(t)\right]_0^\infty + \int_0^\infty f(t) \, s e^{-st} \, dt$$

$$L\left[\frac{df(t)}{dt}\right] = [0 - f(0)] + s\int_0^\infty f(t) \, e^{-st} \, dt = sF(s) - f(0)$$

$$L\left[\frac{df(t)}{dt}\right] = sF(s) - f(0)$$

Derivative

$$L^{-1}[sF(s) - f(0)] = \frac{df(t)}{dt}$$

Since we often work with deviation variables, $f(0) = 0$ in many cases.

In general, you should be able to show the following (see student exercise 1):

$$L\left[\frac{d^n f(t)}{dt^n}\right] = s^n F(s) - s^{n-1} f(0) - s^{n-2} f'(0) - \dots - s f^{(n-2)}(0) - f^{(n-1)}(0)$$

nth Order Derivative

n initial conditions are needed $f(0), \dots, f^{(n-1)}(0)$

7.3.5   Integrals

This is often used in process control, since many controllers use information about the integral of the error between the desired value (setpoint) and the measured value:

$$L\left[\int_0^t f(t) \, dt\right] = \int_0^\infty \left[\int_0^t f(t) \, dt\right] e^{-st} \, dt$$

Again, integrate by parts, using $u = e^{-st} \, dt$ and $v = \int_0^t f(t)dt$, to find

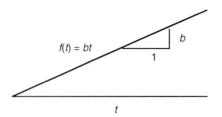

FIGURE 7.4   Ramp function.

$$L\left[\int_0^t f(t)\,dt\right] = \frac{1}{s}F(s)$$ *Integral*

7.3.6 Ramp Function

Consider the following ramp function:

$$f(t) = b\,t$$

as depicted in Figure 7.4.

You should be able to show (see exercise 2) that

$$L\,[bt] = \frac{b}{s^2}$$

Ramp

$$L^{-1}[1/s^2] = t$$

7.3.7 Pulse

Consider the pulse function in Figure 7.5, which consists of a step from 0 to A at $t = 0$, and a step back to 0 at $t = t_p$. Find the Laplace transfer function for this pulse.

There are two ways to solve this problem.

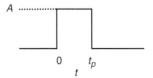

FIGURE 7.5   Pulse function.

ONE METHOD

The pulse function is defined over the following two time intervals:

$$f(t) = A \text{ for } 0 < t < t_p$$
$$f(t) = 0 \text{ for } t > t_p$$

and we can write the Laplace transform as:

$$F(s) = \int_0^\infty f(t)\, e^{-st}\, dt = \int_0^{t_p} f(t)\, e^{-st}\, dt + \int_{t_p}^\infty f(t)\, e^{-st}\, dt$$

$$= \int_0^{t_p} A e^{-st}\, dt + \int_{t_p}^\infty 0 e^{-st}\, dt = -\frac{A}{s}\, [e^{-st}]_0^{t_p}$$

$$F(s) = -\frac{A}{s}\, [e^{-t_p s} - 1] = \frac{A}{s}\, [1 - e^{-t_p s}]$$

The use of $A = 1$ is the unit pulse.

$$L[\text{unit pulse of duration } t_p] = \frac{1}{s}\, [1 - e^{-t_p s}] \qquad\qquad \textit{Unit Pulse}$$

A SECOND METHOD

Consider that the pulse is simply the sum of two step changes, as shown in Figure 7.6.

That is, it is the sum of a positive step change at $t = 0$ and a negative step change at $t = t_p$. Let $f_1(t)$ represent the step change at $t = 0$, and $f_2(t)$ represent the negative step change at $t = t_p$.

$$f(t) = f_1(t) + f_2(t)$$

$$F(s) = F_1(s) + F_2(s) = L[f(t)] = L[f_1(t) + f_2(t)]$$

but notice that $f_2(t) = -f_1(t - t_p)$

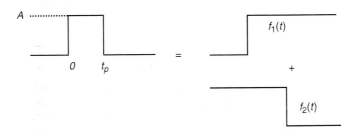

FIGURE 7.6   Pulse function.

and that (from the step function): $F_1(s) = \dfrac{A}{s}$

and (from the delay function): $F_2(s) = e^{-t_p s}\left(-F_1(s)\right)$

$$= -e^{-t_p s}\dfrac{A}{s}$$

So we can write $F(s) = F_1(s) + F_2(s) = \dfrac{A}{s}[1 - e^{-t_p s}]$

which is consistent with the previous derivation.

7.3.8 Unit Impulse

In Figure 7.7, consider the pulse function as the pulse time is decreased, but the pulse area remains the same, as shown by the dashed lines below.

The unit impulse function is a special case of the pulse function, with zero width ($t_p \to 0$) and unit pulse area (so $A = 1/t_p$). Taking the limit and applying L' Hopital's rule:

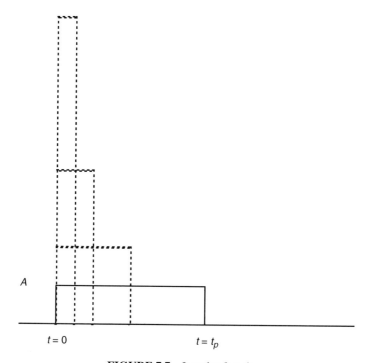

$t = 0$ $t = t_p$

FIGURE 7.7   Impulse function.

$$L[\delta] = \lim_{t_p \to 0} \frac{1}{t_p s} [1 - e^{-t_p s}] = \lim_{t_p \to 0} \frac{-1}{s} [-s e^{-t_p s}] = 1$$

$L[\delta] = 1$	*Unit Impulse*

7.3.9 Review

Thus far we have derived the Laplace transform of a number of functions. For example, we found:

$$L[e^{-at}] = \frac{1}{s + a}$$

If we have a Laplace domain function, such as $1/(s + a)$, we can "invert" it to the time domain. For example,

$$L^{-1}\left[\frac{1}{s + a}\right] = e^{-at}$$

Although the student should be able to derive Laplace transforms of any time domain function, that is not our major objective. Our major objective is to use Laplace transforms as a tool to solve dynamic problems. The Laplace transforms of many time-domain functions have been derived and compiled in various tables and handbooks. Already, we can construct a table of eight (exponential, step, time-delay, derivative, integral, ramp, pulse, and impulse) time-domain functions along with their Laplace domain functions. Additional Laplace transforms are provided in Table 7.1 in Section 7.6.

7.4 FINAL AND INITIAL VALUE THEOREMS

The following theorems are useful for determining limiting values in dynamics studies. They will be used frequently to find the short-term and long-term behavior. The *long term* (final value) of a time domain function can be found by analyzing the Laplace domain behavior in the limit as the s variable approaches zero. The initial value of a time domain function can be found by analyzing the Laplace domain behavior in the limit as s approaches infinity.

Final Value Theorem

$$\lim_{t \to \infty} f(t) = \lim_{s \to 0} [s\, F(s)] \tag{7.3}$$

Initial Value Theorem

$$\lim_{t \to 0} f(t) = \lim_{s \to \infty} [s\, F(s)] \tag{7.4}$$

If we have transformed a time domain function to the s domain, we can still find out the value of the time domain function as it goes to steady-state ($t \to \infty$) by finding the value of the Laplace domain function as $s \to 0$. An application of the final and initial value theorems is shown in Example 7.1.

EXAMPLE 7.1   Application of Final and Initial Value Theorems to the Exponential Function

Consider the exponential function:

$$f(t) = e^{-at}$$

which had the Laplace transform:

$$F(s) = \frac{1}{s + a}$$

Final Value Theorem. We first find:

$$\lim_{s \to 0} s F(s) = \lim_{s \to 0} \frac{s}{s + a} = 0$$

which checks with

$$\lim_{t \to \infty} f(t) = \lim_{t \to \infty} e^{-at} = 0$$

as long as a is positive.

Initial Value Theorem. We first find

$$\lim_{s \to \infty} s F(s) = \lim_{s \to \infty} \frac{s}{s + a} = 1$$

which checks with

$$\lim_{t \to 0} f(t) = \lim_{t \to 0} e^{-at} = 1$$

which is satisfied for any finite a.

One point not often made in textbooks is that the final value theorem only holds for stable systems ($a > 0$).

7.5   APPLICATION EXAMPLES

The following is a checklist for solving dynamics problems using Laplace transforms.

Step 1.   Start with a linear ordinary differential equation and initial conditions.
Step 2.   Transform each of the time domain functions to the Laplace domain, generally by using a table of Laplace transforms.

Step 3. Use algebraic manipulations to solve for the transformed variable. The *partial fraction expansion* approach is particularly useful.

Step 4. "Invert" to the time domain, by using a table of Laplace transforms.

7.5.1 Partial Fraction Expansion

The partial fraction expansion approach is based on representing a ratio of two polynomials as a sum of simpler terms. Let N(s) and D(s) represent numerator and denominator polynomials, respectively.

$$\frac{N(s)}{D(s)} = \frac{C_1}{D_1(s)} + \frac{C_2}{D_2(s)} + \ldots \frac{C_n}{D_n(s)}$$

C_i are constants and D_i are lower order (typically 1) polynomials.

The four-step procedure is used in each of the following examples. The partial fraction expansion is first used in Example 7.3.

EXAMPLE 7.2   Homogeneous First-order Problem

Step 1. Consider the simple *homogeneous* (unforced) first-order problem:

$$\frac{dx}{dt} + 2x = 0 \tag{7.5}$$

subject to the initial condition:

$$x(0) = 4 \tag{7.6}$$

Step 2. Recall the following transforms:

$$L\left[\frac{dx}{dt}\right] = s\,X(s) - x(0)$$

$$L[ax] = aL[x] = a\,X(s)$$

Then we can take the Laplace transform of (7.3) and (7.4) as:

$$L\left[\frac{dx}{dt}\right] + 2L[x] \qquad = 0$$

$$s\,X(s) - x(0) + 2\,X(s) = 0$$

$$s\,X(s) - 4 + 2\,X(s) \quad = 0 \tag{7.7}$$

Step 3. Solving (7.7) for $X(s)$:

$$X(s) = \frac{4}{s + 2} \tag{7.8}$$

Step 4.   Inverting each element back to the time domain:

$$L^{-1}[X(s)] = x(t) \tag{7.9}$$

$$L^{-1}\left[\frac{4}{s+2}\right] = 4\,e^{-2t} \tag{7.10}$$

and the solution is

$$x(t) = 4\,e^{-2t} \tag{7.11}$$

Indeed, using the method in Example 7.2, we can show that the general first-order equation:

$$\frac{dx}{dt} + a\,x = 0$$

with initial condition $x(0)$

has the solution $\qquad\qquad x(t) = x(0)\,e^{-at}$

which, of course, is the same solution obtained by separating the variables and integrating. The real power of Laplace transforms is in solving heterogeneous problems, as illustrated in Example 7.3.

EXAMPLE 7.3   Illustration of the Partial Fraction Expansion Technique

Step 1.   Consider the simple heterogeneous first-order problem:

$$\frac{dx}{dt} + 2\,x = 4.5 \tag{7.12}$$

with the initial condition

$$x(0) = 4 \tag{7.13}$$

Step 2.   Taking the Laplace transform of each element:

$$s\,X(s) - x(0) + 2\,X(s) = \frac{4.5}{s}$$

which can be written (since $x(0) = 4$):

$$(s+2)\,X(s) = 4 + \frac{4.5}{s}$$

Step 3.   Solving for the transformed variable

$$X(s) = \frac{4}{s+2} + \frac{4.5}{s(s+2)} \tag{7.14}$$

We would like to invert (7.14) to the time domain, however we do not know how to invert the last term $4.5/s(s + 2)$.

We will use the approach known as a *partial fraction expansion*. That is, write:

$$\frac{4.5}{s(s + 2)} = \frac{A}{s} + \frac{B}{s + 2} \tag{7.15}$$

to find A, first multiply (7.15) by s:

$$\frac{4.5}{s + 2} = A + \frac{Bs}{s + 1}$$

then set $s = 0$ and solve for A:

$$A = 2.25$$

To find B, first multiply (7.15) by $s + 2$:

$$\frac{4.5}{s} = \frac{A(s + 2)}{s} + B$$

and set $s = -2$ to solve for B:

$$B = -2.25$$

which yields:

$$\frac{4.5}{s(s + 2)} = \frac{2.25}{s} + \frac{-2.25}{s + 2} \tag{7.16}$$

and we can write (7.15) as:

$$X(s) = \frac{4}{s + 2} + \frac{2.25}{s} + \frac{-2.25}{s + 2} \tag{7.17}$$

Step 4.   Inverting element by element in (7.17) we find

$$x(t) = 4 e^{-2t} + 2.25 + -2.25 e^{-2t}$$

or

$$x(t) = 1.75 e^{-2t} + 2.25 \tag{7.18}$$

the reader should verify that this solution satisfies the initial conditions and the differential equation.

Examples 7.4 and 7.5 provide additional illustration of the partial fraction expansion technique.

EXAMPLE 7.4   Find the Inverse Laplace Transform of $1/(s + a)(s + b)$

Write

$$\frac{1}{(s + a)(s + b)} = \frac{A}{s + a} + \frac{B}{s + b} \tag{7.19}$$

Multiply (7.19) by $s + a$, set $s = -a$ to find:

$$A = \frac{1}{-a + b}$$

Multiply (7.19) by $s + b$, set $s = -b$ to find:

$$B = \frac{1}{a - b}$$

Therefore,

$$\frac{1}{(s + a)(s + b)} = \left[\frac{1}{b - a}\right]\left[\frac{1}{s + a}\right] + \left[\frac{1}{a - b}\right]\left[\frac{1}{s + b}\right] \qquad (7.20)$$

and we can take the inverse Laplace transform of each function on the righthand side of (7.20)

$$L^{-1}\left[\frac{1}{(s + a)(s + b)}\right] = \left[\frac{1}{b - a}\right]e^{-at} + \left[\frac{1}{a - b}\right]e^{-bt} \qquad (7.21)$$

Notice that this technique fails if $a = b$.

The method in the previous example failed if the roots of the Laplace domain function were equal. The following example shows how to perform a partial fraction expansion for *repeated roots*.

EXAMPLE 7.5   Consider the Following Transfer Function with *Repeated Roots*

$$\frac{1}{(s + a)^2(s + b)} \qquad (7.22)$$

Expand (7.22) in the following fashion:

$$\frac{1}{(s + a)^2(s + b)} = \frac{A}{s + a} + \frac{B}{(s + a)^2} + \frac{C}{s + b} \qquad (7.23)$$

Here we cannot multiply by $s + a$ and set $s = -a$, because we would find unbounded terms. First, multiply (7.23) by $(s + a)^2$:

$$\frac{1}{s + b} = A(s + a) + B + \frac{C(s + a)^2}{s + b}$$

and set $s = -a$ to find

$$B = \frac{1}{b - a}$$

Multiply (7.23) by $s + b$ and set $s = -b$ to find

$$C = \frac{1}{(a - b)^2}$$

Notice that we have solved for two of the coefficients of (7.23). Now, we can solve for one equation in one unknown, by setting s = any value. For simplicity, choose $s = 0$. From (7.23):

$$\frac{1}{a^2b} = \frac{A}{a} + \frac{1}{(b-a)a^2} + \frac{1}{(a-b)^2b}$$

we can reduce the solution for A to $A = \dfrac{-1}{(a-b)^2}$

We have solved for A, B, and C in (7.23), so we can perform an element-by-element inversion of (7.23) to find the time domain function:

$$L^{-1}\left[\frac{\left[\dfrac{-1}{(a-b)^2}\right]}{(s+a)} + \frac{\left[\dfrac{1}{(b-a)}\right]}{(s+a)^2} + \frac{\left[\dfrac{1}{(a-b)^2}\right]}{(s+b)}\right] =$$

$$\frac{-1}{(a-b)^2}\,e^{-at} + \frac{1}{b-a}\,t\,e^{-at} + \frac{1}{(a-b)^2}\,e^{-bt}$$

and we can write:

$$L^{-1}\left[\frac{1}{(s+a)^2(s+b)}\right] = \frac{-1}{(a-b)^2}\,e^{-at} + \frac{1}{b-a}\,t\,e^{-at} + \frac{1}{(a-b)^2}\,e^{-bt} \qquad (7.24)$$

As an alternative, we can find a common denominator for the righthand side of (7.23) and write:

$$\frac{1}{(s+a)^2(s+b)} = \frac{A(s+a)(s+b) + B(s+b) + C(s+a)^2}{(s+a)^2(s+b)} \qquad (7.25)$$

then expand the numerator and solve for the coefficients A, B, and C such that the righthand side is equal to the lefthand side. See student exercise 13.

The previous examples were for ODEs with real roots. This next example is a problem with complex roots.

EXAMPLE 7.6   A Second-order System with Complex Roots

Step 1.   Consider the homogeneous problem:

$$\frac{d^2x}{dt^2} + \frac{dx}{dt} + x = 0 \qquad (7.26)$$

with the initial conditions:

$$\dot{x}(0) = x(0) = 1 \qquad (7.27)$$

Step 2.   From the table of Laplace transforms:

$$L\left[\frac{d^n x}{dt^n}\right] = s^n\,X(s) - s^{n-1}x(0) - s^{n-2}\,\dot{x}(0) - \dots - s\,x^{(n-2)}(0) - x^{(n-1)}(0)$$

So, for a second derivative:

$$L\left[\frac{d^2x}{dt^2}\right] = s^2 X(s) - s\,x(0) - \dot{x}(0)$$

and, for a first derivative:

$$L\left[\frac{dx}{dt}\right] = s\,X(s) - x(0)$$

We can now write the Laplace transform of (7.26) as:

$$s^2 X(s) - s\,x(0) - \dot{x}(0) + s\,X(s) - x(0) + X(s) = 0$$

Step 3. Attempting to isolate $X(s)$ on the LHS:

$$(s^2 + s + 1)\,X(s) = s\,x(0) + x(0) + \dot{x}(0)$$

dividing by $(s^2 + s + 1)$:

$$X(s) = \frac{s\,x(0)}{s^2 + s + 1} + \frac{x(0) + \dot{x}(0)}{s^2 + s + 1}$$

and from the initial conditions:

$$X(s) = \frac{s}{s^2 + s + 1} + \frac{2}{s^2 + s + 1} \tag{7.28}$$

the roots of $(s^2 + s + 1)$ are $-1/2 \pm \sqrt{3}/2\,j$ (from the quadratic formula):

$$(s^2 + s + 1) = \left(s + \frac{1}{2} + \frac{\sqrt{3}}{2}j\right)\left(s + \frac{1}{2} - \frac{\sqrt{3}}{2}j\right)$$

Notice another way that we can write $(s^2 + s + 1)$ is:

$$(s^2 + s + 1) = \left(s + \frac{1}{2}\right)^2 + \left(\frac{\sqrt{3}}{2}\right)^2$$

which means that we can write (7.28) as:

$$X(s) = \frac{s}{\left(s + \frac{1}{2}\right)^2 + \left(\frac{\sqrt{3}}{2}\right)^2} + \frac{2}{\left(s + \frac{1}{2}\right)^2 + \left(\frac{\sqrt{3}}{2}\right)^2}$$

Step 4. Notice from a table of Laplace transforms that:

$$L[e^{-bt} \sin \omega t] = \frac{\omega}{(s + b)^2 + \omega^2} \tag{7.29}$$

$$L[e^{-bt} \cos \omega t] = \frac{s + b}{(s + b)^2 + \omega^2} \tag{7.30}$$

and we should maneuver (7.28) into the form of (7.29) and (7.30).
Notice that we can write (7.30) as:

$$X(s) = \frac{s + \dfrac{1}{2}}{\left(s + \dfrac{1}{2}\right)^2 + \left(\dfrac{\sqrt{3}}{2}\right)^2} + \frac{1.5}{\left(s + \dfrac{1}{2}\right)^2 + \left(\dfrac{\sqrt{3}}{2}\right)^2} \qquad (7.31)$$

and we invert each element of the RHS of (7.31), using (7.29) and (7.30):

$$x(t) = e^{-t/2} \cos \frac{\sqrt{3}}{2} t + \frac{1.5}{\dfrac{\sqrt{3}}{2}} e^{-t/2} \sin \frac{\sqrt{3}}{2} t$$

$$x(t) = e^{-t/2} \cos \frac{\sqrt{3}}{2} t + \sqrt{3} e^{-t/2} \sin \frac{\sqrt{3}}{2} t$$

which has the time domain response shown in Figure 7.8. As we noticed in Chapter 6, complex roots give oscillatory responses. We see in Chapter 9 that this type of response is called under-damped.

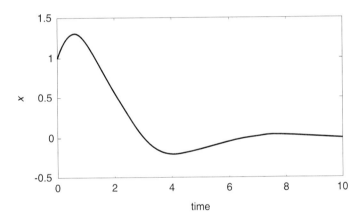

FIGURE 7.8   Oscillatory response due to complex roots.

7.6   TABLE OF LAPLACE TRANSFORMS

For your convenience, selected Laplace transforms are presented in Table 7.1. If you desire to transform a function from the time domain to the Laplace domain, then look for the time domain function in the first column ($f(t)$) and write down the corresponding Laplace domain function in the second column ($F(s)$). Similarly, if you are trying to "invert" a Laplace domain function to the time domain, then look for the Laplace domain function in the second column and write down the corresponding time domain function from the first column.

TABLE 7.1   Laplace Transforms for Selected Time-Domain Functions

$f(t)$	$F(s)$
$\delta(t)$ (unit impulse)	1
$S(t)$ (unit step) $\begin{cases} 0 \text{ for } t < 0 \\ 1 \text{ for } t > 0 \end{cases}$	$\dfrac{1}{s}$
A (constant)	A/s
$f(t-\theta)$ (time delay)	$e^{-\theta s}\, F(s)$
t (ramp)	$\dfrac{1}{s^2}$
t^{n-1}	$\dfrac{(n-1)!}{s^n}$
$\dfrac{df}{dt}$ (derivative)	$sF(s) - f(0)$
$\dfrac{d^n f}{dt^n}$	$s^n F(s) - s^{n-1} f(0) - s^{n-2} f^{(1)}(0) - \ldots$ $\quad - sf^{(n-2)}(0) - f^{(n-1)}(0)$
e^{-at}	$\dfrac{1}{s+a}$
$\dfrac{1}{a_1 - a_2}\left(e^{-a_2 t} - e^{-a_1 t}\right)$	$\dfrac{1}{(s+a_1)(s+a_2)}$
$\dfrac{a_3 - a_1}{a_2 - a_1}\, e^{-a_1 t} + \dfrac{a_3 - a_2}{a_1 - a_2}\, e^{-a_2 t}$	$\dfrac{s+a_3}{(s+a_1)(s+a_2)}$
$1 - e^{-t/\tau}$	$\dfrac{1}{s(\tau s + 1)}$
$\sin \omega t$	$\dfrac{\omega}{s^2 + \omega^2}$
$\cos \omega t$	$\dfrac{s}{s^2 + \omega^2}$
$e^{-at}\sin \omega t$	$\dfrac{\omega}{(s+a)^2 + \omega^2}$
$e^{-at}\cos \omega t$	$\dfrac{s+a}{(s+a)^2 + \omega^2}$
$1 + \dfrac{1}{\tau_2 - \tau_1}\left(\tau_1\, e^{-t/\tau_1} - \tau_2\, e^{-t/\tau_2}\right)$	$\dfrac{1}{s(\tau_1 s + 1)(\tau_2 s + 1)}$
$\left(1 - \dfrac{1}{\sqrt{1-\xi^2}}\, e^{-\xi t/\tau} \sin(\omega t + \Phi)\right)$	$\dfrac{1}{s(\tau^2 s^2 + 2\xi \tau s + 1)}$
$\text{where } \omega = \dfrac{\sqrt{1-\xi^2}}{\tau},\ \Phi = \tan^{-1}\dfrac{\sqrt{1-\xi^2}}{\xi}$	
$1 + \dfrac{\tau_3 - \tau_1}{\tau_1 - \tau_2}\, e^{-t/\tau_1} + \dfrac{\tau_3 - \tau_2}{\tau_2 - \tau_1}\, e^{-t/\tau_2}$	$\dfrac{(\tau_3 s + 1)}{s(\tau_1 s + 1)(\tau_2 s + 1)}$
$1 - \left(1 - \dfrac{\tau_n}{\tau_d}\right) e^{-t/\tau_d}$	$\dfrac{\tau_n s + 1}{s(\tau_d s + 1)}$

SUMMARY

We have defined the Laplace transform and applied it to several functions that commonly appear in the solution of chemical process dynamics problems. Although the Laplace transform concept seems quite abstract at this point, in the chapters that follow you will find it extremely useful in solving differential equation models. The final (7.3) and initial value (7.4) theorems will be useful for checking the long-term (steady-state) behavior and the initial conditions for a particular problem.

A number of examples were provided to illustrate the power of the Laplace transform technique for solving ordinary differential equations. We noted that the technique allows us to convert the ODE problem to an algebraic problem, which is easier for us to solve. After performing algebraic manipulations in the Laplace domain, often with the use of a partial fraction expansion, we then look up inverse transforms to obtain the time domain solution.

In the chapters that follow, we use Laplace transforms to analyze the dynamic behavior of different types of linear process models.

FURTHER READING

Many differential equations and process control textbooks provide details on Laplace transforms. Some examples are:

Boyce, W., & R. DiPrima. (1992). *Ordinary Differential Equations and Boundary Value Problems,* 5th ed. New York: Wiley.

Luyben, W.L. (1990). *Process Modeling, Simulation and Control for Chemical Engineers,* 2nd ed. New York: McGraw-Hill.

Seborg, D.E., T.F. Edgar, & D.A. Mellichamp. (1989). *Process Dynamics and Control.* New York: Wiley.

Stephanopoulos, G. (1984). *Chemical Process Control: An Introduction to Theory and Practice.* Englewood Cliffs, NJ: Prentice Hall.

STUDENT EXERCISES

1–5. The student should derive the Laplace transform for the following functions:

1. $\dfrac{d^n f}{dt^n}$

2. $f(t) = bt$

3. $f(t) = t^2$

4. $L\left[\int_0^t f(t)\, dt\right]$

5. $f(t) = \cos \omega t$

(*Hint:* Although you can solve question 5 using integration by parts, you may wish to use the   Euler identity $\cos \omega t = 1/2\, (e^{j\omega t} + e^{-j\omega t})$.)

6. Find the Laplace transform, $u(s)$, of the following input function:

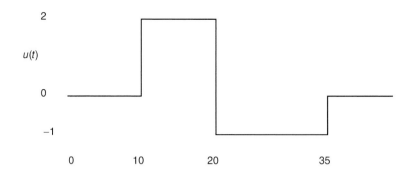

7. Find the Laplace transform of the function $y(t)$ that satisfies the differential equation and initial conditions:

$$\frac{d^3y}{dt^3} + 4\frac{d^2y}{dt^2} + 5\frac{dy}{dt} + 2y = 2$$

$$y(0) = \frac{dy(0)}{dt} = \frac{d^2y(0)}{dt^2} = 0$$

8. Solve the differential equation:

$$\frac{dy}{dt} + 3y = 0$$

$$y(0) = 2.0$$

9. A process input has the following Laplace transform:

$$u(s) = \frac{2}{s^3} - \frac{6}{s^2}e^{-3s}$$

What is the time domain input, $u(t)$? Find this analytically.

Sketch the time domain input.

10. Find the time domain solution $y(t)$ for the Laplace domain transfer function (with $\xi < 1$):

$$Y(s) = \frac{1}{s(\tau^2 s^2 + 2\xi\tau s + 1)}$$

11. Derive the time domain solution $y(t)$ for the Laplace domain transfer function:

$$Y(s) = \frac{(\tau_n s + 1)}{s(\tau_1 s + 1)(\tau_2 s + 1)}$$

12. Derive the time domain solution $y(t)$, for the Laplace domain transfer function:

$$Y(s) = \frac{\tau_n s + 1}{s(\tau_d s + 1)}$$

13. Consider Example 7.5, involving the following transfer function with *repeated roots*:

$$\frac{1}{(s + a)^2(s + b)} = \frac{A}{s + a} + \frac{B}{(s + a)^2} + \frac{C}{s + b}$$

Find a common denominator for the righthand side:

$$\frac{1}{(s + a)^2(s + b)} = \frac{A(s + a)(s + b) + B(s + b) + C(s + a)^2}{(s + a)^2(s + b)}$$

then expand the numerator and solve for the coefficients A, B, and C such that the righthand side is equal to the lefthand side.

TRANSFER FUNCTION ANALYSIS OF FIRST-ORDER SYSTEMS

8

After studying this chapter, the reader should understand:

- The responses of first-order systems to step and impulse inputs.
- How chemical reactions change the time constant of a stirred tank.
- The behavior of an integrating process.
- How to compare the long-term behavior of a nonlinear process with that of a linear process without integrating the nonlinear modeling equations.
- The responses of first-order + time-delay models.
- How to estimate the parameters of first-order and first-order + time-delay transfer functions by applying step input changes.
- The response of a lead/lag model to a step input.

The important sections in this chapter are:

8.1   PERSPECTIVE

One of the powers of the Laplace transform technique is the ease with which it handles heterogeneous (forced input) problems. It is most useful when the models are separate from the type of input imposed (step, ramp, etc.). The models that are developed are called *transfer function* models and will be used frequently in control system design.

Process engineers often learn much about the behavior of a process by changing the inputs and seeing how the outputs respond. The goal of this chapter is to illustrate the typical responses of first-order models to step changes in inputs. Knowledge of these types of responses will allow an engineer to determine a good approximate model for the process, including the best parameter values, based on measured data from the process.

8.2   RESPONSES OF FIRST-ORDER SYSTEMS

The equation for a linear first-order process is generally written in the following form

$$\tau \frac{dy}{dt} + y = k\,u \tag{8.1}$$

where the parameters (τ and k) and variables (y and u) have the following names:

τ = time constant (units of time)

k = process gain (units of output/input)

y = output variable

u = input variable

The model (8.1) is sometimes derived by linearizing a nonlinear model about a given steady-state and then placing the resulting linear model in deviation variable form. For this reason, we assume that the initial conditions are $y(0) = 0$ and $u(0) = 0$. The input, u and the output y are functions of time; $u(t)$ must be specified to solve for $y(t)$.

In the next example, we show how a standard first-order process model arises.

EXAMPLE 8.1   A mixing tank

Assume that a chemical compound, *A*, is in a feedstream entering a mixing tank. Assume that there is no reaction, and that the concentration of *A* has no effect on the density of the fluid (this is true for trace components in water, for example). Also assume that the flowrate is constant and the volume in the tank is constant—this implies that the outlet flowrate is equal to the inlet flowrate, as shown in Figure 8.1. The process is operating at steady-state, then the inlet concentration is suddenly changed to a new value. Find the tank outlet concentration as a function of time.

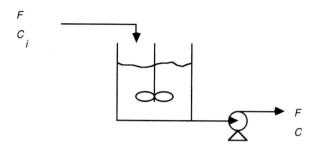

FIGURE 8.1   Mixing tank.

Overall Material Balance

$$\frac{dV}{dt} = F - F = 0 \quad \text{(from problem statement)}$$

Component Material Balance

$$\frac{dVC}{dt} = F\,C_i - F\,C$$

since V is constant:

$$\frac{dC}{dt} = \frac{F}{V}\,C_i - \frac{F}{V}\,C \tag{8.2}$$

First of all, we can solve for the initial steady-state concentration. At steady-state, $dC/dt = 0$, so from (8.2) we find:

$$C_s = C_{is}$$

where C_s is the steady-state tank outlet concentration and C_{is} is the steady-state tank inlet concentration. Now, since $-F/V\,C_{is} + F/V\,C_s = 0$, we can add this to (8.2). Also, since C_s is a constant, $dC/dt = d(C - C_s)/dt$, and we can write:

$$\frac{d(C - C_s)}{dt} = \frac{F}{V}(C_i - C_{is}) - \frac{F}{V}(C - C_s) \tag{8.3}$$

or

$$\frac{V}{F}\frac{d(C - C_s)}{dt} + (C - C_s) = (C_i - C_{is}) \tag{8.4}$$

Equation (8.4) is identical to the first-order equation:

$$\tau\frac{dy}{dt} + y = k\,u \tag{8.1}$$

with $\tau = V/F$, $k = 1$, $y = C - C_s$, $u = C_i - C_{is}$

Notice that the time constant in this case is simply the residence time of the tank, that is, the average amount of time that a molecule stays in the tank.

Notice that for linear systems, we can directly write the deviation variable model directly from the physical model, skipping several intermediate steps. Also, since deviation variables are defined on the basis of a steady-state operating condition, if the process is initially at steady-state, then $y(0) = 0$ and $u(0) = 0$.

Taking the Laplace transform of (8.1) we find:

$$\tau \left[s\, Y(s) - y(0) \right] + Y(s) = k\, U(s)$$
$$\tau s Y(s) + Y(s) = k\, U(s) \qquad (8.5)$$
$$(\tau s + 1)\, Y(s) = k\, U(s)$$

which is most commonly written:

$$Y(s) = \frac{k}{\tau s + 1}\, U(s) \qquad (8.6)$$

or,

$$Y(s) = g(s)\, U(s) \qquad (8.7)$$

where:

$$g(s) = \frac{k}{\tau s + 1} \qquad (8.8)$$

The reader should become familiar with this type of representation. In general terms, $g(s)$ is known as a *transfer function*. In this specific case, $g(s)$ is a first-order transfer function. You will often see a *block diagram* representation of (8.7) as shown in Figure 8.2

One nice thing about (8.6) is that it holds for any first-order process (with zero initial conditions)—we have not had to use any knowledge (yet) about the input u as a function of time. Once we know $u(t)$, we can use Laplace transforms to find $U(s)$ to solve the problem. We will see later that block diagrams and transfer functions are easy to work with, when we have a complex system that is composed of a number of subsystems. Before we deal with such systems, we will first understand the behavior of first-order systems to different types of inputs.

8.2.1   Step Inputs

The most common input forcing function is the step input. For this problem, assume a step input of magnitude ΔU at time $t = 0$. We know that the Laplace transform of a step input is (from Chapter 7):

$$L\,[\Delta U] = \frac{\Delta U}{s} \qquad (8.9)$$

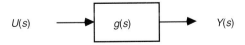

FIGURE 8.2   Block diagram.

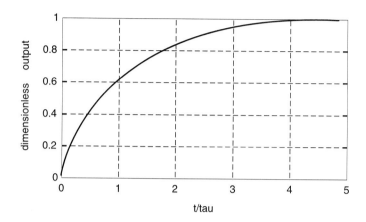

FIGURE 8.3   Dimensionless output step response of a first-order process.

and we can then write (8.6) as:

$$Y(s) = \frac{k}{\tau s + 1} \frac{\Delta U}{s} \tag{8.10}$$

$$Y(s) = \frac{k\Delta U}{s(\tau s + 1)} \tag{8.11}$$

From the table of Laplace transforms in Chapter 7 (the reader should be able to derive this result, using a partial fraction expansion):

$$L^{-1}\left[\frac{1}{s(\tau s + 1)}\right] = 1 - e^{-t/\tau} \tag{8.12}$$

and the solution to (8.11) is then:

$$y(t) = k\Delta U\left[1 - e^{-t/\tau}\right] \tag{8.13}$$

Notice that we can represent the solution of (8.13) with a single plot, by dividing (8.13) by $k\Delta U$ to obtain the dimensionless output:

$$\frac{y(t)}{k\Delta U} = \left[1 - e^{-t/\tau}\right] \tag{8.14}$$

A plot of (8.14) is shown in Figure 8.3, where we have used t/τ as a dimensionless time.

EXAMPLE 8.1   Continued

As a numerical example, consider the case where $V = 5$ ft³, $F = 1$ ft³/min, and the steady-state concentration (inlet and outlet) is 1.25 lbmol/ft³. Consider a step change in inlet concentration from 1.25 lbmol/ft³ to 1.75 lbmol/ft³. Then:

$$U(s) = \frac{\Delta U}{s} = \frac{0.5}{s} \qquad (\Delta u(t) = 1.75 - 1.25 = 0.5 \text{ lbmol/ft}^3)$$

$$Y(s) = \frac{1}{5s + 1} \frac{0.5}{s} \tag{8.15}$$

which has the time domain solution:

$$y(t) = 0.5 \left[1 - e^{-t/5}\right] \tag{8.16}$$

Since we desire to find the actual concentration, we can convert back to the physical variables, from the relationship:

$$y = C - C_s \Rightarrow C(t) = C_s + y(t) \tag{8.17}$$

and (8.17) can be written:

$$C(t) = 1.25 + 0.5 \left[1 - e^{-t/5}\right] \quad \text{lbmol/ft}^3 \tag{8.18}$$

Notice that $C(t \to \infty) = 1.75$, as expected. This can also be obtained by applying the Final Value Theorem to (8.16) and using (8.18). A plot of (8.18) is shown in Figure 8.4.

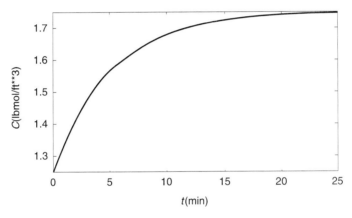

FIGURE 8.4   Transient response of mixing tank.

PARAMETER ESTIMATION FOR FIRST-ORDER PROCESSES

Returning to the general model for a first-order process, we see that there are two parameters of interest: the process gain and the process time constant.

$$y(t) = k\Delta U \left[1 - e^{-t/\tau}\right] \tag{8.13}$$

Process engineers often find process gains and time constants by performing step tests on processes.

GAIN ESTIMATION

We see from (8.14) that after $t \gg \tau$, the $e^{-t/\tau}$ term approaches 0. The value of k can be determined:

$$k = \frac{y(t) \text{ as } t \to (\text{large})}{\Delta U} = \frac{\Delta Y}{\Delta U} \qquad (8.19)$$

that is, the process gain is the change in output (as it approaches a new steady-state) divided by the change in input.

TIME CONSTANT ESTIMATION

We can find the time constant for a first-order process in the following fashion. Apply a step input to the process at $t = 0$. From (8.14), we see that $y(t)$ goes to a value of $k\Delta U$ as $t \to \infty$. When the time is equal to the time constant ($t = \tau$), from (8.13):

$$y(t) = k\Delta U \left[1 - e^{-1}\right] = 0.632 \, k\Delta U$$

that is, the time constant can be determined by finding the time where the output, $y(t)$, is at 63.2% of the ultimate response (new steady-state). This rule is also obvious by looking at Figure 8.3; when $t/\tau = 1$, $y(t)/k\Delta U = 0.632$.

You should be careful, because this is only true for first-order processes with no time-delay and a step input at $t = 0$. If the process is second-order or the input is not a step change, etc., this 63.2% value will not be correct.

You should get in the habit of associating units with all of the variables. Obviously, the process time constant, τ, must have units of time because $e^{-t/\tau}$ must be dimensionless. Also, the process gain, k, must have units of output/input to be dimensionally consistent.

SLOPE METHOD

An alternative method of estimating the time constant is to realize that the initial slope of the output step response for a first-order process is $k\Delta u/\tau$, as shown below. Taking the derivative of (8.13):

$$\frac{dy(t)}{dt} = \frac{k\Delta U}{\tau} \left[e^{-t/\tau}\right]$$

and evaluating at $t = 0$, we find

$$\frac{dy(t = 0)}{dt} = \frac{k\Delta U}{\tau}$$

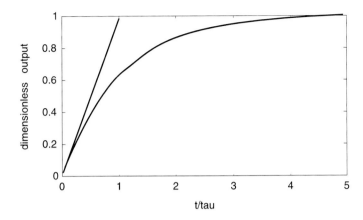

FIGURE 8.5 Slope method for time constant estimation (dimensionless output = $y/k\Delta u$).

If we extrapolate this slope to the final value of the output that is achieved, we find the time constant τ, as shown in Figure 8.5. This is a dimensionless plot, so the intersection at $t/\tau = 1$ indicates an intersection at $t = \tau$ in physical time.

Parameter estimation for first-order processes using a step response is illustrated by the next example.

EXAMPLE 8.2   Parameter Estimation of a First-Order Process

A process operator makes a step change in an input from 20 to 17.5 gal/min (gpm) and finds that the output eventually changes from an initial value of 50 psig to 55 psig, as shown in Figure 8.6 below. Find the process gain and time constant for this system.

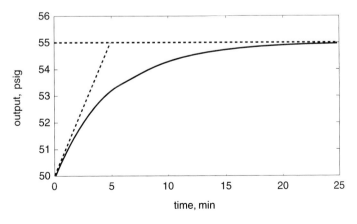

FIGURE 8.6 Slope method for time constant estimation.

We can immediately calculate the process gain from $k = \Delta y/\Delta u = 55 - 50$ psig/17.5 $-$ 20 gpm = -2 psig/gpm. We can calculate the time constant in a number of different ways. One way is to find the time where the output change is 63.2% of the final change. This occurs when the output is $50 + 0.632(5) = 53.2$ psig. From the plot, this occurs at $t = 5$ minutes. Another way to find the time constant is to extrapolate the initial slope of the response to the final value. This occurs at $t = 5$ minutes, as shown. The identified process transfer function is then:

$$g(s) = \frac{-2}{5s + 1}$$

Notice that the gain (-2 psig/gpm) and time constant (5 min) have units associated with them.

8.2.2   Impulse Inputs

Consider a first-order process with an impulse input of magnitude A. The transform of a unit impulse (δ) is 1, so $L[A\delta] = A$. The first order Laplace domain response is:

$$Y(s) = \frac{k}{\tau s + 1}\, U(s) = \frac{kA}{\tau s + 1} \tag{8.20}$$

the time domain response is:

$$y(t) = kA\, e^{-t/\tau} \tag{8.21}$$

Dividing by kA, we find the dimensionless output response shown in the Figure 8.7 below. The prime characteristic of a first-order system is that there is an immediate response to an impulse input.

In practice it is difficult to actually implement an impulse function. A close approximation can be made by implementing a pulse input over a short period of time, as shown in the next example.

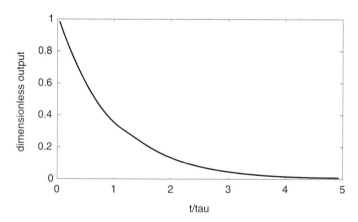

FIGURE 8.7   Impulse Response for a first-order process. The dimensionless output is $y(t)/kA$.

EXAMPLE 8.3   Comparison of Impulse and Pulse Inputs

In the previous example an impulse of magnitude A was applied to the process. Consider a pulse input, where an input value of Δu is applied for t_p units of time, as shown in Figure 8.8. The total applied input is then $A = \Delta u\, t_p$.

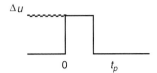

$$\Delta u$$

$$0 \qquad t_p$$

FIGURE 8.8   Pulse input.

From Chapter 7 we find that:

$$U(s) = \frac{\Delta u}{s}[1 - e^{-t_p s}]$$

So, the output for a first-order process with unit gain, is:

$$Y(s) = \frac{\Delta u}{s}[1 - e^{-t_p s}]\frac{1}{\tau s + 1}$$

$$Y(s) = \frac{\Delta u}{s(\tau s + 1)} - \frac{\Delta u\, e^{-t_p s}}{s(\tau s + 1)}$$

which has the time-domain solution (Chapter 7):

$$y(t) = \Delta u\,[1 - e^{-t/\tau}] - \Delta u\,[1 - e^{-(t-t_p)/\tau}]\,H(t)\text{(pulse)} \tag{8.22}$$

where $H(t) = 0$ for $t < t_p$ and 1 for $t \geq t_p$ and the total input applied over the t_p time units is $\Delta u\, t_p$. The impulse response is:

$$y(t) = kA\, e^{-t/\tau} = k\,\Delta u\, t_p\, e^{-t/\tau} \qquad \text{(impulse)} \tag{8.23}$$

The pulse and impulse responses are compared in Figure 8.9 for $t_p = 0.1\tau$ and $A = 1$.

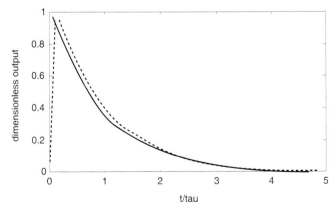

FIGURE 8.9   Comparison of pulse (dashed, $t_p = 0.1\tau$) and impulse (solid) responses.

8.3   EXAMPLES OF SELF-REGULATING PROCESSES

The standard first-order model presented in the previous section is a typical self-regulating process. If the input is changed to another value, the output eventually comes to a new steady-state. Contrast this with non-self-regulating systems where the output continues to change forever after a step input change. Self-regulating behavior is shown by the systems presented in the following example. One key idea to note is that a chemical reaction changes the time constant of a standard mixing tank model.

EXAMPLE 8.4   A CSTR with a First-Order Reaction

Now, extend the Example 8.1 to include a single decomposition reaction. The component material balance is:

$$\frac{dVC}{dt} = F\,C_i - F\,C - k_1\,V\,C \tag{8.24}$$

where k_1 is the reaction rate constant. Since V is constant:

$$\frac{dC}{dt} = -\left(\frac{F}{V} + k_1\right)C + \frac{F}{V}\,C_i \tag{8.25}$$

and we can calculate the steady-state concentrations from $dC/dt = 0$

$$C_s = \frac{\dfrac{F}{V}\,C_{is}}{\dfrac{F}{V} + k_1}$$

The deviation variable form of our dynamic model is

$$\frac{d(C - C_s)}{dt} = -\left(\frac{F}{V} + k_1\right)(C - C_s) + \frac{F}{V}(C_i - C_{is}) \tag{8.26}$$

or

$$\left[\frac{1}{\dfrac{F}{V} + k_1}\right]\frac{d(C - C_s)}{dt} + (C - C_s) = \left[\frac{\dfrac{F}{V}}{\dfrac{F}{V} + k_1}\right](C_i - C_{is}) \tag{8.27}$$

Again, observe that this is simply a first-order ODE with:

$$\tau = \left[\frac{1}{\dfrac{F}{V} + k_1}\right] = \frac{\dfrac{V}{F}}{1 + \dfrac{V}{F}k_1} \quad \text{and} \quad k = \left[\frac{\dfrac{F}{V}}{\dfrac{F}{V} + k_1}\right] = \frac{1}{1 + \dfrac{V}{F}k_1}$$

$$y = C - C_s \qquad\qquad \text{and} \quad u = C_i - C_{is}$$

and therefore, we know the solution for a step change in inlet concentration at $t = 0$.

Notice that the gains and time constants for a stirred tank with reaction are less than those for a stirred tank without reaction. This means that an inlet composition change has a faster dynamic effect in a system with chemical reaction than in a system with just mixing.

Note that the previous examples were linear because the flowrate was constant. If the flowrate were changing (i.e., was considered an input), the models would be nonlinear (actually bilinear), because of the terms where an input multiplies a state variable. The linearization techniques developed in Chapter 5 must then be used before a Laplace transform analysis can be performed. In the following example, linearization must be used because of the second-order reaction term.

EXAMPLE 8.5   A CSTR with a Second-Order Reaction

Here we extend the previous example to include a second-order reaction problem. We will assume that the rate of reaction (per unit volume) is proportional to the square of the concentration of the reacting component. An example would be $A + A \rightarrow B$. As before, we are making the simplifying assumption that the fluid density is not a function of the concentration. Again, assume that C_i is the input. The component material balance is:

$$\frac{dVC}{dt} = F\,C_i - FC - k_2\,V\,C^2 \tag{8.28}$$

where k_2 is the reaction rate constant. Since V is constant,

$$\frac{dC}{dt} = \frac{F}{V}\,C_i - \frac{F}{V}\,C - k_2\,C^2 \tag{8.29}$$

and we can calculate the steady-state concentrations from $dC/dt = 0$

$$k_2\,C_s^2 + \frac{F}{V}\,C_s - \frac{F}{V}\,C_{is} = 0 \tag{8.30}$$

Notice that (8.30) is quadratic in C_s, and will always have one positive and one negative root (the reader should verify this by using the quadratic formula). Obviously, only the positive root makes physical sense.

Now, the problem with obtaining an analytical solution to (8.29) is the nonlinear term. We can use the linearization technique from Chapter 5.

$$\frac{d(C - C_s)}{dt} \approx \left.\frac{\partial f}{\partial C}\right|_{ss} (C - C_s) + \left.\frac{\partial f}{\partial C_i}\right|_{ss} (C_i - C_{is})$$

to find that:

$$\frac{1}{\left(\dfrac{F}{V} + 2\,k_2\,C_s\right)} \frac{d(C - C_s)}{dt} + (C - C_s) = \frac{\dfrac{F}{V}}{\left(\dfrac{F}{V} + 2\,k_2\,C_s\right)} (C_i - C_{is}) \tag{8.31}$$

Again, we have a first-order, linear relationship, where:

$$\text{process gain} = k = \frac{\dfrac{F}{V}}{\left(\dfrac{F}{V} + 2\,k_2\,C_s\right)} = \frac{1}{\left(1 + 2\,k_2\,C_s\dfrac{V}{F}\right)}$$

$$\text{time constant} = \tau = \frac{1}{\left(\dfrac{F}{V} + 2\,k_2\,C_s\right)} = \frac{\dfrac{V}{F}}{\left(1 + 2\,k_2\,C_s\dfrac{V}{F}\right)}$$

Summarizing, the parameters for each of the previous examples are shown in Table 8.1.

TABLE 8.1   Summary of Parameters from Examples

	Ex. 8.1 Mixing Tank No Rxn	Ex. 8.4 CSTR First-Order Rxn	Ex. 8.5 CSTR Second-Order Rxn
Process Gain, k	1	$\dfrac{1}{1 + \dfrac{V}{F}k_1}$	$\dfrac{1}{1 + 2\,k_2\,C_s\dfrac{V}{F}}$
Process Time Constant, τ	$\dfrac{V}{F}$	$\dfrac{\dfrac{V}{F}}{1 + \dfrac{V}{F}k_1}$	$\dfrac{\dfrac{V}{F}}{1 + 2\,k_2\,C_s\dfrac{V}{F}}$

EXAMPLE 8.6   A Numerical Study of Examples 8.1, 8.4, 8.5

Here we will perform a numerical study, using the following values:

$\dfrac{V}{F}$	=	5 min	All cases
k_1	=	0.2 min^{-1}	CSTR with first-order Rxn
k_2	=	0.32 ft^3 lbmol^{-1} min^{-1}	CSTR with second-order Rxn
C_{is}	=	1.25 lbmol ft^{-3}	All cases

Then, we can calculate the following steady-state concentrations:

C_s	=	1.25 lbmol ft^{-3}	Mixing tank with no Rxn
C_s	=	0.625 lbmol ft^{-3}	CSTR with first-order Rxn
C_s	=	0.625 lbmol ft^{-3}	CSTR with second-order Rxn

	Mixing Tank No Rxn	CSTR First-Order Rxn	CSTR Second-Order Rxn
Process Gain, k	1	0.5	0.5
Process Time Constant, τ (min)	5	2.5	2.5

For all of the examples, assume that a step change in the inlet concentration occurs at $t = 0$. That is, C_i changes from 1.25 lbmol ft^{-3} to 1.75 lbmol ft^{-3} at $t = 0$ minutes. In terms of deviation variables, this means that u increases from 0 to 0.5 lbmol ft^{-3} at $t = 0$.

   Recall that the solution for a first-order system with a step input change of magnitude A is:

$$y(t) = kA \left[1 - e^{-t/\tau} \right]$$

and since

$$y(t) = C(t) - C_s$$

our solution is

$$C(t) = C_s + kA \left[1 - e^{-t/\tau} \right] \tag{8.32}$$

For the mixing tank

$$C(t) = 1.25 + 0.5 \left[1 - e^{-t/5} \right] \tag{8.33}$$

For the CSTR with first-order Rxn

$$C(t) = 0.625 + 0.25 \left[1 - e^{-t/2.5} \right] \tag{8.34}$$

For the CSTR with second-order Rxn

$$C(t) = 0.625 + 0.25 \left[1 - e^{-t/2.5} \right] \tag{8.35}$$

Notice that solutions for the mixing tank (8.33) and the CSTR with first-order Rxn (8.34) are *exact* because these systems are inherently linear. The solution to the CSTR with second-order Rxn (8.35) is only approximate, because it is based on a linearized approximation to a nonlinear model.

   The actual response of the nonlinear model (using `ode45`) is compared with the linear solution (8.35) in Figure 8.10. Notice that the initial response is similar, but the long-term re-

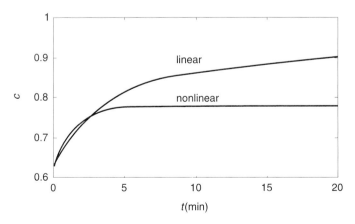

FIGURE 8.10   Reactor concentration response to a step increase in inlet concentration, for a second-order reaction.

sponse of the linear model deviates significantly from the nonlinear model. Indeed, we can calculate the long-term response without doing any numerical integration, as shown below.

Linear Model (8.35) as $t\rightarrow\infty$ $C(\infty) = 0.625 + 0.25 = \mathbf{0.8750}$

Nonlinear Model (8.30) as $t\rightarrow\infty$

$$C_s^2 + \frac{F}{Vk_2}C_s - \frac{F}{Vk_2}C_{is} = 0$$

$$C_s^2 + 0.625\,C_s - 0.625(1.75) = 0$$

The solution that makes physical sense is: $C(\infty) = \mathbf{0.7790}$

In Example 8.6 we were able to find the new steady-state for the nonlinear system by solving a single quadratic equation. For the general case, with a model composed of a set of nonlinear equations, one would need to solve a set of nonlinear algebraic equations. This would be done twice, once to find the initial steady-state, then again to find the final steady-state after a new input change.

EXAMPLE 8.7   First Order + Deadtime

The most common model for process control studies is known as a first-order + deadtime process model, and is written in the following form

$$\tau\frac{dy}{dt} + y = k\,u(t - \theta) \tag{8.36}$$

where θ is known as the time delay. Assume that $y(0) = 0$ and $u(0) = 0$. The input, u and the output y are functions of time; $u(t)$ must be specified to solve for $y(t)$.
To understand how this equation might arise, see Figure 8.11.

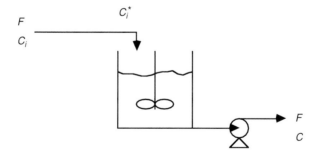

FIGURE 8.11   Mixing tank.

Notice that if the inlet pipe has a significant volume, there will be a delay between a change in the concentration at the inlet pipe and the concentration at the outlet of the pipe. The delay can be calculated as:

$$\theta = \frac{V_p}{F}$$

where V_p is the volume of the pipe. The relationship between the concentration at the exit of the pipe and the inlet of the pipe can be found by:

$$C_i^*(t) = C_i(t - \theta)$$

That is, the concentration at the exit of the pipe is equal to what the concentration at the outlet of the pipe was θ time units in the past. The modeling equation is:

$$\frac{dC}{dt} = -\frac{F}{V}C + \frac{F}{V}C_i^*(t)$$

which can be written:

$$\frac{dC}{dt} = -\frac{F}{V}C + \frac{F}{V}C_i(t - \theta)$$

which is equivalent to (8.36) when written in deviation variable form, where:

$$u = C_i - C_{is} \qquad y = C - C_s \qquad \tau = \frac{V}{F}$$

Taking the Laplace transform of (8.36) we find:

$$\tau\left[s\,Y(s) - y(0)\right] + Y(s) = k\,e^{-\theta s}\,U(s) \qquad (8.37)$$
$$\tau\,sY(s) + Y(s) = k\,e^{-\theta s}\,U(s)$$
$$(\tau s + 1)\,Y(s) = k\,e^{-\theta s}\,U(s)$$

which is most commonly written:

$$Y(s) = \frac{k\,e^{-\theta s}}{\tau s + 1}\,U(s) \qquad (8.38)$$

or,

$$Y(s) = g(s)\,U(s) \qquad (8.39)$$

where:

$$g(s) = \frac{k\,e^{-\theta s}}{\tau s + 1} \qquad (8.40)$$

Assume a step input of magnitude Δu at time $t = 0$. We know that:

$$L\left[\Delta u\right] = \frac{\Delta u}{s} \qquad (8.41)$$

and we can then write (8.38) as:

$$Y(s) = \frac{k\,e^{-\theta s}}{\tau s + 1}\,\frac{\Delta u}{s} \qquad (8.42)$$

$$Y(s) = \frac{k\Delta u\,e^{-\theta s}}{s(\tau s + 1)} \qquad (8.43)$$

$$Y(s) = k\Delta u\,e^{-\theta s}\left[\frac{1}{s} - \frac{\tau}{\tau s + 1}\right] \qquad (8.44)$$

The solution to (8.44) is then (since $L[y(t - \theta)] = e^{-\theta s}Y(s)$):

$$y(t) = 0 \qquad \text{for } 0 \leq t \leq \theta$$

$$y(t) = k\Delta u \left[1 - e^{-(t-\theta)/\tau}\right] \quad \text{for } t \geq \theta \qquad (8.45)$$

Notice that (8.45) is merely a translation of the first-order response by θ time units.

Consider the following plot (Figure 8.12) of the response of a system to a step input change of magnitude 0.5 at time $t = 0$. We see immediately that the time delay is $\theta = 5$ minutes. Since the change in output after a long period of time is $\Delta y = 1 = k \, \Delta u$, we see that $k = 2$ (units of input/output). The process time constant can be determined from the amount of time, after the delay, that it takes for 63.2% of the change to occur. In this case, the time constant is approximately 5 minutes.

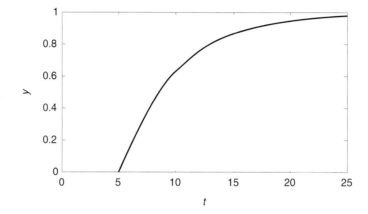

FIGURE 8.12   Response of a first-order + deadtime (5 time units) model to a step input at $t = 0$.

8.4 INTEGRATING PROCESSES

The previous examples were for self-regulating processes. If an input changed, then the process output came to a new steady-state. Another common chemical process is the integrating process, as shown in the example below.

EXAMPLE 8.8   An Integrating System

Consider a water storage tank with inlet and outlet streams that can be independently adjusted. The storage tank has a cross-sectional area of 100 ft^2. Initially, the flow in is equal to the flow out, which is 5 ft^3/min. The initial height of water in the tank is 4 ft and the height of the tank is 10 ft. At 1:00 pm the inlet flowrate is increased to 6 ft^3/min. When does the tank overflow? The material balance (assuming constant density) is

$$\frac{dV}{dt} = F_i - F_o \tag{8.46}$$

where F_i and F_o are the inlet and outlet flowrates, and V is the tank volume. Assuming a constant cross-sectional area:

$$\frac{dh}{dt} = \frac{1}{A} F_i - \frac{1}{A} F_o \tag{8.47}$$

To satisfy steady-state relationships $F_{is} = F_{os}$, so we can use the following deviation variable form:

$$\frac{d(h - h_s)}{dt} = \frac{1}{A}(F_i - F_{is}) - \frac{1}{A}(F_o - F_{os}) \tag{8.48}$$

For simplicity, let's assume that F_o is constant, then:

$$\frac{d(h - h_s)}{dt} = \frac{1}{A}(F_i - F_{is}) \tag{8.49}$$

which has the form:

$$\frac{dy}{dt} = k\,u \tag{8.50}$$

where $y = h - h_s$, $k = 1/A$ and $u = F_i - F_{is}$. Taking Laplace transforms, we find:

$$sY(s) - y(0) = k\,U(s) \tag{8.51}$$

where $y(0) = h(0) - h_s$. Assuming that we are starting from a steady-state, $y(0) = h(0) - h_s = 0$. So we can write (8.51) as:

$$sY(s) = k\,U(s)$$

or,

$$Y(s) = \frac{k}{s} U(s) \tag{8.52}$$

Using the notation Δu for the magnitude of the step increase:

$$U(s) = \frac{\Delta u}{s} \tag{8.53}$$

and

$$Y(s) = k\,\frac{\Delta u}{s^2} \tag{8.54}$$

Taking the inverse Laplace transform:

$$y(t) = L^{-1}[Y(s)] = L^{-1}\left[\frac{k\,\Delta u}{s^2}\right] \tag{8.55}$$

$$y(t) = k\,\Delta u\,t \tag{8.56}$$

Substituting back for the physical variables,

$$h - h_s = \frac{1}{A}\Delta F_i\,t \tag{8.57}$$

or,

$$h = h_s + \frac{1}{A} \Delta F_i t$$

$$h = 4 \text{ ft} + \frac{(6-5) \text{ ft}^3/\text{min}}{100 \text{ ft}^2} t$$

Solving for $h = 10$ ft

$$t = (10 \text{ ft} - 4 \text{ ft}) \frac{100 \text{ ft}^2}{(6-5) \text{ ft}^3/\text{min}} = 600 \text{ minutes}$$

$$= 10 \text{ hours}$$

Since the step change was made at 1:00 pm, the tank will overflow at 11:00 pm. A plot of tank height versus time is shown in Figure 8.13.

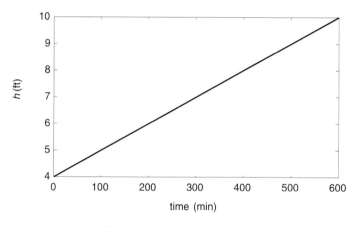

FIGURE 8.13   Integrating system.

Notice in equation (8.52) that the process transfer function has a pole at $s = 0$. This is a characteristic of an integrating system.

8.5   LEAD-LAG MODELS

Some dynamic systems, particularly involved with process control, have the following form for a transfer function model:

$$Y(s) = k \frac{\tau_n s + 1}{\tau_d s + 1} U(s) \tag{8.58}$$

Consider a step input change of magnitude Δu

$$Y(s) = k \frac{\tau_n s + 1}{\tau_d s + 1} \frac{\Delta u}{s} \tag{8.59}$$

The reader should find that time domain response is (see student exercise 11)

$$y(t) = k \, \Delta u \left[1 - \left(1 - \frac{\tau_n}{\tau_d} \right) e^{-t/\tau_d} \right] \tag{8.60}$$

A plot of (8.60) is shown in Figure 8.14, for $k\Delta u = 1$. Notice if $\tau_n > \tau_d$, the immediate increase in the output is greater than the ultimate steady-state increase, while if $\tau_n < \tau_d$, the immediate increase in the output is less than the ultimate steady-state increase.

8.5.1 Simulating Lead/Lag Transfer Functions

We have derived the step response for a lead/lag transfer function. This transfer function does not usually arise in the modeling of a physical system, but it often arises in control system design. Our desire in this section is to show how to convert a lead/lag transfer function to state-space form, so that a general simulation package can be used to integrate the corresponding ordinary differential equation.

Multiplying through by the denominator term in (8.58), we find:

$$(\tau_d s + 1) \, Y(s) = (\tau_n s + 1) \, U(s) \tag{8.61}$$

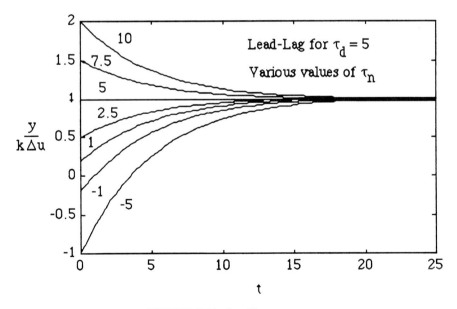

FIGURE 8.14 Lead/lag response.

Using the Laplace transform relationships,

$$\tau_d \left(\frac{dy}{dt} - y(0) \right) + y = \tau_n \left(\frac{du}{dt} - u(0) \right) + u \qquad (8.62)$$

and we know that to obtain the transfer function form, the initial conditions of all variables were assumed to be zero, so:

$$\tau_d \frac{dy}{dt} + y = \tau_n \frac{du}{dt} + u \qquad (8.63)$$

We cannot solve (8.63) by using a general purpose integrator, because it is not in the standard form of $dx/dt = f(x)$. Our goal now is to define a new variable that will allow us to use a standard integrator.

Rearrange (8.63) to find:

$$\tau_d \frac{dy}{dt} - \tau_n \frac{du}{dt} = -y + u \qquad (8.64)$$

We see that we can define a new variable, x, so that:

$$x = \tau_d y - \tau_n u \qquad (8.65)$$

and since τ_d and τ_n are constants, we can take the derivative of (8.65) with respect to time to find:

$$\frac{dx}{dt} = \tau_d \frac{dy}{dt} - \tau_n \frac{du}{dt} \qquad (8.66)$$

Substituting the righthand side of (8.66) for the lefthand side of (8.64), we find:

$$\frac{dx}{dt} = -y + u \qquad (8.67)$$

Now, we must solve (8.65) to find y as a function of x, to obtain:

$$y = \frac{1}{\tau_d} x + \frac{\tau_n}{\tau_d} u \qquad (8.68)$$

which we substitute into (8.67) to find:

$$\frac{dx}{dt} = -\frac{1}{\tau_d} x + \left(1 - \frac{\tau_n}{\tau_d} \right) u \qquad (8.69)$$

and we see that we have the standard state-space form:

$$\dot{\mathbf{x}} = \mathbf{A}\,\mathbf{x} + \mathbf{B}\,\mathbf{u}$$

$$\mathbf{y} = \mathbf{C}\,\mathbf{x} + \mathbf{D}\,\mathbf{u}$$

except that (8.68) and (8.69) consist of scalars:

$$\frac{dx}{dt} = a\,x + b\,u \tag{8.70}$$

$$y = c\,x + d\,u \tag{8.71}$$

where

$$a = -\frac{1}{\tau_d} \qquad b = \left(1 - \frac{\tau_d}{\tau_d}\right)$$

$$c = \frac{1}{\tau_d} \qquad d = \frac{\tau_n}{\tau_d}$$

We will see in Chapter 11 how (8.70) and (8.71) can be used within the context of a block diagram.

SUMMARY

We have studied the response of a number of processes that have denominators of transfer function models that are first-order in the Laplace variable, s. The systems were: first-order, first-order + deadtime, integrating, and lead/lag. Most chemical processes can be represented by a cascade of these types of modes. We found that stirred tank chemical reactors are linear first-order processes, as long as they have first-order kinetics (or no reaction) and the input flowrate is not changing.

For first-order and first-order + time-delay transfer functions, we discussed how to estimate the parameters (which always have units associated with them) by applying a known step input to the process and observing the response. First-order + time-delay models are commonly used in control system design.

In the next chapter we study the transient response behavior of second- and higher-order systems.

STUDENT EXERCISES

1. As a process engineer, you are attempting to estimate the model parameters for a process that you believe is first-order (with no deadtime). At 3:00 pm, you make a step input change to the process. At 4:00 pm, the process output has reached 80% of its final change.

    What is the time constant of the process?

2. Consider a water storage tank with inlet and outlet streams that can be independently adjusted. The storage tank has a cross-sectional area of 100 ft². Initially, the flow in is equal to the flow out, which is 5 ft³/min. The initial height of water in the tank is 4 ft and the height of the tank is 10 ft. At $t = 0$ a ramp increase in the inlet flowrate is made, so that $F_i(t) = 5 + 0.25t$ where the flowrate units are ft³/min.

How long does it take the tank to overflow? Solve using Laplace transforms. Obtain a general expression for systems modeled in deviation variable form by:

$$\frac{dy}{dt} = k\,u(t)$$

where:
$$u(t) = a\,t$$

3. Write a differential equation which corresponds to the following input-output transfer function relationship:

$$y(s) = k\,\frac{\tau_n s + 1}{\tau_d s + 1}\,u(s)$$

4. Consider a chemical reactor that has zero-order kinetics, that is, the rate of reaction per unit volume is a constant (a zero-order kinetic parameter) that does not depend on concentration. Compare this model with that of a stirred tank mixer, and a stirred tank reactor with first-order kinetics. Perform a numerical study, related to Example 8.5, by finding the zero-order parameter that yields the same steady-state concentration as the first-order kinetic model.

5. A process operator makes a step change on an input variable at 2:00 pm and discovers no output response is observed until after 2:10 pm. She finds that the output is 90% of the way to its final steady-state at 2:45 pm. You believe that this is a first order + deadtime process.

time	input	output
1:00 pm	200 lb/hr	100°F
1:30 pm	200 lb/hr	100°F
1:59 pm	200 lb/hr	100°F
2:00 pm	225 lb/hr	100°F
2:10 pm	225 lb/hr	100°F
2:45 pm	225 lb/hr	91°F
after 5:00 pm	225 lb/hr	90°F

(i)   What is the deadtime for this process (show units)?
(ii) What is the time constant for this process (show units)?
(iii) What is the process gain (show units)?

6. As the process engineer for an operating unit in a process plant, you are trying to get a "feel" for the dynamic characteristics of a particular process. You have a discussion with the operator about a process (which you feel is first-order) that uses steam flowrate as an input variable, and process temperature as a measured variable. After the steam flowrate is increased from 1000 lb/hr to 1100 lb/hr (quickly), the process fluid temperature changes from 100°F (the initial steady-state) to 110°F in 30 minutes. The temperature eventually reaches a new steady-state value of 120°F.
(i) Find the process gain (show units).
(ii) Find the process time constant (show units).

7. A process input is:

$$u(t) = 0 \qquad \text{for } t < 0$$
$$u(t) = 1 - e^{-t} \quad \text{for } t > 0$$

The process transfer function is:

$$g(s) = \frac{2.5}{12s + 1}$$

Find the time domain output, $y(t)$. Plot both the input and the output.

8. Consider the mixing process shown below, where a portion of the feed stream bypasses the mixing tank.

   a. Show that the process has a lead/lag transfer function, if the input is C_f and the output is C_3. (*Hint:* Write a dynamic balance around the tank and a static balance around the mixing point (after the tank outlet). Use deviation variable form.)

   b. Let $F = 2$ m³/min, $F_1 = 1$ m³/min, $C_f = 1$ kgmol/m³, and $V = 10$ m³. Find the state-space model and the transfer function representing this system.

   c. Consider a step increase of C_f to 1.5 kgmol/m³. Find the response in C_3 to this change.

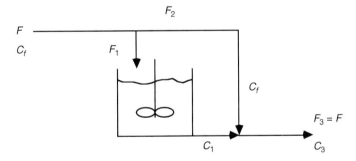

9. *Comparison of Impulse and Pulse Responses.* Consider a tank with constant cross-sectional area, $A_t = 1$ m², and assume that the flow out of the tank is a linear function of the height of liquid in the tank. The steady-state values of tank height and flowrate are 1 meter and 1 m³/hr, respectively. Find the impulse response of tank height if 1 m³ (in addition to the constant steady-state flow) is instantanously dumped in the tank. Compare this with several pulse responses, where the additional 1 m³ is added over 0.05, 0.1, and 0.15 hour periods.

10. Consider a chemical reactor where a step change in coolant flowrate from 10 gal/min to 12 gal/min (at $t = 0$) causes the change in reactor temperature shown in the figure below.

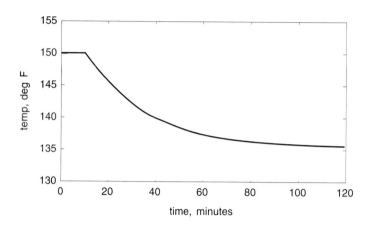

Find the gain, time constant, and time-delay for this system.

11. For step response of the lead/lag transfer function:

$$Y(s) = k\,\frac{\tau_n s + 1}{\tau_d s + 1}\,\frac{\Delta u}{s}$$

Show that time domain response is

$$y(t) = k\,\Delta u\left[1 - \left(1 - \frac{\tau_n}{\tau_d}\right)e^{-t/\tau_d}\right]$$

TRANSFER FUNCTION ANALYSIS 9
OF HIGHER-ORDER SYSTEMS

After studying this chapter, the reader should be able to:

- Understand the dynamic behavior of second-order systems.
- Understand the effect of poles and zeros on the response for higher-order systems.
- Use the Padé approximation for time-delays.
- Understand the concept of inverse response.
- Understand how to simulate transfer function models using ODE solvers that require sets of first-order ODEs.
- Use the MATLAB routine `tf2ss` to convert from transfer function to state-space form.
- Use the MATLAB routines `step` and `impulse`.

The major sections are:

The dynamic behavior of *first-order* systems was studied in Chapter 8. In this chapter, we present results for *higher-order* systems and show how to use standard numerical integration routines for time domain simulation of these models. We first study second-order systems, then generalize our results to higher-order systems.

9.1 RESPONSES OF SECOND-ORDER SYSTEMS

Consider a linear second-order ODE, with constant parameters:

$$a_2 \frac{d^2y}{dt^2} + a_1 \frac{dy}{dt} + a_0 y = b\,u(t) \tag{9.1}$$

This is often written in the form:

$$\tau^2 \frac{d^2y}{dt^2} + 2\zeta\tau \frac{dy}{dt} + y = k\,u(t) \tag{9.2}$$

where (obviously $a_0 \neq 0$):

$$\tau^2 = \frac{a_2}{a_0} \quad 2\zeta\tau = \frac{a_1}{a_0} \quad k = \frac{b}{a_0}$$

where the parameters are:

  k = gain (units of output/input)
  ζ = damping factor (dimensionless)
  τ = natural period (units of time)

We discussed in Chapter 6 that single nth order ODEs do not naturally arise in chemical processes. The second-order model shown in (9.1) or (9.2) generally arises by changing a set of two first-order equations (state-space model) to a single second-order equation. For a given second-order ODE, there are an infinite number of sets of two first-order (state-space) models that are equivalent.

Taking the Laplace transform of (9.2):

$$\tau^2 \left[s^2\,Y(s) - sy(0) - \dot{y}(0) \right] + 2\zeta\tau \left[sY(s) - y(0) \right] + Y(s) = kU(s)$$

where $Y(s)$ indicates the Laplace transformed variable.

Assuming initial conditions are zero, that is $\dot{y}(0) = y(0) = 0$, we find:

$$Y(s) = \frac{k}{\tau^2 s^2 + 2\zeta\tau s + 1}\,U(s) \tag{9.3}$$

which can be represented as:

$$Y(s) = g(s)\,U(s)$$

The *characteristic equation* of the second-order transfer function is $\tau^2 s^2 + 2\zeta\tau s + 1$. We can find the roots (also known as the *poles*) by using the quadratic formula:

TABLE 9.1   Characteristic Behavior of Second-Order Transfer Functions

Case	Damping Factor	Pole Location	Characteristic Behavior
I	$\zeta > 1$	2 real, distinct poles	overdamped
II	$\zeta = 1$	2 real, equal poles	critically damped
III	$\zeta < 1$	2 complex conjugate poles	underdamped

$$p_i = \frac{-2\zeta\tau \pm \sqrt{4\zeta^2\tau^2 - 4\tau^2}}{2\tau^2} \tag{9.4}$$

which yields the following values for the roots:

$$p_1 = -\frac{\zeta}{\tau} + \frac{\sqrt{4\tau^2(\zeta^2 - 1)}}{2\tau^2} = -\frac{\zeta}{\tau} + \frac{\sqrt{\zeta^2 - 1}}{\tau} \tag{9.5}$$

$$p_2 = \qquad\qquad = -\frac{\zeta}{\tau} - \frac{\sqrt{\zeta^2 - 1}}{\tau} \tag{9.6}$$

The following analysis assumes that $\zeta > 0$ and $\tau > 0$. This implies that the real portions of p_1 and p_2 are negative and, therefore, the system is stable. The three possible cases are shown in Table 9.1.

9.1.1   Step Responses

Now, we consider the dynamic response of second-order systems to step inputs ($U(s) = \Delta u/s$):

$$Y(s) = \frac{k}{\tau^2 s^2 + 2\zeta\tau s + 1} \frac{\Delta u}{s} \tag{9.7}$$

where Δu represents the magnitude of the step change.

CASE 1   Overdamped ($\zeta > 1$)

Since $\zeta > 1$, we can see that the two roots will be real and distinct. Also, since we assumed that $\tau > 0$, the system is stable (the roots are less than zero, since we are assured that $\sqrt{\zeta^2 - 1} < \zeta$). We factor the polynomial $\tau^2 s^2 + 2\zeta\tau s + 1$ into the following form:

$$\tau^2 s^2 + 2\zeta\tau s + 1 = (\tau_1 s + 1)(\tau_2 s + 1) \tag{9.8}$$

We see immediately from (9.8) that the poles (values of s where the polynomial = 0) are:

$$p_1 = -1/\tau_1 \quad p_2 = -1/\tau_2 \tag{9.9}$$

from (9.5), (9.6) and (9.8) we find:

$$p_1 = -1/\tau_1 = -\zeta/\tau + \sqrt{(\zeta^2 - 1)}/\tau$$

which gives the following value for the first time constant:

$$\tau_1 = \frac{\tau}{\zeta - \sqrt{\zeta^2 - 1}} \tag{9.10}$$

Also, we find the second pole:

$$p_2 = -1/\tau_2 = -\zeta/\tau - \sqrt{\zeta^2 - 1}/\tau$$

which gives the following value for the second time constant:

$$\tau_2 = \frac{\tau}{\zeta + \sqrt{\zeta^2 - 1}} \tag{9.11}$$

Expanding the righthand side of (9.8),

$$\tau^2 s^2 + 2\zeta\tau s + 1 = \tau_1\tau_2 s^2 + (\tau_1 + \tau_2) s + 1 \tag{9.12}$$

we can write:

$$\tau^2 = \tau_1\tau_2 \quad \text{and} \quad 2\zeta\tau = \tau_1 + \tau_2$$

which lead to the relationships

$$\tau = \sqrt{\tau_1\tau_2} \tag{9.13}$$

$$\zeta = \frac{\tau_1 + \tau_2}{2\sqrt{\tau_1\tau_2}} \tag{9.14}$$

We can derive (see student exercise 1a) the following solution for step responses of overdamped systems

---

Overdamped, $\zeta > 1$

$$y(t) = k\Delta u \left[1 + \frac{\tau_1 e^{-t/\tau_1} - \tau_2 e^{-t/\tau_2}}{\tau_2 - \tau_1} \right] \tag{9.15}$$

where $\quad \tau_1 = \dfrac{\tau}{\zeta - \sqrt{\zeta^2 - 1}} \quad$ and $\quad \tau_2 = \dfrac{\tau}{\zeta + \sqrt{\zeta^2 - 1}}$

Note that, as in the case of first-order systems, we can divide by $k\Delta u$ to develop a dimensionless output. Also, the dimensionless time is t/τ and we can plot curves for dimensionless output as a function of ζ. This is done in Figure 9.1, which includes the critically damped case, as discussed next. Most chemical processes exhibit overdamped behavior. The critically damped step response is also shown in Figure 9.1 (curve with $\zeta = 1$).

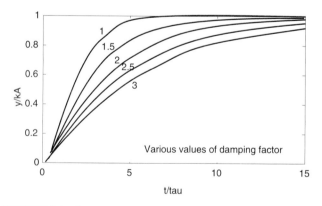

FIGURE 9.1 Step response of a second-order overdamped system.

CASE 2   Critically Damped ($\zeta = 1$)

The transition between overdamped and underdamped is known as critically damped. We can derive the following for the step response of a critically damped system (see student exercise 1b)

> **Critically damped, $\zeta = 1$ [Repeated poles]**
>
> $$y(t) = k\Delta u \left(1 - \left(1 + \frac{t}{\tau}\right) e^{-t/\tau}\right)$$
>
> (9.16)

Notice that the main difference between overdamped (or critically damped) step responses and first-order step responses is that the second-order step responses have an "S" shape with a maximum slope at an inflection point, whereas the first-order responses have their maximum slope initially.

The initial behavior for a step change is really dictated by the *relative order* of the system. The relative order is the difference between the orders of the numerator and denominator polynomials. If the relative order is 1, then output response has a non-zero slope at the time of the step input; the step response of a system with a relative order greater than 1 has a zero slope at the time of the step input.

CASE 3   Underdamped ($\zeta < 1$)

For $\zeta < 1$, from (9.5) and (9.6), we find that the poles are complex:

$$p = -\frac{\zeta}{\tau} \pm \frac{\sqrt{\zeta^2 - 1}}{\tau} = -\frac{\zeta}{\tau} \pm j \frac{\sqrt{(1 - \zeta^2)}}{\tau}$$

which is written in terms of the real and imaginary contributions:

$$p = \alpha \pm j\beta$$

where: $\alpha = -\dfrac{\zeta}{\tau}$ $\beta = \dfrac{\sqrt{(1 - \zeta^2)}}{\tau}$

We can derive the following step response for an underdamped system (see student exercise 1):

> **Underdamped, ($\zeta < 1$) [Complex poles]**
>
> $$y(t) = k\Delta u \left(1 - \frac{1}{\sqrt{1 - \zeta^2}} e^{-\zeta t/\tau} \sin(\beta t + \phi)\right)$$
>
> (9.17)
>
> where $\beta = \dfrac{\sqrt{1 - \zeta^2}}{\tau}$ $\phi = \tan^{-1}\dfrac{\sqrt{1 - \zeta^2}}{\zeta}$

Again, dividing by $k\Delta u$, we can produce the plot shown in Figure 9.2.

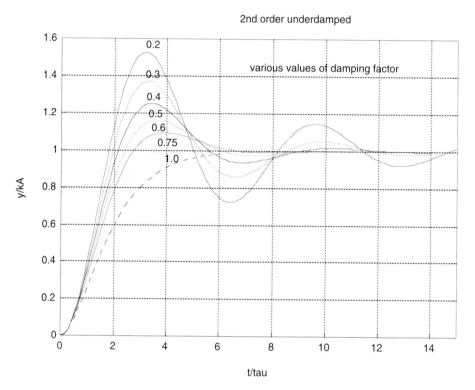

FIGURE 9.2   Step response of a second-order underdamped system as a function of the damping factor (ζ).

A number of insights can be obtained from Figure 9.2 and an analysis of the step response equations. Notice that the poles for the second-order system can be written:

$$p = [-\zeta \pm j \sqrt{(1 - \zeta^2)}]\frac{1}{\tau}$$

Observe that, for smaller ζ, the response is more oscillatory. For $\zeta < 1$, the ratio of the imaginary portion to the real portion of the pole is:

$$\frac{\text{imaginary}}{\text{real}} = \frac{\sqrt{(1 - \zeta^2)}}{\zeta}$$

As the imaginary/real ratio gets larger the response becomes more oscillatory. We also notice that a decreasing τ corresponds to a larger negative value for the real portion. As the real portion becomes larger in magnitude (more negative) the response becomes faster. We use these insights to interpret pole/zero plots in Section 9.3.

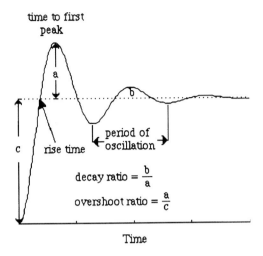

FIGURE 9.3   Step response characteristics of underdamped second-order processes.

9.1.2   Underdamped Step Response Characteristics

The following common measures of underdamped second-order step responses are shown on Figure 9.3 and defined below: (1) rise time, (2) time to first peak, (3) overshoot, (4) decay ratio, (5) period of oscillation.

Rise time. The amount of time it takes to first reach the new steady-state value.

Time to first peak. The time required to reach the first peak. Notice that there are an infinite number of peaks.

Overshoot. The distance between the first peak and the new steady-state. Usually expressed as the overshoot ratio, as shown in the figure.

Decay ratio. A measure of how rapidly the oscillations are decreasing. A *b/a* ratio of 1/4 is commonly called "quarter wave damping".

Period of oscillation. The time between successive peaks.

The following example shows how to use Figure 9.2 to estimate these values.

EXAMPLE 9.1   Underdamped Second-Order System

Consider the following transfer function, subject to a unit step ($\Delta u = 1$) input change (assume time units are minutes):

$$g(s) = \frac{5}{4s^2 + 0.8s + 1}$$

Find the (1) rise time, (2) time to first peak, (3) overshoot, (4) decay ratio, (5) period of oscillation, (6) value of $y(t)$ at the peak time.

Our first step is to calculate the system parameters. We can see that:

$$k = 5$$
$$\tau^2 = 4 \text{ so } \tau = 2$$
$$2\zeta\tau = 0.8 \text{ so } \zeta = \frac{0.8}{2\tau} = \frac{0.8}{4} = 0.2$$

We use Figure 9.2 as the basis for the following calculations.

1. The rise time for $\zeta = 0.2$ is $\frac{t_r}{\tau} \approx 1.8$, so $t_r = 1.8\,\tau = 3.6$ minutes

2. The time to first peak for $\zeta = 0.2$ is $\frac{t_p}{\tau} \approx 3.2$, so $t_p = 3.2\,\tau = 6.4$ minutes

3. The overshoot ratio is $\frac{1.53 - 1}{1} = 0.53$.

4. The decay ratio is $\frac{1.15 - 1}{1.53 - 1} = 0.3$

5. The period of oscillation is $\frac{t_{osc}}{\tau} \approx 9.6 - 3.3$, so $t_{osc} = 6.3\,\tau = 12.6$ minutes.

6. The value of $y(t_p)$ is $\frac{y}{k\Delta u} = 1.53$, so $y = 1.53\,k\Delta u = 1.53(5) = 7.65$.

Although equation (9.17) can be used to solve the previous example, it is often easier to use the dimensionless plot (Figure 9.2).

9.1.3 Impulse Responses

Now, we consider the dynamic response of second-order systems to impulse inputs.

$$Y(s) = \frac{k}{\tau^2 s^2 + 2\zeta\tau s + 1} A\delta$$

where A represents the magnitude of the impulse.

CASE 1   Overdamped ($\zeta > 1$)

The time domain solution for the overdamped case is (see student exercise 2a):

$$y(t)/kA = \frac{1}{\tau} \frac{1}{\sqrt{\zeta^2 - 1}} e^{-\zeta t/\tau} \sinh\left(\sqrt{1 - \zeta^2}\,\frac{t}{\tau}\right)$$

where $y(t)/kA$ is the dimensionless output and t/τ is the dimensionless time.

CASE 2   Critically Damped ($\zeta = 1$)

The impulse response for the critically damped case is (see student exercise 2b):

$$y(t)/kA = \frac{t}{\tau^2} e^{-t/\tau}$$

CASE 3   Underdamped ($\zeta < 1$)

The time domain solution for the underdamped case is (see student exercise 2c):

$$y(t)/kA = \frac{1}{\tau} \frac{1}{\sqrt{1 - \zeta^2}} e^{-\zeta t/\tau} \sin\left(\sqrt{1 - \zeta^2}\, \frac{t}{\tau}\right)$$

The impulse responses as a function of ζ are shown in Figure 9.4.

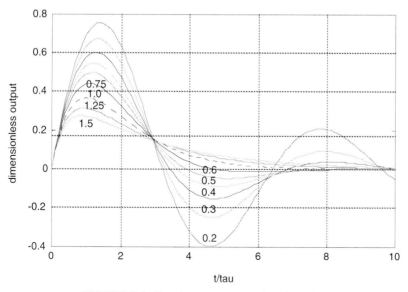

FIGURE 9.4   Impulse response as a function of ζ.

The underdamped responses show characteristic oscillatory behavior.

9.1.4 Response to Sine Inputs

Consider the case where the input is a sine function, with amplitude A and frequency ω:

$$u(t) = A \sin \omega t$$

The Laplace transform is:

$$U(s) = \frac{A\omega}{(s^2 + \omega^2)}$$

when applied to the second-order transfer function and inverted to the time domain, the response after a long period of time is the periodic function:

$$y(t) = \frac{kA}{\sqrt{(1 - \tau^2\omega^2)^2 + (2\zeta\tau\omega)^2}} \sin(\omega t + \phi) \tag{9.18}$$

where:

$$\phi = \tan^{-1}\left(\frac{-2\zeta\tau\omega}{1 - \tau^2\omega^2}\right) \tag{9.19}$$

(see student exercise 3). The amplitude of the output is:

$$\frac{kA}{\sqrt{(1 - \tau^2\omega^2)^2 + (2\zeta\tau\omega)^2}}$$

and the phase angle is ϕ. Often, system behavior is discussed in terms of an amplitude ratio, which is the amplitude of the output divided by the amplitude of the input. The amplitude ratio is:

$$\frac{k}{\sqrt{(1 - \tau^2\omega^2)^2 + (2\zeta\tau\omega)^2}}$$

These relationships are used in Example 9.2.

EXAMPLE 9.2   Sine Forcing of Second-Order Systems

Consider the following system:

$$g(s) = \frac{1}{s^2 + 0.2s + 1}$$

A low frequency sine forcing ($\omega = 0.1$ min^{-1}) yields the input/output response shown in Figure 9.5.

Notice that the output lags slightly behind the input, and the amplitude of the output is slightly smaller than the input amplitude. Contrast this result with the following case of a high frequency input.

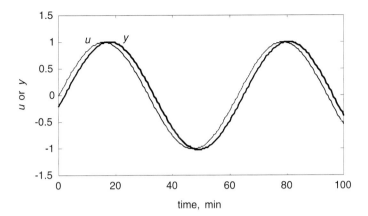

FIGURE 9.5   Low frequency sine input response.

A high frequency sine forcing ($\omega = 5$ min^{-1}) yields the input/output response shown in Figure 9.6.

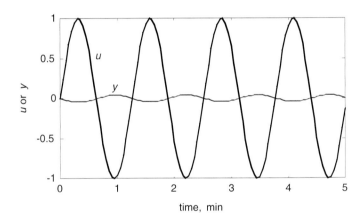

FIGURE 9.6   High frequency sine input response.

Notice that the output lags significantly behind the input, and the amplitude of the output is much smaller than the input amplitude.

A particularly interesting type of behavior that can occur with second-order underdamped systems is known as *resonance peaking,* which occurs in intermediate frequency ranges as shown in Figure 9.7, where a frequency of 1 rad/min is used.

Here the output amplitude is significantly higher than the input amplitude although the input/output gain is 1. At lower (Figure 9.5) and higher (Figure 9.6) frequencies the output had a lower amplitude than the input, while at an intermediate frequency (Figure 9.7) the output had a higher amplitude than the input. This phenomena can only happen in systems with complex roots.

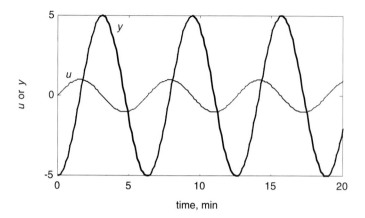

FIGURE 9.7 Resonance peaking phenomenon.

The concept of phase angle is illustrated by Figures 9.5 through 9.7. At low frequencies (Figure 9.5) the output barely lags the input, and therefore has a phase lag of almost 0 deg. At intermediate frequencies (Figure 9.7) the output lags the input by 90°, and at high frequencies (Figure 9.6) the output lags the input by almost 180°. Also note that the notion of "high," "intermediate," and "low" frequencies is relative (dependent on τ). Low, medium, and high frequencies correspond roughly to $\omega\tau = 0.1$, 1, and 10, respectively.

The method of sine-forcing a system is used in the analysis of feedback control systems and is known as frequency response analysis. Bode diagrams are used to plot the amplitude and phase angle as a function of frequency. We do not provide further analysis here, but refer the reader to any textbook on process control for more detail.

9.2 SECOND-ORDER SYSTEMS WITH NUMERATOR DYNAMICS

The previous discussion involved pure second-order systems. Consider now, a second-order system with numerator dynamics with the gain/time constant form:

$$y(s) = \frac{k(\tau_n s + 1)}{(\tau_1 s + 1)(\tau_2 s + 1)} u(s) \tag{9.20}$$

The pole-zero form is:

$$y(s) = \frac{k_{pz}(s - z_1)}{(s - p_1)(s - p_1)} u(s)$$

where:

$$k_{pz} = \frac{k\tau_n}{\tau_1\tau_2} \quad p_1 = -\frac{1}{\tau_1} \quad p_2 = -\frac{1}{\tau_2}$$

The gain/time constant form has the following time domain response to a step input (see student exercise 4):

$$y(t) = k\Delta u\left[1 + \frac{\tau_n - \tau_1}{\tau_1 - \tau_2}e^{-t/\tau_1}\frac{\tau_n - \tau_2}{\tau_2 - \tau_1}e^{-t/\tau_2} \right] \tag{9.21}$$

The reader should show that, if $\tau_n = \tau_2$, the response is the same as a first-order process.

EXAMPLE 9.3   Consider the Following Transfer Function

$$y(s) = \frac{k(\tau_n s + 1)}{(3s + 1)(15s + 1)}u(s) \tag{9.22}$$

The step responses are shown in Figure 9.8. Notice that negative numerator time constants yield a step response that initially decreases before increasing to the final steady-state. This type of response is known as *inverse response* and causes tough challenges for process control systems.

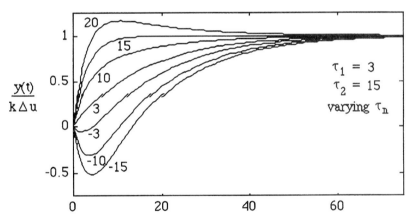

FIGURE 9.8 Step responses of a second-order system with numerator dynamics.

Notice also that a numerator time constant that is greater than the denominator time constant causes overshoot before settling to the final steady-state. Also notice that the inverse response becomes "deeper" as the process zero $(-1/\tau_n)$ approaches a value of zero from the right.

9.3 THE EFFECT OF POLE-ZERO LOCATIONS ON SYSTEM STEP RESPONSES

There are a number of different ways to represent process transfer functions. The "gain-time constant" form is:

$$g(s) = \frac{k(\tau_{n1}s + 1)(\tau_{n2}s + 1)...(\tau_{nm}s + 1)}{(\tau_{d1}s + 1)(\tau_{d2}s + 1)...(\tau_{dn}s + 1)} \tag{9.23}$$

where τ_{ni} is a numerator time constant and τ_{di} is a numerator time constant.

The "polynomial" form is

$$g(s) = \frac{(b_m s^m + b_{m-1}s^{m-1} + ... + b_1 s + b_0)}{a_n s^n + a_{n-1}s^{n-1} + ... + a_1 s + a_0} \tag{9.24}$$

The values of s that cause the numerator of (9.23) or (9.24) to equal zero are known as the "zeros" of the transfer function. The values of s that cause the denominator of (9.23) or (9.24) to equal zero are known as the "poles" of the transfer function.

The "pole-zero" form is:

$$g_{pz}(s) = \frac{k_{pz}(s - z_1)(s - z_2)...(s - z_m)}{(s - p_1)(s - p_2)...(s - p_n)} \tag{9.25}$$

where:

$$k_{pz} = k \frac{\prod_{i=1}^{n} (-p_i)}{\prod_{i=1}^{m} (-z_i)} \tag{9.26}$$

The notation $\prod_{i=1}^{n} (-p_i)$ is shorthand for $(-p_1)(-p_2)...(-p_n)$.

Notice also that the poles are

$$p_i = -\frac{1}{\tau_{di}} \tag{9.27}$$

and the zero is

$$z_i = -\frac{1}{\tau_{ni}} \tag{9.28}$$

and that complex poles (or zeros) must occur in complex conjugate pairs.

EXAMPLE 9.4   Comparison of Various Transfer Function Forms

Consider a transfer function with the following gain-time constant form:

$$g(s) = \frac{2(-10s + 1)}{(3s + 1)(15s + 1)}$$

The polynomial form is:

$$g(s) = \frac{-20s + 2}{45s^2 + 18s + 1}$$

The gain-polynomial form is:

$$g(s) = \frac{2(-10s + 1)}{(45s^2 + 18s + 1)}$$

and the pole-zero form is:

$$g_{pz}(s) = \left(-\frac{4}{9}\right) \frac{\left(s - \frac{1}{10}\right)}{\left(s + \frac{1}{3}\right)\left(s + \frac{1}{15}\right)}$$

The zero is 0.1, and the poles are $-1/3$ and $-1/15$.

Notice that the zero for Example 9.4 is positive. A positive zero is called a right-half-plane (RHP) zero, because it appears in the right half of the complex plane. Right-half-plane zeros have a characteristic *inverse response,* as shown in Figure 9.9.

Also notice that the poles are negative (left-half-plane), indicating a stable process. Right-half-plane poles (positive poles) are unstable. Recall that complex poles will yield an oscillatory response. A pole-zero plot of the transfer function in Example 9.4 is shown in Figure 9.10 (the pole locations are $(-1/3,0)$, $(-1/15,0)$ and the zero location is $(0.1,0)$; the coordinates are (real,imaginary)). For this system, there is no imaginary component and the poles and zeros lie on the real axis (Figure 9.10).

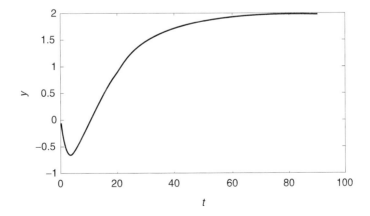

FIGURE 9.9   Inverse response.

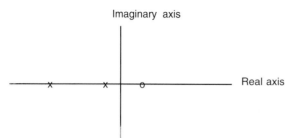

FIGURE 9.10   Pole-zero location plot for Example 9.4 (x-poles, o-zero).

As poles move further to the left they yield a faster response, while increasing the magnitude of the imaginary portion makes the response more oscillatory. This behavior is summarized in Figure 9.11. Recall also that a process with a pole at the origin (and none in the right-half-plane) is known as an *integrating* system, that is the system never settles to a steady-state when a step input change is made.

Multiple right-half-plane zeros cause multiple "changes in direction"; for example, with two RHP zeros, the step response is initially in one direction, switches direction, then switches back to the initial direction.

9.4   PADÉ APPROXIMATION FOR DEADTIME

Recall that the Laplace transfer function for a pure time-delay is $e^{-\theta s}$ where θ is the time-delay. This is an irrational transfer function; an approximation that is rational and often provides an adequate representation of the deadtime is known as the Padé approximation.

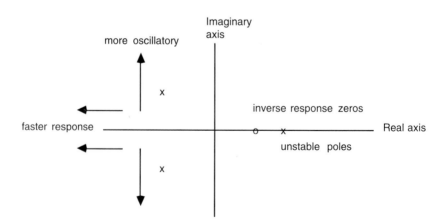

FIGURE 9.11   Effect of pole-zero location on dynamic behavior (x-poles, o-zero). As poles become more negative, the response is faster. As the imaginary/real ratio increases, the response becomes more oscillatory.

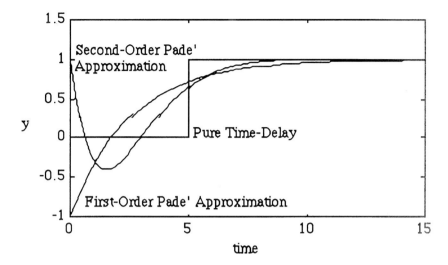

FIGURE 9.12 Comparison of step responses for pure time-delay with first-order and second-order Padé approximations. Deadtime = 5.

The first-order Padé approximation is

$$e^{-\theta s} \approx \frac{1 - \dfrac{\theta}{2}s}{1 + \dfrac{\theta}{2}s} \tag{9.29}$$

The second-order Padé approximation is

$$e^{-\theta s} \approx \frac{1 - \dfrac{\theta}{2}s + \dfrac{\theta^2}{12}s^2}{1 + \dfrac{\theta}{2}s + \dfrac{\theta^2}{12}s^2} \tag{9.30}$$

A comparison of the step responses of first and second-order Padé approximations with pure time delay are shown in Figure 9.12.

EXAMPLE 9.5   Comparison of the Padé Approximations for Deadtime

Consider the following first-order + deadtime transfer function

$$g(s) = \frac{e^{-5s}}{5s + 1}$$

The first-order Padé approximation yields the following transfer function

$$g_1(s) = \frac{-2.5s + 1}{12.5s^2 + 7.5s + 1}$$

and the second-order Padé approximation yields

$$g_2(s) = \frac{2.0833s^2 - 2.5s + 1}{10.4167s^3 + 14.5833s^2 + 7.5s + 1}$$

a comparison of the step responses of $g(s)$, $g_1(s)$ and $g_2(s)$ is shown in Figure 9.13. Notice that the first-order approximation has an inverse response, while the second-order approximation has a "double inverse response." The reader should find that there is a single positive zero for $g_1(s)$ and there are two positive, complex-conjugate zeros of the numerator transfer function of $g_2(s)$.

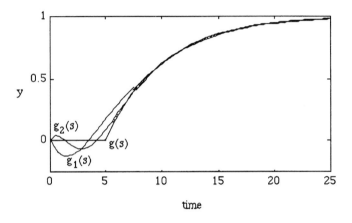

FIGURE 9.13   Comparison of first-order + deadtime response with first- and second-order Padé approximations for deadtime.

Most ordinary differential equation numerical integrators (including `ode45`) require pure differential equations (with no time-delays). If you have a system of differential equations which has time-delays, the Padé approximation can be used to convert them to delay-free differential equations, which can then be numerically integrated. See student exercise 28 as an example.

One of the many advantages to using SIMULINK is that time-delays are easily handled so that no approximation is required.

9.5   CONVERTING THE TRANSFER FUNCTION MODEL TO STATE-SPACE FORM

In this section we show one way to convert the input-output transfer function model to state-space form. Although the Laplace domain is used for analysis, the state-space form will normally be used for time domain simulations. Consider the transfer function relationship:

$$y(s) = \frac{k}{\tau^2 s^2 \zeta \tau + 2\zeta\tau s + 1} u(s)$$

which arises from the following equation:

$$\tau^2 \frac{d^2y}{dt^2} + 2\zeta\tau \frac{dy}{dt} + y = k\,u(t) \tag{9.31}$$

Let: $$x_1 = y \tag{9.32}$$

and: $$x_2 = \dot{x}_1 \tag{9.33}$$

so: $$\ddot{y} = \ddot{x}_1 = \dot{x}_2 \tag{9.34}$$

Divide (9.31) by τ^2 to obtain:

$$\frac{d^2y}{dt^2} + \frac{2\zeta}{\tau} \frac{dy}{dt} + \frac{1}{\tau^2} y = \frac{k}{\tau^2} u(t)$$

which we can write as:

$$\frac{d^2y}{dt^2} = -\frac{2\zeta}{\tau} \frac{dy}{dt} - \frac{1}{\tau^2} y + \frac{k}{\tau^2} u(t)$$

or,

$$\dot{x}_2 = -\frac{2\zeta}{\tau} x_2 - \frac{1}{\tau^2} x_1 + \frac{k}{\tau^2} u(t)$$

and since:

$$\dot{x}_1 = x_2$$

and we can write in the state-space form

$$\begin{bmatrix} \dot{x}_1 \\ \dot{x}_2 \end{bmatrix} = \begin{bmatrix} 0 & 1 \\ -\frac{1}{\tau^2} & -\frac{2\zeta}{\tau} \end{bmatrix} \begin{bmatrix} x_1 \\ x_2 \end{bmatrix} + \begin{bmatrix} 0 \\ \frac{k}{\tau^2} \end{bmatrix} [u] \tag{9.35}$$

$$y = [1 \quad 0] \begin{bmatrix} x_1 \\ x_2 \end{bmatrix} \tag{9.36}$$

The student should show that defining $y = x_2$ leads to the following state-space model:

$$\begin{bmatrix} \dot{x}_1 \\ \dot{x}_2 \end{bmatrix} = \begin{bmatrix} -\frac{2\zeta}{\tau} & -\frac{1}{\tau^2} \\ 1 & 0 \end{bmatrix} \begin{bmatrix} x_1 \\ x_2 \end{bmatrix} + \begin{bmatrix} \frac{k}{\tau^2} \\ 0 \end{bmatrix} [u] \tag{9.37}$$

$$y = [0 \quad 1] \begin{bmatrix} x_1 \\ x_2 \end{bmatrix} \tag{9.38}$$

MATLAB has routines for converting from transfer function form to state-space form (tf2ss) and vice versa (ss2tf). tf2ss is used in Example 9.5.

EXAMPLE 9.5   MATLAB Routine tf2ss

Consider the following second-order system:

$$y(s) = \frac{3}{2s^2 + 0.7071\,s + 1}\,u(s)$$

First, define the numerator and denominator arrays by:

```
num = [ 3 ]
den = [ 2     0.7071     1 ]
```

and enter the command:

```
[a,b,c,d] = tf2ss(num,den)
```

MATLAB returns the state-space matrices:

```
a =     -0.3535 -0.5000
        1.0000       0
b =     1
        0
c =     0          1.5000
d =     0
```

Notice that the state space models in Example 9.5 are different than the matices that are obtained from (9.35) and (9.36) or (9.37) and (9.38), but the different forms would all yield the same results for the output variable via simulation. Remember that a transfer function relates inputs to outputs but does not represent the actual states of the system. There are an infinite number of state-space models that will yield the same input/output model.

After finding the state-space form for a transfer function, we can use any available numerical integrator to solve problems. MATLAB routines of interest include ode45, initial, and step.

9.6   MATLAB ROUTINES FOR STEP AND IMPULSE RESPONSE

MATLAB has routines for step and impulse response of either transfer function models or state-space models. In the following, we show how these routines are used for transfer function models.

9.6.1   step

A quick way to generate step responses is to use the MATLAB function step. This can be used with either a state-space or a Laplace domain model.

Consider the following Laplace domain model:

$$g(s) = \frac{2(10s + 1)}{50\,s^2 + 15\,s + 3}$$

which can be written:

$$g(s) = \frac{20s + 2}{50\,s^2 + 15\,s + 3}$$

The following MATLAB commands are used to generate the response shown in Figure 9.14.

```
num = [20 2];
den = [50 15 3];
[y,x,t] = step(num,den)
plot(t,y)
```

Notice that a time vector is automatically generated, with a length close to the *settling time* of the process.

The same plot could be generated from the state-space form by using:

```
[y,x,t] = step(A,B,C,D,1)
plot(t,y)
```

where *A*, *B*, *C*, and *D* are the state-space matrices and '1' indicates the first input. Although state variables are calculated, only the output variables are of interest.

We could supply an equally spaced time vector and use:

```
[y,x] = step(num,den,t)
```

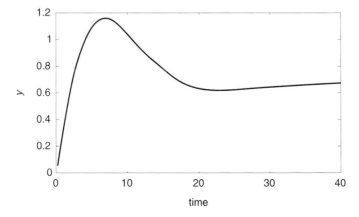

FIGURE 9.14   Step response for the example system.

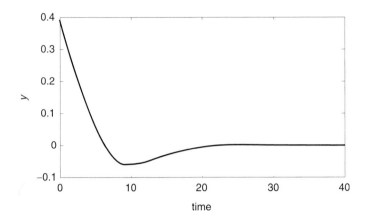

FIGURE 9.15   Impulse response for the example system.

for the step response of a transfer function model. The number of arguments determines whether a transfer function or state space model is used by the step function, and whether the time vector has been specified or not.

9.6.2 `impulse`

The output and time vectors are generated using:

$$[\texttt{y,x,t}] \;=\; \texttt{impulse(num,den)};$$

the plot is obtained from

$$\texttt{plot(t,y)}$$

The plot is shown in Figure 9.15 above. Notice that an impulse has an immediate (discontinuous) effect on the output, because this is a relative order one system.

    We could also supply an equally spaced time vector and use:

$$[\texttt{y,x}] \;=\; \texttt{impulse(num,den,t)};$$

SUMMARY

The step responses of the classical second order system (overdamped, critically damped, and underdamped) were presented. In addition, we showed the effect of numerator dynamics (and particularly right-half-plane zeros) on the response of a second-order system. The Padé approximations for deadtime were presented. You should understand the effect of the location of poles and zeros on the speed and quality of response of a transfer function model. The process gain is simply the ultimate change in output divided by the change in input.

The MATLAB routines used were

    `tf2ss:`   transfer function to state space
    `step:`   step response
    `impulse:` impulse response

Critical concepts from this chapter include:

    damping factor
    natural period
    numerator dynamics
    Padé approximation for time-delay
    relative order

STUDENT EXERCISES

1. Derive the *step* responses for the following second-order systems.
   a. Overdamped
   b. Critically damped
   c. Underdamped

2. Derive the *impulse* responses for the following second-order systems.
   a. Overdamped
   b. Critically damped
   c. Underdamped

3. Consider a sine input with magnitude A and frequency ω. Solve for the time domain value of the output for the following second-order systems.
   a. Overdamped
   b. Critically damped
   c. Underdamped

4. For a second-order system with numerator dynamics, find the step response for the following.
   a. Overdamped.
   b. Underdamped.
   c. Critically damped.

5. A second-order system has the following Laplace transfer function form:

$$Y(s) = \frac{2.5}{25s^2 + 5s + 1} U(s)$$

where the time unit is hours. The initial steady-state value for the output is 20 psig and the input is 4 gpm.

    At $t = 0$, a step input decrease is made, from 4 gpm to 3 gpm.

  a. What is the final value of the output?
  b. When does the output first reach this final value?
  c. What is the minimum value of the output?
  d. When does the output hit this minimum value?
  e. Plot the response.

6. Consider the following second-order ODE:

$$\tau_1 \tau_2 \frac{d^2y}{dt^2} + (\tau_1 + \tau_2) \frac{dy}{dt} + y = k\tau_n \frac{du}{dt} + k u$$

  with the initial conditions $y'(0) = y(0) = u'(0) = u(0) = 0$

  a. Find the Laplace transform of the differential equation. Write this expression in the form of $y(s) = g(s) u(s)$

  b. Now, assume that a step change of magnitude A in the variable u occurs at time = 0. Find the time domain result, $y(t)$.

  c. Now, assume that a step change of magnitude A in the variable u occurs at time = 0. Find the time domain result, $y(t)$, by using a partial fraction expansion and solving for the inverse Laplace transform by hand.

  d. Plot the time domain response, $y(t)$ from part c, using the following parameter values $k = 1$, $\tau_1 = 3$, $\tau_2 = 10$, and try several plots, varying τ_n from 3 to 10.

  e. Plot the time domain response, $y(t)$ from part c, using the following parameter values $k = 1$, $\tau_1 = 3$, $\tau_2 = 10$, and try several plots, varying τ_n from -10 to 0.

7. Consider the following two first-order ODEs:

$$\tau_1 \frac{dx_1}{dt} + x_1 = k_1 u$$

$$\tau_2 \frac{dx_2}{dt} + x_2 = k_2 u$$

  and the static relationship $y = x_1 + x_2$

  where x_1 and x_2 are two state variables, y is the output variable, and u is an input variable.

  a. Show that the two equations can be combined to yield a single ODE in the form of problem 6. Find k and τ_n as a function of k_1, k_2, τ_1, τ_2.

  b. Now, assume that a step change of magnitude Δu in the variable u occurs at time = 0. Find the time domain result, $y(t)$, by using a partial fraction expansion and solving for the inverse Laplace transform by hand.

  c. Plot $x_1(t)$, $x_2(t)$ and $y(t)$ if $\Delta u = 1$, $k_1 = -1$, $k_2 = 2$, $\tau_1 = 3$, and $\tau_2 = 10$.

8. As a process engineer with the Complex Pole Corporation, you are assigned a unit with an exothermic chemical reactor. In order to learn more about the dynamics of the process, you decide to make a step change in the input variable, which is coolant temperature, from 10°C to 15°C. Assume that the reactor was initially at a steady-state. You obtain the following plot for the output variable, which is reactor temperature (notice that the reactor temperature is in °F).

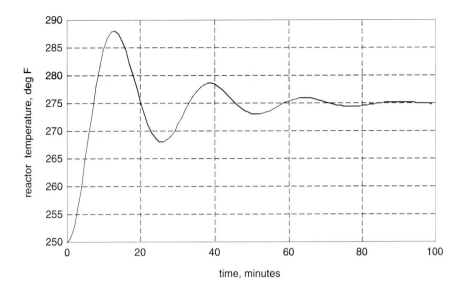

a. What is the value of the process gain? (show units)
b. What is the value of τ? (show units)
c. What is the value of ζ? (show units)
d. What is the decay ratio?
e. What is the period of oscillation? (show units)
f. Write the second-order transfer function.

9. A process is described by the following linear ordinary differential equation:

$$4\frac{d^2y}{dt^2} + 1.2\frac{dy}{dt} + y = 2.5\frac{d^2u}{dt^2}$$

where y is the output and u is the input. Assume that:

$$\frac{dy(0)}{dt} = y(0) = \frac{du(0)}{dt} = 0$$

also, assume that at time $t = 0$, the input begins to increase with the following relationship

$$u(t) = \frac{1}{2}t^2$$

The units for time are minutes.
a. What are the values of the poles of this process (give units)?
b. When does the output of the process reach a maximum value?
c. What is the maximum value of the process output?

10. A process has two poles and one zero. The poles are located at $-1 \pm 0.5j$ and the zero is located at 0.5. Sketch the type of response that you expect to a step change in input. Explain. Find the transfer function and verify these results assuming a gain of one.

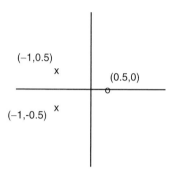

11. Consider the following state-space model (from Module 7):

$$\begin{bmatrix} \dot{x}_1 \\ \dot{x}_2 \end{bmatrix} = \begin{bmatrix} -2.405 & 0 \\ 0.833 & -2.238 \end{bmatrix} \begin{bmatrix} x_1 \\ x_2 \end{bmatrix} + \begin{bmatrix} 7 \\ -1.117 \end{bmatrix} u$$

$$y = \begin{bmatrix} 0 & 1 \end{bmatrix} \begin{bmatrix} x_1 \\ x_2 \end{bmatrix}$$

   a. Find the transfer function $g(s)$ where $y(s) = g(s)\, u(s)$.
   b. Find the poles and zeros.
   c. Plot the response to a unit step input.
   d. Plot the response to a unit impulse input.

12. A process engineer responsible for the operation of a complex chemical reactor has the process operator make a step change in the coolant flowrate from 10 gpm to 15 gpm to the reactor at 2:00 pm. The reactor temperature is initially 150°F at 2:00 pm and drops to a low of 115°F at 2:10 pm. Eventually the reactor temperature comes to a final steady-state temperature of 125°F. Assuming that the response is second-order ($k/\tau^2 s^2 + 2\zeta\tau s + 1$), find k, ζ, τ (show units).

13. The output of a second-order, underdamped system has a rise time of 1 hour, and a maximum value of 15°F (in deviation variables), after a step change at time $t = 0$. After a long period of time, the output is 12°F (again in deviation variables).
   a. What is the value of τ?
   b. What is the value of ζ?
   c. What are the poles? (also, show their location in the complex plane)

14. A step change of magnitude 2 lb/min is applied to the input of a process. The resulting output response, in deviation variables, is shown in the figure below.

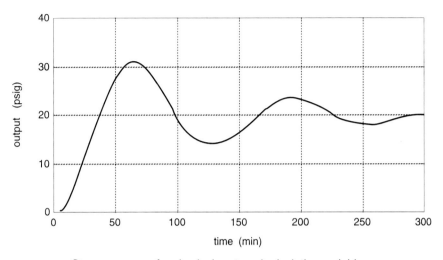

Step response of a physical system, in deviation variables.

a. Find the period of oscillation, rise time, and time to first peak, for this system. Show your work.

b. Find parameters (show the units) in the transfer function, $g(s) = k/(\tau^2 s^2 + 2\zeta\tau s + 1)$, by using the dimensionless plot, Figure 9.2. Show your work.

15. Consider the following third-order transfer function, where β is a parameter. Find the conditions on the parameter β that will give an inverse response.

$$g(s) = \frac{(2s^2 + s + \beta)}{(5s + 1)(3s + 1)(2s + 1)}$$

Show your work and explain your answer.

16. Consider the following transfer function:

$$g(s) = \frac{s^2 + s - 2}{s^2 + 4s + 3}$$

a. Find the poles and zeros for this transfer function.

b. A unit step change is made at $t = 0$. Find the value of the output, using the final and initial value theorems:

  i. After a long time.

  ii. Immediately after the step change.

c. Verify your results in **b** by finding (analytically) the time domain solution.

d. Verify the results in **b** using the MATLAB function `step`.

17. Consider the following state-space model:

$$\begin{bmatrix} \dot{x}_1 \\ \dot{x}_2 \end{bmatrix} = \begin{bmatrix} -6.5 & 2.5 \\ 4 & -6.5 \end{bmatrix} \begin{bmatrix} x_1 \\ x_2 \end{bmatrix} + \begin{bmatrix} 0.00155 \\ 0.00248 \end{bmatrix} u$$

Find the transfer functions relating the input to each output. Find the step response of each output.

18. A unit step change in input is made on a number of processes (I–IV). The resulting outputs are shown in the plot below. Associate each process with a response curve.

Process Transfer Function	Curve (letter) from Plot
I. $g(s) = \dfrac{2e^{-2s}}{2s + 1}$	
II. $g(s) = \dfrac{1}{s^2 + s + 1}$	
III. $g(s) = \dfrac{2s + 1}{s + 1}$	
IV. $g(s) = \dfrac{4(s^2 - 2s + 2)}{(s + 2)^3}$	

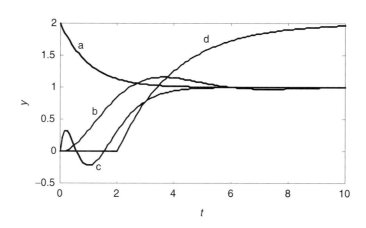

19. Consider a second-order transfer function with numerator dynamics:

$$y(s) = \frac{k(\tau_n s + 1)}{(\tau_1 s + 1)(\tau_2 s + 1)} u(s)$$

let τ_1 represent the smaller denominator time constant. Assume a step change in input. Show that a maximum in $y(t)/k\Delta u$ occurs if $\tau_n > \tau_2$ and that a minimum (indicating inverse response) occurs if $\tau_n < 0$. Also show that there is no extrema in the step response if $0 < \tau_n < \tau_2$. (*Hint:* Realize that a maximum or minimum occurs at $\dot{y}/\Delta u = 0$.)

20. Consider the transfer function $g_p(s) = \dfrac{12s + 2}{3s^2 + 4s + 1}$

a. Write the gain-polynomial form $\quad g_p(s) = \dfrac{k(\tau_n s + 1)}{\tau^2 s^2 + 2\zeta\tau s + 1}$

b. Write the gain-time constant form $\quad g_p(s) = \dfrac{k(\tau_n s + 1)}{(\tau_1 s + 1)(\tau_2 s + 1)}$

c. Write the gain-pole-zero form $\quad g_p(s) = \dfrac{k_1(s - z)}{(s - p_1)(s - p_2)}$

21. The reader should show how the first- and second-order Padé approximations relate to a Taylor series expansion. The Taylor series approximation to a time-delay in the Laplace domain is

$$e^{-\theta s} \approx 1 - \theta s + \frac{\theta^2 s^2}{2!} - \frac{\theta^3 s^3}{3!} + \frac{\theta^4 s^4}{4!} - \frac{\theta^5 s^5}{5!} + \frac{\theta^6 s^6}{6!} - \dots$$

Use long division of the first- and second-order Padé approximations and comment on the number of terms that are consistent with the Taylor series expression.

22. Consider the following interacting tank problem. Assume that the flow between tanks 2 and 1 is linearly proportional (β_1) to the difference in tank heights and that the outlet flow from tank 2 is proportional (β_2) to tank height 2. Develop the transfer function models relating the inlet flowrate to both tank heights.

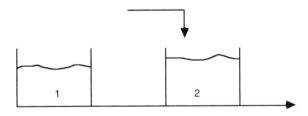

23. Consider an exothermic chemical reactor that has the following transfer function relationship between the inlet flowrate (input) and the reactor temperature (output).

$$g(s) = \frac{2(-2.5s + 1)}{9s^2 + 3s + 1}$$

The units of the input are liter/min and the output is in deg C.

a. Find the values of the zeros and poles. Is this system underdamped or overdamped?

b. For a step input change of +3 liter/min, find how the output changes with time. How much does the temperature decrease before increasing? Compare plots of your analytical solution with those obtained using the MATLAB function `step`.

c. What is the ultimate change in temperature after a long period of time?

d. If the steady-state input and output values (in physical terms) are 10 liter/min and 75°C respectively, what are the physical values of the results in **b** and **c**?

    e. If a step decrease in the input of −3 liter/min is made, what would be the results in **b, c,** and **d**?

24. Consider a CSTR with a first-order irreversible reaction $A \longrightarrow B$. The modeling equations are:

$$\frac{dC_A}{dt} = -\left(\frac{F}{V} + k\right) C_A + \frac{F}{V} C_{Af}$$

$$\frac{dC_B}{dt} = k\, C_A - \frac{F}{V} C_B$$

The following parameters and steady-state input values characterize this system:

$$\frac{F}{V} = 0.2\ \text{min}^{-1}$$

$$k = 0.2\ \text{min}^{-1}$$

$$C_{Afs} = 1.0\ \frac{\text{gmol}}{\text{liter}}$$

The input is C_{Af} and the output is C_B. You should be able to show that the steady-state values of C_A and C_B are 0.5 gmol/liter.

    a. Show that the transfer function relating the feed concentration of A to the concentration of B is:

$$y(s) = \frac{0.5}{(5s + 1)(2.5s + 1)}$$

where the gain is $\dfrac{\text{gmol B/liter}}{\text{gmol A/liter}}$, and the time unit is minutes.

    b. At time $t = 0$, the input begins to vary in a sinusoidal fashion with amplitude 0.25 and frequency 0.5 min^{-1}; that is,

$$u(t) = 0.25 \sin(0.5\ t)$$

Using Laplace transforms, find how the output varies with time.

    c. Compare your results in **b** with the integration of the modeling equations using the MATLAB integration routine `ode45`. Remember to use the correct initial conditions. Also, remember that the transfer function results are in deviation variable form and must be converted back to physical variable values.

    d. Discuss how the amplitude of the output changes if the input frequency is changed to 5 min^{-1}.

25. Often higher-order process transfer functions are approximated by lower-order transfer functions. Consider the following second-order transfer function:

$$g(s) = \frac{1}{(2s + 1)(s + 1)}$$

Find the value of τ in a first-order transfer function, $1/(\tau s + 1)$, which best approximates the step response of this second-order transfer function, in a least-squares sense.

(*Hint:* Define an error as a function of time as $e(t) = y_2(t) - y_1(t)$, where y_2 and y_1 are the step responses of the second- and first-order responses respectively. Find t which minimizes $e^2(t)$ when $t \rightarrow$ inf.)

26. Consider a critically damped second-order system:

$$g(s) = \frac{1}{(\tau s + 1)(\tau s + 1)}$$

   a. For a unit step input change ($\Delta u = 1$), find the time at which the rate of change of the output is greatest (i.e., find the inflection point).

   b. Compare this rate of change with a unit step response of a first-order system with the following transfer function:

$$g(s) = \frac{1}{(2\tau s + 1)}$$

   c. Plot the step responses for *a* and *b,* for $\tau = 1$. Compare and contrast the responses.

27. Pharmacokinetics is the study of how drugs infused to the body are distributed to other parts of the body. The concept of a compartmental model is often used, where it is assumed that the drug is injected into compartment 1. Some of the drug is eliminated (reacted) in compartment 1, and some of it diffuses into compartment 2 (the rest accumulates in compartment 1). Similarly, some of the drug that diffuses into compartment 2 diffuses back into compartment 1, while some is eliminated by reaction and the rest accumulates in compartment 2. Assuming that the rates of diffusion and reaction are directly proportional to the concentration of drug in the compartment of interest, the following balance equations arise:

$$\frac{dx_1}{dt} = -(k_{10} + k_{12})\, x_1 + k_{21}\, x_2 + u$$

$$\frac{dx_2}{dt} = k_{12}\, x_1 - (k_{20} + k_{21})\, x_2$$

where x_1 and x_2 = drug concentrations in compartments 1 and 2 (μg/kg patient weight), and u = rate of drug input to compartment 1 (scaled by the patient weight, μg/kg min).

Experimental studies (of the response of the compartment 1 concentration to various drug infusions) have led to the following parameter values for the drug atracurium, which is a muscle relaxant:

$$(k_{10} + k_{12}) = 0.26 \text{ min}^{-1}$$

$$(k_{20} + k_{21}) = 0.094 \text{ min}^{-1}$$

$$k_{12}k_{21} \qquad = 0.015 \text{ min}^{-1}$$

a. Find the poles and zeros of the transfer function that relate the input, u, to the output, x_1.

b. Find the response of the concentration in compartment 1, x_1, to a *step* input of 1 µg/kg min. What is the value at 10 minutes? What is the value after a long period of time?

c. Find the response of the concentration in compartment 1, x_1, to an *impulse* input of 10 µg/kg. What is the value at $t = 0$? What is the value at 10 minutes?

28. Consider the following delay-differential equations:

$$\frac{dx_1}{dt} = -x_2(t - \theta) + u(t)$$

$$\frac{dx_2}{dt} = -2 x_1$$

using the first-order Padé approximation for deadtime, write the corresponding (approximate) pure differential equations. (*Hint:* define a new variable $x_3 = x_2(t - \theta)$.)

Solve the equations using ode45, for an initial condition of 0 in all states, and a value of 1 for the input.

MATRIX TRANSFER FUNCTIONS

10

Chapter 6 presented simple examples for transforming a state-space model to a single nth order differential equation. Once the single differential equation was obtained, the methods of characteristics and undetermined coefficients (Chapter 6) or Laplace transforms (Chapters 7–9) could be used to obtain a solution. A general method for converting a state-space model directly to the Laplace domain is presented in this chapter. With the transfer function representation, one can easily obtain the corresponding single nth order differential equation. After studying this chapter, the reader should be able to:

- Convert a state-space model to a transfer function model analytically.
- Convert a state-space model to a transfer function model using the MATLAB routine ss2tf.
- Discuss interesting effects from pole-zero cancellation.

The major sections are:

The goal of this chapter is to take a general state-space model:

$$\dot{x} = A\,x + B\,u$$

$$y = C\,x + D\,u$$

and convert it to the matrix transfer function form:

$$Y(s) = G(s)\,U(s)$$

and use this model to solve for the responses of each output to each input. We will also use this technique to easily find the nth order differential equation corresponding to each output variable.

10.1 A SECOND-ORDER EXAMPLE

Consider the following two-state, single-input, single-output model:

$$\frac{dx_1}{dt} = a_{11}\,x_1 + a_{12}\,x_2 + b_{11}\,u \tag{10.1}$$

$$\frac{dx_2}{dt} = a_{21}\,x_1 + a_{22}\,x_2 + b_{21}\,u \tag{10.2}$$

$$y = c_{11}\,x_1 + c_{12}\,x_2 + d_{11}\,u \tag{10.3}$$

Taking Laplace transforms of (10.1) through (10.3), we find:

$$s\,X_1(s) - x_1(0) = a_{11}\,X_1(s) + a_{12}\,X_2(s) + b_{11}\,U(s) \tag{10.4}$$

$$s\,X_2(s) - x_2(0) = a_{21}\,X_1(s) + a_{22}\,X_2(s) + b_{21}\,U(s) \tag{10.5}$$

$$Y(s) = c_{11}\,X_1(s) + c_{12}\,X_2(s) + d_{11}\,U(s) \tag{10.6}$$

Assuming $x_1(0) = x_2(0) = 0$, and rearranging:

$$(s - a_{11})\,X_1(s) - a_{12}\,X_2(s) = b_{11}\,U(s) \tag{10.7}$$

$$(s - a_{22})\,X_2(s) - a_{21}\,X_1(s) = b_{21}\,U(s) \tag{10.8}$$

In order to generalize this procedure later, we write (10.7) and (10.8) in matrix form:

$$\left\{ \begin{bmatrix} s & 0 \\ 0 & s \end{bmatrix} - \begin{bmatrix} a_{11} & a_{12} \\ a_{21} & a_{22} \end{bmatrix} \right\} \begin{bmatrix} X_1(s) \\ X_2(s) \end{bmatrix} = \begin{bmatrix} b_{11} \\ b_{21} \end{bmatrix} U(s)$$

or,

$$\left\{ s\begin{bmatrix} 1 & 0 \\ 0 & 1 \end{bmatrix} - \begin{bmatrix} a_{11} & a_{12} \\ a_{21} & a_{22} \end{bmatrix} \right\} \begin{bmatrix} X_1(s) \\ X_2(s) \end{bmatrix} = \begin{bmatrix} b_{11} \\ b_{21} \end{bmatrix} U(s) \tag{10.9}$$

and (10.6) is written in matrix form:

$$Y(s) = [c_{11} \quad c_{12}] \begin{bmatrix} X_1(s) \\ X_2(s) \end{bmatrix} + d_{11} \, U(s) \qquad (10.10)$$

We see (10.9) is of the form:

$$(s\mathbf{I} - \mathbf{A}) \, \mathbf{X}(s) = \mathbf{B} \, U(s) \qquad (10.11)$$

with the solution for $\mathbf{X}(s)$:

$$\mathbf{X}(s) = (s\mathbf{I} - \mathbf{A})^{-1} \mathbf{B} \, U(s) \qquad (10.12)$$

and writing (10.10) as:

$$Y(s) = \mathbf{C} \, \mathbf{X}(s) + \mathbf{D} \, U(s) \qquad (10.13)$$

combining (10.12) and (10.13):

$$Y(s) = [\mathbf{C} \, (s\mathbf{I} - \mathbf{A})^{-1} \mathbf{B} + \mathbf{D}] \, U(s) \qquad (10.14)$$

recall that often $\mathbf{D} = \mathbf{0},$ in which case (10.15) is written:

$$Y(s) = [\mathbf{C} \, (s\mathbf{I} - \mathbf{A})^{-1} \mathbf{B}] \, U(s) \qquad (10.15)$$

or,

$$Y(s) = \mathbf{G}(s) \, U(s) \qquad (10.16)$$

In this example, since there is a single input and a single output, $\mathbf{G}(s)$ is a single transfer function, which we call $g(s)$. The transfer function is the ratio of a numerator and a denominator polynomial:

$$g(s) = \frac{N(s)}{D(s)} \qquad (10.17)$$

The reader should show that the polynomials in (10.17), based on (10.15) are (see student exercise 4):

$$N(s) = n_1 \, s + n_0 \qquad (10.18a)$$

$$D(s) = s^2 + d_1 \, s + d_0 \qquad (10.18b)$$

where the polynomial coefficients, in terms of the matrix coefficients, are:

$$n_1 = c_{11} \, b_{11} + c_{12} \, b_{21} \qquad (10.19a)$$

$$n_0 = c_{11} \, [a_{12} \, b_{21} - a_{22} \, b_{11}] + c_{12}[a_{21} \, b_{11} - a_{11} \, b_{21}] \qquad (10.19b)$$

$$d_1 = a_{11} + a_{22} \qquad (10.19c)$$

$$d_0 = a_{11} \, a_{22} - a_{12} \, a_{21} \qquad (10.19d)$$

Since the input-output relationship is written:

$$\mathbf{Y}(s) = \frac{N(s)}{D(s)} \mathbf{U}(s)$$

We can further write:

$$D(s)\,\mathbf{Y}(s) = N(s)\,\mathbf{U}(s)$$

or,

$$[s^2 + d_1 s + d_0]\,\mathbf{Y}(s) = [n_1 s + n_0]\,\mathbf{U}(s)$$

The corresponding differential equation is:

$$\frac{d^2y}{dt^2} + d_1\frac{dy}{dt} + d_0\,y = n_1\frac{du}{dt} + n_0\,u$$

We now have an automated procedure to find the transfer function for a single-input, single-output, two-state system. An example is shown below.

Example 10.1   Linear Bioreactor Model

Consider a linearized model of a bioreactor, with the second-state variable (substrate concentration) measured and with dilution rate (F/V) as the input variable.

The state-space matrices are

$$\mathbf{A} = \begin{bmatrix} 0 & 0.9056 \\ -0.7500 & -2.5640 \end{bmatrix}$$

$$\mathbf{B} = \begin{bmatrix} -1.5302 \\ 3.8255 \end{bmatrix}$$

$$\mathbf{C} = [0 \ \ 1]$$

$$\mathbf{D} = 0$$

Using the following steps to find $\mathbf{G}(s) = \mathbf{C}\,(s\mathbf{I} - \mathbf{A})^{-1}\mathbf{B}$:

$$(s\mathbf{I} - \mathbf{A}) = \begin{bmatrix} s & -0.9056 \\ 0.7500 & s + 2.5640 \end{bmatrix}$$

Recalling the simple method for inverting a 2×2 matrix, we find:

$$(s\mathbf{I} - \mathbf{A})^{-1} = \begin{bmatrix} s + 2.5640 & 0.9056 \\ -0.7500 & s \end{bmatrix} \frac{1}{s^2 + 2.5640\,s + 0.67920}$$

$$C(s\mathbf{I} - \mathbf{A})^{-1} = [\,-0.7500\ s\,] \frac{1}{s^2 + 2.5640\,s + 0.67920}$$

$$C(s\mathbf{I} - \mathbf{A})^{-1}B = [\,-0.7500\ s\,] \begin{bmatrix} -1.5302 \\ 3.8255 \end{bmatrix} \frac{1}{s^2 + 2.5640\,s + 0.67920}$$

$$C(s\mathbf{I} - \mathbf{A})^{-1}B = \frac{3.8255\ s + 1.14765}{s^2 + 2.5640\ s + 0.67920}$$

so,

$$Y(s) = \frac{3.8255\ s + 1.14765}{s^2 + 2.5640\ s + 0.67920}\ U(s)$$

and we easily find that

$$\frac{d^2 y}{dt^2} + 2.5640\ \frac{dy}{dt} + 0.67920\ y = 3.8255\ \frac{du}{dt} + 1.14765\ u$$

We generalize this procedure in Section 10.2.

10.2 THE GENERAL METHOD

Consider a general state-space model with n states, m inputs, and r outputs (see Chapter 5):

$$\frac{dx_1}{dt} = a_{11}\ x_1 + a_{12}\ x_2 + \dots + a_{1n}\ x_n + b_{11}\ u_1 + \dots + b_{1m}\ u_m$$

$$\frac{dx_n}{dt} = a_{n1}\ x_1 + a_{n2}\ x_2 + \dots + a_{nn}\ x_n + b_{n1}\ u_1 + \dots + b_{nm}\ u_m$$

$$y_1 = c_{11}\ x_1 + c_{12}\ x_2 + \dots + a_{1n}\ x_n + d_{11}\ u_1 + \dots + b_{1m}\ u_m$$

(10.20)

$$y_r = c_{r1}\ x_1 + c_{r2}\ x_2 + \dots + c_{rn}\ x_n + d_{r1}\ u_1 + \dots + b_{rm}\ u_m$$

which can be written in matrix form as:

$$\begin{bmatrix} \dot{x}_1 \\ \cdot \\ \cdot \\ \dot{x}_n \end{bmatrix} = \begin{bmatrix} a_{11} & a_{12} & \cdot & a_{1n} \\ \cdot & \cdot & \cdot & \cdot \\ \cdot & \cdot & \cdot & \cdot \\ a_{n1} & a_{n2} & \cdot & a_{nn} \end{bmatrix} \begin{bmatrix} x_1 \\ \cdot \\ \cdot \\ x_n \end{bmatrix} + \begin{bmatrix} b_{11} & b_{12} & \cdot & b_{1m} \\ \cdot & \cdot & \cdot & \cdot \\ \cdot & \cdot & \cdot & \cdot \\ b_{n1} & b_{n2} & \cdot & b_{nm} \end{bmatrix} \begin{bmatrix} u_1 \\ \cdot \\ \cdot \\ u_m \end{bmatrix}$$

(10.21)

$$\begin{bmatrix} y_1 \\ \cdot \\ \cdot \\ y_r \end{bmatrix} = \begin{bmatrix} c_{11} & c_{12} & \cdot & c_{1n} \\ \cdot & \cdot & \cdot & \cdot \\ \cdot & \cdot & \cdot & \cdot \\ c_{r1} & c_{r2} & \cdot & c_{rn} \end{bmatrix} \begin{bmatrix} x_1 \\ \cdot \\ \cdot \\ x_n \end{bmatrix} + \begin{bmatrix} d_{11} & d_{12} & \cdot & d_{1m} \\ \cdot & \cdot & \cdot & \cdot \\ \cdot & \cdot & \cdot & \cdot \\ d_{r1} & d_{r2} & \cdot & d_{rm} \end{bmatrix} \begin{bmatrix} u_1 \\ \cdot \\ \cdot \\ u_m \end{bmatrix}$$

which has the form:

$$\dot{\mathbf{x}} = \mathbf{A}\,\mathbf{x} + \mathbf{B}\,\mathbf{u}$$
$$\mathbf{y} = \mathbf{C}\,\mathbf{x} + \mathbf{D}\,\mathbf{u} \qquad (10.22)$$

where the dot over a state variable indicates the derivative with respect to time. Recall from Chapter 5 that the eigenvalues of the Jacobian matrix ($\mathbf{A}$) determine the stability of the system of equations and the "speed" of response. Now, taking the Laplace transform of (10.22):

$$\mathbf{X}(s) = (s\mathbf{I} - \mathbf{A})^{-1}\mathbf{B}\,\mathbf{U}(s)$$
$$\mathbf{Y}(s) = [\mathbf{C}\,(s\mathbf{I} - \mathbf{A})^{-1}\mathbf{B} + \mathbf{D}]\mathbf{U}(s)$$

If $\mathbf{D} = \mathbf{0}$ we can write:

$$\mathbf{Y}(s) = \mathbf{G}(s)\,\mathbf{U}(s)$$

where:

$$\mathbf{G}(s) \quad = \quad \mathbf{C}\,(s\mathbf{I} - \mathbf{A})^{-1}\,\mathbf{B}$$
$$(r \times m) \quad (r \times n)\,(n \times n)\,(n \times m)$$

The transfer function matrix, $\mathbf{G}(s)$, is:

$$\mathbf{G}(s) = \begin{bmatrix} g_{11}(s) & g_{12}(s) & . & . & g_{1m}(s) \\ . & . & . & . & . \\ . & . & . & . & . \\ g_{r1}(s) & g_{r2}(s) & . & . & g_{rm}(s) \end{bmatrix}$$

Notice that $\mathbf{G}(s)$ is square if $r = m$ (number of outputs = number of inputs).

10.3   MATLAB ROUTINE ss2tf

The routine ss2tf can be used to convert a state-space model to a transfer function model. After entering the A, B, C, and D matrices, the command:

[num,den]=ss2tf(A,B,C,D,m)

will generate the numerator and denominator Laplace domain polynomials for the transfer function between input number m and the outputs, in descending order of s.

EXAMPLE 10.2   Example 10.1 Using MATLAB ss2tf

Here we consider the linearized bioreactor model, with two inputs. The first input is dilution rate, the same input used above. The second input is the substrate feed concentration. We will also consider both state 1 and state 2 to be outputs, and modify the C and D matrices so that ss2tf provides the transfer functions between the input and both outputs.

```
» A = [0,0.9056;-0.75,-2.5640]

A =
            0   0.9056
      -0.7500  -2.5640

» B = [-1.5302,0;3.8255,0.3]

B =
      -1.5302        0
       3.8255   0.3000

» C = [1,0;0,1]

C =
       1   0
       0   1

» D = [0,0;0,0]

D =
       0   0
       0   0
```

Input 1

The numerator and denominator polynomials relating the first input to the two outputs are found using the following command:

```
» [num,den]=ss2tf(A,B,C,D,1)

num =
       0   -1.5302   -0.4591
       0    3.8255    1.1476

den =
   1.0000    2.5640    0.6792
```

where the first row of the num matrix is the coefficients of s in the $g_{11}(s)$ polynomial, in decreasing order from left to right. Similarly, the second row of the num matrix is the coefficients of s in the $g_{21}(s)$ polynomial, in decreasing order from left to right:

$$y_1(s) = g_{11}(s)\, u_1(s) = \frac{-1.5302\, s - 0.4591}{s^2 + 2.5640\, s + 0.67920}\, u_1(s)$$

$$y_2(s) = g_{21}(s)\, u_1(s) = \frac{3.8255\, s + 1.14765}{s^2 + 2.5640\, s + 0.67920}\, u_1(s)$$

We realize that the eigenvalues of the **A** matrix and the poles of the transfer functions will be the same. This is verified by the roots and eig commands

```
» roots(den)

ans =
    -2.2640
    -0.3000

» eig(a)

ans =
    -0.3000
    -2.2640
```

We can also write the transfer functions in pole-zero form:

$$y_1(s) = \frac{-1.5302\,(s + 0.3)}{(s + 2.2640)(s + 0.3)}\,u_1(s)$$

$$y_2(s) = \frac{3.8255\,(s + 0.3)}{(s + 2.2640)(s + 0.3)}\,u_1(s)$$

where we have the interesting result that the zero cancels one of the poles to yield first-order systems

$$y_1(s) = \frac{-1.5302}{(s + 2.2640)}\,u_1(s)$$

$$y_2(s) = \frac{3.8255}{(s + 2.2640)}\,u_1(s)$$

which we are more used to seeing in gain-time constant form:

$$y_1(s) = \frac{-0.6759}{0.4417\,s + 1}\,u_1(s)$$

$$y_2(s) = \frac{1.6897}{0.4417\,s + 1}\,u_1(s)$$

We would notice the zero-pole cancellation if we also used the roots command to find the roots of the numerator polynomial

```
»roots(num(1,:))

ans =
   -0.3000

»roots(num(2,:))

ans =
   -0.3000
```

and we see that the root of the numerator polynomial is the same as one of the roots of the denominator polynomial.

Input 2

The numerator and denominator polynomials relating the second input to the two outputs are found using the following command:

```
» [num,den]=ss2tf(a,b,c,d,2)

num =
            0           0      0.2717
            0      0.3000           0

den =
       1.0000      2.5640      0.6792
```

and we have the result that:

$$y_1(s) = g_{12}(s)\, u_2(s) = \frac{0.2717}{s^2 + 2.5640\, s + 0.67920}\, u_2(s)$$

$$y_2(s) = g_{22}(s)\, u_2(s) = \frac{0.3\, s}{s^2 + 2.5640\, s + 0.67920}\, u_2(s)$$

The relationship between the second input and the second output is particularly interesting. The second input has no steady-state effect on the second output, as can be seen from the final value theorem. Assume a step change of magnitude Δu_2 in input 2.

$$y(t \to \infty) = s\, Y(s \to 0) = \frac{0.3\, s}{s^2 + 2.5640\, s + 0.67920} \frac{\Delta u_2}{s} = 0$$

10.3.1   Discussion of the Results from Example 10.2

THE FIRST INPUT

We noticed that the transfer functions with respect to the first input had pole-zero cancellation. This created an input-output relationship where the step response is faster than would be expected, because the slow pole was canceled by the process zero

$$y_1(s) = \frac{-1.5302\,(s + 0.3)}{(s + 2.2640)(s + 0.3)}\, u_1(s)$$

$$y_2(s) = \frac{3.8255\,(s + 0.3)}{(s + 2.2640)(s + 0.3)}\, u_1(s)$$

This can also be seen using the gain-time constant form:

$$y_1(s) = \frac{-0.6759\,(3.3333\, s + 1)}{(0.4417\, s + 1)(3.3333\, s + 1)}\, u_1(s)$$

$$y_2(s) = \frac{1.6897\,(3.3333\, s + 1)}{(0.4417\, s + 1)(3.3333\, s + 1)}\, u_1(s)$$

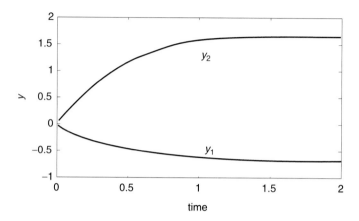

FIGURE 10.1 Unit step change in input 1.

or,

$$y_1(s) = \frac{-0.6759}{0.4417\,s + 1}\,u_1(s)$$

$$y_2(s) = \frac{1.6897}{0.4417\,s + 1}\,u_1(s)$$

The step responses for a unit step input change are shown in Figure 10.1.

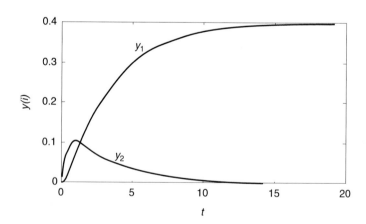

FIGURE 10.2 Unit step change in input 2. Notice that the steady-state value of y_2 does not change.

THE SECOND INPUT

Notice that input 2 does not have a steady-state effect on output 2, only a dynamic effect. This can be seen by using the MATLAB step function, then plotting the results (see Figure 10.2).

```
num =
            0          0      0.2717
            0     0.3000          0

den =
       1.0000     2.5640     0.6792

»[y,x,t]=step(num,den)

plot(t,y)
```

SUMMARY

We have shown how to convert a state-space model to a transfer function model, for multiple inputs and outputs. We have also seen some interesting results regarding pole-zero cancellation. One has to be particularly careful with pole-zero cancellation if a pole is unstable (positive), as will be shown in Section 11.3.

The following MATLAB routines were used:

```
ss2tf:   converts state space to transfer function form
eig:     matrix eigenvalues
roots:   roots of a polynomial
```

STUDENT EXERCISES

1. Compare the step responses of the following three transfer functions:

   a. $g_1(s) = \dfrac{1}{(0.4417\,s + 1)(3.3333\,s + 1)}$

   b. $g_2(s) = \dfrac{1}{0.4417\,s + 1}$

   c. $g_3(s) = \dfrac{1}{3.3333\,s + 1}$

   Which has a faster step response? Why?

2. Consider the following state-space model (a 5-stage absorption column)

$$
\mathbf{A} = \begin{bmatrix}
-0.325 & 0.125 & 0 & 0 & 0 \\
0.2 & -0.325 & 0.125 & 0 & 0 \\
0 & 0.2 & -0.325 & 0.125 & 0 \\
0 & 0 & 0.2 & -0.325 & 0.125 \\
0 & 0 & 0 & 0.2 & -0.325
\end{bmatrix}
$$

$$
\mathbf{B} = \begin{bmatrix}
0.2 & 0 \\
0 & 0 \\
0 & 0 \\
0 & 0 \\
0 & 0.25
\end{bmatrix}
$$

    a. Convert this model to transfer function form, assuming that all of the states are outputs, using `ss2tf`.

    b. Find the response of all of the states to a unit step in input 1. Use the function `step`.

    c. Find the response of all of the states to a unit step in input 2. Use the function `step`.

    d. Compare and contrast the curves from **b** and **c**.

3. Consider the following model for an isothermal CSTR with a single irreversible reaction (see Module 7). Find the transfer function matrix relating both inputs to both states.

$$
\begin{bmatrix}
\dfrac{dx_1}{dt} \\[2mm]
\dfrac{dx_2}{dt}
\end{bmatrix}
=
\begin{bmatrix}
-0.4 & 0 \\
0.2 & -0.2
\end{bmatrix}
\begin{bmatrix}
x_1 \\
x_2
\end{bmatrix}
+
\begin{bmatrix}
0.5 & 0.2 \\
-0.5 & 0
\end{bmatrix}
\begin{bmatrix}
u_1 \\
u_2
\end{bmatrix}
$$

4. For a 2-state, single-input, single-output process, derive the relationships shown in (10.18) and (10.19).

5. For the following state-space model, find the transfer function matrix relating all four inputs to both outputs.

$$
\mathbf{A} = \begin{bmatrix}
-0.4 & 0.3 \\
3 & -4.5
\end{bmatrix}
$$

$$
\mathbf{B} = \begin{bmatrix}
0 & -7.5 & 0.1 & 0 \\
50 & 0 & 0 & 1.5
\end{bmatrix}
$$

$$
\mathbf{C} = \begin{bmatrix}
1 & 0 \\
0 & 1
\end{bmatrix}
$$

6. Consider the following state-space model:

$$
\begin{bmatrix} \dfrac{dx_1}{dt} \\[2mm] \dfrac{dx_2}{dt} \end{bmatrix} =
\begin{bmatrix} -\dfrac{1}{5} & 0 \\[2mm] -\dfrac{1}{10} & \dfrac{1}{2} \end{bmatrix}
\begin{bmatrix} x_1 \\ x_2 \end{bmatrix} +
\begin{bmatrix} \dfrac{7}{5} \\[2mm] \dfrac{2}{25} \end{bmatrix} u
$$

$$
y = x_2
$$

Show that the eigenvalues of the **A** matrix are −1/5 and 1/2, so the system is unstable. Also, plot the step response. Derive the transfer function relating $u(s)$ to $y(s)$ and show that the unstable pole is cancelled by the positive zero. This problem will be analyzed in more detail in Chapter 11.

7. Consider a chemical reactor with bypass, as shown below. Assume that the reaction rate (per unit volume) is first-order ($r = kC_1$) and C_1 is the concentration in the reactor (the reactor is perfectly mixed). Assume that the volume in the reactor (V) and the feed flowrate (F) remain constant. The fraction of feed bypassing the reactor is $(1 - \alpha)F$ and that entering the reactor is αF. Assume that the fraction bypassing the reactor does not change. The inlet concentration (C_{in}) is the input variable and the mixed outlet stream composition (C_2) is the output variable. Write this model in state-space form, using deviation variables.

$$
\dot{\mathbf{x}} = \mathbf{A}\,\mathbf{x} + \mathbf{B}\,\mathbf{u}
$$

$$
\mathbf{y} = \mathbf{C}\,\mathbf{x} + \mathbf{D}\,\mathbf{u}
$$

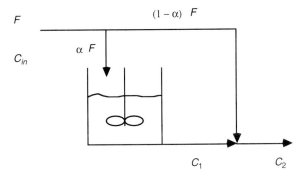

Find the transfer function relating u to y.

For the following parameters, simulate a unit step response.

$$
F = 10 \text{ l/min}, \; V = 100 \text{ l}, \; C_{in} = 1 \text{ gmol/l}, \; \alpha = 0.5, \; k = 0.1 \text{ min}^{-1}.
$$

8. Consider the following set of series and parallel reactions (from Module 7)

$$A \xrightarrow{k_1} B \xrightarrow{k_2} C$$

$$A + A \xrightarrow{k_3} D$$

Material balances on components A and B yield the following two equations:

$$\frac{dC_A}{dt} = \frac{F}{V}(C_{Af} - C_A) + (-k_1 C_A - k_3 C_A^2)$$

$$\frac{dC_B}{dt} = \frac{F}{V}(-C_B) + (k_1 C_A - k_2 C_B)$$

where the rate constants are:

$$k_1 = \frac{5}{6}\,\text{min}^{-1} \qquad k_2 = \frac{5}{3}\,\text{min}^{-1} \qquad k_3 = \frac{1}{6}\,\frac{\text{liters}}{\text{mol min}}$$

and the steady-state feed and reactor concentration of component **A** are:

$$C_{Afs} = 10\,\frac{\text{mol}}{\text{liter}} \qquad C_{As} = 3\,\frac{\text{mol}}{\text{liter}}$$

a. Find the steady-state dilution rate (F/V) and concentration of B (show all units).

b. Linearize and put in state-space form (find the numerical values of the A, B, and C matrices), assuming that the manipulated variables are dilution rate (F/V) and feed concentration and that both states are outputs.

c. Find the eigenvalues (show units).

d. Find the transfer functions relating each output to each input. Find the poles and zeros for each transfer function and make plots of the responses to unit step changes in each input. Comment on your results.

BLOCK DIAGRAMS

11

The objective of this chapter is to introduce block diagram analysis. After studying this chapter, you should be able to:

- Analyze the stability of a block diagram system.
- Understand how inverse response processes can arise.
- Understand potential problems with pole-zero cancellation.
- Write a set of differential equations to simulate systems modeled by transfer functions in series.
- Use the MATLAB routines `series`, `parallel`, `feedback`, `conv`, and `roots`.
- Use SIMULINK for block diagram simulation.

Major sections of this chapter are:

FIGURE 11.1 Block diagram representation.

We have shown how Laplace transforms are used to reduce differential equations to algebraic relationships. Algebraic equations are much easier to manipulate than differential equations. Similarly, block diagrams allow us to easily manipulate complex models that are composed of subsets of simple models.

11.1 INTRODUCTION TO BLOCK DIAGRAMS

Consider a standard first-order process model:

$$\tau \frac{dy(t)}{dt} + y(t) = k\, u(t) \tag{11.1}$$

which has the transfer function form:

$$y(s) = g(s)\, u(s) \tag{11.2}$$

where:

$$g(s) = \frac{k}{\tau s + 1} \tag{11.3}$$

Process engineers usually try and solve problems by sketching diagrams to understand input-output relationships. Process control engineers usually use block diagrams to understand the input-output relationships in a dynamic system. A block diagram representation of (11.2) is shown in Figure 11.1.

We can see that $u(s)$ is the input to the transfer function block and $y(s)$ is the output from the transfer function block. Block diagrams will be particularly useful when analyzing complex dynamic systems, which may be represented as blocks in series or parallel and with feedback. They are particularly useful for feedback control system design and analysis.

11.2 BLOCK DIAGRAMS OF SYSTEMS IN SERIES

Consider now the block diagram representation of two processes in series as shown in Figure 11.2.

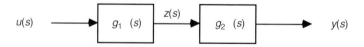

FIGURE 11.2 Block diagram of two processes in series.

The input/output transfer function of Figure 11.2 is:

$$y(s) = g_2(s)\, z(s) = g_2(s)\, g_1(s)\, u(s) \tag{11.4}$$

or,

$$y(s) = g(s)\, u(s) \tag{11.5}$$

where:

$$g(s) = g_2(s)\, g_1(s) \tag{11.6}$$

If the two transfer functions are first-order:

$$g_1(s) = \frac{k_1}{\tau_1 s + 1} \tag{11.7}$$

and

$$g_2(s) = \frac{k_2}{\tau_2 s + 1} \tag{11.8}$$

then the overall process is second order:

$$g(s) = \frac{k}{(\tau_1 s + 1)(\tau_2 s + 1)} \tag{11.9}$$

where:
$$k = k_1 k_2$$

The same idea can be continued for any number of transfer functions in series. The student should notice that the poles of a system composed of many transfer functions in series are simply the poles of each transfer function. This leads to the following conclusion about the *stability of systems with transfer functions in series:*

If a system is composed of transfer functions in series, and if all of those transfer functions are stable, then the overall system is stable.

Also, the zeros of a system of transfer functions in series are simply the zeros of the individual transfer functions.

11.3 POLE-ZERO CANCELLATION

Again, in this section we consider two blocks in series, as shown in Figure 11.2:

$$y(s) = g_2(s)\, z(s) = g_2(s)\, g_1(s)\, u(s) \tag{11.10}$$

If we are not careful, we can overlook possible problems with systems in series, if we look only at the overall input/output relationship. In the next example we show problems with *pole-zero cancellation.*

EXAMPLE 11.1   Lead/Lag in Series with Unstable First-Order System

Consider the following lead/lag in series with an unstable first-order system:

$$g_1(s) = \frac{-2s + 1}{5s + 1} \tag{11.11}$$

$$g_2(s) = \frac{1}{-2s + 1} \tag{11.12}$$

We find that the zeros of $g_1(s)$ cancel the poles of $g_2(s)$

$$y(s) = g_2(s)\, g_1(s)\, u(s) = \left(\frac{-2s + 1}{5s + 1}\right)\left(\frac{1}{-2s + 1}\right) u(s)$$

yielding the transfer function relationship:

$$y(s) = g(s)\, u(s) = \frac{1}{5s + 1}\, u(s) \tag{11.13}$$

We must realize that these transfer functions ultimately represent a physical process. In practice, physical parameters cannot be known perfectly. What this means is that generally the numerator of $g_1(s)$ will not exactly cancel the denominator of $g_2(s)$, in practice.

Consider a realistic case, where $g_2(s)$ has a slight error in the value of the pole

$$g_2(s) = \frac{1}{-2.0001s + 1} \tag{11.14}$$

then we find that

$$y(s) = g_2(s)\, g_1(s)\, u(s) = \left(\frac{-2s + 1}{5s + 1}\right)\left(\frac{1}{-2.0001s + 1}\right) u(s)$$

$$y(s) = g(s)\, u(s) \qquad = \frac{-2s + 1}{-10.0005s^2 + 2.9999s + 1}\, u(s) \tag{11.15}$$

Notice that when we do not have perfect pole/zero cancellation, there is an unstable pole in the input/output relationship, $y(s) = g(s)\, u(s)$. Our goal now is to compare the responses of the two models (11.13) and (11.15). Let $y_1(s)$ represent the output in (11.13) and $y_2(s)$ represent the output in (11.15). Assuming a unit step input, $u(s) = 1/s$,

$$y_1(s) = \frac{1}{s(5s + 1)} \tag{11.16}$$

which has the time domain solution (Chapter 8):

$$y_1(t) = 1 - e^{-t/5} \tag{11.17}$$

Also,

$$y_2(s) = \frac{-2s + 1}{s(-10.0005s^2 + 2.9999s + 1)} \qquad (11.18)$$

which has the time domain solution (Chapter 9):

$$y_2(t) = 1 - \frac{7}{7.0001} e^{-t/5} - \frac{0.0001}{7.0001} e^{t/2.0001} \qquad (11.19)$$

and we can see that, at low t, (11.19) is almost identical to (11.17). As time increases, however, the unstable exponential term in (11.19) begins to dominate. This is shown clearly in Figure 11.3.

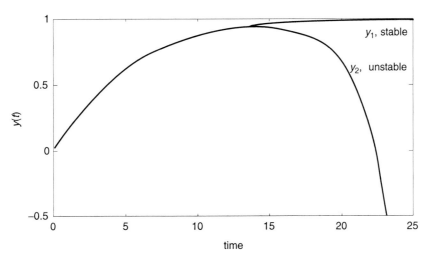

FIGURE 11.3   Comparison of (11.17) and (11.19).

Note that if we had used the state-space form for the model represented by equations (11.10) through (11.12), we would have discovered the instability, even for the perfect parameter case. The following example analyzes the state-space form of Example 11.1.

EXAMPLE 11.1   Continued State-Space Analysis

Refer to $z(s)$ as the output of the lead/lag block. From Chapter 8 we find the following state-space realization of the lead/lag:

$$\frac{dx}{dt} = -\frac{1}{\tau_d} x + \left(1 - \frac{\tau_n}{\tau_d}\right) u$$

$$z = \frac{1}{\tau_d} x + \frac{\tau_n}{\tau_d} u$$

in our case, $g_1(s) = \dfrac{\tau_n s + 1}{\tau_d s + 1} = \dfrac{-2s + 1}{5s + 1}$, so:

$$\frac{dx}{dt} = -\frac{1}{5}x + \frac{7}{5}u \tag{11.20}$$

$$z = \frac{1}{5}x - \frac{2}{5}u \tag{11.21}$$

and the state space realization of the unstable lag is:

$$\frac{dy}{dt} = \frac{1}{2}y - \frac{1}{2}z \tag{11.22}$$

Substituting (11.21) into (11.22), we find

$$\frac{dy}{dt} = \frac{1}{2}y - \frac{1}{10}x + \frac{1}{5}u \tag{11.23}$$

If we use notation

$$x_1 = x$$
$$x_2 = y$$

we can write (11.20) and (11.23) in the following form:

$$\frac{dx_1}{dt} = -\frac{1}{5}x_1 + \frac{7}{5}u \tag{11.24}$$

$$\frac{dx_2}{dt} = \frac{1}{2}x_2 - \frac{1}{10}x_1 + \frac{1}{5}u \tag{11.25}$$

Using the usual state-space notation:

$$\dot{\mathbf{x}} = \mathbf{A}\,\mathbf{x} + \mathbf{B}\,\mathbf{u}$$

$$\mathbf{y} = \mathbf{C}\,\mathbf{x} + \mathbf{D}\,\mathbf{u}$$

we write

$$\begin{bmatrix} \dfrac{dx_1}{dt} \\[2mm] \dfrac{dx_2}{dt} \end{bmatrix} = \begin{bmatrix} -\dfrac{1}{5} & 0 \\[2mm] -\dfrac{1}{10} & \dfrac{1}{2} \end{bmatrix} \begin{bmatrix} x_1 \\ x_2 \end{bmatrix} + \begin{bmatrix} \dfrac{7}{5} \\[2mm] \dfrac{1}{5} \end{bmatrix} u \tag{11.26}$$

$$y = x_2$$

We easily find that the eigenvalues of the A matrix are $-1/5$ and $1/2$. The positive eigenvalue indicates that this system is unstable.

The previous example illustrates the importance of not cancelling an unstable pole with a right-half-plane zero. It also shows how state-space analysis can always be used to address the stability of a system.

11.4   SYSTEMS IN SERIES

The dynamic behavior of chemical processes can often be represented as a series of simple models, such as first-order transfer functions. As an example, consider the following process, which is characterized as n first-order processes with a gain of 1 and a time constant of 5:

$$g(s) = \frac{1}{(5s + 1)^n} \tag{11.27}$$

The step responses for $n = 1$ to 5 are shown in Figure 11.4. Notice the characteristic S-shape for all orders greater than 1 and the additional lag associated with each higher order.

11.4.1   Simulating Systems in Series

Although we analyze processes using transfer functions, to obtain time domain responses we must use a numerical integration package. Consider a system of n first-order processes in series, as shown in Figure 11.5.

Here we write the set of ordinary differential equations that describe this process. The ODE describing the first process is:

$$\frac{dx_1}{dt} = -\frac{1}{\tau_1}x_1 + \frac{k_1}{\tau_1}u \tag{11.28}$$

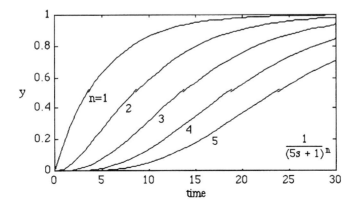

FIGURE 11.4.   Step responses of first-order systems in series.

FIGURE 11.5   *n* processes in series.

Notice that we can think of the output of the first process as the input to the second process:

$$\frac{dx_2}{dt} = -\frac{1}{\tau_2}x_2 + \frac{k_2}{\tau_2}x_1 \tag{11.29}$$

and so on through the *n*th process:

$$\frac{dx_n}{dt} = -\frac{1}{\tau_n}x_n + \frac{k_n}{\tau_n}x_{n-1} \tag{11.30}$$

To solve (11.28) through (11.30) we can use the numerical integration techniques developed in Chapter 4 or the analytical expressions developed in Chapter 6.

We can also write (11.28) through (11.30) in the following state-space form:

$$\begin{bmatrix} \dfrac{dx_1}{dt} \\ \dfrac{dx_2}{dt} \\ \vdots \\ \dfrac{dx_{n-1}}{dt} \\ \dfrac{dx_n}{dt} \end{bmatrix} = \begin{bmatrix} -\dfrac{1}{\tau_1} & 0 & \cdots & 0 & 0 \\ \dfrac{k_2}{\tau_2} & -\dfrac{1}{\tau_2} & \cdots & 0 & 0 \\ \vdots & \vdots & \ddots & & \vdots \\ 0 & 0 & \cdots & -\dfrac{1}{\tau_{n-1}} & 0 \\ 0 & 0 & \cdots & \dfrac{k_n}{\tau_n} & -\dfrac{1}{\tau_n} \end{bmatrix} \begin{bmatrix} x_1 \\ x_2 \\ \vdots \\ \vdots \\ x_{n-1} \\ x_n \end{bmatrix} + \begin{bmatrix} \dfrac{k_1}{\tau_1} \\ 0 \\ \vdots \\ \vdots \\ 0 \\ 0 \end{bmatrix} u \tag{11.26}$$

In Example 11.2 we show how to use the MATLAB routines `series` and `conv` to find a transfer function that represents two blocks in series.

EXAMPLE 11.2   Two Transfer Functions in Series

Consider two processes, $g_1(s)$ and $g_2(s)$, in series, where:

$$g_1(s) = \frac{1.5}{2s + 1}$$

$$g_2(s) = \frac{3}{4s + 1}$$

$$g(s) = g_1(s)\,g_2(s)$$

We use the following MATLAB commands to enter the numerator and denominator polynomials for each transfer function:

```
» num1 = [1.5];

» den1 = [2 1];

» num2 = [3];

» den2 = [4 1];
```

the `series` command generates the numerator and denominator polynomials for the transfer function $g(s)$:

```
»[num,den] = series(num1,den1,num2,den2)

num =

  0  0  4.5000

den =

  8  6  1
```

which indicates that

$$g(s) = \frac{4.5}{8s^2 + 6s + 1}$$

conv

`conv` is used to multiply two polynomials. Using the previous example, we multiply the numerator polynomials to find:

```
» num = conv(num1,num2)

num =

  4.5000
```

and the denominator polynomials to find:

```
» den = conv(den1,den2)

den =

  8  6  1
```

11.5 BLOCKS IN PARALLEL

Sometimes the behavior of a chemical process can be modeled by transfer functions in parallel as shown in Figure 11.6.

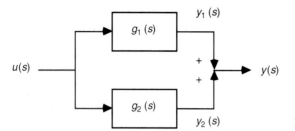

FIGURE 11.6   Systems in parallel.

For this system we can write the total output, $y(s)$, as the sum of two outputs, $y_1(s) + y_2(s)$.

$$y(s) = y_1(s) + y_2(s) \tag{11.31}$$

$$y(s) = [g_1(s) + g_2(s)]\, u(s) \tag{11.32}$$

or,

$$y(s) = g(s)\, u(s) \tag{11.33}$$

where:

$$g(s) = g_1(s) + g_2(s) \tag{11.34}$$

Consider the case where $g_1(s)$ and $g_2(s)$ are first-order transfer functions:

$$g_1(s) = \frac{k_1}{\tau_1 s + 1} \tag{11.35}$$

$$g_2(s) = \frac{k_2}{\tau_2 s + 1} \tag{11.36}$$

so

$$g(s) = \frac{k_1}{\tau_1 s + 1} + \frac{k_2}{\tau_2 s + 1} \tag{11.37}$$

Developing a common denominator, we find:

$$g(s) = \frac{(k_1 + k_2)\left[\left(\dfrac{k_1 \tau_2 + k_2 \tau_1}{k_1 + k_2}\right)s + 1\right]}{(\tau_1 s + 1)\,(\tau_2 s + 1)} \tag{11.38}$$

Notice that (11.38) has the form (see Chapter 9):

$$g(s) = \frac{k(\tau_n s + 1)}{(\tau_1 s + 1)\,(\tau_2 s + 1)} \tag{11.39}$$

where:

$$k = k_1 + k_2 \qquad (11.40)$$

$$\tau_n = \frac{k_1 \tau_2 + k_2 \tau_1}{k_1 + k_2} \qquad (11.41)$$

We will assume that the transfer functions $g_1(s)$ and $g_2(s)$ are stable, so τ_1 and $\tau_2 > 0$. The goal of this section is to show a system where inverse response (discussed in Chapter 9 and Example 9.3) behavior can occur.

11.5.1   Conditions for Inverse Response

Recall that a transfer function will have inverse response only if there is a right-half-plane (positive) zero. Since the zero is $-1/\tau_n$, this system will have inverse response only if $\tau_n < 0$. We find that $\tau_n < 0$ only if k_1 and k_2 are of opposite sign. We can arbitrarily assume that $k_1 > 0$, which means that $k_2 < 0$ is necessary for inverse response. For inverse response, the condition:

$$\tau_n < 0 \quad \text{means that} \quad \frac{k_1 \tau_2 + k_2 \tau_1}{k_1 + k_2} < 0$$

or,

$$\frac{k_1 \tau_2}{k_1 + k_2} < \frac{-k_2 \tau_1}{k_1 + k_2}$$

which yields the following conditions for inverse response.

1. If $k_1 + k_2 > 0$, then $k_1 \tau_2 < -k_2 \tau_1$, which implies that τ_2/τ_1 must be $< -k_2/k_1$ for inverse response.
2. If $k_1 + k_2 < 0$, then $k_1 \tau_2 > -k_2 \tau_1$, which implies that τ_2/τ_1 must be $> -k_2/k_1$ for inverse response.

Physical examples of systems with inverse response include: steam drum level, reboilers in distillation columns, chemical and biochemical reactors. A reason that inverse response behavior is important is that it creates tremendous challenges for tight process control.

We can use the MATLAB routine `parallel` to simulate two systems in parallel, as shown by the next example.

EXAMPLE 11.3   Two systems in parallel

Consider the following system of two first-order processes in parallel (Figure 11.7):

$$g_1(s) = \frac{2}{5s + 1}$$

$$g_2(s) = \frac{-1}{1s + 1}$$

$$g(s) = \frac{2}{5s + 1} + \frac{-1}{1s + 1}$$

```
» num1 = [2];
» den1 = [5    1];
» num2 = [-1];
» den2 = [1    1];
```

The following command is used to find the new transfer function:

```
» [num,den] = parallel(num1,den1,num2,den2)
num =
   0 -3 1
den =
   5 6 1
» [y,x,t] = step(num,den);
» [y1,x1] = step(num1,den1,t);
» [y2,x2] = step(num2,den2,t);
» plot(t,y,t,y1,t,y2)
```

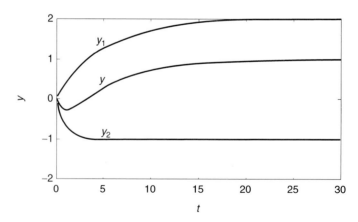

FIGURE 11.7 Two systems in parallel that have an inverse response when added together.

This previous example has shown that inverse response occurs in systems where the gain of the "slow process" (larger time constant) is larger in magnitude (but opposite in sign) than the "fast process" (smaller time constant).

11.6   FEEDBACK AND RECYCLE SYSTEMS

Feedback systems are common in engineering. Examples include chemical and biochemical reactors, where a certain portion of the product stream may be recycled to the feed-stream. Feedback naturally occurs in most "self-regulating" models where, for example, the rate of change of a state variable (say, concentration of A) is a function of the same or another state variable (say, concentration of B).

The entire field of process control is based on the concept and theory of feedback systems. Our goal with this section is to introduce feedback analysis and, in particular, stability analysis of feedback systems. A block diagram of a feedback system is shown in Figure 11.8.

In this figure, the input to the feedback system is $r(s)$ and the output is $y(s)$. Here we develop the relationship between $r(s)$ and $y(s)$.

$$y(s) = g_1(s)\, u(s) \tag{11.44}$$

but

$$u(s) = r(s) + z(s) \tag{11.45}$$

and

$$z(s) = g_2(s)\, y(s) \tag{11.46}$$

So we can write (11.44) as:

$$y(s) = g_1(s)\, (r(s) + g_2(s)y(s)) \tag{11.47}$$

Solving for $y(s)$ we find:

$$y(s) = \frac{g_1(s)}{1 - g_1(s)g_2(s)}\, r(s) \tag{11.48}$$

Notice that we can view this as:

$$y(s) = g_{c1}(s)\, r(s) \tag{11.49}$$

where

$$g_{c1}(s) = \frac{g_1(s)}{1 - g_1(s)g_2(s)} \tag{11.50}$$

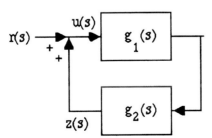

FIGURE 11.8 Feedback diagram.

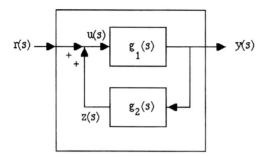

FIGURE 11.9 Equivalent block diagrams.

and we know that if the poles of $g_{cl}(s)$ are stable, then the feedback system is stable. We realize that the two block diagrams shown in Figure 11.9 are equivalent.

EXAMPLE 11.4   Feedback system

Consider two first-order process transfer functions:

$$g_1(s) = \frac{k_1}{\tau_1 s + 1} \tag{11.51}$$

$$g_2(s) = \frac{k_2}{\tau_2 s + 1} \tag{11.52}$$

$$g_{cl}(s) = \frac{g_1(s)}{1 - g_1(s)g_2(s)} = \frac{\dfrac{k_1}{\tau_1 s + 1}}{1 - \left(\dfrac{k_1}{\tau_1 s + 1}\right)\left(\dfrac{k_2}{\tau_2 s + 1}\right)}$$

$$= \frac{k_1(\tau_2 s + 1)}{(\tau_1 s + 1)(\tau_2 s + 1) - k_1 k_2}$$

$$g_{cl}(s) = \frac{k_1(\tau_2 s + 1)}{(\tau_1 \tau_2 s^2 + (\tau_1 + \tau_2)s + 1 - k_1 k_2)} \tag{11.53}$$

and $g_{cl}(s)$ is stable if the roots of $\tau_1 \tau_2 s^2 + (\tau_1 + \tau_2)s + 1 - k_1 k_2$ are stable. We recall from the Routh stability criterion that all of the roots of a quadratic polynomial are negative if the coefficients of the polynomial are positive. If we assume that τ_1 and τ_2 are positive, then (11.53) will be stable if $1 - k_1 k_2$ is positive. For stability, then, $k_1 k_2$ must be less than 1. Let's consider the following numerical example:

$$g_1(s) = \frac{2}{5s + 1}$$

$$g_2(s) = \frac{k_2}{10s + 1}$$

Since $k_1 = 2$, then k_2 must be less than 0.5 for stability.

As a numerical check, let $k_2 = -1$. Solving for the roots of:

$$\tau_1 t_2 s^2 + (\tau_1 + \tau_2)s + 1 - k_1 k_2 = 0$$

we find

$$50\,s^2 + 15\,s + 3 = 0$$

which has the roots (using the quadratic formula)

$$-0.15 \pm 0.1936j$$

(we can verify this result using the MATLAB routine `roots`)

Since the real part of the roots is negative, the system is stable. This is verified in the MATLAB simulation presented in Figure 11.10, where the response of the output to a unit step change in r is presented.

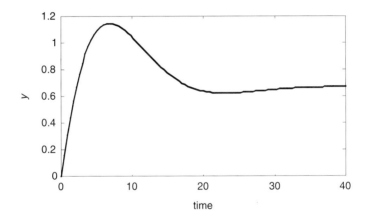

FIGURE 11.10   Step response for the example feedback system.

We can also use the MATLAB `feedback` function to obtain the closed-loop transfer function, as shown below.

$$g_1(s) = \frac{2}{5s + 1}$$

$$g_2(s) = \frac{-1}{10s + 1}$$

```
» num1 = [2];

» num2 = [-1];

» den1 = [5 1];

» den2 = [10 1];
```

and, using the routine `feedback`

```
» [num,den] = feedback(num1,den1,num2,den2,1)

num =

     0 20 2

den =

    50 15 3
```

We use the routine `step` to find the step response:

$$[y,x,t] = step(num,den)$$

which gives the plot shown in Figure 11.10.

11.7 ROUTH STABILITY CRITERION APPLIED TO TRANSFER FUNCTIONS

Recall from Chapter 6 that the purpose of the Routh stability criterion is to determine if a polynomial with the following form has any positive roots:

$$a_n\lambda^n + a_{n-1}\lambda^{n-1} + ... + a_1\lambda + a_o = 0 \qquad (11.54)$$

Since transfer functions that have denominator polynomials in the Laplace transform variable (s) are are the same form as (11.54), we can use Routh analysis to determine the stability of transfer funtions. As before, assume that $a_n > 0$. If $a_n < 0$, then multiply (11.54) by −1. A *necessary* condition for stability is that all of the coefficients in (11.54) must be positive. If any of the coefficients are negative or zero, then at least one pole (root of the characteristic equation) is positive or zero, indicating that the equation is unstable. Even if all of the coefficients are positive, we cannot state that the system is stable. What is needed is a *sufficient* condition for stability. To determine that the system is stable, we must construct the Routh array and use the Routh stability criterion, which provides necessary and sufficient conditions for stability.

Sometimes we simply wish to determine if a particular system is stable or not, without actually evaluating the eigenvalues. This is particularly true if we wish to determine values of system parameters that will cause a system to lose stability. This approach will be useful in performing a bifurcation analysis in later chapters (14 and 15), and in tuning control systems for chemical processes.

11.7.1 Routh Array

If all of the coeffients of the characteristic equation (11.54) are positive, then develop the following Routh array:

Row
1	a_n	a_{n-2}	a_{n-4}	$\cdots$
2	a_{n-1}	a_{n-3}	a_{n-5}	$\cdots$
3	b_1	b_2	b_3	$\cdots$
4	c_1	c_2	$\cdots$	

.   .

$n+1$

where n is the order of the polynomial. Notice that the first two rows consist of the coefficients of the polynomial. The elements of the third row are calculated in the following fashion:

$$b_1 = \frac{a_{n-1} a_{n-2} - a_n a_{n-3}}{a_{n-1}} \qquad b_2 = \frac{a_{n-1} a_{n-4} - a_n a_{n-5}}{a_{n-1}}$$

and so on. Elements of the fourth and larger rows are calculated in a similar fashion:

$$c_1 = \frac{b_1 a_{n-3} - a_{n-1} b_2}{b_1} \qquad c_2 = \frac{b_1 a_{n-5} - a_{n-1} b_3}{b_1}$$

and so on.

A sufficient condition for all roots of the characteristic polynomial to have negative real parts is that all of the elements in the first column of the Routh array are positive.

EXAMPLE 11.5   Routh Array to Determine Closed-Loop Stability

Consider the block diagram of Figure 11.9.

$$y(s) = \frac{g_1(s)}{1 - g_1(s)g_2(s)} r(s)$$

or,

$$y(s) = g_{cl}(s) \, r(s)$$

where:

$$g_{cl}(s) = \frac{g_1(s)}{1 - g_1(s)g_2(s)}$$

And the transfer functions are:

$$g_1(s) = \frac{2}{(5s+1)(3s+1)}$$

$$g_2(s) = \frac{k_2}{10s+1}$$

Our goal is to find k_2 to assure stability of the closed-loop system.
We easily find the transfer function, $g(s)$:

$$g(s) = \frac{2(10s+1)}{(150s^3 + 95s^2 + 18s + 1 - 2k_2)}$$

which has the characteristic polynomial

$$150s^3 + 95s^2 + 18s + 1 - 2k_2$$

which is of the form

$$a_3 s^3 + a_2 s^2 + a_1 s + a_o$$

The Routh array is:

Row
1   150   18
2   95 $1 - 2k_2$
3   b_1 0
4   c_1

The *necessary* condition is that all $a_i > 0$, which is satisfied if $1 - 2k_2 > 0$, or $k_2 < 0.5$.
The sufficient condition is satisfied if all of the coefficients in the first column of the Routh array are positive.

$$b_1 = \frac{a_2 a_1 - a_3 a_o}{a_2} = 18 - \frac{150}{90}(1 - 2k_2) > 0$$

$$c_1 = \frac{b_1 a_0 - a_2 b_2}{b_1} = a_0 = 1 - 2k_2 > 0$$

The b_1 condition is satisfied if $k_2 > -5.2$, while the c_1 condition is the same as the necessary condition. We then have the following restriction on k_2 for stability:

$$-5.2 < k_2 < 0.5$$

11.8 SIMULINK

In the previous sections we have shown how MATLAB routines can be used for block diagram analysis and simulation. The objective of this section is to use the block diagram simulation features of SIMULINK.

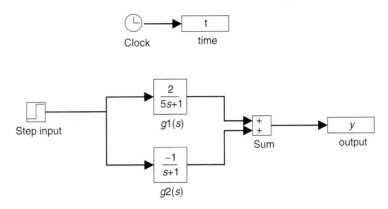

FIGURE 11.11 SIMULINK block diagram for two blocks in parallel.

Consider the block diagram system from Example 11.3. A SIMULINK block diagram is shown in Figure 11.11. Notice the use of step, transfer function, sum, workspace and clock blocks to generate the necessary input and output information.

The parameters menu is used to specify the integration type (LINSIM), final time (30), and mininum (0.01) and maximum (1) step sizes. The results are the same as shown in Figure 11.7. More information on SIMULINK is provided in Module 4 in the final section of the text.

SUMMARY

Block diagram analysis is important because it allows us to think about a system of processes in terms of a combination of the individual processes. We have shown how to analyze the stability of a block diagram system, particularly if there are recycle or feedback processes. We have shown how inverse response processes can arise from systems in parallel. We have also shown potential problems with pole-zero cancellation when analyzing transfer functions in series.

The following MATLAB routines were used:

series:	two models in series (either transfer function or state space)
conv:	multiplies two polynomials
parallel:	two models in parallel (either transfer function or state space)
feedback:	two models in feedback form (either transfer function or state space, and either positive or negative feedback)
roots:	find the roots (zeros) of a polynomial

SIMULINK has also been used for block diagram simulation.

STUDENT EXERCISES

1. Consider a first-order process that has a positive pole (negative time constant), indicating that the process is open-loop unstable.

$$g_1(s) = \frac{1}{-5s + 1}$$

It is desirable to design a feedback compensator $g_2(s)$, so that the feedback system is stable. Assume that $g_2(s)$ is simply a gain:

$$g_2(s) = k_2$$

Find the range of gains that will make the following feedback system stable.

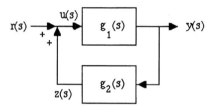

2. Consider the recycle system shown below, where:

$$g_1(s) = \frac{1}{(s - 1)(s + 0.5)}$$

$$g_2(s) = k$$

Find the values of k (if any) that will ensure stability of the system. Show your work and explain your reasoning.

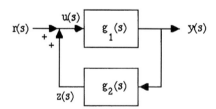

3. Find the analytical solution for a unit step applied to the following process:

$$g(s) = \frac{1}{(5s + 1)^5}$$

4. Consider the recycle system shown below, where:

$$g_1(s) = \frac{1}{(s + 1)}$$

$$g_2(s) = \frac{k_2}{(\tau_2 s + 1)}$$

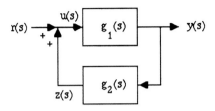

Discuss how the values of k_2 and τ_2 effect the dynamic behavior of y with respect to a unit step input change in r. Use SIMULINK and show compare plots for various values of k_2 and τ_2 to illustrate your points.

LINEAR SYSTEMS SUMMARY   12

One purpose of this chapter is to summarize the techniques that have been developed in Chapters 5 through 11 to solve linear ordinary differential equations. Since the focus has been on initial value problems, we also introduce techniques to solve boundary value ODE problems. Also, since the emphasis has been on continous (differential equation-based) models, another objective is to introduce discrete models. After studying this chapter, the student should be able to:

- Use the characteristic equation method to solve boundary value linear ODE problems
- Select an appropriate technique to solve a particular linear initial value problem
- Formulate linear discrete-time models
- Estimate parameters for linear discrete-time models

The major sections of this chapter are:

12.1 BACKGROUND

Thus far in this text, all of the problems that we have solved have been initial value ordinary differential equations. To solve these problems we simply need to know the initial values of the state variables, and how the inputs change with time. The models are then integrated to find how the states change with time. Ordinary differential equation models may be constrained to satisfy boundary conditions. Boundary value problems often arise when solving for the steady-state behavior of a dynamic system modeled by a partial differential equation. Typically, a boundary value problem has distance as the independent variable and the boundary conditions that must be satisfied are the values of the state variables at different locations (typically at each "end" of the system).

Recall that in Chapter 11 we required n initial conditions to solve an nth order initial value ODE. Similarly, we require n boundary conditions to solve an nth order boundary value ODE. Most chemical processes that can be modeled as second-order boundary value problems (e.g., the reaction-diffusion equation) are two-point boundary value problems. A second-order split boundary value problem has a boundary condition at one end and another boundary condition at the other end. If both boundary conditions were at the front end, then our problem would be an inital value problem. If both boundary conditions were at the rear end, then we would have an initial value problem by simply redefining the independent variable and forming an initial value problem in the opposite direction.

In this chapter, we first cover linear boundary value problems in Section 12.2 and review methods to solve linear initial value problems in Section 12.3. We provide an introduction to discrete-time models in Section 12.4 and show how to estimate parameters for discrete-time models in Section 12.5.

12.2 LINEAR BOUNDARY VALUE PROBLEMS

An analytical solution to boundary value ordinary conditions can be obtained using the method of characteristics when the ODE and the boundary conditions are linear. Consider the following second order ODE

$$a_2 \frac{d^2x}{dz^2} + a_1 \frac{dx}{dz} + a_0 x = 0 \qquad (12.1)$$

where a_0, a_1 and a_2 are constant coefficients, x is the state variable (dependent) and z is the independent variable (often distance). The solution to (12.1) will have the form

$$x = c_1 e^{\lambda_1 z} + c_2 e^{\lambda_2 z} \qquad (12.2)$$

where λ_1 and λ_2 are obtained by rewriting (12.1) as

$$a_2 \lambda^2 + a_1 \lambda + a_0 = 0 \qquad (12.3)$$

using the method discussed in Chapter 6. The constant coefficients (c_1 and c_2) are obtained from the boundary conditions.

EXAMPLE 12.1 Second-order Boundary Value Problem

Consider the following second-order equation:

$$\frac{d^2x}{dz^2} + 4\frac{dx}{dz} + \frac{7}{4}x = 0 \tag{12.4}$$

subject to the boundary conditions at each end:

$$x(z = 0) = 2 \tag{12.5}$$

$$x(z = 1) = 1 \tag{12.6}$$

We solve for the eigenvalues by using the characteristic equation:

$$\lambda^2 + 4\lambda + \frac{7}{4} = 0 \tag{12.7}$$

which yields (from the quadratic formula):

$$\lambda_1 = -3.5 \tag{12.8}$$

$$\lambda_2 = -0.5 \tag{12.9}$$

and the solution is:

$$x = c_1 e^{-3.5z} + c_2 e^{-0.5z} \tag{12.10}$$

Substituting the boundary conditions results in two equations and two unknowns:

$$2 = c_1 + c_2 \tag{12.11}$$

$$1 = c_1 e^{-3.5} + c_2 e^{-0.5} \tag{12.12}$$

which yields:

$$c_1 = 0.36968$$

$$c_2 = 1.63032$$

A plot of the solution, $x = 0.36968\, e^{-3.5z} + 1.63032\, e^{-0.5z}$, is shown in Figure 12.1.

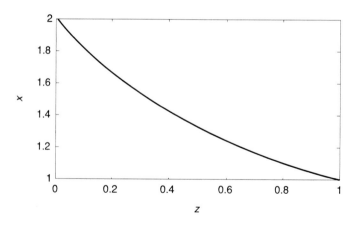

FIGURE 12.1 Solution to Example 12.1.

More generally, the boundary conditions may consist of some function of the state variable and its derivative. The more general linear boundary condition is the form:

$$b_1 \frac{dx}{dz} + b_0 x = d$$

EXAMPLE 12.2   Second-Order Boundary Value Problem

Consider the second-order problem from the previous example:

$$\frac{d^2x}{dz^2} + 4 \frac{dx}{dz} + \frac{7}{4} x = 0 \qquad (12.4)$$

subject to the new boundary conditions,

$$\frac{dx}{dz} + x = 1 \qquad \text{at } z = 0 \qquad (12.13)$$

$$\frac{dx}{dz} = 0 \qquad \text{at } z = 1 \qquad (12.14)$$

Since the solution is:

$$x = c_1 e^{-3.5z} + c_2 e^{-0.5z} \qquad (12.15)$$

then the first derivative with respect to z is:

$$\frac{dx}{dz} = -3.5 \, c_1 e^{-3.5z} - 0.5 \, c_2 e^{-0.5z} \qquad (12.16)$$

and boundary condition (12.13) yields:

$$-3.5 \, c_1 - 0.5 \, c_2 + c_1 + c_2 = 1 \qquad (12.17)$$

while boundary condition (12.14) yields:

$$-3.5 \, c_1 e^{-3.5} - 0.5 \, c_2 e^{-0.5} = 0 \qquad (12.18)$$

Solving these two equations for c_1 and c_2, we obtain the solution:

$$x = -0.37394 \, e^{-3.5z} + 0.13032 \, e^{-0.5z} \qquad (12.19)$$

which is shown in Figure 12.2.

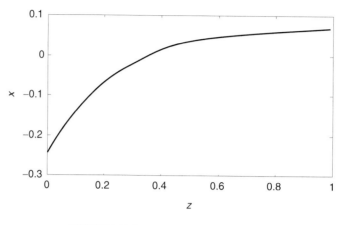

FIGURE 12.2   Solution to Example 12.2.

We have illustrated how the method of characteristics is used to solve linear boundary value problems. The solution to nonlinear boundary value problems generally involves iterative methods. For example, consider a single second-order nonlinear problem with boundary conditions at each end. We know that the second-order equation can be converted to two first-order equations. Typically, one boundary condition will fix an "initial condition" for one of the states. A second initial condition can be iteratively guessed (using a Quasi-Newton method, for example) until the equations, when integrated, yield the correct value for the end boundary condition. This approach is shown in Example 12.3 for the linear system considered in Example 12.1.

EXAMPLE 12.3   Formulating a Boundary Value Problem as an Iterative Initial Value Problem

Consider the second-order boundary value problem:

$$\frac{d^2x}{dz^2} + 4\frac{dx}{dz} + \frac{7}{4}x = 0 \tag{12.4}$$

with the boundary conditions:

$$x(z = 0) = 2 \tag{12.5}$$

$$x(z = 1) = 1 \tag{12.6}$$

It can be shown that (see student exercise 1), by defining $x_1 = x$ and $x_2 = dx/dz$, the following equations are obtained:

$$dx_1/dz = x_2 \tag{12.20}$$

$$dx_2/dz = -\frac{7}{4}x_1 - 4x_2 \tag{12.21}$$

and that one of the initial conditions is

$$x_1(z = 0) = 2 \tag{12.22}$$

We see that $x_2(z = 0)$ must be "guessed," then the two equations can be integrated from $z = 0$ to $z = 1$. The value of x_1 at $z = 1$ is then checked; if $x_1(z = 1)$ is not equal to 1 (within an acceptable tolerance) then values of $x_2(z = 0)$ are iteratively guessed until the final value is satisfied. This method is known as the "shooting method." The reader is encouraged to use this approach to solve exercise 1.

12.3   REVIEW OF METHODS FOR LINEAR INITIAL VALUE PROBLEMS

In Chapters 5 through 11 we presented a number of techniques for solving linear initial value ordinary differential equations. In Chapter 5 we noted that dynamic chemical process models are often formulated as a set of first-order, nonlinear differential equations, where the initial values are known. These equations have the general form:

$$\dot{\mathbf{x}} = \mathbf{f}(\mathbf{x}, \mathbf{u}) \tag{12.23}$$

$$\mathbf{y} = \mathbf{g}(\mathbf{x}, \mathbf{u}) \tag{12.24}$$

where $\mathbf{x}$ is a vector of n state variables, $\mathbf{u}$ is a vector of m input variables, and $\mathbf{y}$ is a vector of r output variables:

$$\dot{x}_1 = f_1(x_1, \ldots, x_n, u_1, \ldots, u_m)$$

$$\cdot \qquad \cdot$$

$$\cdot \qquad \cdot$$

$$\dot{x}_n = f_n(x_1, \ldots, x_n, u_1, \ldots, u_m)$$
$$y_1 = g_1(x_1, \ldots, x_n, u_1, \ldots, u_m)$$

$$\cdot \qquad \cdot$$

$$\cdot \qquad \cdot$$

$$y_r = g_r(x_1, \ldots, x_n, u_1, \ldots, u_m)$$

12.3.1   Linearization

Elements of the linearization matrices are defined in the following fashion:

$$\mathbf{A}_{ij} = \left. \frac{\partial f_i}{\partial x_j} \right|_{\mathbf{x}_s,\mathbf{u}_s} \qquad \mathbf{B}_{ij} = \left. \frac{\partial f_i}{\partial u_j} \right|_{\mathbf{x}_s,\mathbf{u}_s}$$

$$\mathbf{C}_{ij} = \left. \frac{\partial g_i}{\partial x_j} \right|_{\mathbf{x}_s,\mathbf{u}_s} \qquad \mathbf{D}_{ij} = \left. \frac{\partial g_i}{\partial u_j} \right|_{\mathbf{x}_s,\mathbf{u}_s}$$

where $\mathbf{x}_s$, $\mathbf{u}_s$, and $\mathbf{y}_s$ represent the steady-state values of the states, inputs, and outputs, which solve:

$$\mathbf{0} = \mathbf{f}(\mathbf{x}_s,\mathbf{u}_s) \tag{12.25}$$

$$\mathbf{y}_s = \mathbf{g}(\mathbf{x}_s,\mathbf{u}_s) \tag{12.26}$$

After linearization, we have the state space form:

$$\dot{\mathbf{x}}' = \mathbf{A}\,\mathbf{x}' + \mathbf{B}\,\mathbf{u}' \tag{12.27}$$

$$\mathbf{y}' = \mathbf{C}\,\mathbf{x}' + \mathbf{D}\,\mathbf{u}' \tag{12.28}$$

where the deviation variable vectors are:

$$\mathbf{x}' = \mathbf{x} - \mathbf{x}_\mathbf{s} \tag{12.29}$$

$$\mathbf{u}' = \mathbf{u} - \mathbf{u}_\mathbf{s} \tag{12.30}$$

Generally, the (′) notation is dropped and it is understood that the model is in deviation variable form:

$$\dot{\mathbf{x}} = \mathbf{A}\,\mathbf{x} + \mathbf{B}\,\mathbf{u} \tag{12.31}$$

$$\mathbf{y} = \mathbf{C}\,\mathbf{x} + \mathbf{D}\,\mathbf{u} \tag{12.32}$$

Once the model is in this form, a number of techniques can be used.

12.3.2   Direction Solution Techniques

a. Solve the zero-input (perturbation in initial conditions) form (Chapter 5):

$$\mathbf{x}(t) = e^{At}\,\mathbf{x}(0) \tag{12.33}$$

One way the matrix exponential can be solved is

$$e^{At} = \mathbf{V}\,e^{\Lambda t}\,\mathbf{V}^{-1} \tag{12.34}$$

The MATLAB function for matrix exponential is expm.

b. For a constant step input at time zero (Chapter 5):

$$\mathbf{x}(t) = e^{At}\,\mathbf{x}(0) + (e^{At} - \mathbf{I})\,\mathbf{A}^{-1}\,\mathbf{B}\,\mathbf{u}(0) \tag{12.35}$$

c. For inputs that are constant over each time step (from t to $t + \Delta t$) (Chapter 5):

$$\mathbf{x}(t + \Delta t) = e^{A\Delta t}\,\mathbf{x}(t) + (e^{A\Delta t} - \mathbf{I})\,\mathbf{A}^{-1}\,\mathbf{B}\,\mathbf{u}(t) \tag{12.36}$$

which is often written as:

$$\mathbf{x}(k+1) = e^{A\Delta t}\mathbf{x}(k) + (e^{A\Delta t} - \mathbf{I})\,\mathbf{A}^{-1}\,\mathbf{B}\,\mathbf{u}(k) \tag{12.37}$$

where k represents the kth time step. This represents a discrete-time model, which is discussed in more detail in Section 12.4.

12.3.3   Rewrite the State-Space Model as a Single nth Order Ordinary Differential Equation

a. Solve the homogeneous problem (Chapter 6):

$$a_n \frac{d^n x}{dt^n} + a_{n-1} \frac{d^{n-1}x}{dt^{n-1}} + \dots + a_1 \frac{dx}{dt} + a_o x = 0 \tag{12.38}$$

by first writing the characteristic equation:

$$a_n \lambda^n + a_{n-1} \lambda^{n-1} + \dots + a_1 \lambda + a_o = 0 \tag{12.39}$$

and solving for the roots (eigenvalues) of the nth order polynomial. If the roots are distinct, the solution is of the form:

$$x(t) = c_1 e^{\lambda 1 t} + c_2 e^{\lambda 2 t} + \dots + c_n e^{\lambda n t} \tag{12.40}$$

where the coefficients are found using the n initial conditions.

b. Solve the nonhomogeneous problem using the method of undetermined coefficients (Chapter 6):

$$a_n \frac{d^n x}{dt^n} + a_{n-1} \frac{d^{n-1}x}{dt^{n-1}} + \dots + a_1 \frac{dx}{dt} + a_o x = f(u(t)) \tag{12.41}$$

using a three-step procedure.

  i. Solve the *homogeneous* problem to find:

$$x_H(t) \tag{12.42}$$

  ii. Solve for the particular solution by determining the coefficients of a trial function (see Table 6.1, Chapter 6) that satisfy the *nonhomogeneous* equation:

$$x_P(t) \tag{12.43}$$

  iii. Combine the two solutions for:

$$x(t) = x_H(t) + x_P(t) \tag{12.44}$$

c. Use Laplace transforms to solve the nth order equation (most useful for nonhomogeneous equations) (Chapters 7–11):

$$x(s) = \frac{1}{a_n s^n + a_{n-1} s^{n-1} + \dots + a_1 s + a_o} u(s) \tag{12.45}$$

which corresponds to the differential equation:

$$a_n \frac{d^n x}{dt^n} + a_{n-1} \frac{d^{n-1} x}{dt^{n-1}} + \ldots + a_1 \frac{dx}{dt} + a_o x = u(t) \tag{12.46}$$

The more general case is:

$$x(s) = \frac{b_n s^n + b_{n-1} s^{n-1} + \ldots + b_1 s + b_0}{a_n s^n + a_{n-1} s^{n-1} + \ldots + a_1 s + a_o} u(s) \tag{12.47}$$

which corresponds to the differential equation:

$$a_n \frac{d^n x}{dt^n} + a_{n-1} \frac{d^{n-1} x}{dt^{n-1}} + \ldots + a_1 \frac{dx}{dt} + a_o x$$

$$= b_n \frac{d^n u}{dt^n} + b_{n-1} \frac{d^{n-1} u}{dt^{n-1}} + \ldots + b_1 \frac{du}{dt} + b_o u \tag{12.48}$$

For physically realizable systems, $b_n = 0$. Often many of the leading b_i terms are zero. If the leading r terms in the b polynomial are zero, then the system is referred to as relative order r.

12.3.4   Use Laplace Transforms Directly on the State-Space Model

Previously we have assumed that the state-space model has already been converted to a single nth order differential equation. We can also transform the set of n first-order linear state space equations directly using:

$$\mathbf{Y}(s) = [\mathbf{C} \, (s\mathbf{I} - \mathbf{A})^{-1} \, \mathbf{B} + \mathbf{D}] \, \mathbf{U}(s) \tag{12.49}$$

Generally, the Laplace transforms technique is used for nonhomogeneous problems, that is, systems that have an input forcing function (such as a step).

12.4   INTRODUCTION TO DISCRETE-TIME MODELS

Consider the general linear state space model:

$$\begin{bmatrix} \dot{x}_1 \\ \cdot \\ \cdot \\ \cdot \\ \dot{x}_n \end{bmatrix} = \begin{bmatrix} a_{11} & a_{12} & \cdot & a_{1n} \\ \cdot & \cdot & \cdot & \cdot \\ \cdot & \cdot & \cdot & \cdot \\ \cdot & \cdot & \cdot & \cdot \\ a_{n1} & a_{n2} & \cdot & a_{nn} \end{bmatrix} \begin{bmatrix} x_1 \\ \cdot \\ \cdot \\ \cdot \\ x_n \end{bmatrix} + \begin{bmatrix} b_{11} & b_{12} & \cdot & b_{1m} \\ \cdot & \cdot & \cdot & \cdot \\ \cdot & \cdot & \cdot & \cdot \\ \cdot & \cdot & \cdot & \cdot \\ b_{n1} & b_{n2} & \cdot & b_{nm} \end{bmatrix} \begin{bmatrix} u_1 \\ \cdot \\ \cdot \\ \cdot \\ u_m \end{bmatrix}$$

or,

$$\dot{\mathbf{x}} = \mathbf{A}\,\mathbf{x} + \mathbf{B}\,\mathbf{u} \tag{12.31}$$

Recall that the single variable equation:

$$\dot{x} = a\,x + b\,u \tag{12.50}$$

has the solution:

$$x(t) = e^{at}\,x(0) + (e^{at} - 1)\frac{b}{a}\,u(0) \tag{12.51}$$

when $u(t) = \text{constant} = u(0)$.

In a similar fashion, the solution to (12.31), for a constant input ($\mathbf{u}(t) = \mathbf{u}(0)$) from $t = 0$ to t is:

$$\mathbf{x}(t) = \mathbf{\Phi}\,\mathbf{x}(0) + \mathbf{\Gamma}\,\mathbf{u}(0) \tag{12.52}$$

where:

$$\mathbf{\Phi} = e^{At} \tag{12.53}$$

and

$$G = (\mathbf{\Phi} - \mathbf{I})\,\mathbf{A}^{-1}\mathbf{B} \tag{12.54}$$

Equation (12.52) can be used to solve for a system where the inputs change from time step to time step (t to $t+\Delta t$) by using:

$$\mathbf{x}(t + \Delta t) = \mathbf{\Phi}\,\mathbf{x}(t) + \mathbf{\Gamma}\,\mathbf{u}(t) \tag{12.55}$$

More often this is written as:

$$\mathbf{x}(k + 1) = \mathbf{\Phi}\,\mathbf{x}(k) + \mathbf{\Gamma}\,\mathbf{u}(k) \tag{12.56}$$

where k represents the kth time step. The output at time step k is written:

$$\mathbf{y}(k) = \mathbf{C}\,\mathbf{x}(k) + \mathbf{D}\,\mathbf{u}(k) \tag{12.57}$$

The stability of the discrete state-space model is determined by finding the eigenvalues of $\mathbf{\Phi}$. If the magnitude of all of the eigenvalues is less than 1, then the system is stable.

12.4.1   Discrete Transfer Function Models

Continuous time models transfer function models are characterized by the Laplace tranform variable, s. Similarly, for discrete transfer function models, a discrete transform variable, z, is used:

$$\mathbf{Y}(z) = \mathbf{G}(z)\,\mathbf{U}(z) \tag{12.58}$$

where:

$$\mathbf{G}(z) = [\mathbf{C}\,(z\mathbf{I} - \mathbf{A})^{-1}\mathbf{B} + \mathbf{D}] \tag{12.59}$$

For the case of a single input-single output system, $\mathbf{G}(z)$ consists of a numerator and denominator polynomial of the form:

$$\mathbf{g}(z) = \frac{b_n z^n + b_{n-1} z^{n-1} + \ldots + b_1 z + b_o}{a_n z^n + a_{n-1} z^{n-1} + \ldots + a_1 z + a_o} \tag{12.60}$$

The transfer function is normally written in terms of the *backwards shift operator*, z^{-1}. Multiplying the transfer function by z^{-n}/z^{-n}, we find:

$$\mathbf{g}(z) = \frac{b_n + b_{n-1}z^{-1} + \dots + b_1 z^{-n+1} + b_o z^{-n}}{a_n + a_{n-1}z^{-1} + \dots + a_1 z^{-n+1} + a_o z^{-n}} \tag{12.61}$$

The backwards shift operator is defined as:

$$y(k-1) = z^{-1} y(z) \tag{12.62}$$

so $y(k-2) = z^{-2}y(z)$, etc. The discrete transfer function notation:

$$y(z) = \frac{b_n + b_{n-1}z^{-1} + \dots + b_1 z^{-n+1} + b_o z^{-n}}{a_n + a_{n-1}z^{-1} + \dots + a_1 z^{-n+1} + a_o z^{-n}} u(z) \tag{12.63}$$

then represents:

$$(a_n + a_{n-1}z^{-1} + \dots + a_1 z^{-n+1} + a_o z^{-n}) \, y(z)$$
$$= (b_n + b_{n-1}z^{-1} + \dots + b_1 z^{-n+1} + b_o z^{-n}) \, u(z) \tag{12.64}$$

which corresponds to the discrete input/output model:

$$a_n y(k) + a_{n-1}y(k-1) + \dots + a_1 y(k-n) + a_o y(k-n-1)$$
$$= b_n u(k) + b_{n-1}u(k-1) + \dots + b_1 u(k-n) + b_o u(k-n-1) \tag{12.65}$$

Usually we are solving for $y(k+1)$, and without loss of generality we can assume $a_n = 1$.

$$y(k+1) + a_{n-1}y(k) + \dots + a_1 y(k-n+1) + a_o y(k-n)$$
$$= b_n u(k+1) + b_{n-1}u(k) + \dots + b_1 u(k-n-1) + b_o u(k-n) \tag{12.66}$$

Also, for most systems there is not an immediate effect of the input on the output, so $b_n = 0$.

The most common discrete-time model is first-order:

$$y(k+1) + a_o y(k) = b_o u(k) \tag{12.67}$$

or,

$$y(k+1) = -a_o y(k) + b_o u(k)$$

which has the transfer function relationship:

$$y(z) = \frac{b_o z^{-1}}{1 + a_o z^{-1}} \tag{12.68}$$

Physically realizable systems will always have at least a z^{-1} factor (unit time delay) in the numerator.

A first-order discrete system with N additional units of time-delay is written

$$y(k+1) + a_o y(k) = b_o u(k-N) \tag{12.69}$$

or,

$$y(k + 1) = -a_o y(k) + b_o u(k - N)$$

which has the transfer function relationship:

$$y(z) = \frac{b_o z^{-N-1}}{1 + a_o z^{-1}} \tag{12.70}$$

EXAMPLE 12.4   Linear Van de Vusse Reactor Model

Consider a state-space model from the isothermal chemical reactor module (specifically, the Van de Vusse reaction):

$$\mathbf{A} = \begin{bmatrix} -2.4048 & 0 \\ 0.8333 & -2.2381 \end{bmatrix}$$

$$\mathbf{B} = \begin{bmatrix} 7.0000 \\ -1.1170 \end{bmatrix}$$

$$\mathbf{C} = \begin{bmatrix} 0 & 1 \end{bmatrix}$$

$$\mathbf{D} = \begin{bmatrix} 0 \\ 0 \end{bmatrix}$$

For a sample time of 0.1 the discrete state-space model is (using (12.52)–(12.54), or the MATLAB c2d function):

$$\mathbf{\Phi} = \begin{bmatrix} 0.7863 & 0 \\ 0.0661 & 0.7995 \end{bmatrix}$$

$$\mathbf{\Gamma} = \begin{bmatrix} 0.6222 \\ -0.0849 \end{bmatrix}$$

and the discrete input-output model is (using (12.59), or the MATLAB ss2tf function)

$$\mathbf{g}(z) = \frac{-0.0751z^{-1} + 0.1001z^{-2}}{1 - 1.5857z^{-1} + 0.6286z^{-2}}$$

which has poles of 0.7995 and 0.7863 (which have a magnitude less than 1, so the system is stable). The zero of the numerator polynomial is 1.3339.

The step responses of the continuous and discrete systems are compared in Figure 12.3.

For a sample time of 0.75, the discrete state-space model is (using (12.52)–(12.54), or the MATLAB c2d function):

$$\mathbf{\Phi} = \begin{bmatrix} 0.1647 & 0 \\ 0.1096 & 0.1866 \end{bmatrix}$$

$$\mathbf{\Gamma} = \begin{bmatrix} 2.4314 \\ 0.1164 \end{bmatrix}$$

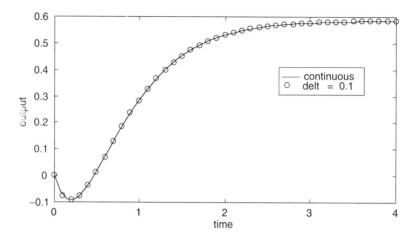

FIGURE 12.3   Step response of continous and discrete ($\Delta t = 0.1$) models.

and the discrete input-output model is (using (12.59), or the MATLAB `ss2tf` function):

$$\mathbf{g}(z) = \frac{0.1564z^{-1} + 0.2408z^{-2}}{1 - 0.3513z^{-1} + 0.0307z^{-2}}$$

which has poles at 0.1866 and 0.1647, indicating stability. The zero of the numerator polynomial is −1.5399.

   A comparison of the step responses of the continuous and discrete models is shown in Figure 12.4. Notice that the discrete sample time is too large to capture the "inverse response" behavior of the continuous system.

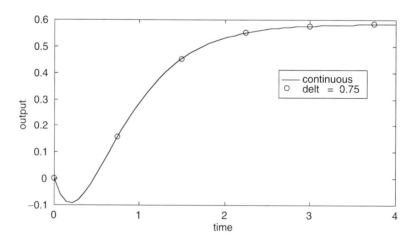

FIGURE 12.4   Step response of continous and discrete ($\Delta t = 0.75$) models.

12.5 PARAMETER ESTIMATION OF DISCRETE LINEAR SYSTEMS

Often when discrete linear models are developed, they are based on experimental system responses rather than converting a continuous model to a discrete model. The estimation of parameters for discrete dynamic models is no different than the linear regression analysis presented in Module 3. Please review Module 3 to understand the notation and ideas behind linear regression.

The measured inputs and outputs are the independent variables, and the dependent variables are the outputs. For simplicity, consider the following single input-single output model:

$$y(k) = -a_1 y(k - 1) - a_0 y(k - 2) + b_1 u(k - 1) + b_0 u(k - 2) \qquad (12.71)$$

Now, for the system of N data points we can write:

$$\mathbf{Y} = \Phi \, \theta \qquad (12.72)$$

where,

$$\mathbf{Y} = \begin{bmatrix} y(1) \\ \cdot \\ \cdot \\ y(N) \end{bmatrix} \quad \Phi = \begin{bmatrix} \varphi(1)^T \\ \cdot \\ \cdot \\ \varphi(N)^T \end{bmatrix} \quad \theta = \begin{bmatrix} -a_1 \\ -a_0 \\ b_1 \\ b_0 \end{bmatrix} \qquad (12.73)$$

$$\varphi(k)^T = [y(k - 1) \;\; y(k - 2) \;\; u(k - 1) \;\; u(k - 2)] \qquad (12.74)$$

The solution to this problem is:

$$\theta = (\Phi^T \Phi)^{-1} \Phi^T \, \mathbf{Y} \qquad (12.75)$$

EXAMPLE 12.5   Parameter Estimation

A unit step input is made to a system at time $t = 0$ ($k = 0$). The sample time is $\Delta t = 0.75$. The step response data are shown below and plotted in Figure 12.5.

k	$y(k)$
0	0
1	0.1564
2	0.4522
3	0.5513
4	0.5770
5	0.5830

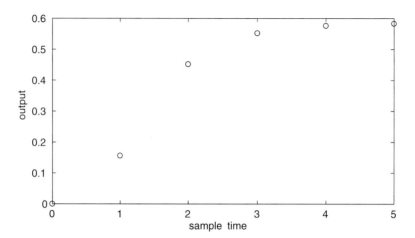

FIGURE 12.5   Step response data (same as Figure 12.4).

$$\mathbf{Y} = \begin{bmatrix} y(1) \\ y(2) \\ y(3) \\ y(4) \\ y(5) \end{bmatrix} = \begin{bmatrix} 0.1564 \\ 0.4522 \\ 0.5513 \\ 0.5770 \\ 0.5830 \end{bmatrix}$$

$$\mathbf{\Phi} = \begin{bmatrix} \varphi(1)^T \\ \cdot \\ \cdot \\ \varphi(5)^T \end{bmatrix} = \begin{bmatrix} y(0) & y(-1) & u(0) & u(-1) \\ \cdot & \cdot & \cdot & \cdot \\ \cdot & \cdot & \cdot & \cdot \\ y(4) & y(3) & u(4) & u(3) \end{bmatrix} = \begin{bmatrix} 0 & 0 & 1 & 0 \\ 0.1564 & 0 & 1 & 1 \\ 0.4522 & 0.1564 & 1 & 1 \\ 0.5513 & 0.4522 & 1 & 1 \\ 0.5770 & 0.5513 & 1 & 1 \end{bmatrix}$$

The solution is:

$$\theta = (\mathbf{\Phi^T \Phi})^{-1} \mathbf{\Phi}^T \mathbf{Y}$$

$$\begin{bmatrix} -a_1 \\ -a_0 \\ b_1 \\ b_0 \end{bmatrix} = \begin{bmatrix} 0.3513 \\ -0.0308 \\ 0.1564 \\ 0.2409 \end{bmatrix}$$

which are the same parameters that we found for the discrete transfer function model that was converted from the continuous model with a sample time of 0.75.

This simple example illustrated the step response of a perfectly modeled system (no measurement noise). The approach can also be applied to a system with arbitrary inputs and with noisy measurements. The data was analyzed in a batch fashion, that is, all of the data were collected before the parameter estimation was performed.

There are other approaches that are useful for estimating model parameters in real time, often using the model parameters to modify feedback control laws. These approaches are beyond the scope of this textbook. The MATLAB System Identification Toolbox is useful for these types of problems. A good reference is the text by Ljung.

SUMMARY

There were multiple objectives to this chapter. The first was to introduce analytical solution techniques for boundary value ordinary differential equations. The second was to provide a concise review of techniques to solve linear initial value ordinary differential equations. The final objective was to introduce discrete-time models and discuss parameter estimation for these models.

For continuous-time models, the eigenvalues of the state-space model must have negative real portions for the system to be stable. Equivalently, the poles of the continuous transfer function models must be negative (the eigenvalues of the state-space model are equal to the poles of the transfer function model). Analogously, the eigenvalues of the discrete state-space model must have a magnitude less than one to be stable. Also, the poles of a discrete transfer function model must have magnitudes less than one to be stable.

Continuous-time input/output (transfer function) models with zeros that are positive exhibit inverse response. Similarly, discrete transfer function models with zeros that have a magnitude greater than one (yet have a negative real portion) exhibit inverse response.

REFERENCES

The following text provides methods to solve boundary value ordinary differential equation models.

Rameriz, W.F. (1989). *Computational Methods for Process Simulation.* Boston: Butterworths.

Many process control textbooks provide coverage of linear discrete-time models. See, for example:

Ogunnaike, B.A., & W.H. Ray. (1994). *Process Dynamics, Modeling and Control.* New York: Oxford.

System identification (model parameter estimation) is covered in the text by Ljung

Ljung, L. (1987). *System Identification—Theory for the User.* Englewood Cliffs, NJ: Prentice-Hall.

STUDENT EXERCISES

1. Consider the following second-order boundary value problem:

$$\frac{d^2x}{dz^2} + 4\frac{dx}{dz} + \frac{7}{4}x = 0$$

where:

$$x(z = 0) = 2$$
$$x(z = 1) = 1$$

Show that, by defining $x_1 = x$ and $x_2 = dx/dz$, the following equations are obtained:

$$dx_1/dz = x_2$$

$$dx_2/dz = -\frac{7}{4}x_1 - 4x_2$$

and that one of the initial conditions is:

$$x_1(z = 0) = 2$$

We see that $x_2(z = 0)$ must be "guessed," then the two equations can be integrated (using ode45) from $z = 0$ to $z = 1$. The value of x_1 at $z = 1$ is then checked; if $x_1(z = 1)$ is not equal to 1 (within an acceptable tolerance) then values of $x_2(z = 0)$ are iteratively guessed until the final value is satisfied. This method is known as the "shooting method". Use fzero to solve for the initial condition that satisfies the end boundary value.

2. Consider the reaction/dispersion equation

$$\frac{\partial C_A}{\partial t} = -v_z\frac{\partial C_A}{\partial z} + \tilde{D}_{AZ}\frac{\partial^2 C_A}{\partial z^2} - k\,C_A$$

let: $C = \dfrac{C_A}{C_{A0}}$ = dimensionless concentration

and: $y = \dfrac{z}{L}$ = dimensionless axial distance

define: $\tau = \dfrac{\tilde{D}_{AZ}t}{L^2}$ = dimensionless time

$P_e = \dfrac{v_z L}{\tilde{D}_{AZ}}$ = Peclet number

and: $D_a = \dfrac{k\,L^2}{\tilde{D}_{AZ}}$ = Damkohler number

to show that: $\dfrac{\partial C}{\partial \tau} = -P_e\dfrac{\partial C}{\partial y} + \dfrac{\partial^2 C}{\partial y^2} - D_a\,C$ (12.2)

Find the dimensionless form of the Danckwerts boundary conditions at steady-state:

$$v_z\,C_{A0} = v_z\,C_A(0^+) - \tilde{D}_{AZ}\frac{dC_A(0^+)}{dz}$$

and:

$$\frac{dC_A(L)}{dz} = 0$$

a. Perform steady-state calculations (analytically) using the Danckwerts boundary conditions for:

    i. $P_e = 1$, $D_a = 1$, 10, 25 (compare on same plot)

    ii. $P_e = 10$, $D_a = 1$, 10, 25, 100 (compare on same plot)

    iii. $P_e = 25$, $D_a = 1$, 10, 25, 100 (compare on same plot)

    iv. $P_e = 100$, $D_a = 1$, 10, 25, 100 (compare on same plot)

3. Consider the following continuous state-space model:

$$A = \begin{bmatrix} -3.6237 & 0 \\ 0.8333 & -2.9588 \end{bmatrix}$$

$$B = \begin{bmatrix} 5.5051 \\ -1.2660 \end{bmatrix}$$

$$C = \begin{bmatrix} 0 & 1 \end{bmatrix}$$

$$D = 0$$

    a. Find the continuous transfer function model.

    b. For a sample time of 0.25, find the discrete state-space and transfer function models.

    c. Compare the step responses of the continuous and discrete models. What do you observe?

4. Consider a unit step change made at $k = 0$, resulting in the output response shown in the plot and table below.

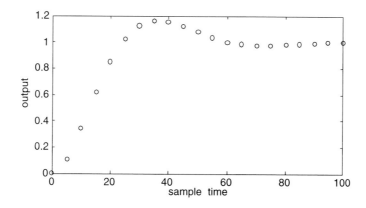

k	0	1	2	3	4	5	6	7	8	9	10	11
y	0	0.1044	0.3403	0.6105	0.8494	1.0234	1.1244	1.1616	1.1531	1.1184	1.0746	1.0336

k	12	13	14	15	16	17	18	19	20
y	1.0023	0.9828	0.9744	0.9742	0.9790	0.9860	0.9929	0.9985	1.0022

Estimate the parameters for a discrete linear model with the form:

$$g(z) = \frac{b_1 z^{-1} + b_0 z^{-2}}{1 - a_1 z^{-1} - a_0 z^{-2}}$$

SECTION IV

NONLINEAR SYSTEMS ANALYSIS

PHASE-PLANE ANALYSIS

13

The objective of this chapter is to introduce the student to phase-plane analysis, which is one of the most important techniques for studying the behavior of nonlinear systems. After studying this chapter, the student should be able to:

- Use eigenvalues and eigenvectors of the Jacobian matrix to characterize the phase-plane behavior
- Predict the phase-plane behavior close to an equilibrium point, based on the linearized model at that equilibrium point
- Predict qualitatively the phase-plane behavior of the nonlinear system, when there are multiple equilibrium points

The major sections of this chapter are:

13.1 BACKGROUND

Techniques to find the transient (time domain) behavior of linear state-space models were discussed in Chapter 5. Recall that the response characteristics (relative speed of response) for unforced systems were dependent on the initial conditions. Eigenvalue/eigenvector analysis allowed us to predict the fast and slow (or stable and unstable) initial conditions. If we plotted the transient responses based on a number of initial conditions, there would soon be an overwhelming number of curves on the transient response plots. Another way of obtaining a feel for the effect of initial conditions is to use a *phase-plane* plot. A phase-plane plot for a two-state variable system consists of curves of one-state variable versus the state variable ($x_1(t)$ versus $x_2(t)$), where each curve is based on a different initial condition. A phase-space plot can also be made for three-state variables, where each curve in 3-space is based on a different initial condition.

Phase-plane analysis is one of the most important techniques for studying the behavior of nonlinear systems, since there is usually no analytical solution for a nonlinear system. It is obviously important to understand phase-plane analysis for linear systems before covering nonlinear systems. Section 13.2 discusses the phase-plane behavior of linear systems and Section 13.3 covers nonlinear systems.

13.2 LINEAR SYSTEM EXAMPLES

Nonlinear systems often have multiple steady-state solutions (see Modules 8 and 9 for examples). Phase-plane analysis of nonlinear systems provides an understanding of which steady-state solution that a particular set of initial conditions will converge to. The local behavior (close to one of the steady-state solutions) can be understood from a linear phase-plane analysis of the particular steady-state solution (equilibrium point).

In this section we show the different types of phase-plane behavior that can be exhibited by linear systems. The phase-plane analysis approach will be shown by way of a number of examples.

EXAMPLE 13.1   A Stable Equilibrium Point (Node Sink)

Consider the system of equations:

$$\dot{x}_1 = -x_1 \tag{13.1}$$

$$\dot{x}_2 = -4\,x_2 \tag{13.2}$$

The reader should find that the solution to (13.1) and (13.2) is:

$$x_1(t) = x_{1o}\,e^{-t} \tag{13.3}$$

$$x_2(t) = x_{2o}\,e^{-4t} \tag{13.4}$$

where x_{1_o} and x_{2_o} are the initial conditions for x_1 and x_2. We could plot x_1 and x_2 as a function of time for a large number of initial conditions (requiring a large number of time domain plots), but the same information is contained on a *phase-plane* plot as shown in Figure 13.1. Each curve corresponds to a different initial condition. Notice that the solutions converge to $(0,0)$ for all initial conditions. The point $(0,0)$ is a stable equilibrium point for the system of equations (13.1) and (13.2)—the plot shown in Figure 13.1 is often called a *stable node*.

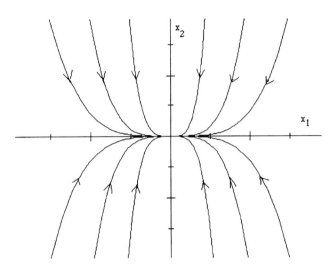

FIGURE 13.1   Phase-plane plot for Example 13.1. The point $x^T = (0,0)$ is a stable node.

EXAMPLE 13.2   An Unstable Equilibrium Point (Saddle)

Consider the system of equations:

$$\dot{x}_1 = -x_1 \tag{13.5}$$

$$\dot{x}_2 = 4\,x_2 \tag{13.6}$$

The student should find that the solution to (13.5) and (13.6) is:

$$x_1(t) = x_{1_o}\,e^{-t} \tag{13.7}$$

$$x_2(t) = x_{2_o}\,e^{4t} \tag{13.8}$$

The phase-plane plot is shown in Figure 13.2. If the initial condition for the x_2 state variable was 0, then a trajectory that reached the origin could be obtained. Notice if the initial condition x_{2_o} is just slightly different than zero, then the solution will always leave the origin. The origin is an unstable equilibrium point, and the trajectories shown in Figure 13.2 represent a *saddle point*.

The x_1 axis represents a stable subspace and the x_2 axis represents an unstable subspace for this problem. The term saddle can be understood if you view the x_1 axis as the line (ridge) between the "horn" and rear of a saddle. A ball starting at the horn could conceptually roll down the saddle and remain exactly on the ridge between the horn and the rear of the saddle. In practice, a small perturbation from the ridge would cause the ball to begin rolling to one "stirrup" or the other. Similarly, a small perturbation in the initial condition from the x_1 axis in Example 13.2 would cause the solution to diverge in the unstable direction.

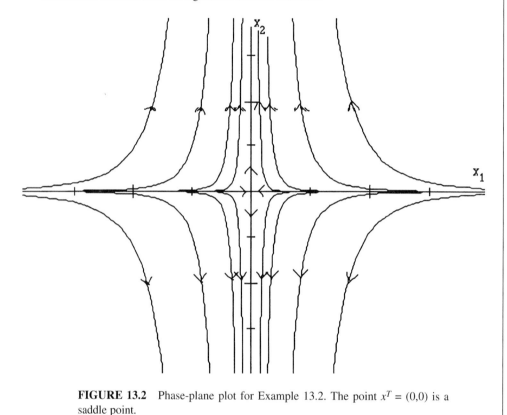

FIGURE 13.2   Phase-plane plot for Example 13.2. The point $x^T = (0,0)$ is a saddle point.

Figures 13.1 and 13.2 clearly show the idea of separatices. A separatrix is a line in the phase-plane that is not crossed by any trajectory. In Figure 13.1 the *separatices* are the coordinate axes. A trajectory that started in any quadrant stayed in that quadrant. This is because the eigenvectors are the coordinate axes. Similar behavior is observed in Figure 13.2, except that the x_1 coordinate axis is unstable.

Solving the equations for Examples 13.1 and 13.2 and the phase-plane trajectories were straight-forward and obvious, because the eigenvectors where simply the coordinate axes. In general, eigenvalue/eigenvector analysis must be used to determine the stable and unstable "subspaces." The eigenvectors are the separatices in the general case.

Example 13.3 shows how eigenvalue/eigenvector analysis is used to find the stable and unstable subspaces, and to define the separatices.

EXAMPLE 13.3   AnotherSaddle Point Problem

Consider the following system of equations:

$$\dot{x}_1 = 2\,x_1 + x_2 \tag{13.9}$$

$$\dot{x}_2 = 2\,x_1 - x_2 \tag{13.10}$$

Using standard state-space notation:

$$\dot{\mathbf{x}} = \mathbf{A}\,\mathbf{x} \tag{13.11}$$

The Jacobian matrix is:

$$A = \begin{bmatrix} 2 & 1 \\ 2 & -1 \end{bmatrix}$$

the eigenvalues are:

$$\lambda_1 = -1.5616 \qquad \lambda_2 = 2.5616$$

and the eigenvectors are:

$$\xi_1 = \begin{bmatrix} 0.2703 \\ -0.9628 \end{bmatrix} \qquad \xi_2 = \begin{bmatrix} 0.8719 \\ 0.4896 \end{bmatrix}$$

Since $\lambda_1 < 0$, ξ_1 is a stable subspace; also, since $\lambda_2 > 0$, ξ_2 is an unstable subspace. A plot of the stable and unstable subspaces is shown in Figure 13.3. These eigenvectors also define the separatices that determine the characteristic behavior of the state trajectories.

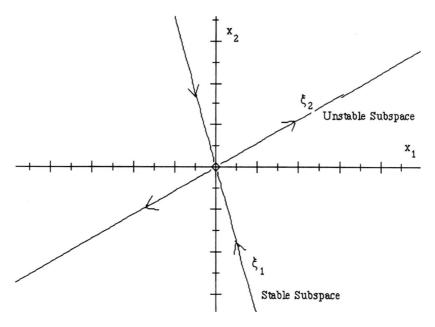

FIGURE 13.3   Stable and unstable subspaces for Example 13.3.

The time domain solution to (13.11) is:

$$\mathbf{x}(t) = e^{At}\,\mathbf{x}(0) \tag{13.12}$$

which is often solved as (see Chapter 5):

$$\mathbf{x}(t) = \mathbf{V}\,e^{\Lambda t}\,\mathbf{V}^{-1}\,x(0) \tag{13.13}$$

which yields the following solution for this system:

$$\mathbf{x}(t) = \begin{bmatrix} 0.2703 & 0.8719 \\ -0.9628 & 0.4896 \end{bmatrix} \begin{bmatrix} e^{-1.5616t} & 0 \\ 0 & e^{2.5616t} \end{bmatrix} \begin{bmatrix} 0.5038 & -0.8972 \\ 0.9907 & 0.2782 \end{bmatrix} \mathbf{x}(0) \tag{13.14}$$

Recall that the solution to (13.14), if $x(0) = \xi_1$, is $x(t) = \xi_1 e^{\lambda_1 t}$, so

$$\text{if } \mathbf{x}(0) = \begin{bmatrix} 0.2703 \\ -0.9628 \end{bmatrix} \text{ then } \mathbf{x}(t) = \begin{bmatrix} 0.2703 \\ -0.9628 \end{bmatrix} e^{-1.5616t}$$

The phase-plane plot is shown in Figure 13.4, where the separatrices clearly define the phase-plane behavior.

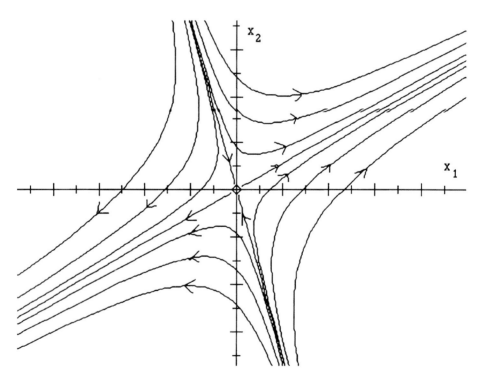

FIGURE 13.4   Phase-plane plot for Example 13.3.

The previous examples were for systems that exhibited stable node or saddle point behavior. In either case, the eigenvalues and eigenvectors where real. Another type of behavior that can occur is a spiral focus (either stable or unstable), which has complex eigenvalues and eigenvectors. Example 13.4 is an unstable focus.

EXAMPLE 13.4   Unstable Focus (Spiral Source)

Consider the following system of equations:

$$\dot{x}_1 = x_1 + 2\,x_2 \tag{13.15}$$

$$\dot{x}_2 = -2\,x_1 + x_2 \tag{13.16}$$

Using standard state-space notation:

$$\dot{\mathbf{x}} = \mathbf{A}\,\mathbf{x}$$

The Jacobian matrix is:

$$A = \begin{bmatrix} 1 & 2 \\ -2 & 1 \end{bmatrix}$$

with eigenvalues $1 \pm 2j$. This system is unstable because the real portion of the complex eigenvalues is positive.

The phase-plane plot is shown in Figure 13.5.

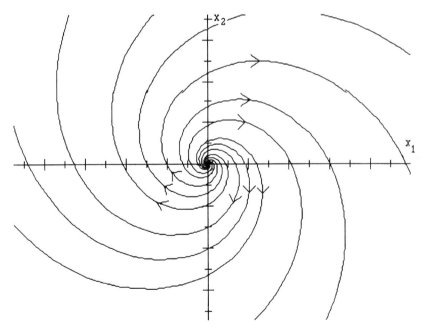

FIGURE 13.5   Example 13.4, unstable focus (spiral source).

Another type of linear system behavior occurs when the eigenvalues have a zero real portion. That is, the eigenvalues are on the real axis. This type of system leads to closed curves in the phase-plane, and is known as center behavior. Example 13.5 illustrates center behavior.

EXAMPLE 13.5   Center

Consider the following system of equations:

$$\dot{x}_1 = -x_1 - x_2 \tag{13.17}$$

$$\dot{x}_2 = 4x_1 + x_2 \tag{13.18}$$

The Jacobian matrix is:

$$A = \begin{bmatrix} -1 & -1 \\ 4 & 1 \end{bmatrix}$$

and the eigenvalues are $0 \pm 1.7321j$. Since the real part of the eigenvalues is zero, there is a periodic solution (sine and cosine), resulting in a phase-plane plot where the equilibrium point is a center, as shown in Figure 13.6.

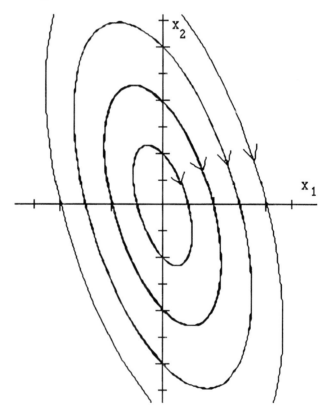

FIGURE 13.6   Example 13.5, eigenvalues with zero real portion are centers.

Examples 13.1 to 13.5 we have provided an introduction to linear system phase-plane behavior. We noted the important role of eigenvectors and eigenvalues, and how these relate to the concept of a separatrix. Section 13.3 provides a generalization of these examples.

13.3 GENERALIZATION OF PHASE-PLANE BEHAVIOR

We wish now to generalize our results for second-order linear systems of the form:

$$\dot{\mathbf{x}} = \mathbf{A}\,\mathbf{x} \tag{13.19}$$

where the Jacobian matrix is:

$$A = \begin{bmatrix} a_{11} & a_{12} \\ a_{21} & a_{22} \end{bmatrix} \tag{13.20}$$

Recall that the eigenvalues are found by solving $\det(\lambda I - A) = 0$:

$$\det(\lambda I - A) = (\lambda - a_{11})(\lambda - a_{22}) - a_{12}\,a_{21} = 0 \tag{13.21}$$

which can be written as:

$$\det(\lambda I - A) = \lambda^2 - \text{tr}(A)\,\lambda + \det(A) = 0 \tag{13.22}$$

The quadratic formula can be used to find the eigenvalues:

$$\lambda = \frac{\text{tr}(A) \pm \sqrt{(\text{tr}(A))^2 - 4\det(A)}}{2} \tag{13.23}$$

or, expressing each eigenvalue separately,

$$\lambda_1 = \frac{\text{tr}(A) - \sqrt{(\text{tr}(A))^2 - 4\det(A)}}{2}$$

and,

$$\lambda_2 = \frac{\text{tr}(A) + \sqrt{(\text{tr}(A))^2 - 4\det(A)}}{2}$$

We notice that at least one eigenvalue will be negative if $tr(A) < 0$. We also notice that the eigenvalues will be complex if $4\det(A) > tr(A)^2$. Remember that the different behaviors resulting from λ_1 and λ_2 are:

Sinks (stable nodes):	Re $(\lambda_1) < 0$ and Re $(\lambda_2) < 0$
Saddles (unstable):	Re $(\lambda_1) < 0$ and Re $(\lambda_2) > 0$
Sources (unstable nodes):	Re $(\lambda 1) > 0$ and Re $(\lambda_2) > 0$
Spirals:	λ_1 and λ_2 are complex complex conjugates. If Re(λ_1) < 0 then stable, if Re$(\lambda_1) > 0$ then unstable.

We can then use Figure 13.7 to find the phase-plane behavior for second-order linear ordinary differential equations as a function of the trace and determinant of A. In Figure

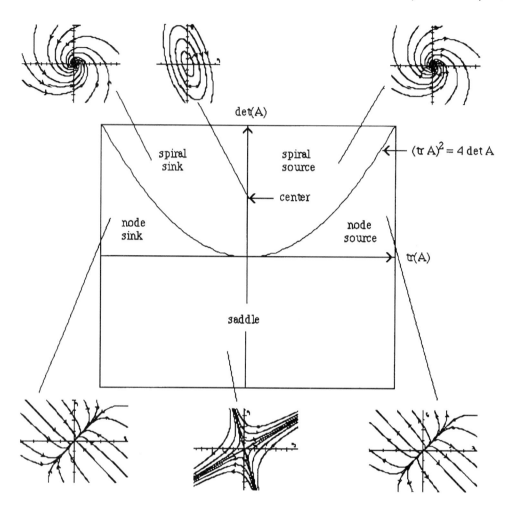

FIGURE 13.7 Dynamic behavior diagram for second-order linear systems.
The *x*-axis is tr(*A*) and the *y*-axis is det(*A*).

13.7, the *x*-axis is the trace of *A* and the *y*-axis is the determinant of *A*. For example, consider Example 13.1, where

$$A = \begin{bmatrix} -1 & 0 \\ 0 & -4 \end{bmatrix}$$
$$\text{tr}(A) = -5$$
$$\det(A) = 4$$

The point (−5,4) lies in the second quadrant in the node sink sector, as expected, since the two real eigenvalues are negative (indicting stable node behavior).

Figure 13.8 shows the phase-plane behavior as a function of the eigenvalue locations in the complex plane. For example, two negative eigenvalues lead to stable node behavior.

13.3.1 Slope Marks for Vector Fields

A qualitative assessment of the phase-plane behavior can be obtained by plotting the *slope marks* for the vector field. Consider a general linear 2-state system

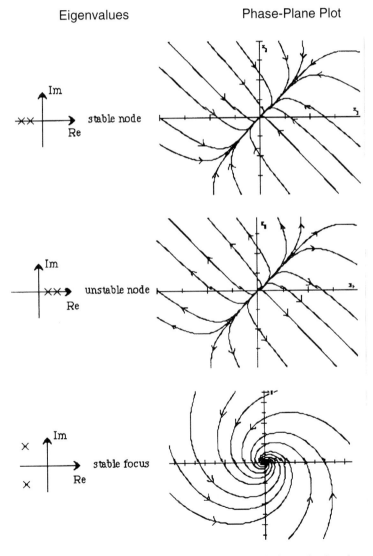

FIGURE 13.8   Phase-plane behavior as a function of Eigenvalue location.

Eigenvalues	Phase-Plane Plot

unstable focus
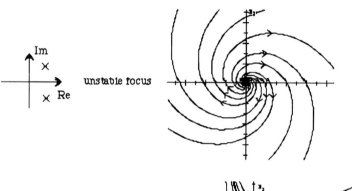

saddle
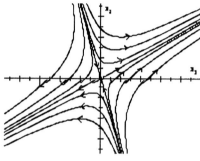

Im
center
Re
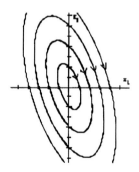

FIGURE 13.8 *Continued*

$$\dot{x}_1 = a_{11} x_1 + a_{12} x_2 \qquad (13.24)$$

$$\dot{x}_2 = a_{21} x_1 + a_{22} x_2 \qquad (13.25)$$

We can divide (13.25) by (13.24) to find how x_2 changes with respect to x_1:

$$\frac{dx_2}{dx_1} = \frac{a_{21} x_1 + a_{22} x_2}{a_{11} x_1 + a_{12} x_2} \qquad (13.26)$$

and we can plot "slope marks" for values of x_1 and x_2 to determine an idea of how the phase plane will look. Let us revisit Example 13.3. The slope marks can be calculated from (13.27):

$$\frac{dx_2}{dx_1} = \frac{2\,x_1 - x_2}{2\,x_1 + x_2} \tag{13.27}$$

Figure 13.9 shows the slope marks for Example 13.3. These are generated by forming a grid of points in the plane, and finding the slope associated with each point; short line segments with the slope calculated are then plotted for each point. Notice that one can use the slope marks to help sketch state variable trajectories, as shown in Figure 13.10. Saddle point behavior found in Example 13.3 is clearly shown in Figure 13.10.

13.3.2 Additional Discussion

Phase-plane analysis can be used to analyze autonomous systems with two state variables. Notice that state variable trajectories cannot "cross" in the plane, as illustrated by the following reasoning. Think of any point of a trajectory as being an initial condition. The model, when integrated from that initial condition, must have a single trajectory. If two trajectories crossed, that would be the equivalent of saying that a single initial condition could have two different trajectories. If a system was non-autonomous (for example, if there was a forcing function that was a function of time) then state variable trajectories could cross, because a model with the same initial conditions but a different forcing function would have different trajectories.

An autonomous (unforced) system with n state variables cannot have trajectories that cross in n-space, but may have trajectories that cross in less than n-space. For exam-

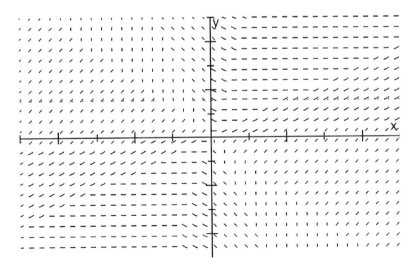

FIGURE 13.9   Slope marks for the vector field of Example 13.3.

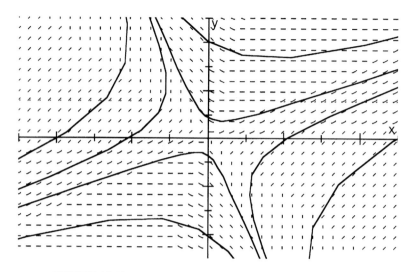

FIGURE 13.10   Slope marks with trajectories for Example 13.3.

ple, a third-order autonomous system cannot have trajectories that cross in 3-space, but the trajectories may cross when placed in a two-dimensional plane.

13.4 NONLINEAR SYSTEMS

In the previous sections we discovered the types of phase-plane behavior that could be observed in linear systems. In this chapter we will find that nonlinear systems will often have the same general phase-plane behavior as the model linearized about the equilibrium (steady-state) point, when the system is close to that particular equilibrium point.

In this section we study two examples. Example 13.6 is based on a simple bilinear model, while Example 13.7 is a classical bioreactor model.

EXAMPLE 13.6   Nonlinear (Bilinear) System

Consider the following system:

$$\frac{dz_1}{dt} = z_2(z_1 + 1) \tag{13.28}$$

$$\frac{dz_2}{dt} = z_1(z_2 + 3) \tag{13.29}$$

which has two steady-state (equilibrium) solutions:

$$\textit{Equilibrium } 1: \text{trivial} \qquad z_{1s} = 0 \qquad z_{2s} = 0$$

$$\textit{Equilibrium } 2: \text{nontrivial} \qquad z_{1s} = -1 \qquad z_{2s} = -3$$

Linearizing (13.28) and (13.29), we find the following Jacobian matrix:

$$A = \begin{bmatrix} z_{2s} & z_{1s} + 1 \\ z_{2s} + 3 & z_{1s} \end{bmatrix}$$

In the following, we analyze the stability of each equilibrium point.

Equilibrium 1 (Trivial)

The Jacobian matrix is: $\qquad A = \begin{bmatrix} 0 & 1 \\ 3 & 0 \end{bmatrix}$

and the eigenvalues are: $\qquad \lambda_1 = -\sqrt{3} \qquad \lambda_2 = \sqrt{3}$

We know from linear system analysis that equilibrium point one is a *saddle point,* since one eigenvalue is stable and the other is unstable.

The stable eigenvector is: $\qquad \xi_1 = \begin{bmatrix} -0.5 \\ 0.866 \end{bmatrix}$

The unstable eigenvector is: $\xi_2 = \begin{bmatrix} 0.5 \\ 0.866 \end{bmatrix}$

The phase-plane of the linearized model around equilibrium point one is a saddle, as shown in Figure 13.11. The linearized model is

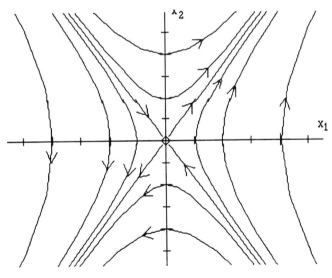

FIGURE 13.11   Phase-plane of the Example 13.6 model linearized around trivial equibrium point. This point is a saddle point.

$$\dot{\mathbf{x}} = \begin{bmatrix} 0 & 1 \\ 3 & 0 \end{bmatrix} \mathbf{x}$$

where $\mathbf{x} = \mathbf{z} - \mathbf{z}_s$

Equilibrium 2 (Nontrivial)

The Jacobian matrix is: $A = \begin{bmatrix} -3 & 0 \\ 0 & -1 \end{bmatrix}$

the eigenvalues are: $\lambda_1 = -3 \qquad \lambda_2 = -1$

So, we know from linear system analysis that equilibrium point two is a *stable node,* since both eigenvalues are stable.

The "fast" stable eigenvector is $\xi_1 = \begin{bmatrix} 1 \\ 0 \end{bmatrix}$

The "slow" stable eigenvector is $\xi_2 = \begin{bmatrix} 0 \\ 1 \end{bmatrix}$

The phase-plane of the linearized model around equilbrium point two is a stable node, as shown in Figure 13.12.

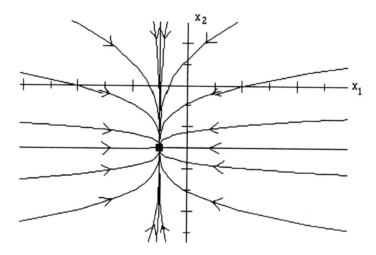

FIGURE 13.12 Phase-plane of the Example 13.6 model linearized around trivial equibrium point. This point is a stable node.

The phase-plane diagram of the nonlinear model is shown in Figure 13.13. Notice how the linearized models capture the behavior of the nonlinear model when close to one of the equilbrium points. Notice, however, that initial conditions inside the "right" saddle "blow up," while initial conditions inside the left saddle are attracted to the stable point. Slope-field marks are shown in Figure 13.14.

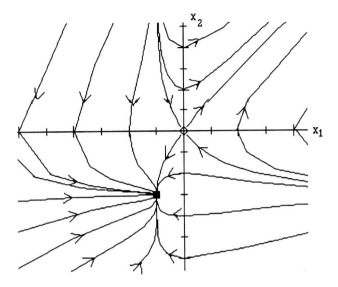

FIGURE 13.13   Phase-plane of Example 13.6. Trajectories (except those of the right side of the saddle) leave the unstable point and are "attracted" to the stable point.

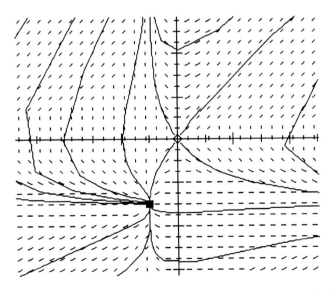

FIGURE 13.14   Slope-field marks and some trajectories for Example 13.6.

EXAMPLE 13.7   Bioreactor with Monod Kinetics

Consider a model for a bioreactor with Monod kinetics (see Module 8):

$$\frac{dx_1}{dt} = (\mu - D)\, x_1 \tag{13.30}$$

$$\frac{dx_2}{dt} = (s_f - x_2)\, D - \frac{\mu x_1}{Y} \tag{13.31}$$

$$\mu = \frac{\mu_{max}\, x_2}{k_m + x_2} \tag{13.32}$$

where:

$$\mu_{max} = 0.53 \qquad k_m = 0.12$$
$$Y = 0.4 \qquad s_f = 4.0$$

x_1 is the biomass concentration and x_2 is the substrate concentration. There are two steady-state (equilibrium) solutions for this set of parameters.

Equilibrium 1: trivial $x_{1s} = 0$ $x_{2s} = 4.0$

Equilibrium 2: nontrivial $x_{1s} = 1.4523$ $x_{2s} = 0.3692$

Linearizing (13.30) and (13.31) we find the following Jacobian matrix:

$$\mathbf{A} = \begin{bmatrix} \mu_s - D_s & x_{1s}\mu_s' \\ -\dfrac{\mu_s}{Y} & -D_s - \dfrac{\mu_s' x_{1s}}{Y} \end{bmatrix}$$

where we have defined $\mu' = \dfrac{\partial \mu}{\partial x_2} = \dfrac{\mu_s}{x_{2s}(k_m + x_{2s})}$

Equilibrium 1 (Trivial)

The Jacobian matrix is: $A = \begin{bmatrix} 0.114563 & 0 \\ -1.286408 & -0.4 \end{bmatrix}$

with eigenvalues of: $\lambda_1 = 0.114563$ and $\lambda_2 = -0.4$
indicating that the steady-state is unstable (it is a saddle point).

The unstable eigenvector is: $\xi_1 = \begin{bmatrix} 0.3714 \\ -0.9285 \end{bmatrix}$

The stable eigenvector is: $\xi_2 = \begin{bmatrix} 0 \\ 1 \end{bmatrix}$

This steady-state is known as the "wash-out" steady-state, because no biomass is produced and the substrate concentration in the reactor is equal to the feed substrate concentration.

Equilibrium 2 (Nontrivial)

The Jacobian matrix is: $A = \begin{bmatrix} 0 & 3.215929 \\ -1 & -8.439832 \end{bmatrix}$

with eigenvalues of $\lambda_1 = -0.4$ and $\lambda_2 = -8.0398$

indicating that the steady-state is a stable node.

The "slow" stable eigenvector is $\xi_1 = \begin{bmatrix} 0.9924 \\ -0.11234 \end{bmatrix}$

The "fast" stable eigenvector is $\xi_2 = \begin{bmatrix} 0.3714 \\ 0.9285 \end{bmatrix}$

The phase-plane plot of Figure 13.15 shows that the trajectories leave unstable point 1 (0,4) and go to stable point 2 (1.4523,0.3692). More detail of the phase-plane around the unstable point is shown in Figure 13.16, while Figure 13.17 shows more detail around the stable point.

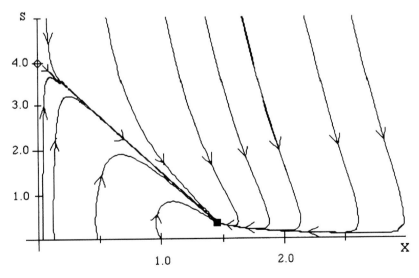

FIGURE 13.15   Phase-plane for bioreactor with Monod kinetics. x is biomass concentration and s is substrate concentration.

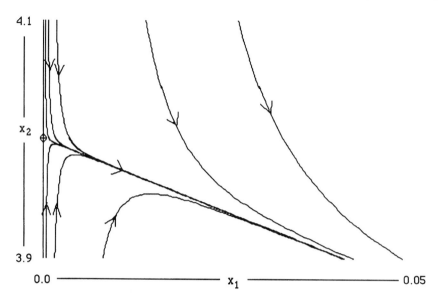

FIGURE 13.16   Phase-plane behavior near the unstable point (0,4) (Equilibrium 1).

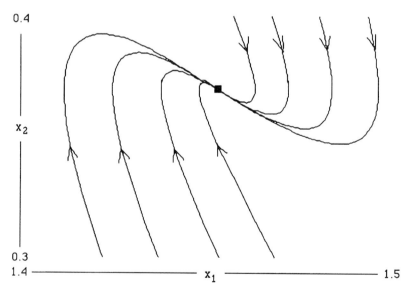

FIGURE 13.17   Phase-plane behavior near the stable point (1.4523, 0.3692) (Equilibrium 2).

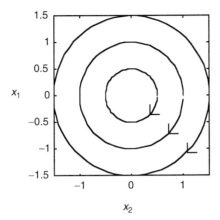

FIGURE 13.18 Example of center behavior.

In the previous examples the system trajectories "left" an unstable point and were "attracted" to a stable point. Another type of behavior that can occur is limit cycle or periodic behavior. This is illustrated in the following section.

13.4.1 Limit Cycle Behavior

In Section 13.2 we noticed that linear systems that had eigenvalues with zero real portion formed centers in the phase plane. The phase-plane trajectories of the systems with centers depended on the initial condition values. An example is shown in Figure 13.18. A somewhat related behavior that can occur in nonlinear systems is known as limit cycle behavior, as shown in Figure 13.19.

The major difference in center (Figure 13.18) and limit cycle (Figure 13.19) behavior is that limit cycles are *isolated* closed orbits. By isolated, we mean that an inital per-

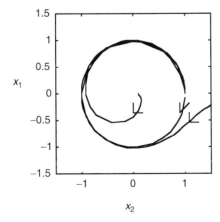

FIGURE 13.19 Example of limit cycle behavior.

turbation from the closed cycle eventually returns to the closed cycle. Contrast that with center behavior, where a perturbation leads to a different closed cycle.

Limit cycle behavior will be discussed in more detail in Chapter 16.

SUMMARY

As noted earlier, phase-plane analysis is a useful tool for observing the behavior of nonlinear systems. We have spent time analyzing autonomous linear systems, because the nonlinear systems will behave like a linear system, in the vicinity of the equilibrium point (where the linear approximation is most valid). A qualitative feel for phase-plane behavior can be obtained by plotting slope marks.

Notice that we have shown examples of nonlinear systems that have multiple equilibrium points (steady-state solutions). Phase-plane analysis can be used to determine regions of initial conditions where a system may converge to one (stable) equilbrium point and regions where the initial conditions may converge to another (stable) equilibrium point.

By sketching the linear behavior around a particular equilibrium point and by using slope marks, we can qualitatively sketch the phase-plane behavior of a given nonlinear system.

Clearly the phase-plane approach is limited to systems with two state variables. Analogous procedures can be used to develop phase-space plots in three dimensions for three-state systems. Linearization and analysis of the locally linear behavior in terms of eigenvalues and eigenvectors can still be used for higher-order systems, but the phase behavior cannot be viewed for these higher-order systems.

FURTHER READING

Strogatz, S.H. (1994). *Nonlinear Dynamics and Chaos.* Reading, MA: Addison Wesley.

STUDENT EXERCISES

Linear Problems

For the following linear systems, use Figure 13.7 to determine the phase-plane behavior. Also, calculate the eigenvalues and use Figure 13.8 to verify your results. Develop your own phase-plane diagrams for any situations not covered in Figures 13.7 and 13.8.

1. $\dot{\mathbf{x}} = \begin{bmatrix} -1 & 0 \\ 1 & -2 \end{bmatrix} \mathbf{x}$

2. $\dot{\mathbf{x}} = \begin{bmatrix} -1 & 3 \\ 2 & -2 \end{bmatrix} \mathbf{x}$

3. $\dot{\mathbf{x}} = \begin{bmatrix} 1 & 3 \\ 2 & 2 \end{bmatrix} \mathbf{x}$

4. $\dot{\mathbf{x}} = \begin{bmatrix} 1 & 0 \\ 1 & 2 \end{bmatrix} \mathbf{x}$

5. $\dot{\mathbf{x}} = \begin{bmatrix} 1 & -2 \\ 1 & 2 \end{bmatrix} \mathbf{x}$

6. $\dot{\mathbf{x}} = \begin{bmatrix} -1 & -2 \\ 1 & -2 \end{bmatrix} \mathbf{x}$

7. Compare $\dot{\mathbf{x}} = \begin{bmatrix} -1 & -0.25 \\ 1 & -2 \end{bmatrix} \mathbf{x}$ with $\dot{\mathbf{x}} = \begin{bmatrix} 1 & -0.25 \\ 1 & 2 \end{bmatrix} \mathbf{x}$

8. $\dot{\mathbf{x}} = \begin{bmatrix} -1 & -0.5 \\ 2 & 1 \end{bmatrix} \mathbf{x}$

9. $\dot{\mathbf{x}} = \begin{bmatrix} -1 & 1 \\ 2 & -2 \end{bmatrix} \mathbf{x}$

10. $\dot{\mathbf{x}} = \begin{bmatrix} -1 & 0 & 0 \\ 0 & -2 & 0 \\ 0 & 0 & -3 \end{bmatrix} \mathbf{x}$

11. $\dot{\mathbf{x}} = \begin{bmatrix} -1 & 0 & 0 \\ 0 & 2 & 0 \\ 0 & 0 & -3 \end{bmatrix} \mathbf{x}$

12. A process engineer has linearized a nonlinear process model to obtain the following state-space model and given it to your boss. Your boss has forgotten everything he learned on dynamic systems and has asked you to study this model using linear system analysis techniques.

$$\dot{\mathbf{x}} = \mathbf{A}\,\mathbf{x}$$

where:

$$\mathbf{A} = \begin{bmatrix} 0 & -1.0 \\ 1.0 & 0.0 \end{bmatrix}$$

with initial conditions $x_1(0) = 0.5$ and $x_2(0) = -0.25$.

  a. What are the eigenvalues of the A matrix? Use both MATLAB and your own analytical solution.

  b. Show a phase-plane plot, placing x_1 on the x-axis and x_2 on the y-axis.

13. Consider a process with a state-space A (Jacobian) matrix that has the following eigenvalues and eigenvectors. Draw the phase-plane plot, clearly showing the direction of the trajectories. The eigenvalues are:

$$\lambda_1 = -1 \qquad \lambda_2 = 1$$

and the eigenvectors are:

$$\xi_1 = \begin{bmatrix} \dfrac{\sqrt{2}}{2} \\ \dfrac{\sqrt{2}}{2} \end{bmatrix} \qquad \xi_2 = \begin{bmatrix} 1 \\ 0 \end{bmatrix}$$

14. An interesting example of phase-plane behavior is presented in the book by Strogatz (1994). He develops a simple model for love affairs, using Romeo and Juliet to illustrate the concepts. Consider the case where Romeo is in love with Juliet, but Juliet is fickle. The more that Romeo loves her, the more that Juliet resists his love. When Romeo becomes discouraged and backs off, Juliet becomes more attracted to him. Let:

$$x_1 = \text{Romeo's love/hate for Juliet}$$

$$x_2 = \text{Juliet's love/hate for Romeo}$$

where positive state variable values indicate love and negative values indicate hate. The model for this relationship is:

$$\frac{dx_1}{dt} = a\,x_2$$

$$\frac{dx_2}{dt} = -b\,x_1$$

where a and b are positive parameters. Show that this model has center behavior and discuss the meaning from a romance perspective.

15. Consider a more general formulation of the Romeo/Juliet problem in 14 above. In this case, let:

$$\frac{dx_1}{dt} = a_{11}\,x_1 + a_{12}\,x_2$$

$$\frac{dx_2}{dt} = a_{21}\,x_1 + a_{22}\,x_2$$

where the parameters a_{ij} can be either positive or negative. The choice of signs specifies the romantic "styles." For each of the following cases (parameters a and b are positive), determine the phase-plane behavior. Interpret the meaning of the results in terms of romantic behavior.

a.
$$\frac{dx_1}{dt} = a\,x_1 + b\,x_2$$

$$\frac{dx_2}{dt} = b\,x_1 + a\,x_2$$

b.
$$\frac{dx_1}{dt} = -a\,x_1 + b\,x_2$$

$$\frac{dx_2}{dt} = b\,x_1 - a\,x_2$$

c.
$$\frac{dx_1}{dt} = a\,x_1 + b\,x_2$$

$$\frac{dx_2}{dt} = -b\,x_1 - a\,x_2$$

Nonlinear Problems

16. As a chemical engineer in the pharmaceutical industry you are responsible for a process that uses a bacteria to produce an antibiotic. The reactor has been contaminated with a protozoan that consumes the bacteria. Assume that predator-prey equations are used to model the system (x_1 = bacteria (prey), x_2 = protozoa (predator)). The time unit is *days*.

$$\frac{dx_1}{dt} = \alpha\,x_1 - \gamma\,x_1 x_2$$

$$\frac{dx_2}{dt} = \varepsilon\,\gamma\,x_1 x_2 - \beta\,x_2$$

a. Show that the nontrivial steady-state values are:

$$x_{1s} = \frac{\beta}{\varepsilon\gamma} \qquad x_{2s} = \frac{\alpha}{\gamma}$$

b. Use the scaled variables, y_1 and y_2,

$$y_1 = \frac{x_1}{x_{1s}} \qquad y_2 = \frac{x_2}{x_{2s}}$$

to find the scaled modeling equations:

$$\frac{dy_1}{dt} = \alpha\,(1 - y_2)\,y_1$$

$$\frac{dy_2}{dt} = -\beta\,(1 - y_1)\,y_2$$

c. Find the eigenvalues of the Jacobian matrix for scale equations, evaluated at $y_1 s$ and $y_2 s$. Realize that $y_1 s$ and $y_2 s$ are 1.0 by definition. Find the eigenvalues in terms of α and β.

d. The parameters are $\alpha = \beta = 1.0$ and the initial conditions are $y_1(0) = 1.5$ and $y_2(0) = 0.75$.

    i. Plot the transient response of y_1 and y_2 as a function of time (plot these curves on the same graph using MATLAB). Using your choice of integration methods, simulate to at least $t = 20$.

    ii. Show a phase-plane plot, placing y_1 on the x-axis and y_2 on the y-axis.

    iii. What is the "peak-to-peak" time for the bacteria? By how much time does the protozoa "lag" the bacteria?

e. Now consider the trivial steady-state $(x_{1s} = x_{2s} = 0)$. Is it stable? Perform simulations when $x_1(0) \neq 0$ and $x_2(0) \neq 0$. What do you find?

f. What if $x_1(0) \neq 0$ and $x_2(0) = 0$?

g. What if $x_1(0) = 0$ and $x_2(0) \neq 0$?

17. Consider the bioreactor model used in Example 13.7 with substrate inhibition rather than Monod kinetics (see Module 8 for more detail)

$$\frac{dx_1}{dt} = (\mu - D)\,x_1$$

$$\frac{dx_2}{dt} = (s_f - x_2)\,D - \frac{\mu x_1}{Y}$$

$$\mu = \frac{\mu_{max}\,x_2}{k_m + x_2 + k_1 x_2^2}$$

where:

$$\mu_{max} = 0.53 \qquad k_m = 0.12$$
$$Y = 0.4 \qquad s_f = 4.0$$
$$k_1 = 0.4545$$

and x_1 is the biomass concentration and x_2 is the substrate concentration. Assume that the steady-state dilution rate is $D_s = 0.3$.

a. Find the steady-state (equilibrium) solutions (Hint: There are three).

b. Analyze the stability of each steady-state. Find the Jacobian, the eigenvalues, and the eigenvectors at each steady-state.

    c. Construct a phase plane plot. What do you observe about the unstable steady-state?

    d. What would you do if it was desirable to operate the reactor at an unstable steady-state?

18. Perform some time domain plots related to the phase-plane plots for Example 13.7. Discuss how these plots relate to the phase-plane results.

19. A chemical reactor that has a single second-order reaction and has an outlet flowrate that is a linear function of height has the following model where the outlet flowrate is linearly related to the volume of liquid in the reactor ($F = \beta V$).

$$\frac{dC}{dt} = \frac{F_{in}}{V}(C_{in} - C) - kC^2$$

$$\frac{dV}{dt} = F_{in} - \beta V$$

The parameters, variables and their steady-state values are shown below:

F_{in} = inlet flowrate (1 liter/min)
C_{in} = inlet concentration (1 gmol/liter)
C = reactor concentration (0.5 gmol/liter)
V = reacto volume (1 liter)
k = reaction rate constant (2 liter/(gmol min))
β = 1 min^{-1}

Perform a phase-plane analysis and discuss your results.

20. Consider two interacting tanks in series, with outlet flowrates that are a function of the square root of tank height. The flow from tank 1 is a function of $\sqrt{h_1 - h_2}$, while the flowrate out of tank 2 is a function of $\sqrt{h_2}$

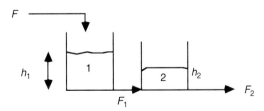

The following modeling equations describe this system:

$$\begin{bmatrix} \dfrac{dh_1}{dt} \\[2mm] \dfrac{dh_2}{dt} \end{bmatrix} = \begin{bmatrix} f_1(h_1,h_2,F) \\ f_2(h_1,h_2,F) \end{bmatrix} = \begin{bmatrix} \dfrac{F}{A_1} - \dfrac{\beta_1}{A_1}\sqrt{h_1 - h_2} \\[3mm] \dfrac{\beta_1}{A_2}\sqrt{h_1 - h_2} - \dfrac{\beta_2}{A_2}\sqrt{h_2} \end{bmatrix}$$

For the following parameter values

$$\beta_1 = 2.5 \frac{\text{ft}^{2.5}}{\text{min}} \qquad \beta_2 = \frac{5}{\sqrt{6}} \frac{\text{ft}^{2.5}}{\text{min}} \qquad A_1 = 5 \text{ ft}^2 \qquad A_2 = 10 \text{ ft}^2$$

and the input $\qquad F = 5 \dfrac{\text{ft}^3}{\text{min}}$

The steady-state height values are

$$h_{1s} = 10 \qquad h_{2s} = 6$$

Perform a phase-plane analysis and discuss your results.

INTRODUCTION TO NONLINEAR DYNAMICS: A CASE STUDY OF THE QUADRATIC MAP

14

This chapter provides an introduction to bifurcation theory and chaos. After studying this chapter, the reader should be able to:

- See the similarity between discrete time dynamic models and numerical methods
- Determine the asymptotic stability of a solution to the quadratic map
- Understand the concept of a bifurcation
- Understand how to find period-2, period-4, . . ., period-n solutions
- Understand the significance of the universal number 4.669196223

When a parameter of a discrete-time model is varied, the number and character of solutions may change—the parameter that is varied is known as a *bifurcation* parameter. For some values of the bifurcation parameter, the dynamic model may converge to a single value after a long value of time, while a small change in the bifurcation parameter may yield periodic (continuous oscillations) solutions. For some discrete equations, values of the parameter may yield solutions that appear random—these are typically "chaotic" solutions. Chaos can occur in a single nonlinear discrete equation, while three autonomous (no explicit dependence on time) ODEs are required for chaos in continuous models.

The major sections in this chapter are:

14.1   BACKGROUND

Many engineers and scientists have assumed (at least, until roughly twenty years ago) that simple models have simple solutions and simple behavior, and that this behavior is predictable. Indeed, the main objective for developing a model is usually to be able to predict behavior or to match observed behavior (measured data). During the past thirty years, a number of scientists and engineers have discovered simple models where the short-term behavior is predictable, but sensitivity to initial conditions make the long-term prediction impossible. By initial conditions, we mean the value of the variables at the beginning of the integration in time. An example is the simple weather prediction model of Lorenz (1963), which is a system of three nonlinear ordinary differential equations; the Lorenz model is covered in more detail in Chapter 17. Another example is the population growth model used by May (1976), which is a single nonlinear discrete time equation. This population model is the topic of this chapter.

The commonly accepted term for the dynamic behavior of a system that exhibits sensitivity to initial conditions is *chaos*. Terms for the branch of mathematics related to chaos include nonlinear dynamics, dynamical systems theory, or nonlinear science. New chaos books, written for a general audience, appear frequently; some of the more interesting ones are referenced at the end of this chapter.

This chapter will not make you an expert on nonlinear dynamics, but it will help you understand what is meant by *sensitivity to initial conditions* and practical limits to long-term predictability.

14.2   A SIMPLE POPULATION GROWTH MODEL

Assume that the population of a species during one time period is a function of the previous time period. Perhaps we are interested in the number of bacteria cells that are growing in a petri dish, or maybe we are concerned about the population of the United States. In either case, the mathematical model is:

$$n_{k+1} = n_k + b_k - d_k \tag{14.1}$$

where n_k =   population at the beginning of time period k

           b_k =   number of births during time period k

           d_k =   number of deaths during time period k

Now assume that the number of births and deaths during time period k is proportional to the population at the beginning of time period k.

$$b_k = \alpha_b \, n_k \tag{14.2}$$

$$d_k = \alpha_d \, n_k \tag{14.3}$$

where α_b and α_d are birth and death constants. Then:

$$n_{k+1} = n_k + \alpha_b \, n_k - \alpha_d \, n_k \tag{14.4}$$

which we can write as:

$$n_{k+1} = n_k + (\alpha_b - \alpha_d) \, n_k \tag{14.5}$$

or, $$n_{k+1} = (1 + r) \, n_k \tag{14.6}$$

where $r = \alpha_b - \alpha_d$. Eqn. (14.6) can be simply written as:

$$n_{k+1} = \alpha \, n_k \tag{14.7}$$

where $\alpha = 1 + r = 1 + \alpha_b - \alpha_d$ (obviously, $\alpha > 0$ for a physical system)
    The analytical solution to (14.7) is:

$$n_k = \alpha^k \, n_0 \tag{14.8}$$

where n_0 is the initial condition.
    From inspection of (14.8) we observe that

if $\alpha < 1$ The population decreases during each time period (converging to 0).
if $\alpha > 1$ The population increases during each time period ($\rightarrow \infty$).
if $\alpha = 1$ The population remains constant during each time period.

These results are also shown in Figure 14.1

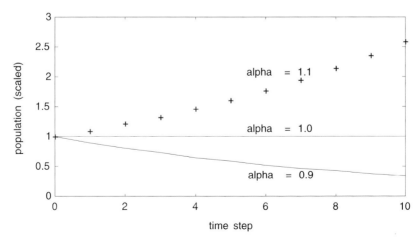

FIGURE 14.1   Simple population growth model.

These results are easily rationalized, since births<deaths, births>deaths, and births = deaths for the three cases. The result for $\alpha > 1$ is consistent with Malthus, who in the nineteenth century predicted an exponential population growth.

The result that the population increases to ∞ for $\alpha > 1$ is a bit unrealistic. In practice, the amount of natural resources available will limit the total population (for the bacteria case, the amount of nutrients or the size of the Petri dish will limit the maximum number of bacteria that can be grown). In the next example, we show a simple model that "constrains" the maximum population.

14.3   A MORE REALISTIC POPULATION MODEL

A common model that has been used to predict population growth is known as the logistic equation or the *quadratic map* (May, 1976).

$$x_{k+1} = \alpha\, x_k \left(1 - x_k\right) \tag{14.9}$$

Here, x_k represents a scaled population variable (see student exercise 3).

Note the similarity of (14.9) with the numerical methods presented in Chapter 3:

$$x_{k+1} = g(x_k) \tag{14.10}$$

Recall that direct substitution is sometimes used to solve a nonlinear algebraic equation. The next guess (iteration $k+1$) for the variable that is being solved for (x) is a function of the current guess (iteration k). Equation (14.9) shows how the population changes from time period to time period—that is, it is a discrete dynamic equation. Since (14.9) is the same form as (14.10), we will learn a lot about the quadratic map from analysis of the direct substitution technique and vice versa. You will also note that many numerical integration techniques (Euler, Runge-Kutta) have the form of (14.10).

Since (14.9) is a discrete dynamic equation, we can determine the steady-state behavior by finding the solution as $k \to \infty$. Writing (14.9) in a more explicit form,

$$x_{k+1} = \alpha\, x_k - \alpha\, x_k^2 \tag{14.11}$$

as we approach a steady-state (*fixed-point*) solution, $x_{k+1} = x_k$, so we can write:

$$x_s = \alpha\, x_s - \alpha\, x_s^2 \tag{14.12}$$

which can be written:

$$\alpha\, x_s^2 - (\alpha - 1)\, x_s = 0 \tag{14.13}$$

We can use the quadratic formula to find the steady-state (fixed-point) solutions:

$$x_s = 0 \text{ and } \frac{\alpha - 1}{\alpha} \tag{14.14}$$

It is easy to see from (14.9) that if the initial population is zero, it will remain at zero. For a non-zero initial condition, one would expect convergence (steady-state) of the population to $(\alpha - 1)/\alpha$. We will use a case study approach to show that the actual long-term

TABLE 14.1   Parameters and Non-zero Solutions for Four Cases

Case	α	x_s-
1	2.95	0.6610
2	3.20	0.6875
3	3.50	0.7143
4	3.75	0.7333

(steady-state) behavior can be quite complex. Table 14.1 shows the α parameter and the non-zero steady-state that is expected from (14.14).

14.3.1   Transient Response Results for the Quadratic (Logistic) Map

Each case presented in Table 14.1 has distinctly different dynamic behavior. As shown in the following sections, case 1 illustrates asymptotically stable behavior, cases 2 and 3 illustrate periodic behavior, and case 4 illustrates chaotic behavior.

ASYMPTOTICALLY STABLE BEHAVIOR

Let x_0 represent the initial condition (the value of the population at the initial time) and x_k represent the population value at time step k. For case 1 we find the following values, using the relationship $x_{k+1} = 2.95\, x_k\, (1 - x_k)$:

Step k	x_k	x_{k+1}
0	0.1	0.2655
1	0.2655	0.5753
2	0.5753	0.7208
3	0.7208	0.5937
4	0.5937	0.7116
5	0.7116	0.6054
.	.	.
∞	0.6610	0.6610

The transient response for case 1 is plotted in Figure 14.2 for an initial condition of 0.1. Notice that the response converges to the predicted steady-state of 0.6610. This type of response for continuous models is usually called *asymptotically stable* behavior since the output converges to the steady-state (fixed-point) solution.

PERIODIC BEHAVIOR

The transient response curve for case 2 is shown in Figure 14.3. The curve oscillates between 0.513 and 0.800, while the predicted result (equation 14.14 and Table 14.1) is

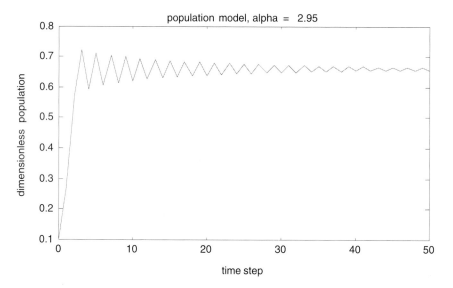

FIGURE 14.2 Transient population response, case 1; converges to single steady-state.

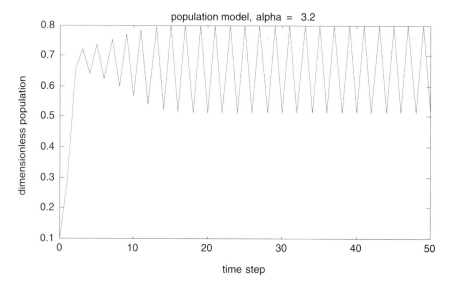

FIGURE 14.3 Transient population response, case 2; oscillates between two values (period-2 behavior).

0.6875. This type of response is known as period-2 behavior. In case 3 the transient response oscillates between 0.383, 0.827, 0.501, and 0.875 as shown in Figure 14.4. This is known as period-4 behavior, since the system returns to the same state value every fourth time step.

CHAOTIC BEHAVIOR

For Case 4, the transient response never settles to a consistent set of values, as shown in Figure 14.5; rather the values appear to be somewhat "random" although a deterministic equation has been used to solve the problem. Figure 14.6 shows that a slight change in initial condition from 0.100 to 0.101 leads to a significantly different point-to-point response—this is known as *sensitivity to initial conditions* and is characteristic of *chaotic* systems.

WHERE WE ARE HEADING

At this point, you are probably wondering how to predict the type of behavior that the quadratic map is going to have. Changes in the α parameter have led to many different types of behavior. The purpose of Section 14.4 is to show how to predict the type of behavior that will be observed using *cobweb* diagrams. Section 14.5 will then introduce bifurcation plots, which reduce the long-term results from many transient plots to a single plot. Section 14.6 introduces linear stability theory for discrete systems. Section 14.7 shows how to find the period-n points.

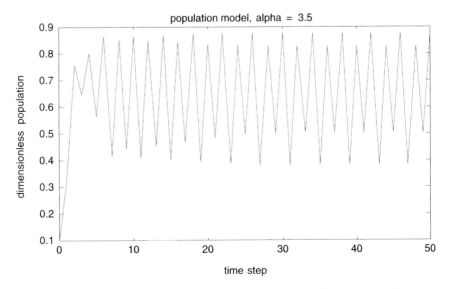

FIGURE 14.4   Transient population response, case 3; oscillates between four values (period-4 behavior).

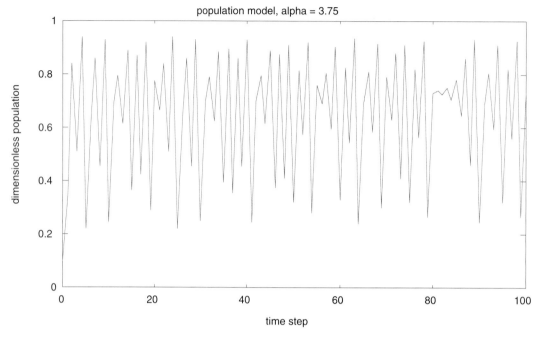

FIGURE 14.5   Transient population response, case 4; chaotic behavior.

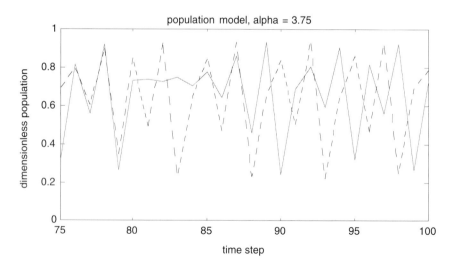

FIGURE 14.6   Transient population response, case 4; chaotic behavior (Solid Line—initial condition of $x_0 = 0.1$. Dashed line—initial condition of $x_0 = 0.101$). This illustrates the sensitivity to initial conditions.

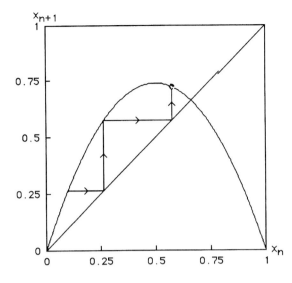

FIGURE 14.7   Cobweb diagram for the quadratic map problem. The initial point is $x_0 = 0.1$.

14.4 COBWEB DIAGRAMS

Insight to the behavior of discrete single-variable systems can be obtained by constructing *cobweb diagrams*. Cobweb diagrams are generated by plotting two curves: (i) $g(x)$ versus x and (ii) x versus x; the solution (fixed-point) is at the intersection of the two curves. For example, consider the case 1 parameter value of $\alpha = 2.95$ and an initial guess, $x_0 = 0.1$. The $x_{n+1} = g(x) = 2.95\, x_n\, (1 - x_n)$ curve is shown as the inverted parabola in Figure 14.7. Since the x_0 value is 0.1, the x_1 value is obtained by first drawing a vertical line to the $g(x)$

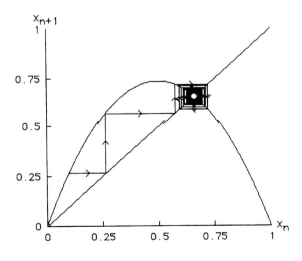

FIGURE 14.8   Case 1 ($\alpha = 2.95$) map, convergence to a single solution ($x = 0.661$); corresponds to the transient response in Figure 14.2.

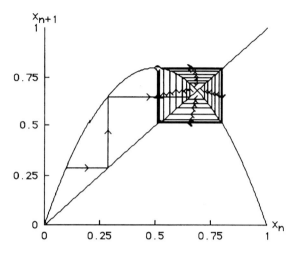

FIGURE 14.9 Case 2 ($\alpha = 3.20$) map, oscillates between $x = 0.5130$ and 0.7995 after initial transient; initial condition of $x_0 = 0.1$—corresponds to the transient response in Figure 14.3.

curve to find $g(x_0) = 0.265$, then drawing a horizontal line to the $x = x$ curve (since $x_1 = g(x_0)$; therefore, $x_1 = 0.265$). A vertical line is drawn to the $g(x)$ curve (to obtain $g(x_1) = 0.575$), then a horizontal line is drawn to the $x = x$ curve (so, $x_2 = 0.575$). These initial steps are shown in Figure 14.7.

Figure 14.8 shows that this process converges to the fixed-point of $x_\infty = 0.661$, for the case 1 parameter value of $\alpha = 2.95$. Figure 14.9 shows that the iterative process eventually "bounces" between two solutions for the case 2 parameter value of $\alpha = 3.2$. This is shown more clearly in Figure 14.10 where an initial guess of $x_0 = 0.5130$ leads to solutions of 0.5130 and 0.7995 (period-2 behavior). Case 3 has period-4 behavior, as shown in Figures 14.11 and 14.12. Case 4 (Figure 14.13) is an example of chaotic behavior, where the sequence of iterates never repeats.

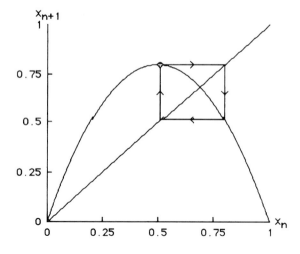

FIGURE 14.10 Case 2 ($\alpha = 3.20$) map, oscillates between $x = 0.5130$ and 0.7995; initial condition of $x_0 = 0.5130$ (period-2 behavior).

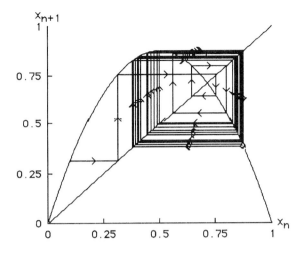

FIGURE 14.11 Case 3 (α = 3.50) map, oscillates between x = 0.3828, 0.8269, 0.5009, and 0.8750. The initial condition is x_0 = 0.1. This corresponds to the transient response in Figure 14.4 (period-4 behavior).

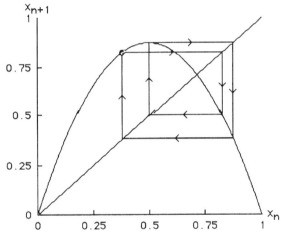

FIGURE 14.12 Case 3(α = 3.50) map, oscillates between 0.3828, 0.8269, 0.5009, and 0.8750, initial condition of x_0 = 0.3828 (period-4 behavior).

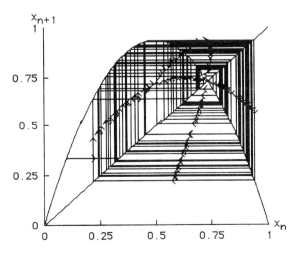

FIGURE 14.13 Case 4 (α = 3.75) map, chaotic behavior; corresponds to the transient response in Figure 14.5.

14.5   BIFURCATION AND ORBIT DIAGRAMS

When a parameter of a discrete-time model is varied, the number and character of solutions may change; the parameter that is varied is known as a *bifurcation* parameter. For the quadratic map, α is a bifurcation parameter. We have seen that somewhere between $\alpha = 2.95$ and 3.2, the behavior of the quadratic map changes from asymptotically stable to period-2 behavior.

A single diagram can be developed that represents the solutions for a large range of α values. We are most interested in the long-term behavior of a system, so for a single α value, we can run a simulation and throw out the initial transient data points (say, the first 250 points). The next points (say, the next 250) should then adequately represent the long-term behavior of the system. We can then move on to another value of α and do the same. This is exactly the technique used to generate Figure 14.14, which is an *orbit diagram* for the quadratic map (see student exercise 7).

14.5.1   Observations from the Orbit Diagram (Figure 14.14)

There is a single steady-state solution until $\alpha = 3$, where a bifurcation to two solutions occurs. The next bifurcation point is $\alpha = 3.44949$, where four solutions emerge. A period-8 bifurcation occurs at $\alpha = 3.544090$, period-16 at $\alpha = 3.564407$, period-32 at $\alpha = 3.568759$, and period-64 at $\alpha = 3.569692$. Chaos occurs at $\alpha = 3.56995$. Notice that there are some interesting "windows" of periodic behavior, after the onset of chaos. For example, at $\alpha = 3.83$ we find a window of period-3 behavior. The period-3 behavior occurs after approximately 60 time steps, with an initial condition of 0.1, as shown in Figure 14.15. This behavior is shown more clearly in Figure 14.16 which is simply the data from

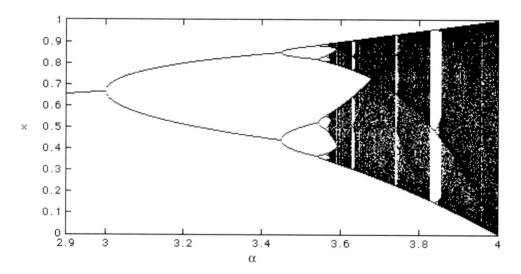

FIGURE 14.14   Orbit diagram for the quadratic map. α is the bifurcation parameter.

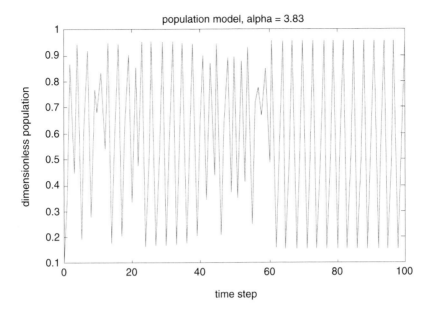

FIGURE 14.15   Period-3 behavior (after initial transient) for $\alpha = 3.83$, initial condition = 0.1. Periodic values are $x = 0.15615$, 0.50466, and 0.957417.

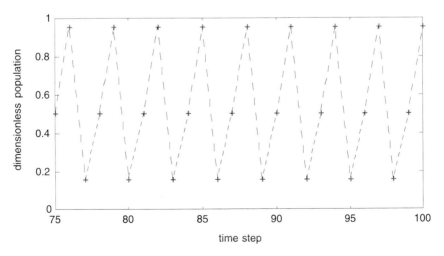

FIGURE 14.16   Period-3 behavior for $\alpha = 3.83$. Values are $x = 0.15615$, 0.50466, and 0.957417.

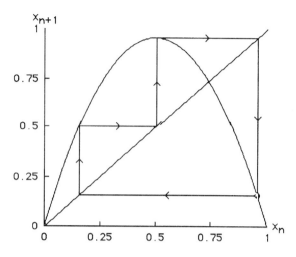

FIGURE 14.17 Period-3 behavior at $\alpha = 3.83$. Values are $x = 0.15615$, 0.50466, and 0.957417.

Figure 14.15 plotted between 75 and 100 time steps. The cobweb diagram of Figure 14.17 also shows the period-3 behavior. Research has shown that period-3 behavior implies chaotic behavior.

14.6 STABILITY OF FIXED-POINT SOLUTIONS

When we performed our case study, we found that case 1 converged to the predicted fixed point, while the other cases had periodic (or chaotic) solutions that were not attracted to the fixed points. We wish now to use an analytical method to determine when the solutions will converge to the fixed-point solution—that is, when is a fixed-point stable? The following stability theorem is identical to the stability theorem used for the numerical analysis in Chapter 3.

Definition   Let x^* represent the *fixed-point* solution of $x^* = g(x^*)$, or $g(x^*) - x^* = 0$.

Theorem x^* is a stable solution of $x^* = g(x^*)$, if $\left| \dfrac{\partial g}{\partial x} \right| < 1$ when evaluated at x^*.

14.6.1 Application of the Stability Theorem to the Quadratic Map

Here we will make use of this theorem to determine the stability of a solution to the quadratic map:

$$g(x) = \alpha x(1 - x) = \alpha x - \alpha x^2 \tag{14.15}$$

So,

$$\frac{\partial g}{\partial x} = \alpha - 2\alpha x = \alpha (1 - 2x)$$

For simplicity in notation, we will use g' to represent $\partial g/\partial x$. Evaluated at x^*, we have:

$$g'(x^*) = \alpha - 2\alpha x^* = \alpha (1 - 2x^*) \qquad (14.16)$$

Therefore, from (14.16) and the stability theorem, if the following condition is satisfied:

$$\left| \alpha (1 - 2x^*) \right| < 1 \qquad (14.17)$$

Then: x^* is a stable solution.

Remember from (14.14) that there are two solutions to $x^* = \alpha x^*(1-x^*)$:

$$x^* = 0 \quad \text{or} \quad x^* = \frac{\alpha - 1}{\alpha}$$

Momentarily we will generalize the stability results for any value of α. First, we will study the four specific cases.

CASE 1 $\alpha = 2.95$

At one fixed point solution, $x^* = 0$, we find:

$$\left| g'(x^*) \right| = \left| \alpha (1 - 2x^*) \right| = 2.95 > 1$$

which indicates that the fixed point is unstable.

At the other fixed point solution, $x^* = \alpha - 1/\alpha = 2.95 - 1/2.95 = 0.6610$, we find:

$$\left| g'(x^*) \right| = \left| \alpha (1 - 2x^*) \right| = \left| 2.95 (1 - 2(0.6610)) \right| = \left| -0.9499 \right|$$
$$= 0.9499 < 1$$

which assures that the second fixed-point is stable.

For $\alpha = 2.95$, we expect the numerical solution to converge to the stable fixed point, $x^* = 0.6610$, since the other fixed point ($x^* = 0$) is unstable.

The stability results for cases 1 through 4 are compared in Table 14.2. Notice that case 1 is the only one of the four cases where there exists a stable solution. The reader should verify that an initial guess of x arbitrarily close to zero (but not exactly 0), will not converge to zero for any of the four cases.

TABLE 14.2   Stability Results for Cases 1–4

Case	α	x^*	$\left| g'(x^*) \right|$	condition	x^*	$\left| g'(x^*) \right|$	condition
1	2.95	0	2.95	unstable	0.6610	0.9499	stable
2	3.20	0	3.20	unstable	0.6875	1.2000	unstable
3	3.50	0	3.50	unstable	0.7143	1.5000	unstable
4	3.75	0	3.75	unstable	0.7333	1.7500	unstable

14.6.2 Generalization of the Stability Results for the Quadratic Map

Notice that we have been quite limited in our study, since we have only considered four cases with $2.95 \leq \alpha \leq 3.75$. Now we will consider the general results for any $\alpha > 0$.

Again, recall that there are two fixed-point solutions for a given value of α

$$x^* = 0 \quad \text{or} \quad x^* = \frac{\alpha - 1}{\alpha}$$

At the risk of complicating our notation, let

$$x_0^* = 0 \quad \text{and} \quad x_1^* = \frac{\alpha - 1}{\alpha}$$

Our goal is to determine how the stability of x_0^* or x_1^* changes as a function of α.

STABILITY OF x_0^* AS A FUNCTION OF α

Since $x_0^* = 0$ and $g'(x_0^*) = \alpha - 2\alpha 0 = \alpha$
then $\left| g'(x_0^*) \right| = \left| \alpha \right|$

Also, since $|g'| < 1$ is required for stability, then x_0^* is a stable solution only as long as $-1 < \alpha < 1$. Otherwise, x_0^* is unstable (recall that $\alpha < 0$ does not make physical sense).

STABILITY OF x_1^* AS A FUNCTION OF α

Since $x_1^* = (\alpha - 1)/\alpha$ and $g'(x_1^* = \alpha - 2\alpha\ (\alpha - 1)/\alpha = \alpha - 2(\alpha - 1) = -\alpha + 2$
then $\left| g'(x_1^*) \right| = \left| -\alpha + 2 \right|$
which indicates stability for $1 < \alpha < 3$. Otherwise, x_1^* is unstable.

These results are shown on the bifurcation diagram of Figure 14.18 for $0 < \alpha < 4$. Generally, solid lines will be used to represent stable solutions and dotted lines will be used to represent unstable solutions. As discussed above, a change of stability for x_0^* occurs at $\alpha = 1$. Also, changes of stability for x_1^* occur at $\alpha = 1$ and $\alpha = 3$. The values of α where the stability characteristics change are known as bifurcation points. The bifurcation that occurs at $\alpha = 1$ is commonly known as a "transcritical" bifurcation (see Chapter 15)—an exchange of stability between the two solutions has occured.

Notice that Figure 14.18 is a *bifurcation diagram* based on a linear stability analysis. It differs from an *orbit diagram* (such as Figure 14.14), because it does not show the periodic behavior obtained from solving the nonlinear algebraic equation for the population growth model. An orbit diagram cannot display unstable solutions, however.

14.6.3 The Stability Theorem and Qualitative Behavior

The theorem states that if $|\partial g/\partial x|$ at the fixed point, the fixed point is stable. Further, if $\partial g/\partial x$ is negative, then the fixed point solution is oscillatory. If $\partial g/\partial x$ is positive, the be-

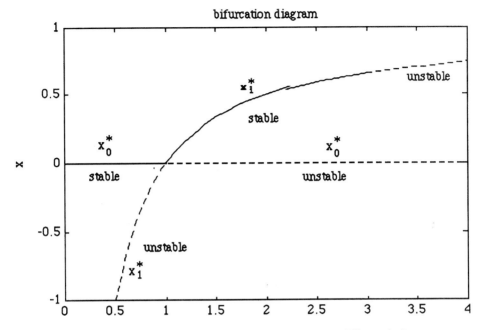

FIGURE 14.18   Bifurcation diagram based on linear stability analysis.

havior is monotonic. We can then develop the following table of results from the stability theorem

$\dfrac{\partial g}{\partial x}$	stability	response
< -1	unstable	oscillatory
$-1 < \dfrac{\partial g}{\partial x} < 0$	stable	oscillatory
$0 < \dfrac{\partial g}{\partial x} < 1$	stable	monotonic
> 1	unstable	monotonic

Although the linear stability analysis is useful for determining if a fixed-point is stable, it cannot be used directly to understand possible periodic behavior. This is the topic of the next section.

14.7 CASCADE OF PERIOD-DOUBLINGS

We have noted that there appears to be a series of period doublings in route to chaos. The limitation to the method presented in Section 14.6 showed that it could predict that a particular fixed point was unstable, but could not identify the type of periodic behavior that might occur. In this section we will show how to find these period-doubling bifurcation points and the respective branches shown in Figure 14.13.

14.7.1 Period-2

When period doubling occurs, the population value at time step k is equal to the value at time step $k - 2$. This can be represented by

$$x_k = x_{k-2} \tag{14.18}$$

or

$$x_{k+2} = x_k$$

using the notation

$$x_{k+1} = g(x_k) \tag{14.19}$$

then

$$x_{k+2} = g(x_{k+1}) \tag{14.20}$$

$$x_{k+2} = g(g(x_k)) \tag{14.21}$$

$$x_{k+2} = g^2(x_k) \tag{14.22}$$

Warning: Do not confuse the $g^2(x_k)$ notation with that of the square of the operator $[g(x_k)]^2$.

For the quadratic map, we can develop the relationship shown in (14.22):

$$x_{k+2} = \alpha\, x_{k+1}\, (1 - x_{k+1}) \tag{14.23}$$

and substituting $x_{k+1} = \alpha\, x_k\, (1 - x_k)$ into (14.23), we find:

$$x_{k+2} = \alpha\, [\alpha\, x_k\, (1 - x_k)]\, [1 - (\alpha\, x_k(1 - x_k))] \tag{14.24}$$

Since (from 14.19):

$$x_{k+2} = x_k$$

we can write (14.24) as

$$x_k = \alpha\, [\alpha\, x_k\, (1 - x_k)]\, [1 - (\alpha\, x_k(1 - x_k))] \tag{14.25}$$

Expanding (14.25),

$$x_k = \alpha^2 \left[-\alpha x_k^4 + 2\alpha x_k^3 - (1 + \alpha)x_k^2 + x_k \right] \qquad (14.26)$$

or

$$g^2(x_k) = \alpha^2 \left[-\alpha x_k^4 + 2\alpha x_k^3 - (1 + \alpha)x_k^2 + x_k \right] \qquad (14.27)$$

Notice that there are four solutions to the fourth-order polynomial. We can find the solutions graphically by plotting $g^2(x)$ versus x, as shown in Figure 14.19 (for $\alpha = 3.2$).

If you closely observe the plot, you will find the following four solutions for period-2 behavior:

$$x^* = 0, 0.5130, 0.6875, \text{ and } 0.7995$$

We can see graphically that the solutions $x^* = 0$ and 0.6875 are unstable, since the slope of $g^2(x^*)$ is greater than 1. (A period-2 solution is stable if $|\partial(g^2(x))/\partial x| < 1$.) Also, notice that the solution for $x = g(x)$ will always appear as one of the solutions for $x = g^2(x)$. If a solution for $x = g(x)$ is unstable, it will also be unstable for $g^2(x)$. We can see that $x = 0.6875$ is the solution for both $x = g(x)$ and $x = g^2(x)$, by observing Figure 14.20.

At this point it is worth showing the results of x versus $g(g(x))$ for case 1, which we know has a single, asymptotically stable solution. Figure 14.21 shows that there is a single stable solution of $x = 0.6610$. This makes sense, because as $k \to \infty$, we know that $x_k = 0.6610$; this means that $x_{k+2} = x_{k+1} = x_k = 0.6610$.

We can see from Figure 14.22 that $\alpha = 3.0$ is a bifurcation point, since absolute values of the slope of $g(x)$ and $g^2(x) = 1$ at $x = 0.66667$. Figure 14.22 is clearly a transition point between Figure 14.21 and Figure 14.20.

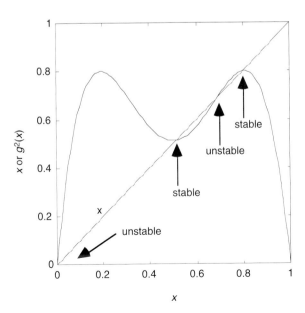

FIGURE 14.19 Plot of $g(g(x))$ versus x to find the period-2 values for $\alpha = 3.2$.

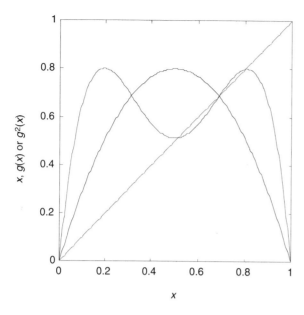

FIGURE 14.20   Plots of $g(x)$ and $g^2(x)$ versus x for $\alpha = 3.2$.

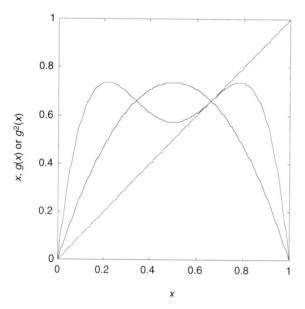

FIGURE 14.21   Plots of $g(x)$ and $g(g(x))$ versus x for $\alpha = 2.95$. No period-2 behavior.

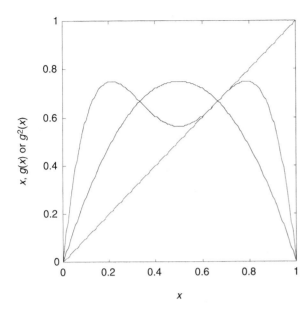

FIGURE 14.22   Plot of $g(x)$ and $g^2(x)$ versus x for $\alpha = 3.0$.

14.7.2   Period-4

When period-4 behavior occurs, the population value at time step k is equal to the value at time step $k - 4$. This can be represented by:

$$x_{k+4} = x_k \qquad (14.28)$$

Using the same arguments that we used for period-2 behavior, we can find that since $x_{k+1} = g(x_k)$,

$$x_{k+4} = g(x_{k+3}) \qquad (14.29)$$
$$= g(g(x_{k+2}))$$
$$= g(g(g(x_{k+1})))$$
$$= g(g(g(g(x_k))))$$
$$x_{k+4} = g^4(x_k) \qquad (14.30)$$

We can obtain the solutions to (14.30) by plotting $g^4(x)$ versus x as shown in Figure 14.23 for $\alpha = 3.5$. Again, do not confuse $g^4(x)$ with $[g(x)]^4$. $g^4(x)$ is an eighth-order polynomial with eight solutions as shown in Figure 14.23.

Figure 14.24 shows that the solutions for $x = g(x)$ and $x = g^2(x)$ are also solutions (although unstable) for $x = g^4(x)$.

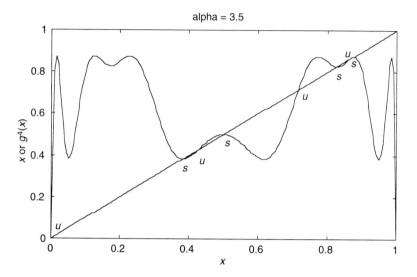

FIGURE 14.23   Plots of $g^4(x)$ versus x to find the period-4 values for $\alpha = 3.5$.

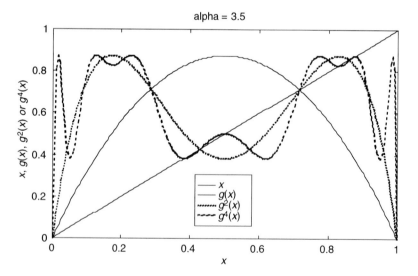

FIGURE 14.24   Plots of $g(x)$, $g^2(x)$ and $g^4(x)$ versus x for $\alpha = 3.5$.

14.7.3   Period-n

By analogy to the period-2 and period-4 behavior, we can see that for any period n, we have the following relationship

$$x_{k+n} = x_k \qquad (14.31)$$

$$x_{k+n} = g^n(x_k) \qquad (14.32)$$

Note that $g^n(x_k)$ will be a polynomial that is order $2n$, and there will be $2n$ solutions, n of which are stable.

14.7.4   Feigenbaum's Number

The quadratic map exhibits a period doubling route to chaos. As the bifurcation parameter α is increased, model goes through a series of period doublings (period-2, period-4, period-8, period-16, etc.). Feigenbaum noticed that the quadratic map had a consistent change in the bifurcation parameter between each period doubling. Indeed, he found that any "one-hump" (see any plot of $g(x)$ for the quadratic map) model will have a cascade of bifurcations which will yield the *Feigenbaum number*. The Feigenbaum number is calculated by comparing α values at each successive bifurcation point in the following fashion

$$\lim_{i \to \infty} \frac{\alpha_i - \alpha_{i-1}}{\alpha_{i+1} - \alpha_i} = 4.669196223 \qquad (14.33)$$

where α_i represents the parameter value at the ith period doubling point, where the period is $n = 2^i$. To obtain a rough estimate of the Feigenbaum number, use the values of α for period-16 (2^4), period-32 (2^5) and period-64 (2^6)

$$\frac{\alpha_5 - \alpha_4}{\alpha_6 - \alpha_5} = \frac{3.568759 - 3.564407}{3.569692 - 3.568759} = 4.6645$$

which is close to 4.6692.

A summary of the bifurcation points is provided in Table 14.3.

Chaos occurs when the period is ∞ (state sequence never repeated) at $\alpha = 3.56995$.

TABLE 14.3   Values of α at Bifurcation Points

i	period	α
1	2	3.0
2	4	3.44949
3	8	3.544090
4	16	3.564407
5	32	3.568759
6	64	3.569692
∞	∞	3.56995

14.8 FURTHER COMMENTS ON CHAOTIC BEHAVIOR

We have used the quadratic map (a model of population growth) to introduce you to non-linear dynamic behavior. This model consisted of a single discrete nonlinear equation. Dynamic behavior similar to period-2 can result from a set of two nonlinear ordinary differential equations. Examples of period behavior in continuous systems include the Lotka-Volterra model used to predict the populations of predator and prey species. The change in a bifurcation parameter that causes a *limit-cycle* to form in a 2 ODE system is known as a *Hopf bifurcation,* and will be covered in Chapter 16. Chaos is possible in a set of three autonomous nonlinear ordinary differential equations. This behavior was discovered by Lorenz in a simple (reduced-order) model of a weather system (really a model of natural convection heat transfer) and will be detailed in Chapter 17. Lorenz coined the phrase "butterfly effect" to describe a system of equations that is sensitive to initial conditions (hence chaotic). He stated conceptually that a butterfly flapping its wings in Troy, New York could cause a monsoon in China several months later (or something similar!).

Some of the earliest results of what is now known as chaos were really discovered by Poincaré in the late nineteenth century, involving the three-body problem. He found that it was easy to determine the planetary motions due to gravity in a system with two bodies, but when three bodies were considered, the system of equations became noninte-grable—leading to the possibility of chaos.

We see turbulence throughout our daily lives, from the water flowing from our faucets, to the effect of wind blowing through our hair as we ride our bicycles, to the boiling water on our stoves. Many researchers have tried to model turbulence by adding stochastic (random) terms to our models of physical behavior. It has only been realized in the past three decades that a good physical (nonlinear) model can simulate the effects of turbulence through chaos.

Numerical methods are used to solve the vast majority of chemical process models. Angelo Lucia (see references) has found that chaos can occur in the solution of some thermodynamic equations of state if the numerical methods are not formulated carefully. It is likely that many people have obtained similarly bad solutions before reformulating them correctly.

SUMMARY

A lot of material has been presented in this chapter. You may be wondering how discrete maps and bifurcation theory ties in with applications in chemical engineering. There are at least two important reasons for studying this material:

- We have shown that the quadratic map problem is conceptually identical to numerical methods that can be used to solve a nonlinear algebraic equation. Since the quadratic map problem exhibits exotic behavior under certain values of the parameter α, this tells us that a poorly posed numerical method may have similar problems. *Be careful when using numerical methods!*

- We noticed that a discrete population growth model is represented by the quadratic map problem. This population growth model is a simple example of a discrete dynamic system, which was modeled by a nonlinear algebraic equation. In the future, we will be studying continuous dynamic systems, that is, systems that are modeled by ordinary differential equations (ODEs). It turns out that nonlinear ODEs can have dynamic properties that are similar to the discrete population model. For example, exothermic chemical reactors can exhibit bifurcation behavior and continuous oscillations in temperature and composition. One main difference is that a system modeled by a set of autonomous ODEs must have at least three equations before chaotic behavior occurs. Chaotic behavior can occur in a discrete model with only one equation.

REFERENCES AND FURTHER READING

The primary reference for the behavior of the quadratic map is (this has been reprinted in a number of sources) is by May.

May, R.M. (1976). Simple mathematical models with very complicated dynamics, *Nature* 261: 459–467.

The general field of chaos theory was introduced to much of the public in the popular book by Gleick:

Gleick, J. (1987). *Chaos: Making a New Science.* New York: Viking.

Lorenz is given credit for the discovery of "sensitivity to initial conditions":

Lorenz, E.N. (1963). Deterministic nonperiodic flows. *Journal of Atmospheric Science,* 20: 130–141.

Software for the Macintosh® that was packaged with the following book was used to generate some of the quadratic map diagrams. This book also does an excellent job of discussing the quadratic map and dynamical systems theory.

Tufillaro, N.B., T. Abbott, & J. Reilly. (1992). *An Experimental Approach to Nonlinear Dynamics and Chaos.* Redwood City, CA: Addison-Wesley.

Chaos can appear in numerical solutions to chemical engineering problems, such as phase equilibrium calculations, as shown in the following paper:

Lucia, A., X. Guo, P.J. Richey, & R. Derebail. (1990). *Simple process equations, fixed point methods, and chaos.* American Institute of Chemical Engineers Journal (*AIChE J.*), 36(5): 641–654.

The book by Strogatz is an excellent introduction to nonlinear dynamics:

Strogatz, S.H. (1994). *Nonlinear Dynamics and Chaos.* Reading, MA: Addison Wesley.

STUDENT EXERCISES

1. Why are the results for the simple quadratic map problem important to understand, when chemical and environmental process models are obviously much more complex (based on ODEs)?

2. Use MATLAB to generate transient responses for the quadratic map, for various values of α. Explore regions of single steady-state solutions, as well as regions of periodic and chaotic behavior. Use Figure 14.14 to try and find regions of periodic behavior in the midst of chaotic behavior.

3. Derive the scaled logistic equation (14.9) from the following unscaled model for population growth.

$$n_{k+1} = n_k + r\left(1 - \frac{n_k}{L}\right) n_k$$

where r and L are constants (*Hint:* Define the scaled variable, $x = nr/(1 + r)L$). What is the physical significance of L?

4. Consider the "constant harvesting" model for population growth, where γ is a term that accounts for a constant removal rate per unit time period (e.g., hunting deer or removing cells from a petri dish),

$$x_{k+1} = \alpha x_k (1 - x_k) - \gamma$$

How does γ effect the equilibrium population values? (Show calculation, and consider stability of the equilibrium.)

   Let $\alpha = 3.2$. What γ values are required for 0 (the trivial solution) to be a stable equibrium solution? What γ values are required for a stable nontrivial solution?

5. Consider the "proportional harvesting" model for population growth, where the removal rate per unit time period is proportional to the amount of population

$$x_{k+1} = \alpha x_k (1 - x_k) - \gamma x_k$$

How does γ effect the equilibrium population values? (Show calculation, and consider stability of the equilibrium.)

   Let $\alpha = 3.2$. What γ values are required for 0 (the trivial solution) to be a stable equibrium solution? What γ values are required for a stable nontrivial solution?

6. Consider the period-2 behavior that occurs at a value of $\alpha = 3.2$. Show that the values of $x = 0$ and $x = 0.6875$ are unstable. (*Hint:* Let $h(x) = g(g(x))$ and show that $|h'(x)| \geq 1$ at those values.)

7. Using MATLAB construct the orbit diagram (Figure 14.14) for the quadratic map.

8. Find the (real) fixed points of $x_{k+1} = \sqrt{x_k}$ and analyze their stability. Also, develop a cobweb diagram for this problem.

9. Consider the nonlinear algebraic equation, $f(x) = -x^2 - x + 1 = 0$. Using the direct substitution method, formulated as $x = -x^2 + 1 = g(x)$, the iteration sequence is

$$x_{k+1} = g(x_k) = -x_k^2 + 1$$

Try several different initial conditions and show whether these converge, diverge or oscillate between values. Discuss the stability of the two solutions $x^* = 0.618$ and $x^* = 1.618$, based on an analysis of $g'(x^*)$. Develop a cobweb diagram for this system.

10. Consider the nonlinear algebraic equation, $f(x) = -x^2 - x + 1 = 0$. Using Newton's method,

$$x_{k+1} = x_k - \frac{f(x_k)}{f'(x_k)}$$

write the iteration sequence in the form of:

$$x_{k+1} = g(x_k)$$

Try several different initial conditions and show whether these converge, diverge, or oscillate between values. Discuss the stability of the two solutions $x^* = 0.618$ and $x^* = 1.618$, based on an analysis of $g'(x^*)$. Develop a cobweb diagram for this system.

11. Consider the scaled Lotka-Volterra (predator (y_2)-prey (y_1)) equations, where

$$\frac{dy_1}{dt} = \alpha\,(1 - y_2)\,y_1$$

$$\frac{dy_2}{dt} = -\beta\,(1 - y_1)\,y_2$$

The parameters are $\alpha = \beta = 1.0$ and the initial conditions are $y_1(0) = 1.5$ and $y_2(0) = 0.75$. The time unit is days. Integrate these equations numerically (using ode45, for example) to show the periodic behavior.

12. The Henon map is a discrete model that can exhibit chaos:

$$x_1\,(k + 1) = x_2(k) + 1 - a\,x_1(k)^2$$
$$x_2(k + 1) = b\,x_1(k)$$

For a value of $b = 0.3$, perform numerical simulations for various values of a. Try to find values of a (try $a > 0.3675$) that yield stable period-2 behavior. Show that chaos occurs at approximately $a = 1.06$.

13. Read the paper by Lucia et al. (1990) and use cobweb diagrams to show different types of periodic behavior that can occur when direct substitution is used to find the volume roots of the SRK equation-of-state for the multicomponent mixture (CH_4, C_2H_4, and C_3H_6O).

APPENDIX: MATLAB M-FILES USED IN THIS MODULE

```
  function [time,x] = pmod(alpha,xinit,n)
% population model (quadratic map), pmod.m
% 29 August 1993 (c) B.W Bequette
% revised 20 Dec 96
% input data:
%   alpha : growth parameter (between 0 and 4)
%   n : number of time steps
%   xinit : initial population (between 0 and 1)
%
  clear x; clear k; clear time;
  x(1) = xinit;
  time(1)= 0;
  for k = 2:n+1;
    time(k) = k-1;
    x(k) = alpha*x(k-1)*(1-x(k-1));
  end
% run this file by entering the following in the command
  window
%   [time,x] = pmod(alpha,xinit,n);
% with proper values for alpha, xinit and n
% then enter the following
%   plot(time,x)
```

```
  function [x,g,g2,g3,g4] = gn_qmap(alpha);
%
% finds g(x), g^2(x), g^3(x) and g^4(x) functions for
% the quadratic map problem
%
% (c) B.W. Bequette
% 23 july 93
% modified 12 Aug 93
% revised 20 Dec 96
%
  x = zeros(201,1);
  g = x; g2 = x; g3 = x; g4 = x;
%
  for i=1:201;
  x(i) = (i-1)*0.005;
  g(i) = alpha*x(i)*(1-x(i));
  g2(i) = alpha*g(i)*(1-g(i));
```

```
        g3(i) = alpha*g2(i)*(1-g2(i));
        g4(i) = alpha*g3(i)*(1-g3(i));
        end
% can plot, for example
% plot(x,x,x,g,x,g2,'—',x,g4,'-.')
```

BIFURCATION BEHAVIOR OF SINGLE ODE SYSTEMS 15

The goal of this chapter is to introduce the student to the concept of bifurcation behavior, applied to systems modeled by a single ordinary differential equation. Chapters 16 and 17 will involve systems with more than one state variable.

After studying this chapter, the student should be able to

- Determine the bifurcation point for a single ODE
- Determine the stability of each branch of a bifurcation diagram
- Determine the number of steady-state solutions near a bifurcation point

The major sections in this chapter are:

15.1 MOTIVATION

Nonlinear systems can have "exotic" behavior such as multiple steady-states and transitions from stable conditions to unstable conditions. In Chapter 14 we presented the *quadratic map* (logistic map or population model), which showed how a discrete-time system could move from a single stable steady-state to periodic behavior as a single parameter was varied. This would be considered a dynamic bifurcation of a discrete-time system, where the behavior changed from asymptotically stable to periodic.

In this chapter we introduce bifurcation behavior of continuous-time systems. A steady-state bifurcation occurs if the number of steady-state solutions changes as a system parameter is changed. If the qualitative (stable vs. unstable) behavior of a system changes as a function of a parameter, we also refer to this as bifurcation behavior. This chapter deals with systems modeled by a single ordinary differential equation.

Bifurcation analysis is particularly important for complex systems such as chemical and biochemical reactors. Although only single variable examples are used in this module, the same types of bifurcation behavior are also observed in chemical and biochemical reactors.

15.2 ILLUSTRATION OF BIFURCATION BEHAVIOR

Here a simple polynomial equation will be used to illustrate what is meant by *bifurcation* behavior. Assume that the following cubic polynomial equation describes the steady-state behavior of a system.

$$f(x,\mu) = \mu x - x^3 = 0$$

The solution can be obtained by plotting the function and finding the values of x where $f(x,\mu) = 0$. A plot of this function for $\mu = -1$, 0 and 1 is shown in Figure 15.1 below. We see that the number of real solutions ($f(x,\mu) = 0$) for $\mu = -1$ is one, while the number of real solutions for $\mu = 1$ is three. The curve for $\mu = 0$ is a transition between the two cases. We say that $\mu = 0$ is a bifurcation point for this system, because the number of real solutions changes from one to three at this point.

We will see in the next section that this behavior is characteristic of a pitchfork bifurcation. We will also find that the number of solutions is always three for this problem; sometimes two of the solutions are complex, and other times the solutions are all the same value.

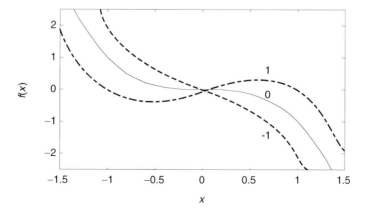

FIGURE 15.1 Polynomial behavior as a function of μ.

15.3   TYPES OF BIFURCATIONS

The types of bifurcations that will be presented by way of examples include: (i) pitchfork, (ii) saddle-node, and (iii) transcritical. We will also cover a form of *hysteresis* behavior and show that it involves two saddle-node bifurcations. Before we cover these specific bifurcations, we will present the general analysis approach.

     Consider the general dynamic equation:

$$\dot{x} = f(x,\mu) \tag{15.1}$$

where x is the state variable and μ is the bifurcation parameter. The steady-state solution (also known as an equilibrium point) of (15.1) is:

$$0 = f(x,\mu) \tag{15.2}$$

A bifurcation point is where the both the function and its first derivative are zero:

$$f(x,\mu) = \frac{\partial f}{\partial x} = 0 \tag{15.3}$$

Notice that the first-derivative is also the Jacobian for the single-equation model. Also, the eigenvalue is simply the Jacobian for a single equation system, so the eigenvalue is 0 at a bifurcation point. The number of solutions of (15.2) can be determined from *catastrophe theory*. Equation (15.2) has k solutions, if the following criteria are satisfied:

$$f(x,\mu) = 0 = \frac{\partial f}{\partial x} = \frac{\partial^2 f}{\partial x^2} = \ldots = \frac{\partial^{k-1} f}{\partial x^{k-1}} = 0 \tag{15.4}$$

and

$$\frac{\partial^k f}{\partial x^k} \neq 0 \tag{15.5}$$

In Example 15.1 this method is applied to a system that exhibits a pitchfork bifurcation.

EXAMPLE 15.1   Pitchfork Bifurcation

Consider the single variable system shown previously in Section 15.2.

$$\dot{x} = f(x,\mu) = \mu x - x^3 \tag{15.6}$$

The equilibrium point is:

$$f(x,\mu) = 0 = \mu x_e - x_e^3$$

the solutions to $f(x,\mu) = 0$ are

$$x_{e0} = 0$$

$$x_{e1} = \sqrt{\mu}$$

$$x_{e2} = -\sqrt{\mu}$$

Notice that if $\mu < 0$, then $x_e = 0$ is the only physically meaningful (real) solution, since $\sqrt{\mu}$ is complex if $\mu < 0$.

The Jacobian is

$$\left.\frac{\partial f}{\partial x}\right|_{x_e,\mu_e} = -3\,x_e^2 + \mu_e$$

Since the Jacobian is a scalar, then the eigenvalue is equal to the Jacobian:

$$\lambda = -3\,x_e^2 + \mu_e$$

If $\lambda < 0$, then the system is stable. If $\lambda > 0$, then the system is unstable. Now, we can find the stability of the system, as a function of the bifurcation parameter, μ.

I. $\mu < 0$.   The only real equilibrium solution is $x_{e0} = 0$, so the value of the eigenvalue is:

$$\lambda = \mu_e$$

which is stable, since $\mu < 0$.

II. $\mu > 0$.   For this case, there are three real solutions; we will analyze each one separately. We use the notation x_{e0}, x_{e1}, and x_{e2} to indicate the three different solutions.

   a. $x_{e0} = 0$
   $$\lambda = -3\,x_e^2 + \mu_e = \mu_e = \text{unstable}$$

   b. $x_{e1} = \sqrt{\mu}$
   $$\lambda = -3\,x_e^2 + \mu_e$$
   $$= -3\,\mu_e + \mu_e = -2\,\mu_e = \text{stable}$$

   c. $x_{e2} = -\sqrt{\mu}$
   $$\lambda = -3\,x_e^2 + \mu_e$$
   $$= -3\,\mu_e + \mu_e = -2\,\mu_e = \text{stable}$$

It is common to plot the equilibrium solutions on a bifurcation diagram, as shown in Figure 15.2. For $\mu < 0$, there is a single real solution, and it is stable. For $\mu > 0$ there are three real solutions; two are stable and one is unstable. A solid line is used to represent the stable solutions, while a dashed line indicates the unstable solution. Notice that a change in the number of equilibrium solutions and the type of dynamic behavior occured at $\mu = 0$—the *bifurcation point*. The bifurcation point satisfies the conditions in (15.3):

$$f(x_e,\mu_e) = \mu_e x_e - x_e^3 = 0$$

and

$$\left.\frac{\partial f}{\partial x}\right|_{x_e,\mu_e} = -3\,x_e^2 + \mu_e = 0$$

The state and parameter values that satisfy these conditions simultaneously are:

$$\mu_e = 0$$

and

$$x_e = 0$$

The higher-order derivatives at the bifurcation point are:

$$\left.\frac{\partial^2 f}{\partial x^2}\right|_{x_e,\mu_e} = 6x = 0$$

and

$$\left.\frac{\partial^3 f}{\partial x^3}\right|_{x_e,\mu_e} = 6 \neq 0$$

This analysis indicates that the number of solutions is three in the vicinity of the bifurcation point (see (15.4) and (15.5)).

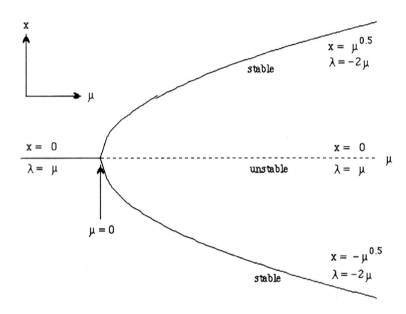

FIGURE 15.2   Pitchfork Bifurcation Diagram—Example 15.1.

It should be noted that there are actually three solutions to the steady-state equation throughout the entire range of μ values. For $\mu < 0$, two of the solutions for x_e are complex and one is real. For $\mu = 0$, all three solutions for x_e are zero. For $\mu > 0$, all three solutions for x_e are real.

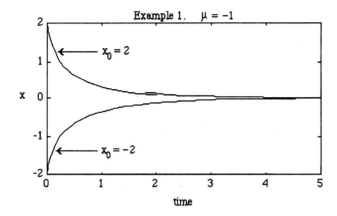

FIGURE 15.3 Transient response for Example 15.1, $\mu = -1$.

15.3.1 Dynamic Responses

Figure 15.3 shows the transient response for $\mu = -1$ for two different initial conditions; both initial conditions converge to the equilibrium solution of $x = 1$. Figure 15.4 shows the transient response for $\mu = 1$ for two different initial conditions; the final steady-state obtained depends on the intial condition. Notice that an initial condition of $x_0 = 0$ would theoretically stay at $x = 0$ for all time, however, a small perturbation (say 10^{-9}) would eventually cause the solution to go to one of the two stable steady-states.

Example 15.1 illustrates pitchfork bifurcation behavior, where a single real (and stable) solution changes to three real solutions. Two of the solutions are stable, while one is unstable. It is easy to find cases where a (subcritical) pitchfork occurs, that is, where a single unstable solution branches to two unstable and one stable solution. For example, consider the system

$$\dot{x} = f(x,\mu) = \mu x + x^3$$

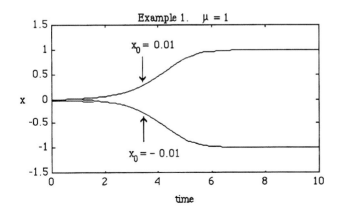

FIGURE 15.4 Transient response for Example 15.1, $\mu = 1$. The final steady-state reached depends on the initial condition.

The reader is encouraged to find the bifurcation behavior of this system shown below (see student exercise 5).

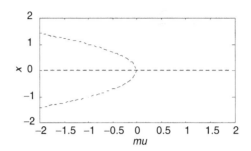

Also, a perturbation of the pitchfork diagram can occur with the following system:

$$\dot{x} = f(x,\mu,u) = u + \mu x - x^3$$

which can have a diagram of the form shown below (see student exercise 7).

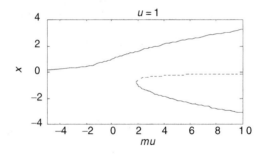

EXAMPLE 15.2   Saddle-Node Bifurcation (Turning Point)

Consider the single variable system:

$$\dot{x} = f(x,\mu) = \mu - x^2 \tag{15.7}$$

The equilibrium point is:

$$f(x,\mu) = 0 = \mu - x_e^2$$

The two solutions are:

$$x_{e1} = \sqrt{\mu}$$
$$x_{e2} = -\sqrt{\mu}$$

The Jacobian (and eigenvalue) is:

$$\left.\frac{\partial f}{\partial x}\right|_{x_e,\mu_e} = -2\,x_e = \lambda$$

The bifurcation conditions, (15.4) and (15.5), are satisfied for:

$$\mu_e = x_e = 0$$

The second derivative is:

$$\frac{\partial^2 f}{\partial x^2} = -2 \neq 0$$

which indicates that there are two solutions in the vicinity of the bifurcation point. Now, we can find the stability of the system, as a function of the bifurcation parameter, μ.

I. $\mu < 0$. From $x_{ei} = \pm\sqrt{\mu}$, we see that there is no real solution for $\mu < 0$.

II. $\mu > 0$. There are now two real solutions; we will analyze the stability of each one.

    a. For solution 1:

$$x_{e1} = \sqrt{\mu}$$

the eigenvalue is

$$\lambda = -2\,x_e = -2\sqrt{\mu_e}$$

which is stable.

    b. For solution 2:

$$x_{e2} = -\sqrt{\mu}$$

the eigenvalue is

$$\lambda = -2\,x_e = -2\left(-\sqrt{\mu_e}\right) = 2\sqrt{\mu_e}$$

which is unstable.

The bifurcation diagram (saddle-node) is shown in Figure 15.5.

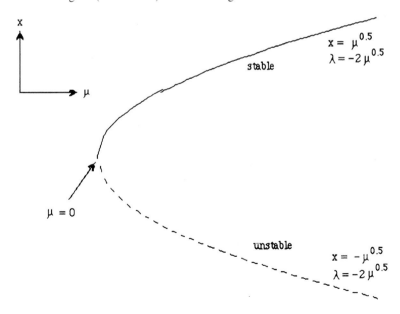

FIGURE 15.5 Saddle-node bifurcation diagram, Example 15.2.

Notice that there are actually two steady-state solutions for x_e throughout the entire range of μ. For $\mu < 0$ both solutions for x_e are complex; for $\mu = 0$ both solutions for x_e are 0; for $\mu > 0$ both solutions for x_e are real.

Dynamic Responses.   Transient response curves for $\mu = 1$ are shown in Figure 15.6, for two different initial conditions. Initial conditions $x_0 > -1$ converge to a steady-state of $x = 1$, while $x_0 < -1$ approach $x = -\infty$. It should be noted that a consistent physical (or chemical) -based model will not exhibit this sort of unbounded behavior, since the variables will have some physical meaning and will therefore be bounded.

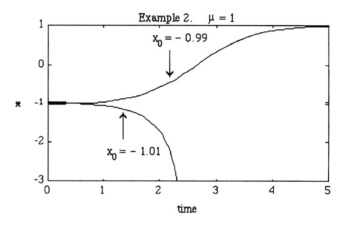

FIGURE 15.6   Transient response for Example 15.2, $\mu = 1$. Initial conditions of $x_0 > -1$ converge to a steady-state of $x = 1$, while $x_0 < -1$ "blows up".

EXAMPLE 15.3   Transcritical Bifurcation

Consider the single variable system:

$$\dot{x} = f(x,\mu) = \mu x - x^2 \qquad (15.8)$$

The equilibrium point is:

$$f(x_e,\mu) = 0 = \mu x_e - x_e^2$$

The solutions are

$$x_{e1} = 0$$

$$x_{e2} = \mu$$

The Jacobian is

$$\left. \frac{\partial f}{\partial x} \right|_{x_e,\mu_e} = \mu - 2 x_e$$

The eigenvalue is also

$$\lambda = \mu - 2\,x_e$$

The bifurcation point is $f(x,\mu) = \partial f/\partial x = 0$, which occurs at $\mu = x_e = 0$. The second derivative is:

$$\frac{\partial^2 f}{\partial x^2} = -2 \neq 0$$

which indicates that there are two equilibrium solutions. Now, we can find the stability of the system, as a function of the bifurcation parameter, μ.

I. $\mu < \boldsymbol{0}$

    a. One solution is:

$$x_{e1} = 0$$

        with an eigenvalue:

$$\lambda = \mu - 2\,x_e = \mu_e$$

        which is stable (since μ_e is negative).

    b. This equilibrium solution is:

$$x_{e2} = \mu_e$$

        which has the eigenvalue:

$$\lambda = \mu - 2x_e = \mu_e - 2\mu_e = -2\mu_e$$

        which is unstable (since μ_e is negative).

II. $\mu > \boldsymbol{0}$

    a. One solution is

$$x_{e1} = 0$$

        which has the eigenvalue:

$$\lambda = \mu - 2\,x_e = \mu_e$$

        which is unstable.

    b. Another solution is:

$$x_{e2} = \mu_e$$

        which has the eigenvalue

$$\lambda = \mu - 2\,x_e = \mu_e - 2\,\mu_e = -2\,\mu_e$$

        which is stable.

These results are shown in the bifurcation diagram of Figure 15.7, which illustrates that the number of real solutions has not changed; however, there is an exchange of stability at the bifurcation point.

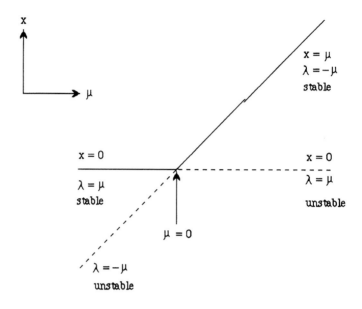

FIGURE 15.7   Transcritical bifurcation, Example 15.3.

Dynamic Responses.   Transient response curves for the transcritical bifurcation are shown in Figures 15.8 and 15.9. Notice that the transient behavior is a strong function of the initial condition for the state variable. For some initial conditions the state variable eventually settles at a stable steady-state, while for other initial conditions the state variable blows up.

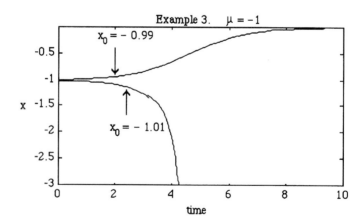

FIGURE 15.8   Transient response for Example 15.3, $\mu = -1$. Notice the importance of initial conditions.

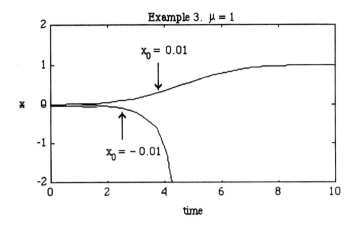

FIGURE 15.9   Transient response for Example 15.3, $\mu = 1$. Notice the importance of initial conditions.

The next example is significantly different from the previous examples. Here we allow two parameters to vary and determine their effects on the system behavior.

EXAMPLE 15.4   Hysteresis Behavior

Consider the system:

$$\dot{x} = f(x,\mu) = u + \mu x - x^3 \qquad (15.9)$$

which has two parameters (u and μ) that can be varied. We think of u as an adjustable input (manipulated variable) and μ as a design-related parameter. We will construct steady-state input-output curves by varying u and maintaining μ constant. We will then change μ and see if the character of the input-output curves (x versus u) changes. We first work with the case $\mu = -1$.

I. $\mu = -1$.   The equilibrium point (steady-state solution) is:

$$f(x_e,\mu) = 0 = u - x_e - x_e^3 \qquad (15.10)$$

The steady-state input-output diagram, obtained by solving (15.10) is shown in Figure 15.10. This curve is generated easily by first generating an x_e vector, then solving $u = x_e + x_e^3$.
   The stability of each point is found from:

$$\left.\frac{\partial f}{\partial x}\right|_{x_e,u_e} = -1 - 3\,x_e^2$$

which is always negative, indicating that there are no bifurcation points and that all equilibrium points are stable for this system. Contrast this result with that for $\mu = 1$, shown next.

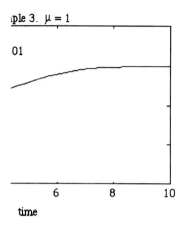

ple 3. μ = 1

01

6   8   10

time

FIGURE 15.10   Input-output diagram for Example 15.4 for $\mu = -1$.

II. μ = 1.   The equilibrium point (steady-state solution) is:

$$f(x_e,u) = 0 = u + x_e - x_e^3 \qquad (15.11)$$

Notice that this is a cubic equation that has three solutions for x_e for each value of u. For example, consider $u = 0$.

At $u = 0$:

$$x_e - x_e^3 = 0$$

so,

$$x_e = 1, 0, \text{ or } 1.$$

The stability of each solution can be determined from the Jacobian:

$$\left.\frac{\partial f}{\partial x}\right|_{x_e,u_e} = 1 - 3\,x_e^2$$

The eigenvalue is then $\lambda = 1 - 3\,x_e^2$. For the three solutions, we find:

$x_e = -1,$	$\lambda = 1 - 3 = -2$	which is stable.
$x_e = 0,$	$\lambda = 1$	which is unstable.
$x_e = 1,$	$\lambda = 1 - 3 = -2$	which is stable.

Now we can vary the input, u, over a range of values and construct a steady-state input-output curve. These results are shown on the diagram of Figure 15.10 (the easiest way to generate this figure is to create an x_e vector, and then solve $u_e = -x_e + x_e^3$. See student exercise 2).

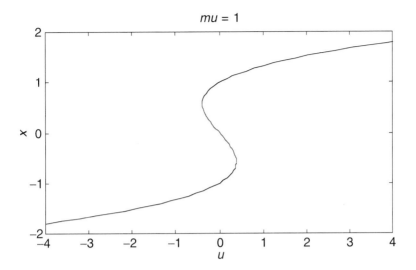

FIGURE 15.11   Input-output diagram for Example 15.4 with $\mu = 1$.

Notice that Figure 15.11 contains two saddle-node (or turning point) bifurcation points (see Example 15.2). The bifurcation (singular) points can be determined from the solution of:

$$\left.\frac{\partial f}{\partial x}\right|_{x_c,u_c} = 0 = 1 - 3\,x_c^2 \tag{15.12}$$

The bifurcation points are then:

$$x_c^2 = \frac{1}{3}$$

or,

$$x_c = \pm\frac{1}{\sqrt{3}} \tag{15.13}$$

which can be seen to be the x values at the upper and lower *turning points*. Substituting (15.13) into (15.11), we find that the bifurcation points occur at the input values of $u_c = \pm\ 2/3\sqrt{3}$, as shown in Figure 15.10. Notice that for $u < -2/3\sqrt{3}$ or $u > 2/3\sqrt{3}$ there is only a single, stable solution, while for $-2/3\sqrt{3} < u < 2/3\sqrt{3}$ there are three solutions; two are stable and one is unstable.

We have referred to the behavior of this example as hysteresis behavior—now let us show why.

***Starting at Low Values of* u.**   Notice that if we begin with a low value of u (say, -3) a single, stable, steady-state value is achieved. If we increase u a slight amount (to say, -2.9), we will achieve a slightly higher steady-state value for x. As we keep increasing u, we will continue to achieve a new stable steady-state value for x for each u. This continues until $u = 2/3\sqrt{3}$, where we find that the stable solution "jumps" to the top curve. Again, as we slowly increase u, the stable steady-state solution remains on the top curve.

***Starting at High Values of* u.**   Notice that if we begin with a high value of u (say, 3) a single stable steady-state value is achieved. If we decrease u a slight amount (to say, 2.9), we will achieve a slightly lower steady-state value for x. As we keep decreasing u, we will continue to achieve a new stable steady-state value for x for each u. This continues until $u = -2/3\sqrt{3}$ where we find that the stable solution "jumps" to the bottom curve. As we slowly decrease u further, the stable steady-state solution remains on the bottom curve.

This is termed hysteresis behavior, because the trajectory (path) taken by the state variable (x) depends on how the system is started-up. A jump discontinuity occurs at each "limit" or "turning" point (the saddle-node bifurcation points).

Discussion.   Notice that there is a significant difference between the input-output behavior exhibited in Figures 15.10 and 15.11. For $\mu = -1$ (Figure 15.9), there is monotonic relationship between the input (u) and the output (x). For $\mu = 1$ (Figure 15.10), there is a region of multiplicity behavior, where there are three values of the output (x) for a single value of the input (u). There has been a qualitative change in the behavior of this system as μ varies from -1 to 1. The value of μ where this occurs is a *hystersis* bifurcation point. At this point the following conditions are satisfied (since there are three solutions in the vicinity of the bifurcation point):

$$f(x,\mu) = 0 = \frac{\partial f}{\partial x} = \frac{\partial^2 f}{\partial x^2} = 0$$

and

$$\frac{\partial^3 f}{\partial x^3} \neq 0$$

The equations are:

$$\dot{x} \qquad = f(x,\mu) = u + \mu x - x^3 = 0$$

$$\frac{\partial f}{\partial x}\bigg|_{x_e,x_e} = \mu - 3x_e^2 \qquad\qquad = 0$$

$$\frac{\partial^2 f}{\partial x^2}\bigg|_{x_e,u_e} = -6x_e \qquad\qquad = 0$$

$$\frac{\partial^3 f}{\partial x^3}\bigg|_{x_e,u_e} = -6 \qquad\qquad \neq 0$$

It is easy to show that, for a value of $u = 0$, the bifurcation conditions are satisfied at:

$$x_e = \mu_e = 0$$

The steady-state input-output curve for this situation is found by solving:

$$f(x_e,u) = 0 = u - x_e^3$$

which yields the plot in Figure 15.12, which is clearly a transition between Figures 15.10 and 15.11.

A three-dimensional plot of x versus u as a function of μ is shown in Figure 15.13. The behavior represented by this diagram is commonly known as a *cusp catastrophe*. At low values

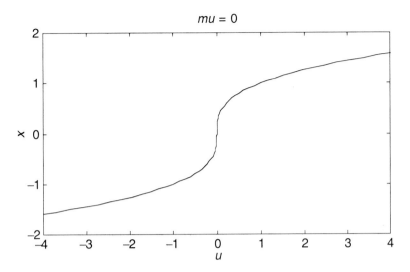

FIGURE 15.12   Input-output diagram for Example 15.4.

of μ we observe monotonic input-output behavior, with a transition to multiplicity (hysteresis) behavior at $\mu = 0$.

The turning points in Figure 15.13 can be projected to the μ-u plane to find the bifurcation diagram shown in Figure 15.14. A saddle-node (turning point) bifurcation occurs all along the boundary of the regions, except at the "cusp point" ($\mu = 0$, $u = 0$), where a codimension-2 bifurcation occurs. The term "codimension-2" means that two parameters (μ,u) are varied to

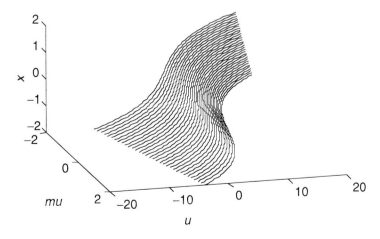

FIGURE 15.13   "Cusp catastrophe" diagram for Example 15.4.

achieve this bifurcation (Strogatz, 1994). The reader is encouraged to construct this diagram (see student exercise 6).

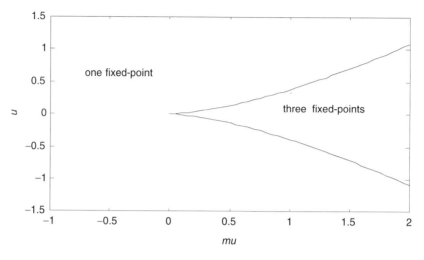

FIGURE 15.14   Two-parameter (μ,u) bifurcation diagram for Example 15.4.

SUMMARY

We have studied the bifurcation behavior of some example single nonlinear ordinary differential equations of the form $\dot{x} = f(x,\mu)$, where x is the state variable and μ is the bifurcation parameter. The equilibrium (steady-state) points are found by solving $f(x_e,\mu_e) = 0$. The stability is determined by finding the eigenvalue, λ, which is simply the Jacobian, $\partial f/\partial x \, |_{x_e,\mu_e}$, for a single equation system. If λ is negative, the equilibrium point is stable. If λ is positive, the equilibrium point is unstable.

A bifurcation diagram is drawn by plotting the equilibrium value of the state variable as a function of the bifurcation parameter. If the equilibrium point is stable ($\lambda = \partial f/\partial x \, |_{x_e,\mu_e} < 0$), a solid line is drawn. If the equilibrium point is unstable, a dashed line is drawn. The bifurcation points can be found by solving for $\partial f/\partial x \, |_{x_e,\mu_e} = 0$ where $f(x_e,\mu_e) = 0$.

These same techniques can also be applied to systems of several equations, particularly if the equations can be reduced to a single steady-state algebraic equation (in a single state variable). This can be done for many simple chemical and biochemical reactor problems.

REFERENCES AND FURTHER READING

The following texts provide nice introductions to bifurcation behavior:

Hale, J.K., & H. Kocak. (1991). *Dynamics and Bifurcations*. New York: Springer-Verlag.

Jackson, E.A. (1991). *Perspectives of Nonlinear Dynamics*. Cambridge, UK: Cambridge University Press.

Strogatz, S.H. (1994). *Nonlinear Dynamics and Chaos*. Reading, MA: Addison-Wesley.

STUDENT EXERCISES

1. For the system in Example 15.4:

$$\dot{x} = f(x,\mu) = u + \mu x - x^3$$

with $u = 0$ and $\mu = 1$, perform transient response simulations (using MATLAB) to show that the final steady-state obtained depends on the initial condition.

2. For the system in Example 15.4, we found that there are ranges of u where there are three equilibrium solutions for x (when $\mu = 1$). When solving for the roots of a cubic polynomial, either a complex analytical solution (see any math handbook) or a root solving routine (such as the MATLAB routine roots) must be used. Show how x can be considered the independent variable and u the dependent variable to obtain an easier analysis of this problem. Then, simply plot x versus u.

3. For the system in Example 15.4:

$$\dot{x} = f(x,\mu) = u + \mu x - x^3$$

with $\mu = 1$, show that the saddle-node bifurcation conditions are satisfied at the "turning points."

4. Consider the constant harvesting model of population growth (Hale & Kocak, 1991):

$$\dot{x} = f(x,k,c,h) = k x - c x^2 - h$$

where all of the parameters are positive. h is the rate of harvesting, while k and c are intrinsic growth rate parameters.

   The problem is, for fixed k and c, to determine the effect of the harvesting on the population. Since the population density cannot be negative, we are interested in solutions where $x \geq 0$. For a positive initial population density (x_0) the population is exterminated if there is a finite value of t such that $x = 0$. Without finding explicit solutions of the differential equation, show the following:

   a. If h satisfies $0 < h \leq k^2/4c$, then there is a threshold value of the initial size of the population such that if the initial size is below the threshold value, then the population is exterminated. If the initial size is above the threshold value, then the population approaches an equilibrium (steady-state) point.

   b. If h satisfies $h > k^2/4c$, then the population is exterminated regardless of its initial size.

   c. Comment on the physical ramifications of parts a and b. Should models be used by State Fish and Game authorities to determine proper hunting and fishing limits?

5. Show that the following system exhibits a pitchfork bifucation, with three real solutions (one stable, two unstable) for $\mu < 0$ and a single unstable real solution for $\mu > 0$.

$$\dot{x} = f(x,\mu) = \mu x + x^3$$

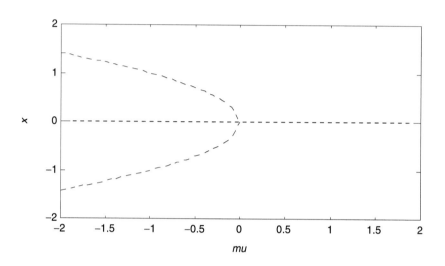

6. Consider the system shown in Example 15.4:

$$\dot{x} = f(x,\mu) = u + \mu x - x^3$$

Develop the cusp bifurcation diagram shown below. Find the values of u and μ on the boundaries between the one and three fixed-point solution behavior.

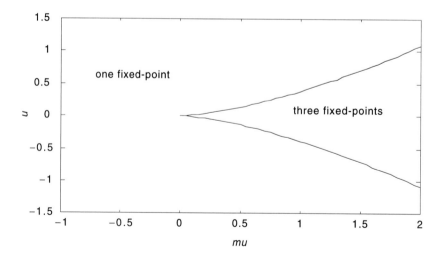

7. Consider the system shown in Example 15.4:

$$\dot{x} = f(x,\mu,u) = u + \mu x - x^3$$

For a value of $u = 1$, develop the steady-state bifurcation diagram shown below. Find the values of x and μ where the saddle-node (turning point) bifurcation occurs. Notice that this is a perturbation of a pitchfork bifurcation. This type of behavior can occur, for example, in exothermic chemical reactors when the feed flowrate is varied while maintaining a constant jacket temperature (a so-called *isola* forms).

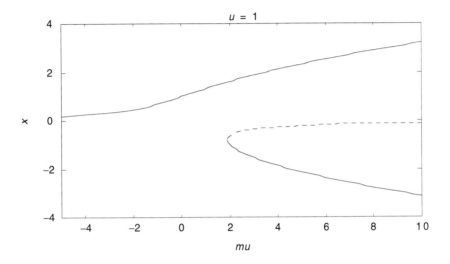

APPENDIX

```
% cusp diagram
%
% b.w. bequette
% 15 dec 96
%
% solves the problem
%   f(x,u,mu) = u + mu*x - x^3 = 0
% with x varying between -2 and 2
% mu varying from -2 to 2 and
% whatever u's result
%
  clear x;
  clear u;
  clear mu;
  x = -2:0.05:2;
  u = x.^3 +2*x;
  plot3(u,-2*ones(size(u)),x,'w')
  hold on
```

```
mu = -1.875:0.125:2;
for i = 1:32;
u = x.^3 - mu(i).*x;
plot3(u,mu(i)*ones(size(u)),x,'w')
end
hold off
»view(15,-30)
```

BIFURCATION BEHAVIOR OF TWO-STATE SYSTEMS

16

The goal of this chapter is to introduce the reader to limit cycle behavior and the Hopf bifurcation. After studying this chapter, the reader should be able to

- Find that many of the same types of bifurcations that occur in single-state systems also occur in two-state systems (pitchfork, saddle-node, transcritical)
- Understand the difference between limit cycles (nonlinear behavior) and centers (linear behavior)
- Distinguish between stable and unstable limit cycles
- Determine the conditions for a Hopf bifurcation (formation of a limit cycle)
- Discuss the differences between subcritical and supercritical Hopf bifurcations

The major sections in this chapter are:

16.1 Background
16.2 Single-Dimensional Bifurcations in the Phase-Plane
16.3 Limit Cycle Behavior
16.4 The Hopf Bifurcation

16.1 BACKGROUND

In Chapter 15 we presented the bifurcation behavior of single-state systems. We found that a number of interesting bifurcation phenomena could occur in these systems, including transcritical, pitchfork, and saddle-node bifurcations. We find in this chapter that these types of bifurcations can also occur in higher-order systems. This is the subject of Section 16.2. In Section 16.3 we review limit cycle behavior, which was initially presented in Chapter 13 (phase-plane analysis). In Section 16.4 we present a type of bifurcation that can only occur in second- and higher-order systems. In a Hopf bifurcation, we find that a stable node can *bifurcate* to a stable limit cycle if a parameter is varied; this is an example of a supercritical Hopf bifurcation. This phenomena has been shown to occur in a number of chemical and biochemical reactors. Before turning to the interesting Hopf bifurcation phenomena, we will discuss single dimensional bifurcations in the phase plane.

16.2 SINGLE-DIMENSIONAL BIFURCATIONS IN THE PHASE-PLANE

Consider the two-variable system (notice that the two equations are decoupled):

$$\dot{x}_1 = f_1(x,\mu) = \mu x_1 - x_1^3 \tag{16.1}$$

$$\dot{x}_2 = f_2(x,\mu) = -x_2 \tag{16.2}$$

The equilibrium (steady-state or fixed-point) solution is:

$$\mathbf{f}(\mathbf{x},\mu) = \begin{bmatrix} \mu x_{1e} - x_{1e}^3 \\ -x_{2e} \end{bmatrix} = \begin{bmatrix} 0 \\ 0 \end{bmatrix}$$

There are three solutions to $\mathbf{f}(\mathbf{x},\mu) = \mathbf{0}$. The trivial solution is:

$$\mathbf{x}_e = \begin{bmatrix} x_{1e} \\ x_{2e} \end{bmatrix} = \begin{bmatrix} 0 \\ 0 \end{bmatrix}$$

and the two nontrivial solutions are:

$$\mathbf{x}_e = \begin{bmatrix} x_{1e} \\ x_{2e} \end{bmatrix} = \begin{bmatrix} \sqrt{\mu} \\ 0 \end{bmatrix}$$

and

$$\mathbf{x}_e = \begin{bmatrix} x_{1e} \\ x_{2e} \end{bmatrix} = \begin{bmatrix} -\sqrt{\mu} \\ 0 \end{bmatrix}$$

Notice that only the trivial solution exists for $\mu < 0$, since we will assume that equilibrium values must be real (not complex).

We can determine the stability of each equilibrium point from the Jacobian, which is:

$$\mathbf{A} = \begin{bmatrix} \mu - 3x_{1e}^2 & 0 \\ 0 & -1 \end{bmatrix}$$

which has the following eigenvalues:

$$\lambda_1 = \mu - 3x_{1e}^2$$

$$\lambda_2 = -1$$

Since the second eigenvalue is always stable, the stability of each equilibrium point is determined by the first eigenvalue. Here we consider three cases, $\mu < 0$, $\mu = 0$, and $\mu > 0$.

I $\mu < 0$

The only equilibrium solution is the trivial solution $(x_{1e} = 0)$, so:

$$\lambda_1 = \mu - 3x_{1e}^2 = \mu$$

which is stable, since $\mu < 0$.

II $\mu = 0$

The equilibrium solution is $x_{1e} = 0$, so:

$$\lambda_1 = \mu - 3x_{1e}^2 = 0$$

which is stable; the system can be shown to exhibit a slow approach to equilibrium by observing the analytical solution to the differential equations.

III $\mu > 0$

The eigenvalue for the trivial solution $(x_{1e} = 0)$:

$$\lambda_1 = \mu - 3x_{1e}^2 = \mu$$

is unstable since $\mu > 0$.

The eigenvalues for the nontrivial solutions $(\pm\sqrt{\mu})$ are:

$$(\text{for } x_{1e} = \sqrt{\mu})\lambda_1 = \mu - 3x_{1e}^2 = \mu - 3\mu = -2\mu$$

and

$$(\text{for } x_{1e} = \sqrt{\mu})\lambda_1 = \mu - 3x_{1e}^2 = \mu - 3\mu = -2\mu$$

So the nontrivial solutions are stable for $\mu > 0$. This means that a saddle point (trivial solution) is bounded by two stable nodes for this case, since the three solutions are:

$$x_e = \begin{bmatrix} x_{1e} \\ x_{2e} \end{bmatrix} = \begin{bmatrix} -\sqrt{\mu} \\ 0 \end{bmatrix} \text{ with } \lambda = \begin{bmatrix} \lambda_1 \\ \lambda_2 \end{bmatrix} = \begin{bmatrix} -2\mu \\ -1 \end{bmatrix} = \text{stable node}$$

$$x_e = \begin{bmatrix} x_{1e} \\ x_{2e} \end{bmatrix} = \begin{bmatrix} 0 \\ 0 \end{bmatrix} \text{ with } \lambda = \begin{bmatrix} \lambda_1 \\ \lambda_2 \end{bmatrix} = \begin{bmatrix} \mu \\ -1 \end{bmatrix} = \text{saddle point}$$

$$x_e = \begin{bmatrix} x_{1e} \\ x_{2e} \end{bmatrix} = \begin{bmatrix} \sqrt{\mu} \\ 0 \end{bmatrix} \text{ with } \lambda = \begin{bmatrix} \lambda_1 \\ \lambda_2 \end{bmatrix} = \begin{bmatrix} -2\mu \\ -1 \end{bmatrix} = \text{stable node}$$

We notice the following phase-plane diagrams (Figure 16.1) as μ goes from negative to positive.

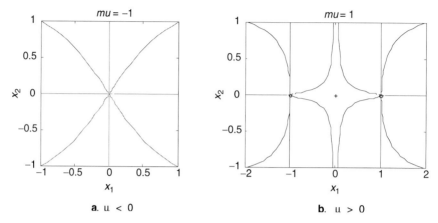

FIGURE 16.1 Pitchfork bifurcation behavior in the plane. There is a single stable node for $\mu < 0$, and two stable (o) nodes and a saddle point (+, unstable) for $\mu > 0$.

16.3 LIMIT CYCLE BEHAVIOR

In Chapter 13 we noticed that linear systems that had eigenvalues with zero real portion formed centers in the phase-plane. The phase-plane trajectories of the systems with centers depended on the initial condition values, as shown in Figure 16.2 below. Different initial conditions lead to different closed-cycles.

In this section, and the rest of this chapter, we are interested in limit cycle behavior, as shown in Figure 16.3. The major difference in center (Figure 16.2) and limit cycle

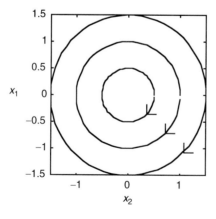

FIGURE 16.2 Example of center behavior.

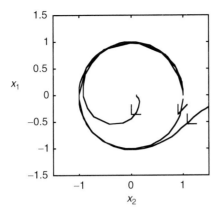

FIGURE 16.3   Example of limit cycle behavior.

(Figure 16.3) behavior is that limit cycles are *isolated* closed orbits. By isolated, we mean that a perturbation in initial conditions from the closed cycle eventually returns to the closed cycle (if it is stable). Contrast that with center behavior, where a perturbation in initial condition leads to a different closed cycle.

EXAMPLE 16.1   A Stable Limit Cycle

Consider the following system of equations, based on polar coordinates:

$$\dot{r} = r\,(1 - r^2) \tag{16.3}$$

$$\dot{\theta} = -1 \tag{16.4}$$

Notice that these equations are decoupled, that is, the value of $r(t)$ is not required to find $\theta(t)$ and vice versa. The second equation indicates that the angle is constantly decreasing. The stability of this system is then determined from an analysis of the first equation.

The steady-state solution of the first equation yields two possible values for r. The *trivial* solution is $r = 0$ and the *nontrivial* solution is $r = 1$.

The Jacobian of the first equation is:

$$\frac{\partial f}{\partial r} = 1 - 3\,r^2$$

We see then that the trivial solution ($r = 0$) is unstable, because the eigenvalue is positive ($+1$). The nontrivial solution is stable, because the eigenvalue is -2. Any trajectory that starts out close to $r = 0$ will move away, while any solution that starts out close to $r = 1$ will move towards $r = 1$. The time domain behavior for x_1 is shown in Figure 16.4. Notice that we have converted the states to rectangular coordinates ($x_1 = r \cos \theta$, $x_2 = r \sin \theta$). The phase-plane behavior is shown in Figure 16.5. Initial conditions that are either "inside" or "outside" the limit cycle converge to the limit cycle.

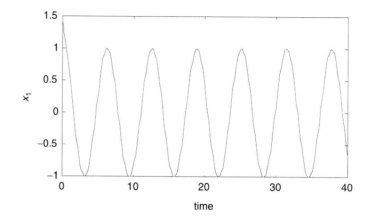

FIGURE 16.4   Stable limit cycle behavior (Example 16.1).

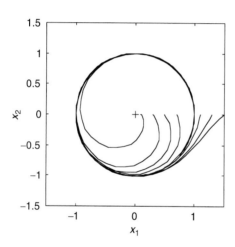

FIGURE 16.5   Stable limit cycle behavior (Example 16.1).

The previous example was for a stable limit cycle. It is also possible for a limit cycle to be unstable, as shown in Example 16.2.

EXAMPLE 16.2   An Unstable Limit Cycle

Consider the following system of equations, based on cylindrical coordinates

$$\dot{r} = -r(1 - r^2) \tag{16.5}$$

$$\dot{\theta} = -1 \tag{16.6}$$

Again, notice that these equations are decoupled. The second equation indicates that the angle is constantly decreasing. The stability of this system is then determined from an analysis of the first equation.

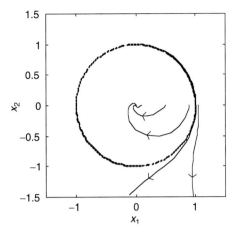

FIGURE 16.6   Phase-plane behavior for an unstable limit cycle.

The steady-state solution of the first equation yields two possible values for r. The trivial solution is $r = 0$ and the nontrivial solution is $r = 1$.

The Jacobian of the first equation is:

$$\frac{\partial f}{\partial r} = -1 + 3\,r^2$$

We see then that the trivial solution is stable, because the eigenvalue is negative (-1). The non-trivial solution is unstable, because the eigenvalue is positive ($+2$). Any trajectory that starts out less than $r = 1$ will converge to the origin, while any solution that starts out greater than $r = 1$ will increase at an exponential rate. This leads to the phase-plane behavior shown in Figure 16.6. The time domain behavior is shown in Figure 16.7.

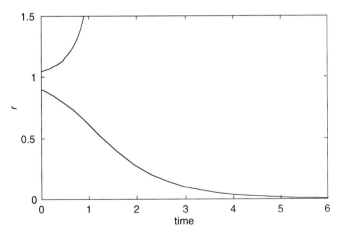

FIGURE 16.7   Time domain behavior for an unstable limit cycle. An initial condition of $r(0) = 0.9$ converges to 0, while an initial condition of $r(0) = 1.05$ blows up.

Examples 16.1 and 16.2 have shown the existence of two different types of limit cycles. In the first case (16.1) the limit cycle was stable, meaning that all trajectories were "attracted" to the limit cycle. In the second case (16.2) the limit cycle was stable, and all trajectories were "repelled" from the limit cycle. Although both of these examples yielded limit cycles that were circles in the plane, this will not normally be the case. Usually the limit cycle forms more of an ellipse. Now that we have covered limit cycle behavior, we are ready to determine what types of system parameter changes will cause limit cycle behavior to occur. That is the subject of the next section.

16.4 THE HOPF BIFURCATION

In Chapter 15 we studied systems where the number of steady-state solutions changed as a parameter was varied. The point where the number of solutions changed was called the bifurcation point. We also found that an exchange of stability generally occurred at the bifurcation point.

A *Hopf* bifurcation occurs when a limit cycle forms as a parameter is varied. In the next example we show a *supercritical* Hopf bifurcation, where the system moves from a stable steady-state at the origin to a stable limit cycle (with an unstable origin) as a parameter is varied.

EXAMPLE 16.3   Supercritical Hopf Bifurcation

Consider the system:

$$\dot{x}_1 = x_2 + x_1 \left(\mu - x_1^2 - x_2^2 \right) \tag{16.7}$$

$$\dot{x}_2 = -x_1 + x_2 \left(\mu - x_1^2 - x_2^2 \right) \tag{16.8}$$

This can be written (see student exercise 4) in polar coordinates as:

$$\dot{r} = r \left(\mu - r^2 \right) \tag{16.9}$$

$$\dot{\theta} = -1 \tag{16.10}$$

Since these equations are decoupled, the stability is determined from the stability of:

$$\dot{r} = f(r) = r \left(\mu - r^2 \right)$$

the Jacobian is:

$$\frac{\partial f}{\partial r} = \mu - 3\, r^2$$

The equilibrium (steady-state) point is $f(r) = 0$, which yields,

$$r \left(\mu - r^2 \right) = 0$$

which has three solutions:

$$r = 0 \text{ (trivial solution)}$$

$$r = \sqrt{\mu}$$

$$r = -\sqrt{\mu} \text{ (not physically realizable)}$$

For $\mu < 0$, only the trivial solution ($r = 0$) exists. For $\mu < 0$,

$$\frac{\partial f}{\partial r} = \mu$$

which is stable, since $\mu < 0$.

For $\mu = 0$, all of the steady-state solutions are $r = 0$, and the Jacobian is:

$$\frac{\partial f}{\partial r} = -3 r^2$$

which is stable, but has slow convergence to $r = 0$.

For $\mu > 0$, the trivial solution ($r = 0$) is unstable, because:

$$\frac{\partial f}{\partial r} = \mu$$

The nontrivial solution ($r = \sqrt{\mu}$) is stable because:

$$\frac{\partial f}{\partial r} = \mu - 3 r^2 = -2\mu$$

and we find the following phase-plane plots shown in Figure 16.8.

The bifurcation-diagram for this system is shown in Figure 16.9.

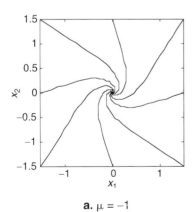

a. $\mu = -1$

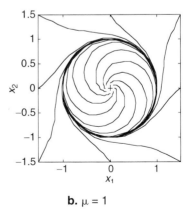

b. $\mu = 1$

FIGURE 16.8   Phase-plane plots. As μ goes from -1 to 1, the behavior changes from stable node to a stable limit cycle.

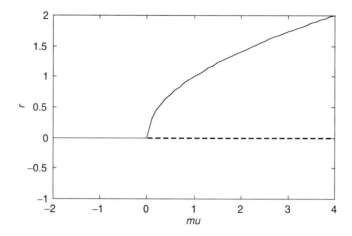

FIGURE 16.9   Bifurcation diagram. Indicates that the origin ($r = 0$) is stable when $\mu < 0$. When $\mu > 0$ the origin becomes unstable, but a stable limit cycle (with radius $r = \sqrt{\mu}$) emerges.

Here we analyze this system in rectangular ($x_1 - x_2$) coordinates. The only steady-state (fixed-point or equilibrium) solution to (16.7) and (16.8) is:

$$x_e = \begin{bmatrix} x_{1e} \\ x_{2e} \end{bmatrix} = \begin{bmatrix} 0 \\ 0 \end{bmatrix}$$

Linearizing (16.7) and (16.8):

$$\frac{\partial f_1}{\partial x_1} = \mu - 3\,x_1^2 - x_2^2 \qquad \frac{\partial f_1}{\partial x_2} = 1 - 2\,x_1\,x_2$$

$$\frac{\partial f_2}{\partial x_1} = -1 - 2\,x_1 x_2 \qquad \frac{\partial f_2}{\partial x_2} = \mu - x_1^2 - 3\,x_2^2$$

We find the Jacobian matrix:

$$\mathbf{A} = \begin{bmatrix} \mu - 3\,x_{1e}^2 - x_{2e}^2 & 1 - 2\,x_{1e}\,x_{2e} \\ -1 - 2\,x_{1e}\,x_{2e} & \mu - x_{1\,e}^2 - 3\,x_{2e}^2 \end{bmatrix}$$

which is, for the equilibrium solution of the origin:

$$\mathbf{A} = \begin{bmatrix} \mu & 1 \\ -1 & \mu \end{bmatrix}$$

The characteristic polynomial, from $\det(\lambda\mathbf{I} - \mathbf{A}) = 0$, is:

$$\lambda^2 - 2\mu\lambda + \mu^2 + 1 = 0$$

which has the eigenvalues (roots):

$$\lambda = \mu \pm \frac{\sqrt{4\mu^2 - 4(\mu^2 + 1)}}{2} = \mu \pm 1j$$

We see that when $\mu < 0$, the complex eigenvalues are stable (negative real portion); when $\mu = 0$, the eigenvalues lie on the imaginary axis; and when $\mu > 0$, the complex eigenvalues are unstable (positive real portion). The transition of eigenvalues from the left-half plane to the right-half plane is shown in Figure 16.10.

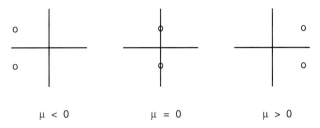

$\mu < 0$ $\mu = 0$ $\mu > 0$

FIGURE 16.10   Location of eigenvalues in complex plane as a function of μ. A Hopf bifurcation occurs as the eigenvalues pass from the lefthand side to the righthand side of the complex plane.

Example 16.3 was for a *supercritical* Hopf bifurcation, where a stable limit cycle was formed. We leave it as an exercise for the reader (student exercise 6) to show the formation of a *subcritical* Hopf bifurcation, where an unstable limit cycle is formed.

We have found that the Hopf bifurcation occurs when the real portion of the complex eigenvalues became zero. In Example 16.3 the eigenvalues crossed the imaginary axis with zero slope, that is, parallel to the real axis. In the general case, the eigenvalues will cross the imaginary axis with non-zero slope.

We should also make it clearer how an analysis of the characteristic polynomial of the Jacobian (A) matrix can be used to identify when a Hopf bifurcation can occur. For a two-state system, the characteristic polynomial has the form:

$$a_2(\mu)\, \lambda^2 + a_1(\mu)\, \lambda + a_0(\mu) = 0$$

where the polynomial parameters, a_i, are shown to be a function of the bifurcation parameter, μ (It should also be noted that it is common for $a_2 = 1$). Assume that the $a_i(\mu)$ parameters do not become 0 for the same value of μ. It is easy to show that a Hopf bifurcation occurs when $a_1(\mu) = 0$ (see student exercise 7).

16.4.1   Higher Order Systems ($n > 2$)

Thus far we have discuss Hopf bifurcation behavior of two-state systems. Hopf bifurcations can occur in any order system ($n \geq 2$); the key is that two complex eigenvalues cross the imaginary axis, while all other eigenvalues remain negative (stable). This is shown in Figure 16.11 for the three state case.

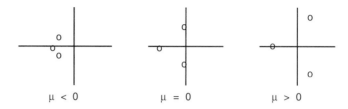

FIGURE 16.11   Location of eigenvalues in complex plane as a function of μ. A Hopf bifurcation occurs as the eigenvalues pass from the lefthand side to the righthand side of the complex plane.

SUMMARY

In this chapter we have shown that the same bifurcations that occured in single-state systems (saddle-node, transcritical, and pitchfork) also occur in systems with two or more states. We have also introduced the Hopf bifurcation, which occurs when complex eigenvalues pass from the left-half plane to the right-half plane, as the bifurcation parameter is varied. A Hopf bifurcation can also occur in systems with more than two states. For a supercritical Hopf bifurcation, two complex conjugate eigenvalues cross from the left-half to the right-half plane, while all of the other eigenvalues remain stable (in the left-half plane).

FURTHER READING

The following sources provide general introductions to bifurcation theory:

> Hale, J., & H. Kocak (1991). *Dynamics and Bifurcations.* New York: Springer-Verlag.
>
> Strogatz, S.H. (1994). *Nonlinear Dynamics and Chaos.* Reading, MA: Addison-Wesley.

The following textbook shows a complete example of the occurance of Hopf bifurcations in a 2-state exothermic chemical reactor model:

> Varma, A., & M. Morbidelli. (1997). *Mathematical Methods in Chemical Engineering.* New York: Oxford University Press.

STUDENT EXERCISES

1. Show that the two-variable system

$$\dot{x}_1 = f_1(x,\mu) = \mu - x_1^2$$
$$\dot{x}_2 = f_2(x,\mu) = -x_2$$

exhibits saddle-node behavior in the phase plane.

2. Show that the two-variable system

$$\dot{x}_1 = f_1(x,\mu) = \mu x_1 - x_1^2$$
$$\dot{x}_2 = f_2(x,\mu) = -x_2$$

exhibits transcritical behavior in the phase plane.

3. Show that the two-variable system:

$$\dot{x}_1 = f_1(x,\mu) = u + x_1 - x_1^3$$
$$\dot{x}_2 = f_2(x,\mu) = -x_2$$

exhibits hysteresis behavior in the x_1 state variable. This means that, as u is varied, x_{1e} follows an S-shaped curve, which exhibits the ignition/extinction behavior shown in Chapter 15.

4. Show that:

$$\dot{x}_1 = x_2 + x_1 \left(\mu - x_1^2 - x_2^2 \right)$$
$$\dot{x}_2 = -x_1 + x_2 \left(\mu - x_1^2 - x_2^2 \right)$$

can be written:

$$\dot{r} = r \left(\mu - r^2 \right)$$
$$\dot{\theta} = -1$$

if $x_1 = r \cos \theta$ and $x_2 = r \sin \theta$.

5. Consider a generalization of Example 16.3, which was a supercritical Hopf bifurcation (a stable limit cycle):

$$\dot{r} = r \left(\mu - r^2 \right)$$
$$\dot{\theta} = \omega + b r^2$$

Discuss how ω affects the direction of rotation. Also, discuss how b relates the frequency and amplitude of the oscillations.

6. Consider the following system, which undergoes a subcritical Hopf bifurcation:

$$\dot{r} = \mu r + r^3 - r^5$$
$$\dot{\theta} = \omega + b r^2$$

Show that, for $\mu < 0$ an unstable limit cycle lies in between a stable limit cycle and a stable attractor at the origin. What happens when $\mu = 0$ and $\mu > 0$?

7. Show that the condition for a Hopf bifurcation for the following characteristic equation

$$a_2(\mu) \lambda^2 + a_1(\mu) \lambda + a_0(\mu) = 0$$

is $a^1(\mu) = 0$. This is easy to do if you realize that a Hopf bifurcation occurs when the roots have zero real portion and write the polynomial in factored form.

Relate this condition to the Jacobian matrix, $A(\mu)$, realizing that:

$$\lambda^2 - \text{tr}(A(\mu))\,\lambda + \det(A(\mu)) = 0$$

8. Consider a Hopf bifurcation of a three-state system. Realizing that one pole is negative (and real) and that the other two poles are on the imaginary axis, relate the Hopf bifurcation to the coefficients of the characteristic polynomial of the Jacobian matrix are:

$$a_3(\mu)\,\lambda^3 + a_2(\mu)\,\lambda^2 + a_1(\mu)^l + a_0(\mu) = 0$$

You can assume, without loss of generality, that $a^3(\mu) = 1$. How do the conditions on the polynomial coefficients relate to the conditions on the Jacobian matrix (trace, determinant, etc.)?

INTRODUCTION TO CHAOS: THE LORENZ EQUATIONS

17

The objective of this chapter is to present the Lorenz equations as an example of a system that has chaotic behavior with certain parameter values. After studying this chapter, the reader should be able to:

- Understand what is meant by chaos (extreme sensitivity to initial conditions)
- Understand conceptually the physical system that the Lorenz equations attempt to model.
- Understand how the system behavior changes as the parameter r is varied.

The major sections in this chapter are:

17.1 INTRODUCTION

In Chapter 14 we presented the quadratic map (logistic equations) and found that the transient behavior of the population varied depending on the growth parameter. Recall that when the qualitative behavior of a system changes as a function of a certain parameter, we refer to the parameter as a *bifurcation* parameter. As the growth parameter was varied, the population model went through a series of period-doubling behavior, finally becoming chaotic at a certain value of the growth parameter. At that time we noted that chaos is possible with one discrete nonlinear equation, but that chaos could only occur in continuous (ordinary differential equation) models with three or more equations (assuming the model is autonomous). In this chapter we study a continuous model that has probably received the most attention in the study of chaos—the Lorenz equations. Before we write the equations, it is appropriate to give a brief historical perspective on the Lorenz model. For a more complete history, see the book *Chaos* by James Gleick (1987).

17.2 BACKGROUND

In 1961, Edward Lorenz, a professor of Meteorology at MIT, was simulating a reduced-order model of the atmosphere, which consisted of twelve equations. Included were functional relationships between temperature, pressure, and wind speed (and direction) among others. He performed numerical simulations and found recognizable patterns to the behavior of the variables, but the patterns would never quite repeat. One day he decided to examine a particular set of conditions (parameter values and initial conditions) for a longer period of time than he had previously simulated. Instead of starting the entire simulation over, he typed in a set of initial conditions based on results from midway through the previous run, started the simulation and walked down the hall for a cup of coffee. When he returned, he was shocked to find that his simulation results tracked the previous run for a period of time, but slowly began to diverge, so that after a long period of time there appeared to be no correlation between the runs. His first instinct was to check for a computer error; when he found none, he realized that he had discovered a very important aspect of certain types of nonlinear systems—that of extreme *sensitivity to initial conditions*. When he had entered the new initial conditions, he had done so only to a few decimal places, whereas several more decimal places were carried internally in the calculations. This small difference in the initial conditions built up over a period of time, to the point where the two runs did not look similar. This discovery led to the realization that long-term prediction of certain systems (such as the weather) will never be possible, no matter how many equations are used and how many variables are measured.

In order to learn more about the behavior of these types of systems, he reduced his model of the atmosphere to the fewest equations that could describe the bare essentials—this required three equations. Here we discuss the "physics" of the three equations, while Section 17.3 presents the equations and discusses the equilibrium solutions and stabilty of the equations.

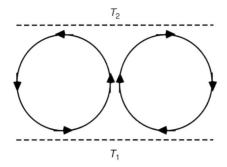

FIGURE 17.1 Convection rolls due to a temperature gradient in a fluid where density decreases as a function of temperature $(T_1 > T_2)$.

Consider a fluid maintained between two parallel plates, as shown in Figure 17.1. When the top plate temperature (T_2) is equal to the bottom plate temperature (T_1), there is no flow and the system is in equilibrium. Now, slowly increase the bottom temperature. At low temperature differences, there is still no flow because the viscous forces are greater than the buoyancy forces (the tendency for the less dense fluid at the bottom to move toward the top and the more dense fluid to move toward the bottom). Finally, at some critical temperature difference, the buoyancy forces overcome the viscous forces and the fluid begins to move and form convection rolls. As the temperature difference is increased, the fluid movement becomes more and more vigorous. Although the following point may be less clear to the reader, for some systems there is a value of temperature difference that will cause the smooth convection rolls to break up and become turbulent or chaotic.

One can think of the speed of these convection rolls as wind speed in a miniature "weather model" and the direction of the convection rolls as wind direction.

In the next section, we analyze the Lorenz equations, which attempt to model the flow pattern of Figure 17.1.

17.3 THE LORENZ EQUATIONS

The Lorenz equations are:

$$\dot{x}_1 = \sigma(x_2 - x_1) \tag{17.1}$$
$$\dot{x}_2 = r\,x_1 - x_2 - x_1 x_3 \tag{17.2}$$
$$\dot{x}_3 = b\,x_3 + x_1 x_2 \tag{17.3}$$

Notice that the only nonlinear terms are the bilinear terms $x_1 x_3$ in (17.2) and $x_1 x_2$ in (17.3).

The state variables have the following physical significance:

x_1 = proportional to the intensity (speed) of the convective rolls

x_2 = proportional to the temperature difference between the ascending and descending currents

x_3 = proportional to distortion of the vertical temperature profile from linearity

Three parameters, σ, r, and b have the following physical significance:

σ = Prandtl number (ratio of kinematic viscosity to thermal conductivity)
r = ratio of the Rayleigh number, Ra, to the critical Rayleigh number, Ra_c
b = a geometric factor related to the aspect ratio (height/width) of the convection roll

The Rayleigh number is: $$Ra = \frac{g\alpha H^3 \Delta T}{vk}$$

where:

α = coefficient of expansion
H = distance between plates
g = gravitational acceleration
ΔT = temperature difference between the plates $(T_1 - T_2)$
v = kinematic viscosity
k = thermal conductivity

For a fixed geometry and fluid, Ra is a dimensionless measure of the temperature difference between the plates. For $0 \le r < 1$ ($Ra < Ra_c$) the temperature difference is not large enough for the buoyancy forces to overcome the viscous forces and cause motion. For $r > 1$ ($Ra > Ra_c$) the temperature difference is large enough to cause motion.

17.3.1 Steady-State (Equilibrium) Solutions

The Lorenz equations have three steady-state (equilibrium) solutions under certain conditions. First, we present the trivial solution, then the nontrivial solutions. In Section 17.4 we will determine the stability of each equilibrium solution.

TRIVIAL SOLUTION

By inspection we find that the trivial solution to (17.1)–(17.3) is $x_{1s} = x_{2s} = x_{3s} = 0$ (17.4)

The trivial condition corresponds to no convective flow of the fluid.

NONTRIVIAL SOLUTIONS

From (17.1) we find that:

$$x_{1s} = x_{2s} \tag{17.5}$$

Substituting (17.5) into (17.3), we find that $x_{3s} = \frac{1}{b}x_{1s}^2$, or:

$$x_{1s} = \pm \sqrt{b\, x_{3s}} \tag{17.6}$$

TABLE 17.1   Summary of the Equilibrium Solutions

State Variable	Trivial Solution	Nontrivial 1 ($r > 0$ required)	Nontrivial 2 ($r > 0$ required)
x_{1s}	0	$\sqrt{b(r-1)}$	$-\sqrt{b(r-1)}$
x_{2s}	0	$\sqrt{b(r-1)}$	$-\sqrt{b(r-1)}$
x_{3s}	0	$r-1$	$r-1$

Substituting (17.6) into (17.2) at steady-state, we find:

$$x_{3s} = r - 1 \tag{17.7}$$

and substituting (17.7) into (17.6) and using the results of (17.5):

$$x_{1s} = x_{2s} = \pm\sqrt{b(r-1)} \tag{17.8}$$

For real solutions to (17.8), condition $r \geq 1$ must be satisfied. This means that for $r < 1$, there is only one real solution (the trivial solution), while for $r > 1$ there are three real solutions. This is an example of a pitchfork bifurcation. For $Ra < Ra_c$, there is no convective flow. The equilibrium behavior is summarized in Table 17.1.

17.4   STABILITY ANALYSIS OF THE LORENZ EQUATIONS

Linearizing (17.1)–(17.3) around the steady-state, we find the following Jacobian matrix:

$$A = \begin{bmatrix} -\sigma & \sigma & 0 \\ r - x_{3s} & -1 & -x_{1s} \\ x_{2s} & x_{1s} & -b \end{bmatrix} \tag{17.9}$$

which we will analyze to determine the stability of the equilibrium solution.

17.4.1   Stability of the Trivial Solution

For the trivial solution, $x_{1s} = x_{2s} = x_{3s} = 0$, the Jacobian matrix is:

$$A = \begin{bmatrix} -\sigma & \sigma & 0 \\ r & -1 & 0 \\ 0 & 0 & -b \end{bmatrix} \tag{17.10}$$

and the stability is determined by finding the roots of $\det(\lambda I - A) = 0$.

$$\lambda I - A = \begin{bmatrix} \lambda + \sigma & -\sigma & 0 \\ -r & \lambda + 1 & 0 \\ 0 & 0 & \lambda + b \end{bmatrix}$$

$$\det(\lambda I - A) = (\lambda + b) \begin{vmatrix} \lambda + \sigma & -\sigma \\ -r & \lambda + 1 \end{vmatrix} \tag{17.11}$$

$$= (\lambda + b)\left[(\lambda + \sigma)(\lambda + 1) - \sigma r\right]$$

$$\det(\lambda I - A) = (\lambda + b)\left[\lambda^2 + (\sigma + 1)\lambda + \sigma(1 - r)\right]$$

and we see from (17.11) that the eigenvalues are

$$\lambda_1 = -b \tag{17.12}$$

$$\lambda_2 = \frac{-(\sigma + 1) - \sqrt{(\sigma + 1)^2 - 4\sigma(1 - r)}}{2} \tag{17.13}$$

$$\lambda_3 = \frac{-(\sigma + 1) + \sqrt{(\sigma + 1)^2 - 4\sigma(1 - r)}}{2} \tag{17.14}$$

Clearly, the first eigenvalue is always stable, since $b > 0$. It is also easy to show that the second and third eigenvalues can never be complex. The second eigenvalue is always negative and the third eigenvalue is only negative for $r < 1$. We then see the following eigenvalue structure for the *trivial* (no flow) solution

$r < 0$: all eigenvalues are negative, trivial solution is stable

$r > 0$: saddle point (one unstable eigenvalue), trivial solution is unstable

For $r > 1$, then $Ra > Ra_c$, which means that flow will occur. Notice that Ra is proportional to ΔT. This means that once ΔT is increased beyond a certain critical ΔT_c, convective flow will begin.

17.4.2   Stability of the Nontrivial Solutions

Here we find the roots of $\det(\lambda I - A) = 0$ for the nontrivial solutions. Starting with:

$$\lambda I - A = \begin{bmatrix} \lambda + \sigma & -\sigma & 0 \\ x_{3s} - r & \lambda + 1 & x_{1s} \\ -x_{2s} & -x_{1s} & \lambda + b \end{bmatrix} \tag{17.15}$$

and solving $\det(\lambda I - A) = 0$, you should find:

$$\lambda^3 + b_2 \lambda^2 + b_1 \lambda + b_0 \tag{17.16}$$

where:

$$b_2 = -tr(A) = \sigma + b + 1 \tag{17.17}$$
$$b_1 = (r + \sigma)b \tag{17.18}$$
$$b_0 = -\det(A) = 2\sigma b(r - 1) \tag{17.19}$$

Recall that real nontrivial solutions only exist for $r > 1$. The coefficients b_2, b_1, and b_0 are then all positive, satisfying the *necessary condition* for stability. The Routh array must be used to check the *sufficient* condition for stability. As derived in the appendix, the critical r for stability is:

$$r_H = \frac{\sigma (\sigma + b + 3)}{(\sigma - b - 1)} \tag{17.20}$$

If $r > r_H$, then the stability condition is not satisfied. This is an interesting result, because it means that none of the equibrium solutions (trivial, nontrivial 1, nontrivial 2) is stable for $r > r_H$. The subscript H is used in (17.20), because a Hopf bifurcation forms at that value (see student exercise 1). If a *supercritical* Hopf bifurcation occurred, a stable limit cycle would form, yielding periodic behavior for the nontrivial solutions. It turns out that a *subcritical* Hopf birfurcation is formed, that is, the limit cycle is unstable (see Strogatz for a nice discussion). Since there are no stable equilibrium points for $r > r_H$, and no stable limit cycles, the solution "wanders" in phase space, never repeating the same trajectory. This behavior is known as *chaos* and the solution is said to be a *strange attractor*.

17.4.3   Summary of Stability Results

We have seen that for $r < 1$ there is only one real solution, the trivial solution, and it is stable. When $r = 1$ there is a pitchfork bifurcation, yielding three real solutions for $r > 1$. The trivial solution is unstable for $r > 1$, while the nontrivial solutions are stable for $1 < r < r_H$. This behavior is shown clearly by the bifurcation diagram shown in Figure 17.2. The formation of the unstable limit cycle at $r = r_H$ is discussed by Strogatz.

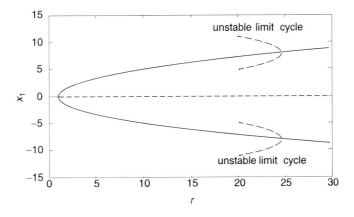

FIGURE 17.2   Bifurcation diagram for the Lorenz equations. Based on parameters in Section 17.5.

17.5 NUMERICAL STUDY OF THE LORENZ EQUATIONS

Lorenz used the following values to illustrate the chaotic nature of the equations

$$\sigma = 10$$
$$b = \frac{8}{3}$$
$$r = 28$$

from (17.22) we calculate that $r_H = 470/19 = 24.74$, indicating that all of the equilibrium points are unstable, since the value of $r = 28$ is greater than r_H.

Before we continue with the set of parameters that Lorenz used to illustrate chaotic behavior, we will first perform simulations for two other cases. In the first, we show a set of conditions where the trivial solution is stable. In the second, we show a set of conditions where the nontrivial solutions are stable.

17.5.1 Conditions for a Stable Trivial (No Flow) Solution

We have found that the trivial steady-state is stable for $0 \leq r < 1$. Here we use the σ and b parameters used by Lorenz, but set $r = 0.5$ for a stable trivial steady-state.

$$\sigma = 10 \qquad b = \frac{8}{3} \qquad r = 0.5$$

As in future simulations, we assume an initial condition of $x_0 = [0 \quad 1 \quad 0]^T$. A time domain plot for all three-state variables is shown in Figure 17.3, and a phase-plane plot (x_3 vs. x_1) is shown in Figure 17.4. The convergence to equilibrium occurs rapidly.

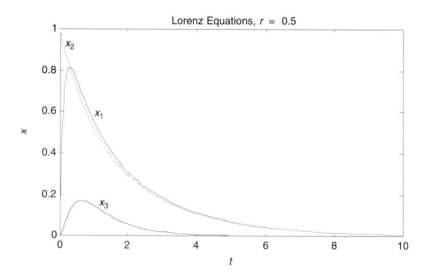

FIGURE 17.3   Lorenz equations under conditions for a stable trivial solution.

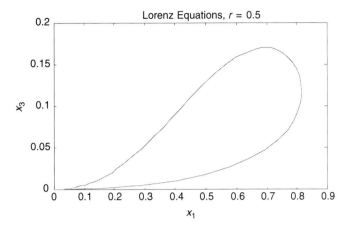

FIGURE 17.4 Phase-plane under stable conditions for the trivial solution
$(x_1 - x_3$ plane).

17.5.2   Stable Nontrivial Solutions

We have found that the nontrivial steady-states are stable for $1 < r < r_H$. Here we use the
σ and b parameters used by Lorenz, but set $r = 10$ (recall that $r_H = 24.74$ for these values
of σ and b) to show that the nontrivial steady-states are stable.

$$\sigma = 10 \qquad b = \frac{8}{3} \qquad r = 21$$

For the trivial steady-state, $\mathbf{x} = [0 \quad 0 \quad 0]^T$, the eigenvalues are:

$$\lambda_1 = -\ 2.67$$
$$\lambda_2 = -20.67$$
$$\lambda_3 = \ \ \ \ 9.67$$

as expected, the trivial steady-state is unstable (a saddle point).
   The nontrivial steady-states, $\mathbf{x} = [\sqrt{160/3} \quad \sqrt{160/3} \quad 20]^T$ and $\mathbf{x} = [-\sqrt{160/3}$
$-\sqrt{160/3} \quad 20]^T$, have eigenvalues of:

$$\lambda_1 = -13.4266$$
$$\lambda_2 = -0.1200 + 8.9123\,j$$
$$\lambda_3 = -0.1200 - 8.9123\,j$$

verifying that the nontrivial steady-states are stable.
   Time domain plots for $\mathbf{x}_1$ are shown in Figure 17.5, for $r = 21$ and two different initial conditions. Notice that convergence to a particular equilibrium point depends on the
initial condition, that is, plot a converges to one equilibrium point, while plot b converges
to a different equilibrium point. Also notice that plot b exhibits what is known as *transient*

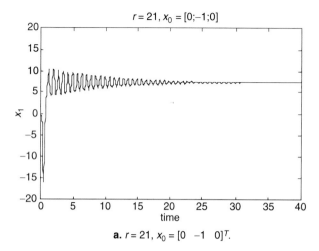

a. $r = 21$, $x_0 = [0 \quad -1 \quad 0]^T$.

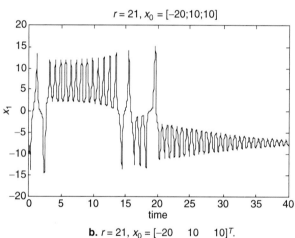

b. $r = 21$, $x_0 = [-20 \quad 10 \quad 10]^T$.

FIGURE 17.5 Lorenz equations under conditions for stable nontrivial solutions.

chaos. The initial trajectory appears chaotic, but eventually the trajectory converges to an equilibrium point. In other systems the system can exhibit transient chaos and settle into periodic behavior.

The phase plane diagram of Figure 17.6 also clearly shows the effect of two different initial conditions. In curve *a* the trajectory almost immediately goes to the equilibrium point on the right (positive value of x_1). In curve *b* the trajectory first winds around the left equilibrium point, switches to the right equilibrium point, and (after going back and forth a few times) eventually winds around the left equibrium point, slowly converging.

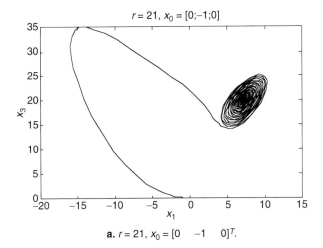

a. $r = 21$, $x_0 = [0 \quad -1 \quad 0]^T$.

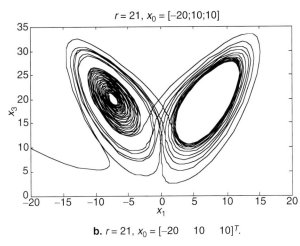

b. $r = 21$, $x_0 = [-20 \quad 10 \quad 10]^T$.

FIGURE 17.6   Phase plane under stable conditions for the nontrivial solutions.

17.5.3   Chaotic Conditions

The parameters used for this case are

$$\sigma = 10 \qquad b = \frac{8}{3} \qquad r = 28$$

Recall that all of the equilibrium points are unstable, since the value of $r = 28$ is greater than r_H.

For the trivial steady-state, $\mathbf{x} = [0 \quad 0 \quad 0]^T$, the eigenvalues are:

$$\lambda_1 = -2.67$$
$$\lambda_2 = -22.83$$
$$\lambda_3 = 11.83$$

as expected, the trivial steady-state is unstable (a saddle point).

For the nontrivial steady-states, $\mathbf{x} = [\sqrt{72} \quad \sqrt{72} \quad 27]^T$ and $\mathbf{x} = [-\sqrt{72} \quad -\sqrt{72} \quad 27]^T$, the eigenvalues are:

$$\lambda_1 = -13.8546$$
$$\lambda_2 = 0.0940 - 10.1945j$$
$$\lambda_3 = 0.0940 + 10.1945j$$

indicating that the nontrivial steady-states are unstable. Notice that all of the steady-state operating points are unstable. This means that the curves in the three-dimensional "phase-cube" plots will not asymptotically approach any single equilibrium point. The curves may exhibit periodic-type behavior, where the three-dimensional equivalent of a limit cycle is reached. The curves could even have "quasi-periodic" behavior, where the oscillations appear to have two frequency components. It turns out for this set of parameters that the curves never repeat. The curves have a *strange attractor* because they stay in a certain region of three-space, but never intersect or repeat. This is known as chaotic behavior.

Figure 17.7 shows the Lorenz behavior for the x_1 variable under unstable (chaotic) conditions. The initial condition is $x_0 = [0 \quad 1 \quad 0]^T$. Plots of the other states (x_2 and x_3) are similiar.

Figure 17.7 was a time domain plot for x_1 under chaotic conditions. More interesting results are also shown in the following phase-plane diagram (Figure 17.8). Notice that the trajectory will spend some time "winding around" one equilibrium point, before jumping to the other side and winding around the other equilibrium point for a while. This process goes on forever, with the trajectory never crossing itself (in 3-space).

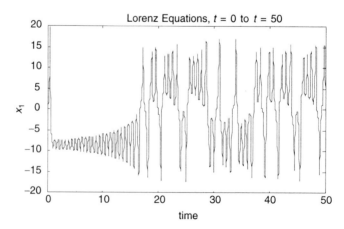

FIGURE 17.7   Transient response of x_1 under chaotic conditions.

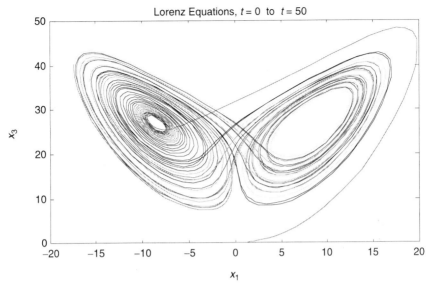

FIGURE 17.8   Phase-plane of Lorenz equations under chaotic conditions.

The development of the curve in Figure 17.8 is shown more clearly in the phase-plane plots in Figure 17.9, which show varies "pieces" of time.

A three-dimensional plot (*phase cube*) of this trajectory is shown in Figure 17.10.

The reader is encouraged to perform simulations of the Lorenz equations, to be understand concepts such as sensitivity to initial conditions. MATLAB has a demo titled `lorenz.m` (simply enter `lorenz` in the command window) that traces a three-dimensional plot of solutions to the Lorenz equations. Each new run uses a new set of random initial conditions. If you write an m-file to simulate the Lorenz equations using `ode45`, remember to use a name that is different than `lorenz.m`, to avoid conflicts with the MATLAB demo.

17.6   CHAOS IN CHEMICAL SYSTEMS

The Lorenz equations provide a nice example of chaos, because the equations are reasonably simple to analyze. An even simpler set of equations was developed by Rossler to demonstrate chaotic behavior (see student exercise 3). Chaos has also been shown to appear in models of chemical process systems, particularly exothermic chemical reactors. Reactors that are forced periodically (jacket temperature is a sine wave, for example) have been shown to exhibit chaos. Also, a series of reactors with heat integration can exhibit chaotic behavior. It appears that chaotic reactors may have had low amplitude "oscillations" (say in temperature) that may have been interpreted as measurement and process *noise* in the past. A comprehensive review of nonlinear dynamic behavior in chemical reactors is provided in the article by Razon and Schmitz (1987).

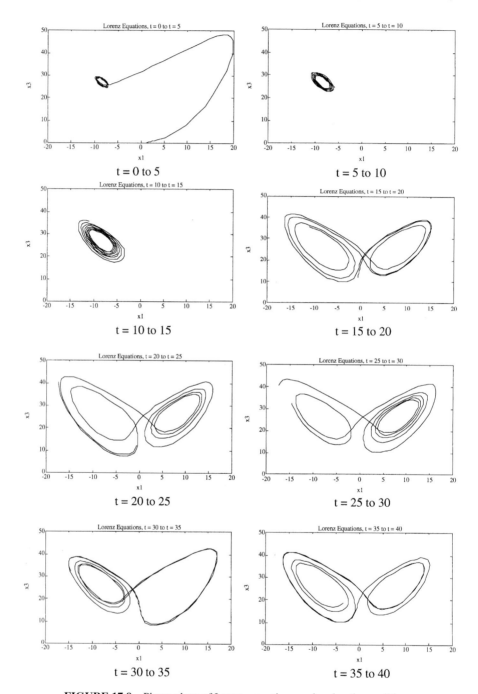

FIGURE 17.9   Phase-plane of Lorenz equations under chaotic conditions.

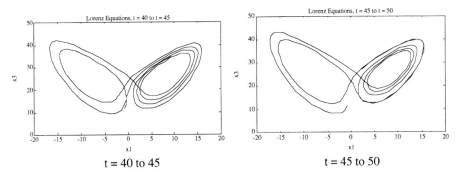

t = 40 to 45   t = 45 to 50

FIGURE 17.9   *Continued*

17.7   OTHER ISSUES IN CHAOS

Chaos is a complex field with many books and conferences devoted to this simple topic. Clearly, it is impossible to give this topic adequate coverage in a single chapter. Our goal is to provide an introduction to, and motivation for, the topic. The reader is encouraged to consult the many books and articles available on the topic.

$r = 28, \quad x_0 = [0;1;0]$

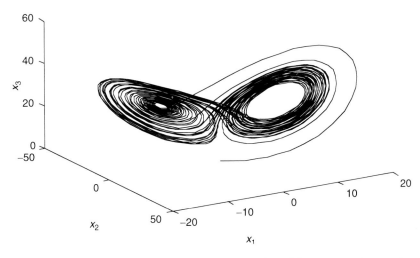

FIGURE 17.10   Three-dimensional phase space plot of Lorenz equations under chaotic conditions.

Issues that may be of particular interest include:

- How does one detect chaos experimentally? One method is to use experimental data to calculate *Lyapunov exponents*. See Strogatz for example.
- Chaos can be used to encode secret messages. See Cuomo and Oppenheim (1993), who used ideas presented by Pecora and Carroll (1990).

SUMMARY

We have presented an introduction to chaotic behavior by studying the Lorenz convective flow equations. A number of chemical processes have been shown to exhibit similar behavior. It is necessary to have three nonlinear autonomous differential equations before chaos can occur. Although not shown here, chaos can occur in a system of two nonlinear *nonautonomous* equations (that is, if some type of periodic input forcing is used). Also, we saw in Chapter 14 that chaos can occur in a single discrete nonlinear equation (the quadratic map, or logistic equation).

REFERENCES AND FURTHER READING

A nice book on the history of chaos, written for the general public, is by Gleick.

Gleick, J. (1987). *Chaos: Making a New Science.* New York: Viking.

The first paper to develop the notion of sensitivity to initial conditions is by Lorenz.

Lorenz, E.N. (1963). Deterministic nonperiodic flow. *Journal of Atmospheric Sciences,* 20: 130–141.

A number of introductory-level textbooks provide nice introductions to chaos. These include:

Strogatz, S.H. (1994). *Nonlinear Dynamics and Chaos.* Reading, MA: Addison-Wesley.

A review of nonlinear dynamic behavior (including Chaos) of chemical reactors is provided by:

Razon, L.F., & R.A. Schmitz. (1987). Multiplicities and instabilities in chemically reacting systems—A Review. *Chem. Eng. Sci.,* 42(5): 1005–1047.

A large number of papers on chaos have been collect in the following book:

Hao, Bai-Lin. (Ed.). (1990). *Chaos II.* Singapore: World Scientific Press.

Papers that develop a way of encoding secret messages using chaos are:

Pecora, L.M., & T.L. Carroll. (1990). Synchronization in chaotic systems. *Physical Review Letters,* 64: 821.

Cuomo, K.M., & A.V. Oppenheim. (1993). Circuit implementation of synchronized chaos, with applications to communications. *Physical Review Letters.* 71: 65.

STUDENT EXERCISES

1. Consider the following parameter values for the Lorenz equations:

$$\sigma = 10$$
$$r = r_c = 470/19 = 24.74$$
$$b = \frac{8}{3}$$

For the *nontrivial solution*, show that a supercritical Hopf bifurcation occurs at this value of r. That is, for $r < r_c$, all eigenvalues are stable, for $r = r_c$, two eigenvalues are on the imaginary axis, and for $r > r_c$, two eigenvalues have crossed into the right half plane.

2. Show the sensitivity to initial conditions of the Lorenz equations. Run two simulations with the parameter values shown in the numerical study

$$\sigma = 10$$
$$r = 28$$
$$b = 8/3$$

For the first simulation use the initial condition $x_0 = [0 \quad 1 \quad 0]^T$. For the second simulation use the initial condition $x_0 = [0 \quad 1.01 \quad 0]^T$. When do the simulations begin to diverge?

Run some more simulations with smaller perturbations in the initial conditions. Also, make perturbations in the initial conditions for the other state variables. What do you find?

3. Consider the Rossler equations (see Strogatz, for example):

$$\dot{x}_1 = -x_2 - x_3$$
$$\dot{x}_2 = x_1 + a\,x_2$$
$$\dot{x}_3 = b + x_3(x_1 - c)$$

which have a single nonlinear term. Let the parameters a and b be constant with a value of 0.2. Use simulations to show that this system has period-1 (limit cycle), period-2, and period-4 behavior for $c = 2.5$, 3.5, and 4, respectively. Show that chaotic behavior occurs for $c = 5$.

4. The Henon map is a discrete model that can exhibit chaos:

$$x_1(k + 1) = x_2(k) + 1 - a\,x_1(k)^2$$
$$x_2(k + 1) = b\,x_1(k)$$

For a value of $b = 0.3$, perform numerical simulations for various values of a. Try to find values of a (try $a > 0.3675$) that yield stable period-2 behavior. Show that chaos occurs at approximately $a = 1.06$.

APPENDIX

Stability analysis of the nontrivial steady-state using the Routh array:

Row

1	1	b_1
2	b_2	b_0
3	$\dfrac{b_2 b_1 - b_0}{b_2}$	0
4	b_0	

Where $b_0 = 2\,\sigma b(r - 1)$ $b_1 = (r + \sigma)b$ $b_2 = \sigma + b + 1$

Since the nontrivial steady-state only exists for $r \geq 1$, then b_0 is always positive. It also follows that b_1 and b_2 will always be positive. The only entry from the Routh array that we must check is the first column in row 3. This entry will be positive if:

$$b_2 b_1 - b_0 > 0 \quad \text{or} \quad b_2 b_1 > b_0 \tag{A-1}$$

Making the substitutions for parameter values in the coefficients:

$$(\sigma + b + 1)(r + \sigma)b > 2\,\sigma b(r - 1) \tag{A-2}$$

After some algebra, this can be written:

$$(-\sigma + b + 1)\,r > -\sigma(\sigma + b + 3) \tag{A-3}$$

or,

$$(\sigma - b - 1)\,r < \sigma(\sigma + b + 3) \tag{A-4}$$

and, assuming that $\sigma > b + 1$, the condition on r for stability is:

$$r < \frac{\sigma(\sigma + b + 3)}{(\sigma - b - 1)} \tag{A-5}$$

Notice from (A–3) that if $\sigma < b + 1$, then any r satisfies the requirement for stability. We will often define the critical value, r_c:

$$r_c = \frac{\sigma(\sigma + b + 3)}{(\sigma - b - 1)} \tag{A-6}$$

If $r > r_c$, then the system is unstable.

SECTION IV

REVIEW AND LEARNING MODULES

INTRODUCTION TO MATLAB

MODULE

1

The purpose of this module[1] is to review MATLAB for those who have used it before and to provide a brief introduction to MATLAB for those who have not used it before. This is a hands-on tutorial introduction. After using this tutorial, you should be able to:

- Enter matrices
- Make plots
- Write script files
- Perform matrix operations
- Use MATLAB functions
- Write function files

The major sections of this module are:

[1]Good references include *Introduction to MATLAB for Engineers and Scientists* by D.M. Etter (Prentice-Hall, 1996), and *Mastering MATLAB* by D. Hanselman and B. Littlefield (Prentice-Hall, 1996).

415

M1.1 BACKGROUND

MATLAB is an interactive program for numerical computation and data visualization. It was originally developed in FORTRAN as a MATrix LABoratory for solving numerical linear algebra problems. The original application may seem boring (except to linear algebra enthusiasts), but MATLAB has advanced to solve nonlinear problems and provide detailed graphics. It is easy to use, yet very powerful. A few short commands can accomplish the same results that required a major programming effort only a few years ago.

M1.2 USING THIS TUTORIAL

This tutorial provides a brief overview of essential MATLAB commands. *You will learn this material more quickly if you use MATLAB interactively as you are reviewing this tutorial.* The MATLAB commands will be shown in the following font style:

```
Monaco font
```

the prompt for a user input is shown by the double arrow

```
»
```

MATLAB has an extensive on-line help facility. For example, type `help pi` at the prompt

```
» help pi

PI    PI = 4*atan(1) = 3.1415926535897....
```

so we see that MATLAB has the number π "built-in". As another example

```
» help exp

EXP   EXP(X) is the exponential of the elements of X, e to
the X.
```

sometimes you do not know the exact command to perform a particular operation. In this case, one can simply type

```
» help
```

and MATLAB will provide a list of commands (and m-files, to be discussed later) that are available. If you do not know the exact command for the function that you are after, another useful command is `lookfor`. This command works somewhat like an index. If you did not know the command for the exponential function was `exp`, you could type

```
» lookfor exponential
```

```
EXP          Exponential.
EXPM         Matrix exponential.
EXPM1 Matrix exponential via Pade' approximation.
EXPM2 Matrix exponential via Taylor series approximation.
EXPM3 Matrix exponential via eigenvalues and eigenvectors.
EXPME Used by LINSIM to calculate matrix exponentials.
```

M1.3 ENTERING MATRICES

The basic entity in MATLAB is a rectangular matrix; the entries can be real or complex. Commas or spaces are used to delineate the separate values in a matrix. Consider the following vector, x (recall that a vector is simply a matrix with only one row or column):

```
» x = [1,3,5,7,9,11]

x =
     1     3     5     7     9    11
```

Notice that a row vector is the default in MATLAB (in contrast to the default column vector used by this and most other texts). We could have used spaces as the delimiter between columns:

```
» x = [1   3   5   7   9   11]

x =
     1     3     5     7     9    11
```

There is a faster way to enter matrices or vectors that have a consistent pattern. For example, the following command creates the previous vector:

```
» x = 1:2:11

x =
     1     3     5     7     9    11
```

Transposing a row vector yields a column vector (' is the transpose command in MAT-LAB):

```
» y = x'

y =
        1
        3
        5
        7
        9
       11
```

If we want to make x a column-vector, we use a semicolon as the delimeter between rows:

```
» x = [1;3;5;7;9;11]

x =
        1
        3
        5
        7
        9
       11
```

To make x a row vector again, we use the transpose:

```
» x = x'
```

Say that we want to create a vector z, which has elements from 5 to 30, by 5s:

```
» z = 5:5:30

z =

        5    10    15    20    25    30
```

If we wish to suppress printing, we can add a semicolon (;) after any MATLAB command:

```
» z = 5:5:30;
```

The z vector is generated, but not printed in the command window. We can find the value of the third element in the z vector, z(3), by typing

```
» z(3)

ans =
     15
```

Notice that a new variable, ans, was defined automatically.

M1.4 THE MATLAB WORKSPACE

We can view the variables currently in the workspace by typing:

```
» who

Your variables are:

ans     x       y       z

leaving 621420 bytes of memory free.
```

More detail about the size of the matrices can be obtained by typing:

```
» whos
              Name          Size          Total       Complex

               ans        1 by 1            1           No
                 x        1 by 6            6           No
                 y        6 by 1            6           No
                 z        1 by 6            6           No

Grand total is (19 * 8) = 152 bytes,

leaving 622256 bytes of memory free.
```

We can also find the size of a matrix or vector by typing:

```
» [m,n]=size(x)

m =
      1

n =
      6
```

where m represents the number of rows and n represents the number of columns. If we do not put place arguments for the rows and columns, we find:

```
» size(x)

ans =
     1      6
```

Since x is a vector, we can also use the length command

```
» length(x)

ans =
     6
```

It should be noted that MATLAB is case sensitive with respect to variable names. An X matrix can coexist with an x matrix.

You may wish to quit MATLAB but save your variables so you don't have to re-type or recalculate them during your next MATLAB session. To save all of your variables, use:

```
» save file_name
```

(Saving your variables does not remove them from your workspace; only `clear` can do that)

You can also save just a few of your variables:

```
» save file_name x y z
```

To load a set of previously saved variables:

```
» load file_name
```

Sometimes it is desirable to clear all of the variables in a workspace. This is done by simply typing:

```
» clear
```

More frequently, you may wish to clear a particular variable, such as x

```
» clear x
```

This is particularly true if you are performing a new calculation of x and the new vector is shorter than the old vector. If the new vector has length n, then all of the elements of the new x greater than x(n) will contain values of the previous x vector.

M1.5 COMPLEX VARIABLES

Both i and j represent the imaginary number, $\sqrt{-1}$, by default:

```
» i

ans =
          0 + 1.0000i

» j

ans =
          0 + 1.0000i

» sqrt(-3)

ans =
          0 + 1.7321i
```

Note that these variables (i and j) can be redefined (as the index in a for loop, for example), as will be shown later.

Matrices can be created where some of the elements are complex and the others are real:

```
» a = [sqrt(4), 1;sqrt(-4), -5]

a =
    2.0000                 1.0000
          0 + 2.0000i   -5.0000
```

Recall that the semicolon designates the end of a row.

M1.6 SOME MATRIX OPERATIONS

Matrix multiplication is straightforward:

```
» b = [1 2 3;4 5 6]

b =
     1        2        3
     4        5        6
```

Using the a matrix that was generated in Section 1.5:

```
» c = a*b
```

```
c =
    6.0000              9.0000              12.0000
  -20.0000 + 2.0000i  -25.0000 + 4.0000i  -30.0000 + 6.0000i
```

Notice again that MATLAB automatically deals with complex numbers.

Sometimes it is desirable to perform an *element-by-element* multiplication rather than matrix multiplication. For example, $d(i,j) = b(i,j)*c(i,j)$ is performed by using the . * command

```
»  d = c.*b
```

```
d =
   1.0e+02 *
   0.0600              0.1800              0.3600
  -0.8000 + 0.0800i   -1.2500 + 0.2000i   -1.8000 + 0.3600i
```

(Notice the scaling that is performed when the numbers are displayed.)

Similarly, element by element division, $b(i,j)/c(i,j)$, can be performed using . /:

```
» e = b./c
```

```
e =
   0.1667              0.2222              0.2500
  -0.1980 - 0.0198i   -0.1950 - 0.0312i   -0.1923 - 0.0385i
```

Other matrix operations include: (i) taking matrix to a power and (ii) the matrix exponential. These are operations on a square matrix:

```
» f = a^2
```

```
f =
   4.0000 + 2.0000i   -3.0000
        0 - 6.0000i   25.0000 + 2.0000i
```

```
» g = expm(a)
```

```
g =
   7.2232 + 1.8019i    1.0380 + 0.2151i
  -0.4302 + 2.0760i   -0.0429 + 0.2962i
```

M1.7 PLOTTING

For a standard solid line plot, simply type:

```
» plot(x,z)
```

where *x* and *z* have been generated using

```
» x = 1:2:11;
» z = 5:5:30;
```

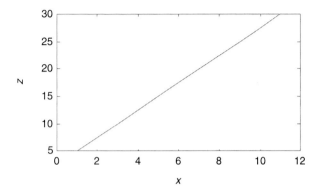

Axis labels are added by using the following commands:

```
» xlabel('x')
```

```
» ylabel('z')
```

For more plotting options, type:

```
» help plot
```

```
PLOT   Plot vectors or matrices. PLOT(X,Y) plots vector X versus
       vector Y. If X or Y is a matrix, then the vector is plotted
       versus the rows or columns of the matrix, whichever lines
       up. PLOT(X1,Y1,X2,Y2) is another way of producing multiple
       lines on the plot. PLOT(X1,Y1,':',X2,Y2,'+') uses a
       dotted line for the first curve and the point symbol +
       for the second curve. Other line and point types are:

            solid    -       point  .     red        r
            dashed   --       plus   +      green      g
            dotted   :       star   *     blue       b
            dashdot  -.      circle o     white      w
                            x-mark x     invisible i
                                         arbitrary c1, c15, etc.

       PLOT(Y) plots the columns of Y versus their index. PLOT(Y)
       is equivalent to PLOT(real(Y),imag(Y)) if Y is complex.
       In all other uses of PLOT, the imaginary part is ignored.
       See SEMI, LOGLOG, POLAR, GRID, SHG, CLC, CLG, TITLE, XLABEL
       YLABEL, AXIS, HOLD, MESH, CONTOUR, SUBPLOT.
```

If we wish to plot discrete points, using + as a symbol, we can use the following:

```
»    plot(x,z,'+')
```

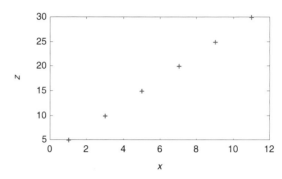

Text can be added directly to a figure using the `gtext` command.

    `gtext('string')` displays the graph window, puts up a cross-hair, and waits for a mouse button or keyboard key to be pressed. The cross-hair can be positioned with the mouse. Pressing a mouse button or any key writes the text string onto the graph at the selected location.

    Consider now the following equation:

$$y(t) = 4\,e^{-0.1t}$$

We can solve this for a vector of t values by two simple commands:

```
» t = 0:1:50;
» y = 4*exp(-0.1*t);
```

and we can obtain a plot by typing:

```
» plot(t,y)
```

Notice that we could shorten the sequence of commands by typing:

```
» plot(t,4*exp(-0.1*t))
```

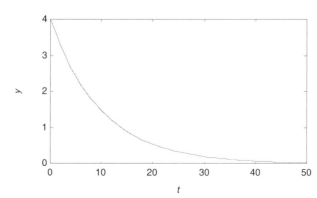

We can plot the function $y(t) = t\,e^{-0.1t}$ by using:

```
» y = t.*exp(-0.1*t);
» plot(t,y)
» gtext('hey, this is the peak!')
» xlabel('t')
» ylabel('y')
```

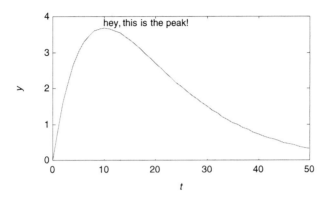

axis('square') will place the plot in a square box, while axis('normal') will change back to a normal aspect ratio.

You can also explicitly set the upper and lower bounds on the plot with:

axis([xlow xhigh ylow yhigh])

For this example we would use:

```
» v = [0   50   0   4];

» axis(v);
```

Multiple curves can be placed on the same plot in the following fashion.

plot(t,4*exp(-0.1*t),t,t.*exp(-0.1*t),'--')

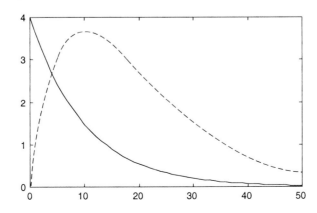

The subplot command can be used to make multiple plots.

```
»subplot(2,1,1), plot(t,4*exp(-0.1*t))
»subplot(2,1,2), plot(t,t.*exp(-0.1*t))
```

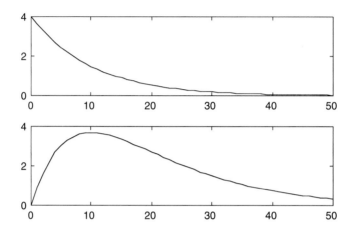

Here, subplot(i,j,k) means that there are i "rows" of figures, j "columns" of figures, and the current plot is the kth figure (counting right to left and top to bottom).

To return to single plots, simply enter subplot(1,1,1).

M1.8 MORE MATRIX STUFF

A matrix can be constructed from two or more vectors. If we wish to create a matrix v, which consists of two columns, the first column containing the vector x (in column form) and the second column containing the vector z (in column form), we can use the following:

```
» v = [x',z']

v =
       1       5
       3      10
       5      15
       7      20
       9      25
      11      30
```

If we wished to look at the first column of v, we could use:

```
» v(:,1)

ans =
       1
       3
       5
       7
       9
      11
```

If we wished to look at the second column of *v*, we could use:

```
» v(:,2)

ans =
       5
      10
      15
      20
      25
      30
```

And we can construct the same plot as before, by using (' -- ' gives a dotted line):

```
» plot(v(:,1),v(:,2),'--')
```

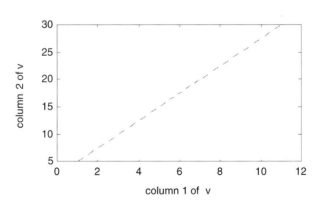

M1.9 FOR LOOPS AND IF-THEN STATEMENTS

A `for` loop in MATLAB is similar to a DO loop in FORTRAN. The main difference is that the FORTRAN DO loop must have an integer index variable; `for` does not have this restriction. An example of a `for` loop that is virtually identical to a DO loop is:

```
» for k = 1:5001;
  t(k) = (k-1)*0.01;
  y(k) = sin(t(k));
end
```

Another way of implementing the same loop is increment t from 0 to 50 in increments of 0.01:

```
» k = 0
» for t = 0:0.01:50;
  k = k + 1;
  y(k) = sin(t);
end
```

The developers of MATLAB highly recommend that you use the vectorized version of the above for loops:

```
t = 0:0.01:50;
y = sin(t);
```

since the computation time for this method is over 200 times faster than the nonvectorized methods.

M1.10 M-FILES

Thus far we have shown the interactive features of MATLAB by entering one command at a time. One reason that MATLAB is powerful is that it is a language, and programs of MATLAB code can be saved for later use. There are two ways of generating your own MATLAB code: (1) script files and (2) function routines.

1.10.1   Script Files

A script file is simply a sequence of commands that could have been entered interactively in the MATLAB command window. When the sequence is long, or must be performed a number of times it is much easier to generate a script file.

The following example is for the quadratic map population growth model:

$$x_{k+1} = \alpha \, x_k \, (1 - x_k)$$

where x_k represents the value of the population (dimensionless) at the kth time step. We have titled the file popmod.m and stored it in our directory.

```
% popmod.m
% population model, script file example
%
  clear x,k
  n       = input('input final time step ');
  alpha   = input('input alpha ');
  xinit   = input('input initial population ');
  x(1)    = xinit;
  time(1) = 0;
  for  k = 2:n+1;
    time(k) = k-1;
    x(k)      = alpha*x(k-1)*(1-x(k-1));
  end
  plot(time,x)
%  end of script file example
```

Notice that we have used the MATLAB `input` function to prompt the user for data. Also note that a percent sign (%) may be used to put comments in a script or function file. Any text after a % is ignored.

The file is run by simply entering:

```
» popmod
```

in the MATLAB command window.

M1.10.2 Function Routines

A more powerful way of solving problems is to write MATLAB function routines. Function routines are similar to subroutines in FORTRAN. Consider the previous example.

```
  function [time,x] = pmod(alpha,xinit,n)
% population model example, pmod.m
  clear time; clear x; clear k;
  x(1)    = xinit;
  time(1) = 0;
  for  k = 2:n+1;
    time(k) = k-1;
    x(k)      = alpha*x(k-1)*(1-x(k-1));
  end
%  end of function file example
```

We can now "run" this function routine (using `alpha` = 2.8, `xinit` = 0.1, `n` = 30) by typing

```
» [tstep,xpop]=pmod(2.8,0.1,30);
» plot(tstep,xpop)
```

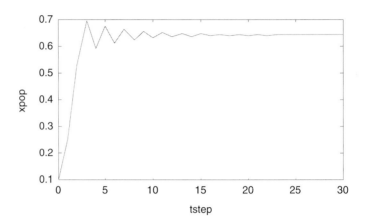

This function routine can also be called by other function routines. This feature leads to "structured programming"; structured programs are easy to follow and debug.

MATLAB has many built-in function routines that you will use throughout this course. The most commonly used routines are `fzero`, `fsolve`, and `ode45`, to solve a single nonlinear algebraic equation, multiple nonlinear algebraic equations, and a set of nonlinear differential equations, respectively.

M1.11 DIARY

When preparing homework solutions, it is often necessary to save the sequence of commands and output results in a file to be turned in with the homework. The `diary` command allows this.

`diary file_name` causes a copy of all subsequent terminal input and most of the resulting output to be written on the file named `file_name`. `diary off` suspends it. `diary on` turns it back on. `diary`, by itself, toggles the diary state. Diary files may be edited later with a text editor to add comments or remove mistaken entries.

Often the consultants wish to see a diary file of your session to assist them in troubleshooting your MATLAB problems.

M1.12 TOOLBOXES

MATLAB toolboxes are a collection of function routines written to solve specialized problems. The Signals and Systems Toolbox is distributed with the Student Edition of MATLAB. The newest edition of Student MATLAB also contains the Symbolic Algebra toolbox, which is a collection of routines that allows one to obtain analytical solutions to algebraic and differential calculus problems. One toolbox often used in chemical process control is the Control Systems Toolbox.

M1.13 LIMITATIONS TO STUDENT MATLAB

The Student Edition of MATLAB has a few limitations compared with the Professional version. For example, Student MATLAB (4.0) is limited to 8192-element arrays.

M1.14 CONTACTING MATHWORKS

MATHWORKS has a homepage on the World Wide Web. The URL for this homepage is:

```
http://www.mathworks.com/
```

You can find answers to frequently asked questions (FAQ) on this homepage. Also, there are a number of technical notes that give more information on using MATLAB function routines. Suggestions are made for modifying these routines for specific problems.

There is also a MATLAB newsgroup (bulletin board) that has up-to-date questions (usually supplied by MATLAB users) and answers (supplied by other users as well as MATHWORKS personnel). The newgroup is:

```
comp.soft-sys.matlab
```

Before posting questions on this newsgroup it is a good idea to read the FAQ from the MATHWORKS.

Summary of Commonly Used Commands

axis	axis limis for plots
clear	removes all variables from workspace
clc	clears command window
diary	save the text of a MATLAB session
end	end of loop
exp	exponential function
for	generates loop structure
format	output display format
function	user generated function
gtext	place text on a plot
help	help function
hold	holds current plot and allows new plot to be placed on current plot
if	conditional test
length	length of a vector

`lookfor`	keyword search on help variables
`plot`	plots vectors
`size`	size of the array (rows, columns)
`subplot`	multiple plots in a figure window
`who`	view variables in workspace
`whos`	view variables in workspace, with more detail (size, etc.)
`*`	matrix multiplication
`'`	transpose
`;`	suppress printing (also - end of row, when used in matrices)
`.*`	element by element multplication
`./`	element by element division
`:`	denotes a column or row in a matrix. also creates a vector
`%`	placed before comment statements

Frequently Used MATLAB Functions

Function	*Use*	*Chapter*
`eig`	eigenvalues, eigenvectors	5
`fsolve`	solve algebraic equations	3
`fzero`	solve a single algebraic equation	3
`impulse`	impulse response	9
`ode45`	integrate set of ordinary differential equations	4
`polyfit`	least squares fit of a polynomial	Module 3 (linear regression)
`ss2tf`	convert state space to transfer function model	10
`step`	step response	5, 9
`tf2ss`	convert transfer function to state space model	9

STUDENT EXERCISES

1. Plot the following three curves on (i) a single plot and (ii) multiple plots (using the `subplot` command): 2 cos(t), sin(t) and cos(t)+sin(t). Use a time period such that two or three peaks occur for each curve. Use solid, dashed, and '+' symbols for the different curves. Use roughly 25-50 points for each curve.

2. **a.** Calculate the rank, determinant and matrix inverse of the following matrices (use `help rank`, `help det`, and `help inv`):

$$A = \begin{bmatrix} 1 & 2 & 1 \\ -1 & -2 & -1 \\ 2 & 4 & 2 \end{bmatrix}$$

$$B = \begin{bmatrix} 1 & 2 & 1 \\ -1 & 4 & -1 \\ 2 & 4 & 2 \end{bmatrix}$$

$$C = \begin{bmatrix} 1 & 2 & 1 \\ -1 & 4 & -1 \\ 2 & 4 & 5 \end{bmatrix}$$

3. Find CC^{-1}, where:

$$C = \begin{bmatrix} 1 & 2 & 1 \\ -1 & 4 & -1 \\ 2 & 4 & 5 \end{bmatrix}$$

4. **a.** Calculate x^Tx, and
   b. Calculate xx^T, where:

$$x = \begin{bmatrix} 1 \\ 2 \\ 3 \\ 4 \end{bmatrix}$$

5. Find the eigenvalues of the matrix:

$$D = \begin{bmatrix} -1 & 0 & 0 & 2 \\ 1 & -2 & 0 & 6 \\ 1 & 3 & -1 & 8 \\ 0 & 0 & 0 & -2 \end{bmatrix}$$

6. Consider the expression:

$$-KA - A^TK - Q - KBR^{-1}B^TK = 0$$

with

$$A = \begin{bmatrix} 0 & 3 \\ 2 & -1 \end{bmatrix} \qquad B = \begin{bmatrix} 1 \\ 4 \end{bmatrix}$$

$$Q = \begin{bmatrix} -11.896 & -20.328 \\ -17.192 & -18.856 \end{bmatrix} \qquad K = \begin{bmatrix} 7 & 3 \\ 5 & 2 \end{bmatrix}$$

Solve for **R**.

7. Find the solutions to the equation $f(x) = 3x^3 + x^2 + 5x - 6 = 0$. Use `roots` and `fzero`.

8. Integrate the following equations from $t = 0$ to $t = 5$:

$$\frac{dx_1}{dt} = -x_1 + x_2$$

$$\frac{dx_2}{dt} = -x_2$$

with the initial condition $x_1(0) = x_2(0) = 1$. Use `ode45` and plot your results.

9. Write your own function file for the following equation:

$$k(T) = a \exp\left(b - \frac{c}{T} - \ln dT - e\,T + f\,T^2\right)$$

for $a = 3.33$, $b = 13.148$, $c = 5639.5$, $d = 1.077$, $e = 5.44 \times 10^{-4}$, $f = 1.125 \times 10^{-7}$
T is in units of Kelvin.
Plot k as a function of T for temperatures from 373 to 600 K. (we suggest 50 points)

10. Find $\hat{V}\left(\dfrac{\text{cm}^3}{\text{gmol}}\right)$ for the following equation of state:

$$P = \frac{RT}{\hat{V} - b} - \frac{a}{T^{0.5}\hat{V}\,(\hat{V} + b)}$$

for $P = 13.76$ bar, $b = 44.891$ cm^3/gmol, $T = 333$ K, $a = 1.56414 \times 10^8$ cm^6 bar/gmol^2 $K^{0.5}$, $R =$ ideal gas constant in appropriate units.

REVIEW OF MATRIX ALGEBRA

MODULE
2

The purpose of this module is to provide a concise review of matrix operations and linear algebra.

After studying this module the reader should be able to:

- Multiply matrices
- Find the transpose of a matrix
- Find the eigenvalues and eigenvectors of a matrix
- Understand concepts of rank and singularity

The major sections of this module are:

In this module we review only the essentials of matrix operation necessary to understand material presented in the chapters of this textbook. For more detailed explanations and more advanced concepts, please consult any textbook on matrices or linear algebra. MATLAB is very useful for performing matrix operations. Module 1 provides a review of MATLAB.

M2.1 MOTIVATION AND NOTATION

In the study of dynamic systems it is common to use matrix notation. The use of matrix notation allows the compact representation of a system composed of many variables and equations.

Matrices are two-dimensional arrays that contain scalar elements that can be either real or complex. In this module most of our examples involve real matrices. Consider the following matrix, which consists of n rows and m columns:

$$\mathbf{A} = \begin{bmatrix} a_{11} & a_{12} & . & . & a_{1m} \\ a_{21} & a_{22} & . & . & a_{2m} \\ . & . & . & . & . \\ a_{n1} & a_{n2} & . & . & a_{nm} \end{bmatrix}$$

where a_{ij} represents the scalar element in the ith row and jth column of matrix $\mathbf{A}$.

It is a good habit to denote the numbers of rows and columns beneath the matrix as (n,m) or $(n \times m)$. For example, the following is a 2×3 matrix:

$$\mathbf{A} = \begin{bmatrix} 1 & 2 & 3 \\ 4 & 5 & 6 \end{bmatrix}$$
$$(2 \times 3)$$

A *vector* is a special case of a two-dimensional matrix. Normally, the use of the term vector implies a column vector. The following is an example of a *column vector* of length 3 (a 3×1 matrix), where v_i is an element of the vector:

$$\mathbf{v} = \begin{bmatrix} v_1 \\ v_2 \\ v_3 \end{bmatrix}$$

A *row vector* of length three is

$$\mathbf{w} = \begin{bmatrix} w_1 & w_2 & w_3 \end{bmatrix}$$

The convention of this textbook is to use lower-case bold letters to represent vectors and upper-case bold letters to represent matrices. Unless stated otherwise, it is assumed that a vector is a column vector.

M2.2 COMMON MATRIX OPERATIONS

M2.2.1 Matrix Addition

Two matrices can be added by simply adding the individual elements of the matrix. The matrices must have the same number of rows and columns. The operation:

$$\mathbf{C} = \mathbf{A} + \mathbf{B}$$

is simply element-by-element addition to form a new matrix:

$$c_{ij} = a_{ij} + b_{ij}$$

For example,

$$\mathbf{C} = \begin{bmatrix} 1 & 2 & 3 \\ 4 & 5 & 6 \end{bmatrix} + \begin{bmatrix} 7 & 8 & 9 \\ 10 & 11 & 12 \end{bmatrix} = \begin{bmatrix} 8 & 10 & 12 \\ 14 & 16 & 18 \end{bmatrix}$$

Clearly the order of matrices does not matter for the addition operation ($\mathbf{A} + \mathbf{B} = \mathbf{B} + \mathbf{A}$).

M2.2.2   Matrix Multiplication

Consider an $n \times m$ matrix $\mathbf{A}$ and an $r \times n$ matrix $\mathbf{B}$. An $r \times m$ matrix $\mathbf{C}$ is defined as the product of $\mathbf{B}$ times $\mathbf{A}$:

$$\begin{array}{ccc} \mathbf{C} & = & \mathbf{B} \quad \mathbf{A} \\ (r \times m) & & (r \times n)(n \times m) \end{array}$$

Notice that the number of columns of $\mathbf{B}$ must be equal to the number of rows of $\mathbf{A}$. The element in the ith row and jth column of $\mathbf{C}$ is:

$$C_{ij} = B_{i1}A_{1j} + B_{i2}A_{2j} + \ldots + B_{in}A_{nj}$$

$$C_{ij} = \sum_{k=1}^{n} B_{ik}A_{kj} \qquad (M2.1)$$

Notice that we can view this as an operation on row and column vectors within the matrices. That is, the ith element in the jth column of $\mathbf{C}$ is equal to the scalar value that is obtained from "multiplying" the ith row of $\mathbf{B}$ times the jth column of $\mathbf{A}$.

$$\begin{array}{ccc} \mathbf{C} & = & \mathbf{B} \quad \mathbf{A} \\ (r,m) & & (r,n) \quad (n,m) \end{array}$$

ijth element ith row jth column

For example,

$$\mathbf{C} = \begin{bmatrix} 7 & 8 \\ 9 & 10 \\ 11 & 12 \end{bmatrix} \begin{bmatrix} 1 & 2 & 3 \\ 4 & 5 & 6 \end{bmatrix}$$

(rows,cols)   (3,2)   (2,3)

$$= \begin{bmatrix} 7(1) + 8(4) & 7(2) + 8(5) & 7(3) + 8(6) \\ 9(1) + 10(4) & 9(2) + 10(5) & 9(3) + 10(6) \\ 11(1) + 12(4) & 11(2) + 12(5) & 11(3) + 12(6) \end{bmatrix}$$

(3,3)

$$= \begin{bmatrix} 39 & 54 & 69 \\ 49 & 68 & 87 \\ 59 & 82 & 105 \end{bmatrix}$$

Clearly the order of the matrices are important in multiplication. In the example above, **BA** is a consistent multiplication, while **AB** is not. Even for square matrices (where the number of rows is equal to the number of columns), $\mathbf{AB} \neq \mathbf{BA}$ in general.

M2.2.3 Transpose

Let **D** be a transpose of the matrix **A**, then:

$$\mathbf{D} = \mathbf{A}^T \tag{M2.2}$$

where (^{T}) represents the transpose operation. The ij element of **D** is the ji element of **A**

$$d_{ij} = a_{ji}$$

Columns become rows and rows become columns. For example,

$$\mathbf{A} = \begin{bmatrix} 1 & 2 & 3 \\ 4 & 5 & 6 \end{bmatrix} \quad \mathbf{A}^T = \begin{bmatrix} 1 & 4 \\ 2 & 5 \\ 3 & 6 \end{bmatrix}$$

It can be shown that the following property holds:

$$(\mathbf{ABC})^T = \mathbf{C}^T \mathbf{B}^T \mathbf{A}^T$$

As noted earlier, we normally think of vectors as being column vectors. The transpose of a column vector is a row vector. Consider an n-dimensional vector **x**. The transpose of **x** is $\mathbf{x}^T$.

$$\mathbf{x} = \begin{bmatrix} x_1 \\ x_2 \\ x_3 \\ \cdot \\ \cdot \\ x_n \end{bmatrix} \quad \text{and} \quad \mathbf{x}^T = \begin{bmatrix} x_1 & x_2 & x_3 & . & . & x_n \end{bmatrix}$$

The vector transpose is often used to calculate a scalar function of two vectors. For example, if $\mathbf{w}$ and $\mathbf{z}$ are column vectors, of length n, the operation $\mathbf{w}^T\mathbf{z}$ results in a scalar value:

$$\begin{bmatrix} w_1 & w_2 & w_3 & . & . & w_n \end{bmatrix} \begin{bmatrix} z_1 \\ z_2 \\ z_3 \\ . \\ . \\ z_n \end{bmatrix} = \sum_{j=1}^{n} w_j z_j$$

Notice that multiplication in the opposite order, $\mathbf{z}\mathbf{w}^T$, results in an $n \times n$ matrix:

$$\begin{bmatrix} z_1 \\ z_2 \\ z_3 \\ . \\ . \\ z_n \end{bmatrix} \begin{bmatrix} w_1 & w_2 & w_3 & . & . & w_n \end{bmatrix} = \begin{bmatrix} z_1 w_1 & z_1 w_2 & z_1 w_3 & . & z_1 w_n \\ z_2 w_1 & z_2 w_2 & z_2 w_3 & . & z_2 w_n \\ . & . & . & . & . \\ z_n w_1 & z_n w_2 & z_n w_3 & . & z_n w_n \end{bmatrix}$$

$\quad\quad (n,1) \quad\quad\quad\quad (1,n) \quad\quad\quad\quad\quad\quad\quad\quad (n,n)$

M2.2.4 Diagonal Matrices

Diagonal matrices are commonly used in weighted least squares (regression) problems. Matrix $\mathbf{Q}$ is diagonal if

$$\begin{aligned} Q_{ij} &= q_i \text{ for } i = j \\ &= 0 \text{ for } i \neq j \end{aligned}$$

For example, $\mathbf{Q} = \text{diag}(1,3,12)$ represents the following matrix:

$$\mathbf{Q} = \begin{bmatrix} 1 & 0 & 0 \\ 0 & 3 & 0 \\ 0 & 0 & 12 \end{bmatrix}$$

IDENTITY MATRIX

An identity matrix, $\mathbf{I}$, is a diagonal matrix with ones on the diagonal and zeros off-diagonal.

$$\begin{aligned} \mathbf{I}_{ij} &= 1 \text{ if } i = j \\ &= 0 \text{ if } i \neq j \end{aligned}$$

The 5×5 identity matrix is:

$$\mathbf{I} = \text{diag}(1,1,1,1,1) = \begin{bmatrix} 1 & 0 & 0 & 0 & 0 \\ 0 & 1 & 0 & 0 & 0 \\ 0 & 0 & 1 & 0 & 0 \\ 0 & 0 & 0 & 1 & 0 \\ 0 & 0 & 0 & 0 & 1 \end{bmatrix}$$

M2.3   SQUARE MATRICES

A number of the important matrix operations used in this textbook involve square (number of rows equal number of columns) matrices.

M2.3.1   Trace

The trace of an $n \times n$ (square) matrix is simply the sum of its diagonal elements,

$$\text{tr } \mathbf{A} = \sum_{i=1}^{n} a_{ii} \qquad \text{(M2.3)}$$

M2.3.2   Determinant

One of the most important properties of a square matrix is the determinant. Consider a 2×2 matrix:

$$\mathbf{A} = \begin{bmatrix} a_{11} & a_{12} \\ a_{21} & a_{22} \end{bmatrix}$$

The determinant of $\mathbf{A}$ is

$$\det \mathbf{A} = a_{11}a_{22} - a_{21}a_{12}$$

Sometimes the det $\mathbf{A}$ is written as $|\mathbf{A}|$. An algorthim for finding the determinant of a larger matrix involves matrix cofactors and is shown below.

M2.3.3   Minors and Cofactors of a Matrix

The matrix formed by deleting the ith row and the jth column of the matrix $\mathbf{A}$ is the *minor* of the element a_{ij}, denoted $\mathbf{M}_{ij}$. Consider the 3×3 matrix:

$$\mathbf{A} = \begin{bmatrix} a_{11} & a_{12} & a_{13} \\ a_{21} & a_{22} & a_{23} \\ a_{31} & a_{32} & a_{33} \end{bmatrix}$$

The minor of a_{12} of the matrix is obtained by deleting the first row and the second column. In this case:

$$\mathbf{M}_{12} = \begin{bmatrix} a_{21} & a_{23} \\ a_{31} & a_{33} \end{bmatrix}$$

The determinant of the resulting matrix is:

$$\det \mathbf{M}_{12} = a_{21}a_{33} - a_{31}a_{23} \qquad \text{(M2.4)}$$

The *cofactor* is the "signed" value of the minor. That is, the cofactor is defined as:

$$c_{ij} = (-1)^{i+j} \det \mathbf{M}_{ij} \qquad \text{(M2.5)}$$

The cofactor c_{12} of the 3×3 example matrix is:

$$c_{12} = (-1)^{i+j} \det \mathbf{M}_{12} = (-1)^{1+2} (a_{21}a_{33} - a_{31}a_{23}) = -a_{21}a_{33} + a_{31}a_{23}$$

The determinant of a matrix can be found by expanding about any row or column (this approach is sometimes called the Laplace expansion).

(i) Expanding around any column j:

$$\det \mathbf{A} = \sum_{i=1}^{n} a_{ij} c_{ij} \tag{M2.6}$$

(ii) Expanding around any row i:

$$\det \mathbf{A} = \sum_{j=1}^{n} a_{ij} c_{ij} \tag{M2.7}$$

As an example, consider the 3×3 matrix $\mathbf{A}$. The expansion around row 1 is

$$\det \mathbf{A} = a_{11}c_{11} + a_{12}c_{12} + a_{13}c_{13}$$
$$\det \mathbf{A} = a_{11}(-1)^{1+1} \det \mathbf{M}_{11} + a_{12}(-1)^{1+2} \det \mathbf{M}_{12} + a_{13}(-1)^{1+3} \det \mathbf{M}_{13}$$
$$\det \mathbf{A} = a_{11}(a_{22}a_{33} - a_{32}a_{33}) - a_{12}(a_{21}a_{33} - a_{31}a_{23}) + a_{13}(a_{21}a_{32} - a_{31}a_{22})$$

M2.3.4 Matrix Inversion

The inverse of a square matrix $\mathbf{A}$, is called $\mathbf{A}^{-1}$ and is defined as:

$$\mathbf{A}\,\mathbf{A}^{-1} = \mathbf{I} \tag{M2.8}$$

The matrix inverse is conceptually useful when finding $\mathbf{x}$ to solve the following problem:

$$\mathbf{A}\,\mathbf{x} = \mathbf{y}$$

Multiplying each side (on the left) by $\mathbf{A}^{-1}$, we find:

$$\underset{(n \times n)(n \times n)(n \times 1)}{\mathbf{A}^{-1} \quad \mathbf{A} \quad \mathbf{x}} \quad = \quad \underset{(n \times n)(n \times 1)}{\mathbf{A}^{-1} \quad \mathbf{y}}$$

which yields:

$$\mathbf{I}\,\mathbf{x} = \mathbf{A}^{-1}\,\mathbf{y}$$

or,

$$\mathbf{x} = \mathbf{A}^{-1}\,\mathbf{y}$$

An algorithm for finding the inverse of an $n \times n$ matrix is known as Cramer's rule:

$$\mathbf{A}^{-1} = \frac{\text{adj } \mathbf{A}}{\det \mathbf{A}} \tag{M2.9}$$

where the *adjoint* matrix of $\mathbf{A}$ (adj $\mathbf{A}$) is the transpose of the matrix of cofactors of $\mathbf{A}$.

$$\text{adj } \mathbf{A} = \{c_{ij}\}^{T} \tag{M2.10}$$

In practice, this procedure is not used because the computational time is quite large for large matrices. Generally, a method such as Gaussian Elimination or LU decomposition is used.

EXAMPLE M2.1   Inversion of a 2 × 2 Matrix Using Cramer's Rule

$$\mathbf{A} = \begin{bmatrix} a_{11} & a_{12} \\ a_{21} & a_{22} \end{bmatrix}$$

$$\mathbf{A}^{-1} = \frac{\text{adj } \mathbf{A}}{\det \mathbf{A}}$$

$$\text{adj } \mathbf{A} = \{c_{ij}\}^T = \begin{bmatrix} c_{11} & c_{12} \\ c_{21} & c_{22} \end{bmatrix}^T = \begin{bmatrix} c_{11} & c_{21} \\ c_{12} & c_{22} \end{bmatrix}$$

where the cofactors are

$$c_{11} = (-1)^{1+1} \det \mathbf{M}_{11} = a_{22}$$

$$c_{12} = (-1)^{1+2} \det \mathbf{M}_{12} = -a_{21}$$

$$c_{21} = (-1)^{2+1} \det \mathbf{M}_{21} = -a_{12}$$

$$c_{22} = (-1)^{2+2} \det \mathbf{M}_{22} = a_{11}$$

$$\text{adj } \mathbf{A} = \{c_{ij}\}^T = \begin{bmatrix} c_{11} & c_{12} \\ c_{21} & c_{22} \end{bmatrix}^T = \begin{bmatrix} c_{11} & c_{21} \\ c_{12} & c_{22} \end{bmatrix} = \begin{bmatrix} a_{22} & -a_{12} \\ -a_{21} & a_{11} \end{bmatrix}$$

$$\mathbf{A}^{-1} = \frac{\text{adj } \mathbf{A}}{\det \mathbf{A}} = \begin{bmatrix} a_{22} & -a_{12} \\ -a_{21} & a_{11} \end{bmatrix} \frac{1}{\det \mathbf{A}}$$

$$= \begin{bmatrix} a_{22} & -a_{12} \\ -a_{21} & a_{11} \end{bmatrix} \frac{1}{a_{11}a_{22} - a_{21}a_{12}}$$

The formula for the inverse of a 2 × 2 matrix should be committed to memory:

$$\mathbf{A} = \begin{bmatrix} a & b \\ c & d \end{bmatrix}$$

$$\mathbf{A}^{-1} = \begin{bmatrix} d & -b \\ -c & a \end{bmatrix} \frac{1}{\det \mathbf{A}} = \begin{bmatrix} \dfrac{d}{\det \mathbf{A}} & \dfrac{-b}{\det \mathbf{A}} \\[2ex] \dfrac{-c}{\det \mathbf{A}} & \dfrac{a}{\det \mathbf{A}} \end{bmatrix}$$

where $\det \mathbf{A} = ad - bc$

That is, swap the diagonal terms, take the negative of the off-diagonal terms, and divide by the determinant.

M2.3.5   Other Issues in Matrix Inversion

A *singular* matrix cannot be inverted. A matrix $\mathbf{A}$ is singular if det $\mathbf{A} = 0$. Also, a singular matrix has a *rank* that is lower than the dimension of the matrix. The rank of a matrix is the number of independent rows or columns. For example, the matrix:

$$\mathbf{A} = \begin{bmatrix} 1 & 2 \\ 2 & 4 \end{bmatrix}$$

has rank = 1, because the second row is linearly dependent on the first row. The matrix is singular, because

$$\det \mathbf{A} = 4 - 4 = 0$$

We note that the matrix inverse of a singular matrix does not exist. For this example, note the division by 0 in the following calculation

$$\mathbf{A}^{-1} = \begin{bmatrix} 4 & -2 \\ -2 & 4 \end{bmatrix} \frac{1}{\det \mathbf{A}} = \begin{bmatrix} 4 & -2 \\ -2 & 4 \end{bmatrix} \frac{1}{0} = \text{undefined}$$

The rank of a square matrix is equal to the number of nonzero eigenvalues. Eigenvalues of a square matrix are discussed next.

M2.3.6   Eigenvalues and Eigenvectors

Eigenvalue/eigenvector analysis will be important when studying dynamic systems. Eigenvalues determine how "fast" a dynamic system responds, and the associated eigenvector indicates the "direction" of that response. This material is used in Chapter 5.

Eigenvalue/eigenvector analysis can only be performed on square matrices. The eigenvector/eigenvalue problem is:

$$\mathbf{A}\,\xi = \lambda\,\xi \tag{M2.11}$$

where:

    $\mathbf{A}$ = matrix (square, number of rows = number of columns)

    λ = eigenvalue (scalar)

    ξ = eigenvector (vector)

Normally, we will use the notation ξ_i to represent the eigenvector that is associated with eigenvalue λ_i. There are n eigenvalues and n eigenvectors associated with an $n \times n$ matrix.

The interesting thing about equation (M2.11) is that a matrix times a vector is equal to a scalar times a vector. A scalar multiplying a vector does not change its "direction" in n-dimensional space, only its magnitude. In this case, a matrix $\mathbf{A}$ times the eigenvector ξ yields the eigenvector ξ back, scaled by λ.

Equation (M2.11) can be written as

$$(\lambda\mathbf{I} - \mathbf{A})\xi = 0 \tag{M2.12}$$

where λ must be a scalar value for which $\lambda\mathbf{I} - \mathbf{A}$ is singular, otherwise $\xi = 0$ (the trivial vector). A requirement for $\lambda\mathbf{I} - \mathbf{A}$ to be singular is $\det(\lambda\mathbf{I} - \mathbf{A}) = 0$, where the notation $\det(\lambda\mathbf{I} - \mathbf{A})$ is used to represent the determinant of $\lambda\mathbf{I} - \mathbf{A}$.

EIGENVALUES

The eigenvalues of an $n \times n$ matrix $\mathbf{A}$ are the n scalars that solve:

$$\det(\lambda\mathbf{I} - \mathbf{A}) = 0 \tag{M2.13}$$

Equation (M2.13) is also known as the *characteristic polynomial* for the matrix $\mathbf{A}$; the eigenvalues are the roots of the characteristic polynomial. For an $n \times n$ matrix $\mathbf{A}$, the characteristic polynomial will be nth order, so there will be n roots (n eigenvalues).

EIGENVECTORS

The corresponding n eigenvectors (ξ_i) can be found from

$$\mathbf{A}\,\xi_i = \lambda_i\,\xi_i \tag{M2.14}$$

ξ_i is the eigenvector associated with the eigenvalue λ_i.

For example, consider the general 2×2 matrix:

$$\mathbf{A} \quad = \begin{bmatrix} a_{11} & a_{12} \\ a_{21} & a_{22} \end{bmatrix}$$

$$\lambda\mathbf{I} - \mathbf{A} \quad = \begin{bmatrix} \lambda - a_{11} & -a_{12} \\ -a_{21} & \lambda - a_{22} \end{bmatrix}$$

$$\det(\lambda\mathbf{I} - \mathbf{A}) = (\lambda - a_{11})(\lambda - a_{22}) - a_{12}\,a_{21} = 0$$

$$\det(\lambda\mathbf{I} - \mathbf{A}) = \lambda^2 - [a_{11} + a_{22}]\,\lambda + [a_{11}a_{22} - a_{21}a_{12}] = 0 \tag{M2.15}$$

Equation (M2.15) can also be written as:

$$\det(\lambda\mathbf{I} - \mathbf{A}) = \lambda^2 - \mathrm{tr}(\mathbf{A})\,\lambda + \det(\mathbf{A}) = 0 \tag{M2.16}$$

where:

$$\mathrm{tr}(\mathbf{A}) = a_{11} + a_{22}$$
$$\det(\mathbf{A}) = a_{11}a_{22} - a_{21}a_{12}$$

Equation (M2.16) is the characteristic polynomial for a 2×2 matrix.

It can easily be shown that, if $\mathrm{tr}(\mathbf{A}) < 0$ and $\det(\mathbf{A}) > 0$, the roots (eigenvalues) of (M2.16) will be negative. Also, if $\det(\mathbf{A}) = 0$, then one root (eigenvalue) will be zero.

The reader should derive the following characteristic polynomial for a 3×3 matrix (see student exercise 7).

The characteristic equation for a 3×3 matrix is:

$$\det(\lambda \mathbf{I} - \mathbf{A}) = \lambda^3 - \text{tr}(\mathbf{A})\,\lambda^2 + \mathbf{M}\,\lambda - \det(\mathbf{A}) = 0 \qquad \text{(M2.17)}$$

where:

$$\text{tr}(\mathbf{A}) = a_{11} + a_{22} + a_{33}$$

$$\mathbf{M} = \det \mathbf{M}_{11} + \det \mathbf{M}_{22} + \det \mathbf{M}_{33}$$

$$= a_{22}a_{33} - a_{32}a_{23} + a_{11}a_{33} - a_{31}a_{13} + a_{11}a_{22} - a_{12}a_{21}$$

$$\det \mathbf{A} = a_{11}(a_{22}a_{33} - a_{21}a_{12}) - a_{12}(a_{21}a_{33} - a_{31}a_{23}) + a_{13}(a_{21}a_{32} - a_{31}a_{22})$$

The Routh stability criterion (Chapters 5 and 9) can be used to find the conditions on the polynomial coefficients that will yield negative roots (eigenvalues). The necessary condition is that all of the polynomial coefficents are positive. The necessary condition is then $\text{tr}(\mathbf{A}) < 0$, $M > 0$, and $\det(\mathbf{A}) < 0$. The reader can use the Routh stability criterion to determine the sufficient conditions for negative roots.

EXAMPLE M2.2   Eigenvalue/Eigenvector Calculation

Consider the following matrix:

$$\mathbf{A} = \begin{bmatrix} 2 & 1 \\ 2 & -1 \end{bmatrix}$$

$$\det(\lambda \mathbf{I} - \mathbf{A}) = \lambda^2 - [2 - 1]\lambda + [2(-1) - 2(1)] = 0$$

$$\lambda^2 - \lambda - 4 = 0$$

From the quadratic equation:

$$\lambda = \frac{1 \pm \sqrt{1 - 4(-4)}}{2} = \frac{1}{2} \pm \frac{\sqrt{17}}{2}$$

so,

$$\lambda_1 = -1.5616$$

and

$$\lambda_2 = 2.5616$$

The first eigenvector is found from (M2.14):

$$\mathbf{A}\,\xi_1 = \lambda_1\,\xi_1 \qquad \text{(M2.18)}$$

where we have used the notation:

$$\xi_1 = \begin{bmatrix} v_{11} \\ v_{21} \end{bmatrix}$$

Applying equation (M2.18),

$$\begin{bmatrix} 2 & 1 \\ 2 & -1 \end{bmatrix} \begin{bmatrix} v_{11} \\ v_{21} \end{bmatrix} = -1.5616 \begin{bmatrix} v_{11} \\ v_{21} \end{bmatrix}$$

So the following two equations must be satisfied:

$$2\, v_{11} + v_{21} = -1.5616\, v_{11}$$
$$2\, v_{11} - v_{21} = -1.5616\, v_{21}$$

which yield:

$$v_{21} = -3.5615\, v_{11}$$

If we wish, we can arbitrarily select $v_{11} = 1.0$ and $v_{21} = 3.5615$

$$\xi_1 = \begin{bmatrix} v_{11} \\ v_{21} \end{bmatrix} = \begin{bmatrix} 1.0 \\ -3.5615 \end{bmatrix} \tag{M2.19}$$

Most computer packages will report eigenvectors that have a unit norm (length = 1). If we divide (M2.19) by $\sqrt{(v_{11})^2 + (v_{21})^2} = \sqrt{1 + 12.6841} = 3.6992$, we find that:

$$\xi_1 = \begin{bmatrix} v_{11} \\ v_{21} \end{bmatrix} = \begin{bmatrix} 0.2703 \\ -0.9628 \end{bmatrix}$$

An equivalent solution is:

$$\xi_1 = \begin{bmatrix} -0.2703 \\ 0.9628 \end{bmatrix}.$$

The reader should show that:

$$\mathbf{A}\, \xi_2 = \lambda_2\, \xi_2$$

leads to:

$$\xi_2 = \begin{bmatrix} v_{12} \\ v_{22} \end{bmatrix} = \begin{bmatrix} 0.8719 \\ 0.4896 \end{bmatrix}$$

Eigenvector problems are easily solved used the `eig` function in MATLAB (see Module 1), as shown below.

```
a =
      2       1
      2      -1

»[v,d]=eig(a)

v =
    0.8719    -0.2703
    0.4896     0.9628

d =
    2.5616         0
         0   -1.5616
```

Column i in the v matrix is the eigenvector associated with the ith eigenvalue (which is the ith diagonal element in the d matrix). The trace and determinant can also be found using MATLAB commands:

```
»  trace(a)

ans =
      1

»  det(a)

ans =
     -4
```

M2.3.7 The Similarity Transform

The *similarity transform* is useful for understanding the dynamic behavior of linear systems (see Chapter 5). Recall the eigenvector/eigenvalue problem for a 2×2 matrix:

$$\mathbf{A}\, \xi_1 = \lambda_1\, \xi_1 \tag{M2.20}$$
$$\mathbf{A}\, \xi_2 = \lambda_2\, \xi_2 \tag{M2.21}$$

Notice that equations (M2.20) and (M2.21) can be written in the following form:

$$\mathbf{A}\,\mathbf{V} = \mathbf{V}\,\Lambda \tag{M2.22}$$

where $\mathbf{V}$ is the matrix of eigenvectors:

$$\mathbf{V} = [\xi_1 \quad \xi_2] = \begin{bmatrix} v_{11} & v_{12} \\ v_{21} & v_{22} \end{bmatrix} \tag{M2.23}$$

where:

$$\xi_1 = \begin{bmatrix} v_{11} \\ v_{21} \end{bmatrix} \quad \text{and} \quad \xi_2 = \begin{bmatrix} v_{12} \\ v_{22} \end{bmatrix}$$

Λ is the matrix of eigenvalues:

$$\Lambda = \begin{bmatrix} \lambda_2 & 0 \\ 0 & \lambda_2 \end{bmatrix} \tag{M2.24}$$

Multiplying (M2.22) on the right side by $\mathbf{V}^{-1}$ we find

$$\mathbf{A} = \mathbf{V}\,\Lambda\,\mathbf{V}^{-1} \tag{M2.25}$$

Equation (M2.25) will be useful when developing analytical solutions for sets of linear differential equations, and can be used for any $n \times n$ (square) matrix.

M2.4   OTHER MATRIX OPERATIONS

M2.4.1   Vector Norms

It is quite natural to think of vectors as having a "length" property. This property is called a vector norm. The most common norm is the Euclidean norm. In two-dimensional space, the Euclidean norm is:

$$\| \mathbf{x} \| = \sqrt{x_1^2 + x_2^2}$$

Sometimes the Euclidian norm is called the 2-norm, $\| \mathbf{x} \|_2$.
   In n-dimensional space,

$$\| \mathbf{x} \|_2 = \sqrt{x_1^2 + x_2^2 + \ldots + x_n^2} = \sqrt{\sum_{j=1}^{n} x_j^2}$$

notice that the sum of the squares can be written as $\mathbf{x}^T\mathbf{x}$, that is:

$$\sum_{j=1}^{n} x_j x_j = \sum_{j=1}^{n} x_j^2 = \mathbf{x}^T\mathbf{x}$$

and the 2-norm is then:

$$\| \mathbf{x} \|_2 = \sqrt{\mathbf{x}^T\mathbf{x}}$$

M2.4.2   Orthogonality

A column vector $\mathbf{w}$ is orthogonal to a column vector $\mathbf{z}$ if $\mathbf{w}^T\mathbf{z} = 0.0$.

For example, consider the 2-D vector $\mathbf{w} = \begin{bmatrix} 1 \\ -0.5 \end{bmatrix}$

and the 2-D vector $\mathbf{z} = \begin{bmatrix} 0.5 \\ 1 \end{bmatrix}$

Then $\mathbf{x}^T\mathbf{z} = \begin{bmatrix} 1 & -0.5 \end{bmatrix} \begin{bmatrix} 0.5 \\ 1 \end{bmatrix} = 1(0.5) + (-0.5)1 = 0.0$

Notice that orthogonal vectors are "perpendicular" to each other, as depicted below.

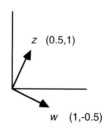

z (0.5,1)

w   (1,-0.5)

M2.4.3 Normalized or Scaled Vectors

Many times, we will scale vectors, so that they are *normalized*. A vector is normalized if it has a norm (length) of 1.0. Normalization is performed by multiplying each element of the vector by 1/vector norm. For a given vector, $\mathbf{x}$, the scaled vector, $\mathbf{u}$ is $\mathbf{u} = \mathbf{x}\, 1/\sqrt{\mathbf{x}^T\mathbf{x}}$

EXAMPLE M2.3   Vector Normalization

Consider the vector $\mathbf{x}^T = [1 \quad 2]$:

Then,
$$\mathbf{x}^T\mathbf{x} = [1 \quad 2]\begin{bmatrix}1\\2\end{bmatrix} = 1 + 4 = 5$$

And,
$$\sqrt{\mathbf{x}^T\mathbf{x}} = \sqrt{5}$$

The normalized vector is
$$\mathbf{u} = \begin{bmatrix}1\\2\end{bmatrix}\frac{1}{\sqrt{5}} = \begin{bmatrix}\dfrac{1}{\sqrt{5}}\\[2mm]\dfrac{2}{\sqrt{5}}\end{bmatrix}$$

A vector norm on R^n is a function $\|\ \| : R^n \to R$ satisfying:

1. $\mathbf{x} \neq 0 \qquad \rightarrow \qquad \|\mathbf{x}\| > 0$
2. $\|\alpha \mathbf{x}\| \qquad = \qquad |\alpha|\,\|\mathbf{x}\| \qquad$ Where α is a scalar
3. $\|\mathbf{x} + \mathbf{y}\| \qquad \leq \qquad \|\mathbf{x}\| + \|\mathbf{y}\| \qquad$ Triangle inequality

Common Vector Norms

1. $\|\mathbf{x}\|_1 = \displaystyle\sum_{i=1}^{n} |x_i| \qquad\qquad$ 1-norm (sum of absolute values)

2. $\|\mathbf{x}\|_2 = \left(\displaystyle\sum_{i=1}^{n} |x_i|^2\right)^{1/2} \qquad$ 2-norm, Euclidean Norm

3. $\|\mathbf{x}\|_p = \left(\displaystyle\sum_{i=1}^{n} |x_i|^p\right)^{1/p} \qquad$ p-norm

4. $\|\mathbf{x}\|_\infty = \max |x_i| \qquad\qquad$ ∞-norm

Notice that the p-norm includes all other vector norms

EXAMPLE M2.4   Comparison of Norms

$$\mathbf{x}^T \quad = [1 \ {-2} \ 3]$$
$$\|\mathbf{x}\|_1 = |1| + |-2| + |3| \qquad\qquad\quad = 6$$
$$\|\mathbf{x}\|_2 = (|1|^2 + |-2|^2 + |3|^2)^{1/2} \quad = 3.742$$
$$\|\mathbf{x}\|_\infty = \max\,[|1|, |-2|, |3|] \qquad\quad = 3$$
$$\|\mathbf{x}\|_{10} = (|1|^{10} + |-2|^{10} + |3|^{10})^{1/10} = 3.005$$

Notice that, as $p \to \infty$, $\|\mathbf{x}\|_p \to \|\mathbf{x}\|_\infty$

SUMMARY

In this module we have provided a concise review of matrix operations. This material should be sufficient for most of the matrix operations used in the textbook. It is particularly important that the reader understand:

- Matrix notation
- How to multiply matrices
- That an $n \times n$ matrix has n *eigenvalues*, which are the n roots to the characteristic equation: $\det(\lambda I - A) = 0$.
- That a matrix A is singular if the $\det(A) = 0$. A singular matrix does not have an inverse. A singular matrix is not full *rank*.
- That the rank of a square matrix is equal to the number of nonzero eigenvalues

STUDENT EXERCISES

For problems 1–3, use $\mathbf{x} = \begin{bmatrix} 3 \\ 1 \\ 7 \end{bmatrix}$

1. Find $\mathbf{x}^T$.
2. Find $\mathbf{x}^T\mathbf{x}$.
3. Find $\| \mathbf{x} \|_2$
4. Perform the indicated matrix operations for the following matrices:

$$\mathbf{A} = \begin{bmatrix} 3 & 2 \\ 1 & 0 \end{bmatrix} \quad \mathbf{B} = \begin{bmatrix} 7 & 6 & 8 \\ 9 & 5 & 4 \end{bmatrix}$$

   a. Find $\mathbf{C} = \mathbf{AB}$
   b. Find $\mathbf{D} = \mathbf{B}^T\mathbf{A}$
   c. Find $\mathbf{E} = \mathbf{A}^{-1}\mathbf{B}$

5. Solve for $\mathbf{x}$ as a function of $\mathbf{y}$ where: $\mathbf{y} = \begin{bmatrix} 1 & 1 \\ 2 & -1 \end{bmatrix} \mathbf{x}$

6. Find the determinant of:

$$\mathbf{A} = \begin{bmatrix} 1 & 0 & 0 & 2 \\ 1 & 2 & 0 & 6 \\ 1 & 3 & 1 & 8 \\ 0 & 0 & 0 & 2 \end{bmatrix}$$

7. Derive the characteristic polynomial $(\det (\lambda I - A))$ for a general 3×3 matrix:

$$\mathbf{A} = \begin{bmatrix} a_{11} & a_{12} & a_{13} \\ a_{21} & a_{22} & a_{23} \\ a_{31} & a_{32} & a_{33} \end{bmatrix}$$

8. Find the determinant, rank, and eigenvalues of the following matrices:

a. $\mathbf{A} = \begin{bmatrix} -1 & 2 \\ 1 & -3 \end{bmatrix}$

b. $\mathbf{B} = \begin{bmatrix} -1 & 2 \\ 1 & -2 \end{bmatrix}$

LINEAR REGRESSION

MODULE
3

The objective of this module is to review linear regression analysis, with a focus on a matrix algebra formulation and MATLAB routines for linear regression. After reviewing this module, the reader should be able to:

- Formulate a matrix algebra solution for a least squares problem
- Use the MATLAB routine `polyfit` to find a polynomial to fit data
- Use the MATLAB routine `polyval` to evaluate a polynomial

The major sections of this module are:

M3.1   Motivation
M3.2   Least Squares Solution for a Line
M3.3   Solution for the Equation of a Line Using Matrix-Vector Notation
M3.4   Generalization of the Linear Regression Technique
M3.5   MATLAB Routines `polyfit` and `polyval`

M3.1 MOTIVATION

The models developed in this text require the values of many parameters such as reaction rate coefficients, etc. It is common to use linear regression (also known as linear least squares analysis) to estimate the values of these parameters. To illustrate the basic principles of linear regression analysis, we first consider the equation of a line. Consider the

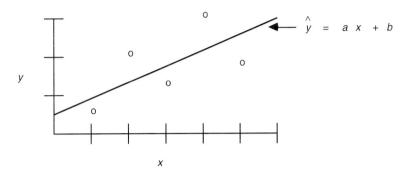

FIGURE M3.1   Experimental data and linear model prediction.

data shown as open circles in Figure M3.1. We would like to find a line that provides a best fit to the data.

Let x_i and y_i represent the independent variable and measured dependent variable for the ith experimental data point. Also, let $\hat{y}_i$ represent a model prediction of the dependent variable, given the experimental value of the independent variable, x_i. The sets of dependent and independent variables can be represented as vectors:

$$\mathbf{x} = \begin{bmatrix} x_1 \\ x_2 \\ . \\ . \\ . \\ x_N \end{bmatrix} \quad \mathbf{y} = \begin{bmatrix} y_1 \\ y_2 \\ . \\ . \\ . \\ y_N \end{bmatrix}$$

which can be written in the following fashion, to minimize space:

$$\mathbf{x}^T = [x_1\, x_2 \ldots x_N]$$
$$\mathbf{y}^T = [y_1\, y_2 \ldots y_N]$$

where ^{T} represents the transpose operation.

Given experimental data, $\mathbf{x}$ and $\mathbf{y}$, we wish to find parameters that allow a model to best fit the data. For a good fit of the data, we would like to minimize the difference between the data points and the model prediction ($\hat{y} = ax + b$). Our first instinct might be to find the model parameters, a and b, so that the sum of the absolute values of the errors is minimized. Here we have defined the error as the difference between the experimental value and the model prediction. That is, our objective is to find the values of a and b such that equation (M3.1) is minimized:

$$|y_1 - \hat{y}_1| + |y_2 - \hat{y}_2| + |y_3 - \hat{y}_3| + |y_4 - \hat{y}_4| + |y_5 - \hat{y}_5| \qquad \text{(M3.1)}$$

The major disadvantage to this approach is that there is not a simple, closed-form, analytical solution to this problem. An alternative approach is to minimize the *sum of the squares* of the errors. That is, find a and b such that (M3.2) is minimized.

$$(y_1 - \hat{y}_1)^2 + (y_2 - \hat{y}_2)^2 + (y_3 - \hat{y}_3)^2 + (y_4 - \hat{y}_4)^2 + (y_5 - \hat{y}_5)^2 \qquad \text{(M3.2)}$$

For N data points, we can represent (M3.2) using the more compact notation in (M3.3):

$$\sum_{i=1}^{N} (y_i - \hat{y}_i)^2 \qquad \text{(M3.3)}$$

Besides the analytical solution that will follow, another advantage to the sum of the squares formulation is that large errors are penalized more heavily than small errors.

M3.2   LEAST SQUARES SOLUTION FOR A LINE

Using optimization notation, we refer to (M3.3) as the *objective function*. We desire to find the values of a and b (known as *decision variables*) that minimize the objective function. Let $f(a,b)$ represent the objective function:

$$f(a,b) = \sum_{i=1}^{N} (y_i - \hat{y}_i)^2 \qquad \text{(M3.4)}$$

and since $\hat{y}_i = ax_i + b$:

$$f(a,b) = \sum_{i=1}^{N} (y_i - ax_i - b)^2 \qquad \text{(M3.5)}$$

We know from calculus the necessary condition for a minimum of a function with respect to a variable. The minimum of $f(a,b)$ is satisfied by (M3.6) and (M3.7).

$$\frac{\partial f(a,b)}{\partial a} = 0 \qquad \text{(M3.6)}$$

$$\frac{\partial f(a,b)}{\partial b} = 0 \qquad \text{(M3.7)}$$

From (M3.6)

$$\frac{\partial f(a,b)}{\partial a} = -2 \sum_{i=1}^{N} (y_i - ax_i - b)(x_i) = 0 \qquad \text{(M3.8)}$$

and from (M3.7):

$$\frac{\partial f(a,b)}{\partial b} = -2 \sum_{i=1}^{N} (y_i - ax_i - b) = 0 \qquad \text{(M3.9)}$$

Removing the constant value -2, we find that (M3.8) and (M3.9) can be written as (M3.10) and (M3.11):

$$\sum_{i=1}^{N} (y_i x_i - ax_i^2 - bx_i) = 0 \qquad \text{(M3.10)}$$

$$\sum_{i=1}^{N} (y_i - ax_i - b) \qquad = 0 \qquad \text{(M3.11)}$$

Removing the parameters a and b from the summations:

$$b \sum_{i=1}^{N} x_i + a \sum_{i=1}^{N} x_i^2 = \sum_{i=1}^{N} y_i x_i \qquad \text{(M3.12)}$$

$$a \sum_{i=1}^{N} x_i + bN = \sum_{i=1}^{N} y_i \qquad \text{(M3.13)}$$

Notice that we have two equations ((M3.12) and (M3.13)) and two unknowns (a and b). We can easily solve for b in terms of a from (M3.13) to obtain:

$$b = \frac{1}{N}\left[\sum_{i=1}^{N} y_i - a \sum_{i=1}^{N} x_i\right] \qquad \text{(M3.14)}$$

Of course, we see that the terms in (M3.14) are merely the mean values of y and x.

$$\bar{y} = \frac{1}{N} \sum_{i=1}^{N} y_i$$

$$\bar{x} = \frac{1}{N} \sum_{i=1}^{N} x_i$$

so that:

$$b = \bar{y} - a\bar{x} \qquad \text{(M3.15)}$$

Substituting (M3.15) into (M3.12) we find:

$$(\bar{y} - a\bar{x})\, N\bar{x} + a \sum_{i=1}^{N} x_i^2 = \sum_{i=1}^{N} y_i x_i \qquad \text{(M3.16)}$$

or,

$$a\left[-N\bar{x}^2 + \sum_{i=1}^{N} x_i^2\right] = \sum_{i=1}^{N} y_i x_i - N\bar{y} \qquad \text{(M3.17)}$$

which leads to:

$$a = \frac{\left(\displaystyle\sum_{i=1}^{N} y_i x_i\right) - N\bar{y}\bar{x}}{\left(\displaystyle\sum_{i=1}^{N} x_i^2\right) - N\bar{x}^2} \qquad \text{(M3.18)}$$

and b can be determined from (M3.15).

M3.3   SOLUTION FOR THE EQUATION OF A LINE USING MATRIX-VECTOR NOTATION

The model prediction of each dependent variable can be written as:

$$\hat{y}_i = ax_i + b \qquad \text{(M3.19)}$$

which can be written in matrix-vector form, for N data points, as:

$$
\begin{bmatrix} \hat{y}_1 \\ \hat{y}_2 \\ . \\ . \\ . \\ \hat{y}_N \end{bmatrix} = \begin{bmatrix} x_1 & 1 \\ x_2 & 1 \\ . & . \\ . & . \\ . & . \\ x_N & 1 \end{bmatrix} \begin{bmatrix} a \\ b \end{bmatrix}
\tag{M3.20}
$$
$$
(N \times 1) \qquad (N \times 2) \quad (2 \times 1)
$$

and we can see that the dimensioning of the matrices and vectors is consistent. Using compact matrix-vector notation, we write (M3.20) as:

$$
\hat{\mathbf{Y}} = \boldsymbol{\Phi} \, \boldsymbol{\theta}
\tag{M3.21}
$$

The objective function is:

$$
f(\boldsymbol{\theta}) = \sum_{i=1}^{N} (y_i - \hat{y}_i)^2
\tag{M3.22}
$$

which can be written as:

$$
f(\boldsymbol{\theta}) = [\mathbf{Y} - \hat{\mathbf{Y}}]^{\mathrm{T}}[\mathbf{Y} - \hat{\mathbf{Y}}]
\tag{M3.23}
$$
$$
(1 \times 1) \quad (1 \times N) \quad (N \times 1)
$$

The general optimization statement is then:

Minimize $\qquad\qquad f(\boldsymbol{\theta}) = [\mathbf{Y} - \hat{\mathbf{Y}}]^{\mathrm{T}}[\mathbf{Y} - \hat{\mathbf{Y}}]$ $\qquad\qquad$ (M3.23)

subject to $\qquad\qquad \hat{\mathbf{Y}} = \boldsymbol{\Phi} \, \boldsymbol{\theta}$ $\qquad\qquad$ (M3.21)

where $f(\boldsymbol{\theta})$ is the objective function, $\boldsymbol{\theta}$ are the decision variables, and (M3.21) is the equality constraint equation. Since the constraint equations are linear and are equality constraints, and the objective function is quadratic, there is an analytical solution to this problem. The solution is (Edgar and Himmelblau, 1988):

$$
\boldsymbol{\theta} = (\boldsymbol{\Phi}^{\mathrm{T}}\boldsymbol{\Phi})^{-1}\boldsymbol{\Phi}^{\mathrm{T}}\,\mathbf{Y}
\tag{M3.24}
$$

M3.4 GENERALIZATION OF THE LINEAR REGRESSION TECHNIQUE

It should be noted that the only requirement for parameter estimation using the least squares solution (M3.24) is that the model must be linear with respect to the parameters. There is no limitation to the functionalities with respect to the independent variables. As an example, consider:

$$
\hat{y} = a\,x_1 + b\,x_2^2 + c\ln x_3 + d\,e^{x_4}
$$

where the x's are the independent variables. Represent the kth data point as:

$$
\hat{y}(k) = a\,x_1(k) + b\,x_2(k)^2 + c\ln x_3(k) + d\,e^{x_4(k)}
$$

which can be written as:

$$\hat{y}(k) = \varphi(k)^{T}\theta$$

where:

$$\varphi(k)^{T} = [x_{1}(k) \quad x_{2}(k)^{2} \quad \ln x_{3}(k) \quad e^{x_{4}(k)}]$$
$$\theta^{T} \quad = [a \quad b \quad c \quad d]$$

Now, for the system of N data points we can write:

$$\hat{\mathbf{Y}} = \mathbf{\Phi}\,\theta$$

where:

$$\hat{\mathbf{Y}} = \begin{bmatrix} \hat{y}(1) \\ \cdot \\ \cdot \\ y(N) \end{bmatrix} \qquad \mathbf{\Phi} = \begin{bmatrix} \varphi(1)^{T} \\ \cdot \\ \cdot \\ \varphi(N)^{T} \end{bmatrix} \qquad \theta = \begin{bmatrix} a \\ b \\ c \\ d \end{bmatrix}$$

The solution to this problem is (M3.24) and the generalization to any system that is linear in the parameters is clear. A common formulation is the least squares fit of a polynomial.

Consider a general nth-order polynomial equation, where p_i is a parameter (coefficient) to be found as a best fit to data:

$$\hat{y} = p_{1}x^{n} + p_{2}x^{n-1} + \ldots + p_{n}x + p_{n+1}$$

The parameter vector is:

$$\theta = \begin{bmatrix} p_{1} \\ p_{2} \\ \cdot \\ \cdot \\ \cdot \\ p_{n} \\ p_{n+1} \end{bmatrix}$$

The matrix of the independent variable functions is:

$$\mathbf{\Phi} = \begin{bmatrix} x(1)^{n} & x(1)^{n-1} & \cdot & \cdot & x(1) & 1 \\ x(2)^{n} & x(2)^{n-1} & \cdot & \cdot & x(2) & 1 \\ \cdot & \cdot & \cdot & \cdot & \cdot & \cdot \\ \cdot & \cdot & \cdot & \cdot & \cdot & \cdot \\ x(N)^{n} & x(N)^{n-1} & \cdot & \cdot & x(N) & 1 \end{bmatrix}$$

The measured variable vector is:

$$\mathbf{Y} = \begin{bmatrix} y(1) \\ y(2) \\ \cdot \\ \cdot \\ y(N) \end{bmatrix}$$

where $x(i)$ and $y(i)$ represent the independent and dependent variables at the ith data point. The solution to this problem is equation (M3.24).

M3.5   **MATLAB ROUTINES** `polyfit` **AND** `polyval`

The MATLAB routine `polyfit` is used to fit data to an nth order polynomial, and the routine `polyval` is used to evaluate an nth order polynomial. Let x = independent variable vector, y = dependent variable vector, and n = order of polynomial. The best-fit polynomial coefficients are found from:

$$p = \texttt{polyfit(x,y,n)}$$

where the elements of the p vector are ordered from the highest power on down. Given a polynomial p and an independent vector x1, the resulting dependent vector y1 can be found from:

$$\texttt{y1 = polyval(p,x1)}$$

We show the use of `polyfit` and `polyval` by way of Example M3.1.

EXAMPLE M3.1   Batch Reactor Example

Consider a batch reactor with a single first-order reaction, $A \rightarrow B$. The model is:

$$\frac{dC_A}{dt} = -kC_A$$

where C_A = concentration of A, k = rate constant, and t = time. Separating variables and integrating:

$$\frac{dC_A}{C_A} = -k\,dt$$

$$\ln C_A = \ln C_{A0} - k\,t$$

where C_{A0} is the initial concentration of A. Notice that this is the equation for a line. If we let

$$\begin{aligned} y &= \ln C_A \\ p(1) &= -k \\ p(2) &= \ln C_{A0} \end{aligned}$$

and we have the form:

$$y = p(1)t + p(2)$$

Now we use `polyfit` to find the best linear fit of the data. The batch reactor data are shown in Table M3.1 and Figure M3.2.

TABLE M3.1   Concentration as a Function of Time

time, min	0	1	2	3	4	5
C_A, kgmol/m^3	8.47	5.00	2.95	1.82	1.05	0.71

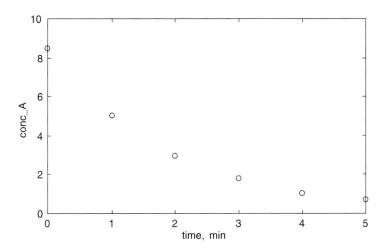

FIGURE M3.2   Batch reactor data. Concentration of A as a function of time.

```
» t = 0:1:5;
» t = t';
» CA = [8.47;5.00;2.95;1.82;1.05;0.71];
» y = log(CA);
» p = polyfit(t,y,1)

p =
  -0.5017    2.1098
```

The same values for the parameters can be obtained from equation (M3.24), by performing the following:

```
» phi = [t ones(6,1)];
» p = inv(phi'*phi)*phi'*y
  p =
  -0.5017
   2.1098
```

The parameters from the linear regression are converted back to the physical parameters:

$$k \quad = -p(1) \quad = 0.502 \text{ min}^{-1}$$
$$C_{A0} = \exp(p(2)) = 8.24 \text{ kgmol/m}^3$$

Now, we wish to compare the experimental data with the best fit line (model). The line is generated using the `polyval` function. The model and experiment are compared in Figure M3.3.

```
>> ymod = polyval(p,t);
>> plot(t,ymod,t,y,'o')
```

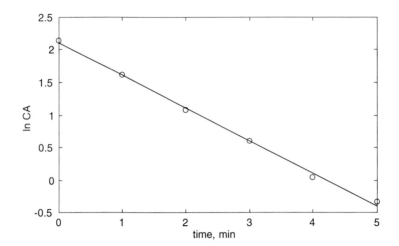

FIGURE M3.3 Semilog plot of concentration data and best fit line.

We also wish to compare the experimental data with the model on a time-concentration plot.

```
>> t1 = 0:0.25:5; % notice that more points are used for smoothness
>> CA_mod = 8.24*exp(-0.502*t1);
>> plot(t1,CA_mod,t,CA,'o')
```

The data and model are compared in Figure M3.4.

The rate constant has been evaluated at a single temperature. It can be evaluated at a number of temperature and linear regression can be used to evaluate the Arrhenius constants (frequency factor and activation energy) as shown in student exercises 3 and 4.

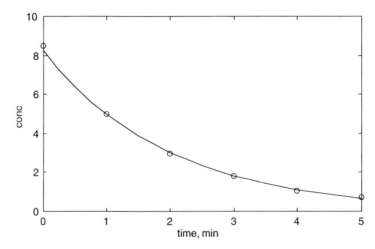

FIGURE M3.4 Concentration data and model as a function of time.

REFERENCES AND FURTHER READING

Good coverage of least squares analysis, and optimization in general, is provided by the following text:

Edgar, T.F., & D.M. Himmelblau. (1988). *Optimization of Chemical Processes.* New York: McGraw-Hill.

STUDENT EXERCISES

1. Note that a different solution for the best fit of a line is obtained if it is assumed that b is known; this becomes a single parameter estimation problem. Derive the following result for the estimate of a if it is assumed that b is known.

$$a = \frac{\left(\sum_{i=1}^{N} y_i x_i\right) - bN\bar{x}}{\left(\sum_{i=1}^{N} x_i^2\right)}$$

2. Use the matrix algebra approach to solve for the slope of a line, if the intercept b is known. You should obtain the same result as problem 1 above.

3. The Arrhenius rate expression is used to find reaction rate constants as a function of temperature:

$$k = A \exp(-E/RT)$$

Taking the natural log (ln) of each side of the Arrhenius rate expression, we find:

$$\ln k = \ln A - (E/R)(1/T)$$

where R is the ideal gas constant (1.987 cal/gmol K). Linear regression analysis can be used to find A and E.

Rate constants as a function of temperature for a first-order decomposition of benzene diazonium chloride are shown below:[1]

k, min^{-1}	0.026	0.062	0.108	0.213	0.43
T, K	313	319	323	328	333

Find A and E using least squares analysis (show units). Show that `polyfit` and the matrix algebra approach (M3.24) yield the same results. Compare model and experiment by (*i*) plotting ln k versus $1/T$ and (*ii*) k versus T.

---

[1]This data is from Example 3.1 in Fogler, H.S. (1992). *Elements of Chemical Reaction Engineering*, 2nd ed. Englewood Cliffs, NJ: Prentice Hall.

4. The Arrhenius rate expression is used to find rate constants as a function of temperature:

$$k = A \exp(-E/RT)$$

Taking the natural log (ln) of each side, we find:

$$\ln k = \ln A - (E/R)(1/T)$$

Linear regression analysis can be used to find A and E.

Rate constants as a function of temperature for a first-order reaction are shown below.

k, min^{-1}	0.0014	0.0026	0.0047	0.0083	0.014	0.023	0.038	0.059	0.090
T, K	300	310	320	330	340	350	360	370	380

Find A and E using least squares analysis (show units). Show that `polyfit` and the matrix algebra approach (M3.24) yield the same results. Compare model and experiment by (i) plotting $\ln k$ versus $1/T$ and (ii) k versus T.

5. The growth rate expression for a biochemical reaction, using a Monod model, is:

$$\mu = \frac{\mu_{max}\, x}{k_m + x}$$

where μ is the specific growth rate, μ_{max} and k_m are parameters, and x is the substrate concentration. The growth rate relationship can be rearranged to:

$$(1/\mu) = (1/\mu_{max}) + (k_m/\mu_{max})(1/x)$$

Data for a particular reactor are shown below. Use linear regression to solve for the parameters (μ_{max} and k_m).

μ, hr^{-1}	0.25	0.31	0.36	0.43	0.45	0.47	0.50	0.52
x, g/liter	0.1	0.15	0.25	0.50	0.75	1.00	1.50	3.00

Show that `polyfit` and the matrix algebra approach (M3.24) yield the same results.

6. The growth rate expression for a biochemical reaction, using a substrate inhibition model, is

$$\mu = \frac{\mu_{max}\, x}{k_m + x + k_1 x^2}$$

where μ is the specific growth rate, μ_{max} and k_m are parameters, and x is the substrate concentration. The growth rate relationship can be rearranged to

$$(1/\mu) = (k_1/\mu_{max})\, x + (k_m/\mu_{max})(1/x) + (1/\mu_{max})$$

Data for a particular reactor are shown below. Use linear regression (M3.24) to solve for the parameters (μ_{max}, k_m, and k_1).

μ, hr^{-1}	0.24	0.27	0.34	0.35	0.35	0.34	0.33	0.22
x, g/liter	0.1	0.15	0.25	0.50	0.75	1.00	1.50	3.00

Note that:

$$
\mathbf{Y} = \begin{bmatrix} 1/\mu(1) \\ 1/\mu(2) \\ . \\ . \\ 1/\mu(8) \end{bmatrix} \quad \mathbf{\Phi} = \begin{bmatrix} x(1) & 1/x(1) & 1 \\ x(2) & 1/x(2) & 1 \\ . & . & . \\ . & . & . \\ x(8) & 1/x(8) & 1 \end{bmatrix} \quad \theta = \begin{bmatrix} k_1/\mu_{\max} \\ k_m/\mu_{\max} \\ 1/\mu_{\max} \end{bmatrix}
$$

and the solution is $\theta = (\mathbf{\Phi}^{\mathbf{T}}\mathbf{\Phi})^{-1}\mathbf{\Phi}^{\mathbf{T}}\mathbf{Y}$

Compare the model and experiment by plotting μ versus x.

INTRODUCTION TO SIMULINK

The purpose of this module is to introduce SIMULINK, a block diagram environment for simulation. After studying this module, the reader should be able to:

- Construct a block diagram for simulation
- Set proper simulation parameters
- Print SIMULINK windows

The major sections of this module are:

M4.1   INTRODUCTION

There are limitations to using the standard MATLAB environment for simulation. Engineers (and particularly control engineers) tend to think in terms of block diagrams. Although MATLAB functions such as `parallel`, `series`, and `feedback` allow the simulation of block diagrams, these functions are not visually pleasing. SIMULINK provides a much more natural environment (a graphical user interface, GUI) for simulating systems that are described by block diagrams. SIMULINK also provides integration

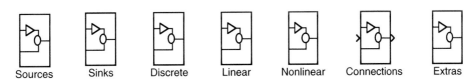

SIMULINK Block Library (Version 1.3c)

FIGURE M4.1   SIMULINK block library.

methods that can handle systems with time-delays (rather than making a Padé approximation for deadtime).

We will illustrate the use of SIMULINK by way of example.

In the MATLAB command window, enter "simulink" (small letters):

```
» simulink
```

The SIMULINK block library window appears, as shown in Figure M4.1.

There are many possible SIMULINK functions (icons) available. You will find it very useful to use the most commonly used blocks, which can be found by double-clicking on the *Extras* icon. The resulting window is shown in Figure M4.2.

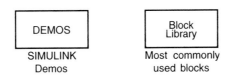

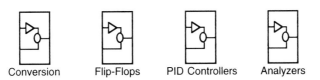

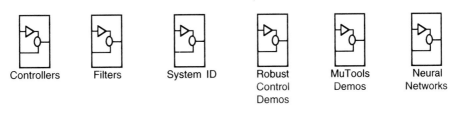

FIGURE M4.2   SIMULINK *Extras* block.

Most commonly used blocks:

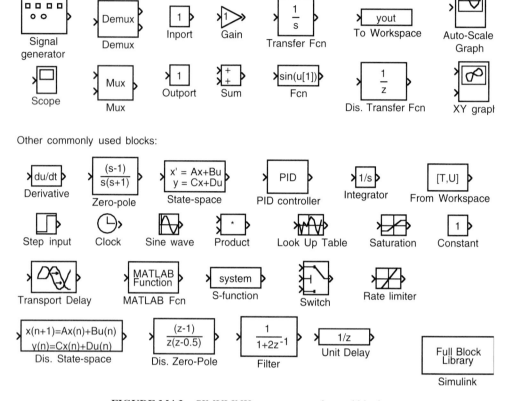

Other commonly used blocks:

FIGURE M4.3   SIMULINK most commonly used blocks.

Double-clicking on the *Block Library* icon then reveals a window for the most commonly used blocks, shown in Figure M4.3. Alternatively, you can simply enter *blocklib* in the MATLAB command window.

M4.2   TRANSFER FUNCTION-BASED SIMULATION

Here we provide an example simulation of a first-order process. The first step is to drag icons from the most commonly used block library, as shown in Figure M4.4. The clock icon is used to create a time vector. The *To Workspace* icons allow the vectors to be written to the MATLAB workspace for later manipulation or plotting.

Notice that each of the variable names, as well as the icon names, can be changed by the user, as shown in Figure M4.5. The variable names are changed by double-clicking on the icon and changing the name in the variable space.

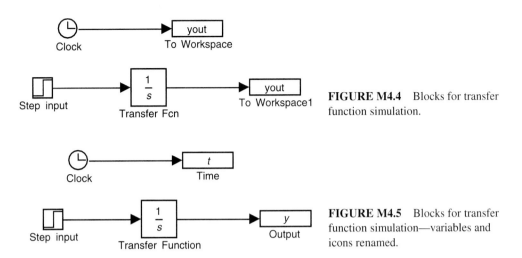

FIGURE M4.4   Blocks for transfer function simulation.

FIGURE M4.5   Blocks for transfer function simulation—variables and icons renamed.

Notice that the default transfer function has a pole at zero, that is, a simple integrator. This is changed by double-clicking on the icon and entering numerator and denominator coefficients. An example of a system with a gain of 2 and a time constant of 5 is shown in Figure M4.6.

The default step input is to step the input from 0 to 1 at $t = 1$. This can easily be changed by double-clicking on the *Step input* icon and changing the specifications.

The simulation parameters (start time, stop time, integration method, etc.) are changed by using the simulation pulldown menu and selecting *Parameters*. It is particularly important to change the stop time, which is 999999 by default (this is because SIMULINK was initially developed for use by electrical engineers who often use a continuous oscilloscope function). Also, you will usually want to change the maximum integration step size; if the step size is too large, the simulation may still be accurate (particularly for linear systems), but the resulting plots may be jagged. The default integration method is RK-45 (Runge-Kutta), which you may wish to change to Linsim to improve the computational speed. You may wish to try several different integration methods, depending on the type of system you are attempting to simulate.

To begin the simulation, simply select *Start* from the simulation pulldown menu. The t and y variables are stored in the MATLAB workspace. The resulting response is

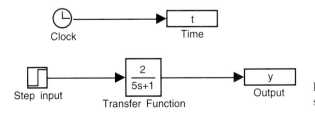

FIGURE M4.6   Transfer function simulation—first-order process.

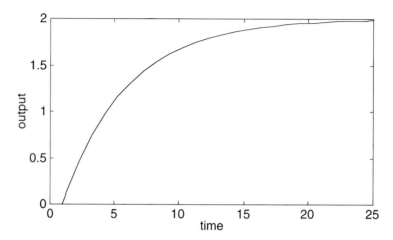

FIGURE M4.7   Simulation result for first-order example with a step input at $t = 1$.

shown in Figure M4.7. Notice that the input was stepped from 0 to 1 at $t = 1$, that is, the default values were used.

A major advantage of SIMULINK over the standard MATLAB integration routines is the ability to handle systems with time delays. This is done using the *Transport Delay* block shown in Figure M4.8. Here we use a time-delay of 5, obtained by modifying the default value of 1. It should also be noted that the user must supply an initial input value for the transport delay block. In this case the initial value is 0, because we are dealing with a system in deviation variable form (that is, all of the initial conditions are 0).

The resulting simulation is shown in Figure M4.9. Notice that no output change occurs before $t = 6$. This is due to the step input changing at $t = 1$, combined with the 5 unit time delay ($1 + 5 = 6$).

Although the block diagram feature of SIMULINK is very handy, it would be time-consuming (and tiring) if every time you wanted to change a set of parameters you had to double-click on the related icons. You can place variable names, such as k, tau, theta, for a first-order + time delay system directly in the blocks, then simply change the values (of k,

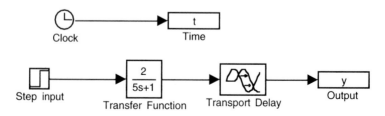

FIGURE M4.8   Transfer function simulation—first-order + time-delay process.

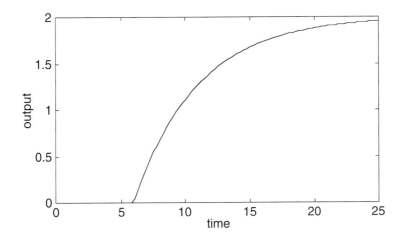

FIGURE M4.9   Simulation result. Step input at $t = 1$ for first-order system with delay of 5.

tau, theta) in the MATLAB command window before performing a new simulation. This block diagram is shown in Figure M4.10.

M4.3   PRINTING SIMULINK WINDOWS

When using a PC or Mac it is easy to print a window directly to the default printer simply by selecting *print* from the *file* pulldown menu. If you are using a Unix-based version of MATLAB/SIMULINK, or you wish to create a postscript file for a window, you need to enter a command directly in the MATLAB command window.

Let's say that you titled the example simulation window "tf_examp" and you want to print the window from a Unix-based workstation. The command that you would use is:

```
print -stf_examp
```

which will send the figure to the default printer.

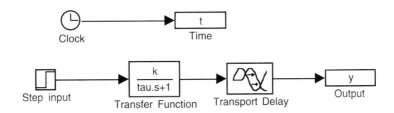

FIGURE M4.10   Transfer function simulation. The values of k, tau, and theta can be entered in the MATLAB command window before beginning the simulation.

SUMMARY

The purpose of this module was to provide a brief introduction to the use of SIMULINK for block diagram programming. Other inputs (sources) such as sine waves or ramps can be used. Also, other "sinks" such as an "XY graph" can be used; we generally recommend that variables be written to the MATLAB workspace, allowing the user to use standard MATLAB plotting functions. There are many options for "systems" to simulate, including "state space" models and *s*-functions which describe nonlinear systems. The user is encouraged to experiment with state space models, and to read the SIMULINK User's Guide for more advanced usage.

EXERCISES

Use SIMULINK to perform the following. Show the SIMULINK diagram used for your simulations.

1. Plot the step response of the following underdamped system

$$g(s) = \frac{3}{25s^2 + 5s + 1}$$

   when a step input of magnitude 2 is made at $t = 5$.

2. Plot the step response of the following two transfer functions by performing a single simulation

$$g_1(s) = \frac{3(-3s + 1)}{25s^2 + 5s + 1}$$

$$g_2(s) = \frac{3(-3s + 1)e^{-5s}}{25s^2 + 5s + 1}$$

   Use a step input change of 2 at $t = 1$.

3. Compare the following first-order + time-delay system

$$g_1(s) = \frac{3e^{-5s}}{10s + 1}$$

   with it's first-order Padé approximation

$$g_2(s) = \frac{3(-22.5s + 1)}{(2.5s + 1)(10s + 1)}$$

   for a step input change of 1 at $t = 0$.

STIRRED TANK HEATERS

MODULE

5

After studying this module, the reader should be able to:

- Develop the nonlinear dynamic model of a perfectly mixed stirred tank heater
- Find the state-space and transfer function form of a linearized stirred tank heater
- Compare the dynamic responses of the nonlinear and linear models
- Use MATLAB for linear and nonlinear simulations
- Analyze phase-plane behavior

The major sections in this module are:

M5.1   Introduction
M5.2   Developing the Dynamic Model
M5.3   Steady-State Conditions
M5.4   State-Space Model
M5.5   Laplace Domain Model
M5.6   Step Responses: Linear versus Nonlinear Models
M5.7   Unforced System Responses: Perturbations in Initial Conditions

M5.1   INTRODUCTION

Mixing vessels are used in many chemical processes. Often these mixing vessels are heated, either by a coil or a jacket surrounding the vessel. For example, a mixing vessel may serve as a chemical reactor, where two or more components are reacted to produce

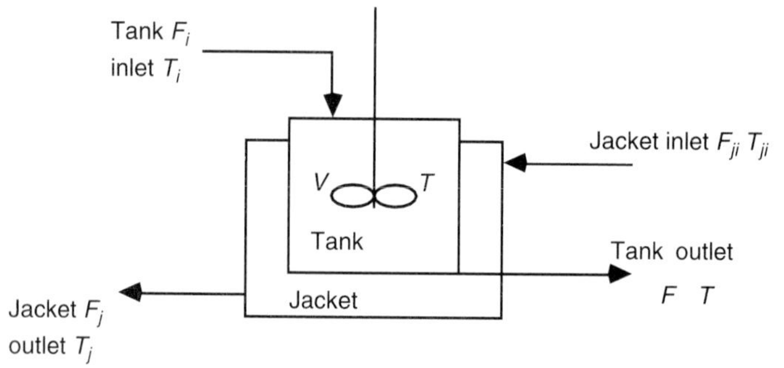

FIGURE M5.1   Jacketed stirred tank heater.

one or more products. Often this reaction must occur at a certain temperature to achieve a desired yield. The temperature in the vessel is maintained by varying the flowrate of a fluid through the jacket or coil.

Consider a stirred tank heater as shown in Figure M5.1, where the tank inlet stream is received from another process unit. The objective is to raise the temperature of the inlet stream to a desired value. A heat transfer fluid is circulated through a jacket to heat the fluid in the tank. In some processes steam is used as the heat transfer fluid and most of the energy transported is due to the phase change of steam to water. In other processes a heat transfer fluid is used where there is no phase change. In this module we assume that no change of phase occurs in either the tank fluid or the jacket fluid.

M5.2   DEVELOPING THE DYNAMIC MODEL

In this section we write the dynamic modeling equations to find the tank and jacket temperatures. This methodology was developed in Chapter 2. We make the following assumptions:

- The volume and liquids are constant with constant density and heat capacity.
- Perfect mixing is assumed in both the tank and jacket.
- The tank inlet flowrate, jacket flowrate, tank inlet temperature, and jacket inlet temperature may change (these are the inputs).
- The rate of heat transfer from the jacket to the tank is governed by the equation $Q = UA(T_j - T)$, where U is the overall heat transfer coefficient and A is the area for heat transfer.

The notation that we use is presented in Table M5.1.

TABLE M5.1   Notation

Variables	Subscripts
A	area for heat transfer
c_p	heat capacity (energy/mass*temp)
F	volumetric flowrate (volume/time)
ρ	density (mass/volume)
T	temperature
t	time
Q	rate of heat transfer (energy/time)
U	heat transfer coefficient (energy/time* area*temp)
V	volume
i	inlet
j	jacket
ji	jacket inlet
ref	reference state
s	steady-state

M5.2.1   Material Balance Around Tank

The first step is to write a material balance around the tank, assuming constant density:

$$\frac{dV\rho}{dt} = F_i\rho - F\rho$$

Also, assuming constant volume ($dV/dt = 0$), we find:

$$F = F_i$$

M5.2.2   Energy Balance Around Tank

The next step is to write an energy balance around the tank.

accumulation = in by flow − out by flow + in by heat transfer + work done on system

$$\frac{dTE}{dt} = F\rho\overline{TE}_i - F\rho\overline{TE} + Q + W_T$$

The next step is to neglect the kinetic and potential energy and write the total work done on the system as a combination of the shaft work and the energy added to the system to get the fluid into the tank and the energy that the system performs on the surroundings to force the fluid out. This allows us to write the energy balance as (see Chapter 2 for the details):

$$\frac{dU}{dt} = F\rho\left(\overline{U}_i + \frac{p_i}{\rho_i}\right) - F\rho\left(\overline{U} + \frac{p}{\rho}\right) + Q + W_s$$

and since $H = U + pV$, we can rewrite the energy balance as:

$$\frac{dH}{dt} - \frac{dpV}{dt} = F\rho\,\overline{H}_i - F\rho\,\overline{H} + Q + W_s$$

Note that $dpV/dt = V\,dp/dt + p\,dV/dt$, and the *volume is constant*. Also, the mean pressure change can be neglected since the density is constant:

$$\frac{dH}{dt} = F\rho\,\overline{H}_i - F\rho\,\overline{H} + Q + W_s$$

Neglecting the work done by the mixing impeller, and assuming single phase and a constant heat capacity, we find:

$$V\rho c_p \frac{dT}{dt} = F\rho c_p(T_i - T) + Q$$

or

$$\frac{dT}{dt} = \frac{F}{V}(T_i - T) + \frac{Q}{V\rho c_p} \tag{M5.1}$$

We must also perform a material and energy balance around the jacket and use the connecting relationship for heat transfer between the jacket and the tank.

M5.2.3 Material Balance Around the Jacket

The material balance around the jacket is (assuming constant density):

$$\frac{dV_j\rho_j}{dt} = F_{ji}\rho_j - F_j\rho_j$$

assuming constant volume ($dV/dt = 0$), we find:

$$F_j = F_{ji}$$

M5.2.4 Energy Balance Around the Jacket

Next, we write an energy balance around the jacket. Making the same assumptions as around the tank:

$$\frac{dT_j}{dt} = \frac{F_j}{V_j}(T_{ji} - T_j) + \frac{Q}{V_j\rho_j c_{pj}} \tag{M5.2}$$

We also have the relationship for heat transfer from the jacket to the tank:

$$Q = UA \, (T_j - T) \tag{M5.3}$$

Substituting (M5.3) into (M5.1) and (M5.2) yields the two modeling equations for this system:

$$\frac{dT}{dt} = \frac{F}{V}(T_i - T) + \frac{UA \, (T_j - T)}{V\rho c_p} \tag{M5.4}$$

$$\frac{dT_j}{dt} = \frac{F_j}{V_j}(T_{ji} - T_j) - \frac{UA \, (T_j - T)}{V_j \rho_j c_{pj}} \tag{M5.5}$$

M5.3 STEADY-STATE CONDITIONS

Before linearizing the nonlinear model to find the state-space form, we must find the state variable values at steady-state. The steady-state is obtained by solving the dynamic equations for $dx/dt = 0$. The steady-state values of the system variables and some parameters for this process are given below.

M5.3.1   Parameters and Steady-State Values

$F_s = 1 \dfrac{\text{ft}^3}{\text{min}}$	$\rho C_p = 61.3 \dfrac{\text{Btu}}{°\text{F ft}^3}$	$\rho_j C_{pj} = 61.3 \dfrac{\text{Btu}}{°\text{F ft}^3}$
$T_{is} = 50°\text{F}$	$T_s = 125°\text{F}$	$V = 10 \text{ ft}^3$
$T_{jis} = 200°\text{F}$	$T_{js} = 150°\text{F}$	$V_j = 1 \text{ ft}^3$

Notice that the values of UA and F_{js} have not been specified. These values can be obtained by solving the two dynamic equations ((M5.4) and (M5.5)) at steady-state.

From $dT/dt = 0$ at steady-state, solve (M5.4) to obtain $UA = 183.9$ Btu/°F min
From $dT_j/dt = 0$ at steady-state, solve (M5.5) to obtain $F_{js} = 1.5$ ft³/min

M5.4 STATE-SPACE MODEL

Here we linearize the nonlinear modeling equations to find the state-space form:

$$\dot{\mathbf{x}} = \mathbf{A}\,\mathbf{x} + \mathbf{B}\,\mathbf{u}$$
$$\mathbf{y} = \mathbf{C}\,\mathbf{x}$$

where the state, input, and output vectors are in deviation from:

$$\mathbf{x} = \begin{bmatrix} T - T_s \\ T_j - T_{js} \end{bmatrix} \quad = \text{state variables}$$

$$\mathbf{u} = \begin{bmatrix} F_j - F_{js} \\ F - F_s \\ T_i - T_{is} \\ T_{jin} - T_{jins} \end{bmatrix} \quad = \text{input variables}$$

$$\mathbf{y} = \begin{bmatrix} T - T_s \\ T_j - T_{js} \end{bmatrix} \quad = \text{output variables}$$

Recall that our two dynamic functional equations are:

$$\frac{dT}{dt} = f_1(T,T_j,F_j,F,T_i,T_{ji}) = \frac{F}{V}(T_i - T) + \frac{UA\,(T_j - T)}{V\rho c_p} \tag{M5.4}$$

$$\frac{dT_j}{dt} = f_2(T,T_j,F_j,F,T_i,T_{jin}) = \frac{F_j}{V_j}(T_{ji} - T_j) - \frac{UA\,(T_j - T)}{V_j\rho_j c_{pj}} \tag{M5.5}$$

The elements of the state-space A matrix are found by: $A_{ij} = \dfrac{\partial f_i}{\partial x_j}$

$$A_{11} = \frac{\partial f_1}{\partial x_1} = \frac{\partial f_1}{\partial (T - T_s)} = \frac{\partial f_1}{\partial T} = -\frac{F_s}{V} - \frac{UA}{V\rho c_p} \quad = -0.4$$

$$A_{12} = \frac{\partial f_1}{\partial x_2} = \frac{\partial f_1}{\partial (T_j - T_{js})} = \frac{\partial f_1}{\partial T_j} = \frac{UA}{V\rho c_p} \quad = 0.3$$

$$A_{21} = \frac{\partial f_2}{\partial x_1} = \frac{\partial f_2}{\partial (T - T_s)} = \frac{\partial f_2}{\partial T} = \frac{UA}{V_j\rho_j c_{pj}} \quad = 3.0$$

$$A_{22} = \frac{\partial f_2}{\partial x_2} = \frac{\partial f_2}{\partial (T_j - T_{js})} = \frac{\partial f_2}{\partial T_j} = -\frac{F_{js}}{V_j} - \frac{UA}{V_j\rho_j c_{pj}} \quad = -4.5$$

The elements of the B matrix are:

$$B_{11} = \frac{\partial f_1}{\partial u_1} = \frac{\partial f_1}{\partial (F_j - F_{js})} = \frac{\partial f_1}{\partial F_j} = 0 \quad\quad = 0$$

$$B_{12} = \frac{\partial f_1}{\partial u_2} = \frac{\partial f_1}{\partial (F - F_s)} = \frac{\partial f_1}{\partial F} = \frac{T_{is} = T_s}{V} = -7.5$$

$$B_{13} = \frac{\partial f_1}{\partial u_3} = \frac{\partial f_1}{\partial (T_i - T_{is})} = \frac{\partial f_1}{\partial T_i} = \frac{F_s}{V} \quad\quad = 0.1$$

$$B_{14} = \frac{\partial f_1}{\partial u_4} = \frac{\partial f_1}{\partial (T_{ji} - T_{jis})} = \frac{\partial f_1}{\partial T_{ji}} = 0 \quad\quad = 0$$

$$B_{21} = \frac{\partial f_2}{\partial u_1} = \frac{\partial f_2}{\partial (F_j - F_{js})} \quad = \frac{\partial f_2}{\partial F_j} = \frac{T_{jis} - T_{js}}{V_j} = 50$$

$$B_{22} = \frac{\partial f_2}{\partial u_2} = \frac{\partial f_2}{\partial (F - F_s)} \quad = \frac{\partial f_2}{\partial F} = 0 \qquad = 0$$

$$B_{23} = \frac{\partial f_2}{\partial u_3} = \frac{\partial f_2}{\partial (T_i - T_{is})} \quad = \frac{\partial f_2}{\partial T_i} = 0 \qquad = 0$$

$$B_{24} = \frac{\partial f_2}{\partial u_4} = \frac{\partial f_2}{\partial (T_{ji} - T_{jis})} \quad = \frac{\partial f_2}{\partial T_{ji}} = \frac{F_{js}}{V_j} \qquad = 1.5$$

Since both states are measured, the C matrix is I. The numerical values for the matrices are:

$$\mathbf{A} = \begin{bmatrix} -0.4 & 0.3 \\ 3 & -4.5 \end{bmatrix}$$

$$\mathbf{B} = \begin{bmatrix} 0 & -7.5 & 0.1 & 0 \\ 50 & 0 & 0 & 1.5 \end{bmatrix}$$

$$\mathbf{C} = \begin{bmatrix} 1 & 0 \\ 0 & 1 \end{bmatrix}$$

M5.5   LAPLACE DOMAIN MODEL

Recall that a transfer function matrix relates the inputs and outputs:

$$\mathbf{y}(s) = \mathbf{G}(s)\,\mathbf{u}(s)$$

The input/output transfer function matrix is found from:

$$\mathbf{G}(s) = \mathbf{C}\,(s\mathbf{I} - \mathbf{A})^{-1}\mathbf{B} \tag{M5.6}$$

Using the MATLAB routine ss2tf (see appendix), we find:

$$\mathbf{G}(s) = \frac{\begin{bmatrix} 15 & (-7.5s - 33.75) & (0.1s + 0.45) & 0.45 \\ (50s + 20) & -22.5 & 0.3 & (1.5s + 0.6) \end{bmatrix}}{s^2 + 4.9s + 0.9}$$

The poles of the characteristic polynomial $(s^2 + 4.9s + 0.9)$ are -0.191 and -4.709. The transfer function relating output 1 (tank temperature) to input 1 (jacket flowrate) is:

$$g_{11}(s) = \frac{15}{s^2 + 4.9s + 0.9} \tag{M5.7}$$

Dividing (M5.7) by 0.9, we find:

$$g_{11}(s) = \frac{16.6667}{1.1111s^2 + 5.4444s + 1} \tag{M5.8}$$

(M5.8) can be factored into:

$$g_{11}(s) = \frac{16.6667}{(0.21236s + 1)(5.23207s + 1)} \qquad (M5.9)$$

which is the gain-time constant form. Any of these forms can be used, however we will generally use the gain-time constant form.

The transfer function relating output 2 (jacket temperature) to input 1 (jacket flowrate) is:

$$g_{21}(s) = \frac{50s + 20}{s^2 + 4.9s + 0.9} \qquad (M5.10)$$

Similarly, we find (dividing (M5.10) by 0.9):

$$g_{21}(s) = \frac{22.2 \, (2.5s + 1)}{1.1111s^2 + 5.4444s + 1} \qquad (M5.11)$$

Also, (M5.11) can be factored into:

$$g_{21}(s) = \frac{22.2 \, (2.5s + 1)}{(0.21236s + 1)(5.2307s + 1)} \qquad (M5.12)$$

M5.6   STEP RESPONSES: LINEAR VERSUS NONLINEAR MODELS

In this section we will compare the step responses of the nonlinear model with those predicted by the linearized model. For comparison purposes, we will plot the physical variables.

For the linear model, which is in deviation variable form, we convert to physical values by realizing that:

$$x_1(t) = T(t) - T_s$$
$$x_2(t) = T_j(t) - T_{js}$$

So, for the linear model:

$$T(t) = T_s + x_1(t)$$

$$T_j(t) = T_{js} + x_2(t)$$

Also,

$$u_1(t) = F_j(t) - F_{js}$$

We will consider step changes of different magnitudes and different directions for this system.

M5.6.1   Small Increase in Jacket Temperature

Consider a step change of 0.1 (from the steady-state value of 1.5 to 1.6 ft^3/min) in the jacket flowrate. The responses of the nonlinear and linear models are shown in Figure M5.2. For both the tank temperature and the jacket temperature, the linear model closely approximates the nonlinear process.

M5.6.2   Large Increase in Jacket Flowrate

Consider a step change of 1.0 (from 1.5 to 2.5 ft^3/min) in jacket flowrate. The responses of the linear and nonlinear models are shown in Figure M5.3. The linear response is quali-

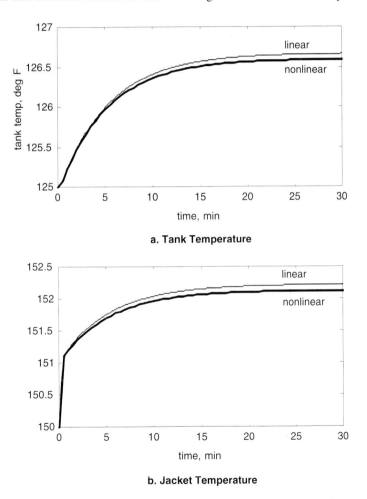

a. Tank Temperature

b. Jacket Temperature

FIGURE M5.2   Response to small increase in jacket flowrate (0.1 ft^3/min).

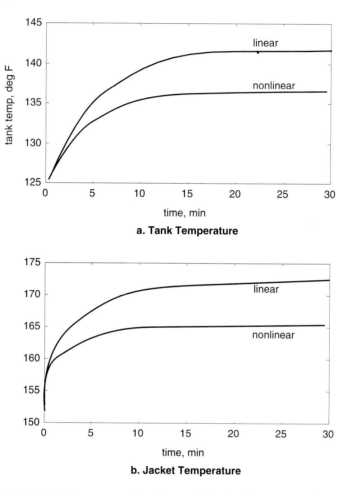

FIGURE M5.3   Reponse to a large increase in jacket flowrate (1.0 ft³/min).

tatively the same as the nonlinear response, but the magnitude of change is different. The gain (change in output/change in input) of the linear model is *greater* than the gain of the nonlinear model.

M5.6.3   Large Decrease in Jacket Flowrate

Consider a step decrease of −1.0 (from 1.5 to 0.5 ft³/min) in jacket flowrate. The responses of the linear and nonlinear models are shown in Figure M5.4. Notice that gain of the linear model is now *less* than the gain of the nonlinear model. This is the opposite effect from what we observed for a large increase in the jacket flowrate.

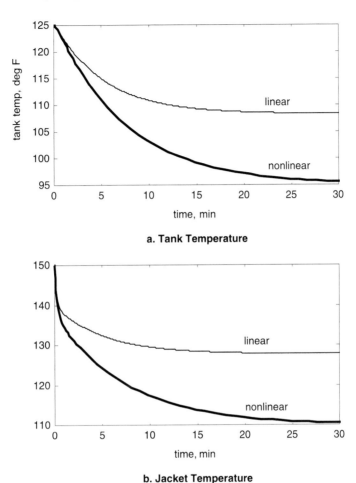

a. Tank Temperature

b. Jacket Temperature

FIGURE M5.4 Response to a large decrease in jacket flowrate (-1.0 ft^3/min)

M5.6.4 Small Decrease in Jacket Flowrate

Consider a step change of -0.1 (from 1.5 to 1.4 ft^3/min) in jacket flowrate. The responses of the linear and nonlinear models are shown in Figure M5.5. Since the the change in input was small, the linear model provides a good approximation to the nonlinear model.

M5.6.5 Remarks on Step Response Behavior

For the nonlinear model, the gain (change in output/change in input) varied as a function of both the input magnitude and direction. If small input changes were made, the gain did

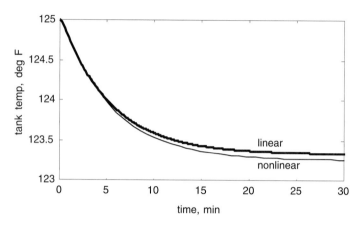

a. Tank Temperature

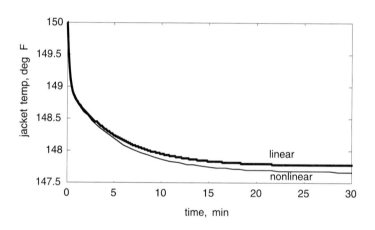

b. Jacket Temperature

FIGURE M5.5   Response to a small decrease in jacket flowrate (-0.1 ft^3/min).

not change much from the linear model case. For large input changes, the gain of the non-linear system was less than the linear system for increases in jacket flowrate, but more than the linear system for decreases in jacket flowrate.

The response of the jacket temperature is faster than that of the tank temperature. This makes sense from a physical point of view, because the jacket volume is one-tenth of the heater volume. Also, the jacket flowrate has a direct effect of jacket temperature and an indirect effect on tank temperature. Notice that the numerator time constant partially "cancels" the slow denominator time constant for the transfer function relating jacket flowrate to jacket temperature.

M5.7   UNFORCED SYSTEM RESPONSES: PERTURBATIONS IN INITIAL CONDITIONS

The eigenvalues of the A matrix in the state-space model provide information about stability and the relative speed of response. The eigenvectors provide information about the directional dependence of the speed of response. The MATLAB `eig` routine is used for eigenvalue/eigenvector calcuations. Recall that a positive eigenvalue is unstable, while a negative eigenvalue is stable. A large magnitude eigenvalue is "faster" than a small magnitude eigenvalue.

The eigenvalues for this system are (see appendix):

$$\lambda_1 = -0.1911 \quad \text{slow}$$
$$\lambda_2 = -4.7089 \quad \text{fast}$$

The eigenvectors are:

$$v_1 = \begin{bmatrix} 0.8207 \\ 0.5714 \end{bmatrix} \quad \text{slow}$$
$$v_2 = \begin{bmatrix} -0.0695 \\ 0.9976 \end{bmatrix} \quad \text{fast}$$

The second eigenvalue and eigenvector tell us that a perturbation in the initial condition of the second state variable will have a fast response. The first eigenvalue and eigenvector tell us that a perturbation in the direction of v_1 will have a slow response.

M5.7.1   Slow Response

Consider an initial perturbation in the slow direction. Let the perturbation be of magnitude 5:

$$x(0) = 5*v_2 = 5*\begin{bmatrix} 0.8207 \\ 0.5714 \end{bmatrix} = \begin{bmatrix} 4.1035 \\ 2.857 \end{bmatrix}$$

The physical state variables (tank and jacket temperature) are:

$$\begin{bmatrix} T(0) \\ T_j(0) \end{bmatrix} = \begin{bmatrix} 125 \\ 150 \end{bmatrix} + \begin{bmatrix} 4.1035 \\ 2.857 \end{bmatrix} = \begin{bmatrix} 129.1035 \\ 152.857 \end{bmatrix}$$

The responses for a perturbation in the slow direction are shown in Figure M5.6.

M5.7.2   Fast Response

Consider an initial perturbation in the fast direction. Let the pertubation be:

$$x(0) = 5*v_2 = 5*\begin{bmatrix} -0.0695 \\ 0.9976 \end{bmatrix} = \begin{bmatrix} -0.3475 \\ 4.988 \end{bmatrix}$$

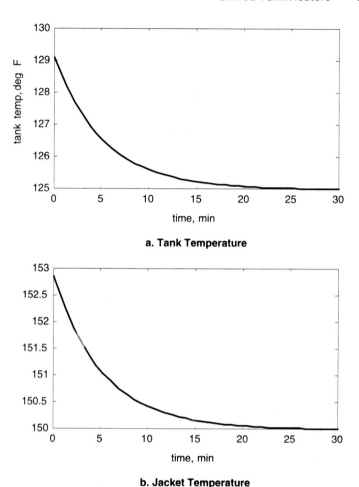

a. Tank Temperature

b. Jacket Temperature

FIGURE M5.6   Initial perturbation in the slow direction.

The physical state variables (tank and jacket temperature) are:

$$\begin{bmatrix} T(0) \\ T_j(0) \end{bmatrix} = \begin{bmatrix} 125 \\ 150 \end{bmatrix} + \begin{bmatrix} -0.3475 \\ 4.988 \end{bmatrix} = \begin{bmatrix} 124.6525 \\ 154.988 \end{bmatrix}$$

The responses for a perturbation in the fast direction are shown in Figure M5.7 (note the time scale change from Figure M5.6).

In Figures M5.6 and M5.7 we have shown the effect of initial condition "direction." This can be illustrated more completely by viewing the phase-plane behavior.

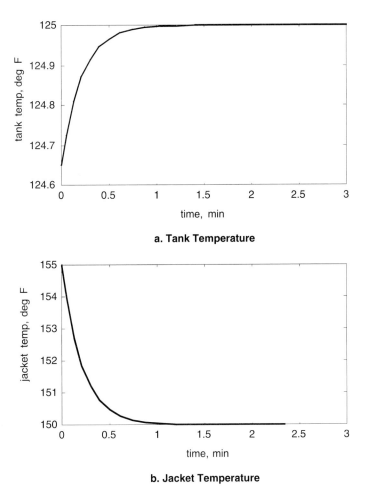

a. Tank Temperature

b. Jacket Temperature

FIGURE M5.7   Initial perturbation in the fast direction.

M5.7.3   Phase-Plane Behavior

The phase-plane behavior is shown in Figure M5.8, where the Tank Temperature is plotted on the x-axis and the Jacket Temperature is plotted on the y-axis. Notice that initial conditions in the $\begin{bmatrix} 4.1035 \\ 2.857 \end{bmatrix}$ direction respond much more slowly than initial conditions in the $\begin{bmatrix} -0.3475 \\ 4.988 \end{bmatrix}$ direction.

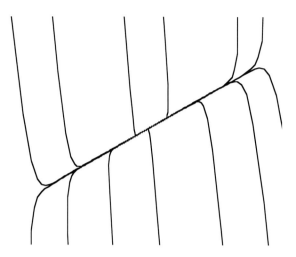

FIGURE M5.8   Phase-plane behavior of the stirred tank heater

SUMMARY

In this module we have developed the dynamic modeling equations for a perfectly mixed stirred tank heater. We solved for the steady-state conditions, linearized to obtain the state-space model, and found the transfer function model. We compared the step responses of the linear and nonlinear models. We also showed the importance of the "direction" of initial conditions. Perturbations in certain directions cause a fast response, while perturbations in other directions yield a slow response.

FURTHER READING

The following text is a good reference for chemical process modeling, with a particularly nice presentation of energy balances:

Denn, M.M. (1986). *Process Modeling,* New York: Longman.

STUDENT EXERCISES

1. In the heater example we performed step tests on the jacket flowrate. Perform step tests on the other inputs: (i) Tank flowrate, (ii) tank inlet temperature, (iii) jacket inlet temperature. Find the transfer functions relating the inputs and outputs. Discuss the linear versus nonlinear effects.

2. Show that a decrease in jacket volume will cause the difference in magnitude between the fast and slow poles to increase.

3. Find the nonlinear steady-state relationships between jacket flowrate and the two temperatures (tank and jacket). Plot the input (jacket flowrate) versus the outputs for the physical parameters given. Discuss how the gain changes with coolant flowrate.

4. Develop a simple model with imperfect mixing on the jacket side. Assume that the jacket can be modeled as two compartments, where the flow is from one compartment to another, as shown in the diagram below.

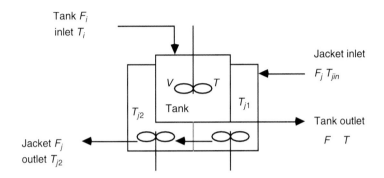

APPENDIX

1. Using MATLAB to convert from state-space to transfer function models—ss2tf

STATE SPACE MODEL

```
a  =
    -0.4000      0.3000
     3.0000     -4.5000

b  =
          0     -7.5000      0.1000           0
    50.0000           0           0      1.5000

c  =
     1         0
     0         1

d  =
     0         0         0         0
     0         0         0         0
```

Find transfer function polynomials for input 1:

```
»[num1,den] = ss2tf(a,b,c,d,1)

num1 =
          0     0.0000    15.0000
          0    50.0000    20.0000

den =
     1.0000     4.9000     0.9000
```

Find transfer function polynomials for input 2:

```
»[num2,den] = ss2tf(a,b,c,d,2)

num2 =
          0    -7.5000   -33.7500
          0    -0.0000   -22.5000

den =
     1.0000     4.9000     0.9000
```

Find transfer function polynomials for input 3:

```
»[num3,den] = ss2tf(a,b,c,d,3)

num3 =
          0     0.1000     0.4500
          0     0.0000     0.3000

den =
     1.0000     4.9000     0.9000
```

Find transfer function polynomials for input 4:

```
»[num4,den] = ss2tf(a,b,c,d,4)

num4 =
          0     0.0000     0.4500
          0     1.5000     0.6000

den =
     1.0000     4.9000     0.9000
```

Eigenvalues and eigenvectors

```
»[v,lambda] = eig(a)

v =
     0.8207    -0.0695
     0.5714     0.9976

lambda =
    -0.1911          0
          0    -4.7089
```

2. MATLAB function routine for nonlinear heater model

```
    function xdot = heater(t,x);
%
%   Dynamics of a stirred tank heater
%   (c) 1994 - B.W. Bequette
%   8 July 94
%
%   x(1)      = T  = temperature in tank
%   x(2)      = Tj = temperature in jacket
%   delFj     =      change in jacket flowrate
%   F         =      Tank flowrate
%   Tin       =      Tank inlet temperature
%   Tji       =      Jacket inlet temperature
%   V         =      Tank volume
%   Vj        =      Jacket volume
%   rhocp     =      density*heat capacity
%   rhocpj    =      density*heat capacity,jacket fluid
%
%   parameter and steady-state variable values are:
%
    F     =    1;
    Fjs   =    1.5;
    Ti    =   50;
    Tji   =  200;
    V     =   10;
    Vj    =    1;
    rhocp = 61.3;
    rhocpj= 61.3;
    UA    = 183.9;
%
    delFj =     -0.1;
    Fj = Fjs + delFj;
    T = x(1);
    Tj= x(2);
%
% odes
%
    dTdt = (F/V)*(Ti - T) + UA*(Tj - T)/(V*rhocp);
    dTjdt = (Fj/Vj)*(Tji - Tj) - UA*(Tj - T)/(Vj*rhocpj);
    xdot(1) = dTdt;
    xdot(2) = dTjdt;
```

Figure M5.7a is obtained using the following commands:

```
x0 =
  124.6525
  154.9880
»[t,x] = ode45('heater',0,5,x0);
»plot(t,x(:,1))
```

ABSORPTION

An Example of a Linear Equilibrium Stage System

MODULE

6

After studying this module, the reader should be able to:

- Understand step response behavior of absorption columns
- Use the MATLAB functions `initial` and `step` for state space simulations
- Use eigenvalue/eigenvector analysis to understand unforced system behavior
- Understand how zeros affect the step response behavior of transfer functions

The major sections of this module are:

M6.1 BACKGROUND

Separations processes play an important role in most chemical manufacturing processes. Streams from chemical reactors often contain a number of components; some of these components must be separated from the other components for sale as a final product, or for use in another manufacturing process. A common example of a separation process is gas absorption, which is normally used to remove a dilute component from a gas stream.

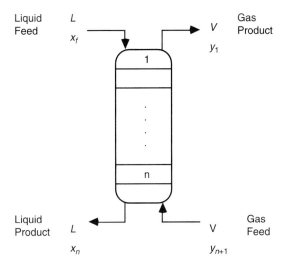

FIGURE M6.1   Gas absorption column, n stages.

Consider the gas absorption column shown in Figure M6.1. Components that enter the bottom of the column in the gas feed stream are absorbed by the liquid stream, so that the gas product stream (leaving the top of the column) is more "pure." An example is the use of a heavy oil stream (liquid) to remove benzene from a benzene/air gas feed stream. Absorption columns often contain "trays" with a liquid layer flowing across the tray; these trays are often modeled as equilibrium stages.

An objective of this module is to develop a dynamic (time dependent) model of a gas absorption column. We will show how to solve for the steady-state stage compositions using matrix algebra. We will then place the dynamic model in the linear state-space form. Dynamic results for stage compositions will be shown for step changes in the inlet feed compositions.

M6.2   THE DYNAMIC MODEL

M6.2.1   Basic Assumptions

We assume that the major component of the liquid stream is "inert" and does not absorb into the gas stream. We also assume that the major component of the gas stream is "inert" and does not absorb into the liquid stream. It is assumed that each stage of the process is an equilibrium stage, that is, the vapor leaving a stage is in thermodynamic equilibrium with the liquid on that stage.

M6.2.2   Definition of Variables

We use the following variable definitions:

$$L = \frac{\text{moles inert liquid}}{\text{time}} \qquad = \text{liquid molar flowrate}$$

$$V = \frac{\text{moles inert vapor}}{\text{time}} \qquad = \text{vapor molar flowrate}$$

$$M = \frac{\text{moles liquid}}{\text{stage}} \qquad = \text{liquid molar holdup per stage}$$

$$W = \frac{\text{moles vapor}}{\text{stage}} \qquad = \text{vapor molar holdup per stage}$$

$$x_i = \frac{\text{moles solute (stage } i\text{)}}{\text{mole inert liquid (stage } i\text{)}}$$

$$y_i = \frac{\text{moles solute (stage } i\text{)}}{\text{mole inert vapor (stage } i\text{)}}$$

M6.2.3   Equilibrium Stage

The concept of an equilibrium stage is important for the development of a dynamic model of the absorption column. An equilibrium stage is represented schematically in Figure M6.2.

M6.2.4   Solute Balance Around Stage i

The total amount of solute on stage i is the sum of the solute in the liquid phase and the gas phase (that is, $Mx_i + Wy_i$). The rate of change of the amount of solute is then $d(Mx_i + Wy_i)/dt$. The component material balance around stage i can now be written (accumulation = in − out):

$$\frac{d(M_{x_i} + W_{y_i})}{dt} = L\,x_{i-1} + V\,y_{i+1} - L\,x_i - V\,y_i \qquad \text{(M6.1)}$$

Since liquid is much more dense than vapor, we can assume that the major contribution to the accumulation term is the Mx_i term. Equation (M6.1) can now be written:

$$\frac{d(Mx_i)}{dt} = L\,x_{i-1} + V\,y_{i+1} - L\,x_i - V\,y_i \qquad \text{(M6.2)}$$

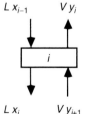

FIGURE M6.2   A typical gas absorption stage.

Our next assumption is that the liquid molar holdup (M) is constant, then we can write (M6.2) as:

$$\frac{dx_i}{dt} = \frac{L}{M} x_{i-1} + \frac{V}{M} y_{i+1} - \frac{L}{M} x_i - \frac{V}{M} y_i \tag{M6.3}$$

The solution to this problem will be easier if we can develop an explicit relationship between vapor phase composition and liquid phase composition. We will assume that the vapor on each stage is in equilibrium with the liquid on that stage.

M6.2.5 Equilibrium Relationships

The simplest relationship is a linear equilibrium relationship:

$$y_i = ax_i \tag{M6.4}$$

where y is the gas phase composition (mole solute/mole inert liquid), x is the liquid phase composition (mole solute/mole inert vapor), i represents the ith stage, and a is an equilibrium parameter.

M6.2.6 The Modeling Equation for Stage i

Substituting the vapor/liquid relationship from (M6.4) into the material balance (M6.3) yields:

$$\frac{dx_i}{dt} = \frac{L}{M} x_{i-1} + \frac{V}{M} ax_{i+1} - \frac{L}{M} x_i - \frac{V}{M} ax_i \tag{M6.5}$$

which can be written in the following form:

$$\frac{dx_i}{dt} = \frac{L}{M} x_{i+1} - \frac{(L + Va)}{M} x_i + \frac{Va}{M} x_{i+1} \tag{M6.6}$$

It should be noted that (M6.6) will yield a matrix with a tridiagonal structure.

M6.2.7 Top Stage

The balance around the top stage (stage 1) yields:

$$\frac{dx_i}{dt} = -\frac{(L + Va)}{M} x_1 + \frac{Va}{M} x_2 + \frac{L}{M} x_f \tag{M6.7}$$

where x_f is known (liquid feed composition).

M6.2.8 Bottom Stage

The balance around the bottom stage (stage n) yields:

$$\frac{dx_n}{dt} = \frac{L}{M} x_{n-1} - \frac{(L + Va)}{M} x_n + \frac{V}{M} y_{n+1} \tag{M6.8}$$

where y_{n+1} is known (vapor feed composition).

EXAMPLE M6.1   Five-stage Absorption Column

The modeling equations (M6.6)–(M6.8) can be written in the following form:

$$\frac{dx_1}{dt} = \qquad -\frac{(L + Va)}{M}x_1 + \frac{Va}{M}x_2 - \frac{L}{M}x_f \quad \text{(stage 1)}$$

$$\frac{dx_2}{dt} = \frac{L}{M}x_1 - \frac{(L + Va)}{M}x_2 + \frac{Va}{M}x_3 \qquad \text{(stage 2)}$$

$$\frac{dx_3}{dt} = \frac{L}{M}x_2 - \frac{(L + Va)}{M}x_3 + \frac{Va}{M}x_4 \qquad \text{(stage 3)}$$

$$\frac{dx_4}{dt} = \frac{L}{M}x_3 - \frac{(L + Va)}{M}x_4 + \frac{Va}{M}x_5 \qquad \text{(stage 4)}$$

$$\frac{dx_5}{dt} = \frac{L}{M}x_4 - \frac{(L + Va)}{M}x_5 \qquad + \frac{V}{M}y_6 \quad \text{(stage 5)}$$

The state variables are x_i (i = 1 to 5), and the input variables are x_f (liquid feed composition) and y_6 (vapor feed composition). *It is assumed that the liquid and vapor flowrates are constant.*

These equations can be written in matrix form as (notice the tridiagonal structure):

$$\begin{bmatrix} \dot{x}_1 \\ \dot{x}_2 \\ \dot{x}_3 \\ \dot{x}_4 \\ \dot{x}_5 \end{bmatrix} = \begin{bmatrix} -\dfrac{(L + Va)}{M} & \dfrac{Va}{M} & 0 & 0 & 0 \\[2ex] \dfrac{L}{M} & -\dfrac{(L + Va)}{M} & \dfrac{Va}{M} & 0 & 0 \\[2ex] 0 & \dfrac{L}{M} & -\dfrac{(L + Va)}{M} & \dfrac{Va}{M} & 0 \\[2ex] 0 & 0 & \dfrac{L}{M} & -\dfrac{(L + Va)}{M} & \dfrac{Va}{M} \\[2ex] 0 & 0 & 0 & \dfrac{L}{M} & -\dfrac{(L + Va)}{M} \end{bmatrix} \begin{bmatrix} x_1 \\ x_2 \\ x_3 \\ x_4 \\ x_5 \end{bmatrix}$$

$$+ \begin{bmatrix} \dfrac{L}{M} & 0 \\[2ex] 0 & 0 \\[2ex] 0 & 0 \\[2ex] 0 & 0 \\[2ex] 0 & \dfrac{V}{M} \end{bmatrix} \begin{bmatrix} x_f \\ y_6 \end{bmatrix} \qquad \text{(M6.9)}$$

which is the so-called state-space form

$$\dot{\mathbf{x}} = \mathbf{A}\,\mathbf{x} + \mathbf{B}\,\mathbf{u} \qquad \text{(M6.10)}$$

M6.3   STEADY-STATE ANALYSIS

We have shown that the form of the dynamic equations for the absorption column are:

$$\dot{\mathbf{x}} = \mathbf{A}\,\mathbf{x} + \mathbf{B}\,\mathbf{u} \qquad\qquad (M6.10)$$

At steady-state the time derivatives are zero, so:

$$0 = \mathbf{A}\,\mathbf{x} + \mathbf{B}\,\mathbf{u} \qquad\qquad (M6.11)$$

The steady-state values of x can be found by solving (M6.11) to find:

$$\mathbf{x} = -\mathbf{A}^{-1}\mathbf{B}\,\mathbf{u} \qquad\qquad (M6.12)$$

EXAMPLE M6.2   Parameters for the Fifth-stage Column

Let the liquid feed flowrate $L = 80$ kgmol inert oil/hr, vapor feed flowrate $V = 100$ kgmol air/hr, equilibrium parameter $a = 0.5$, liquid feed composition $x_f = 0.0$ kgmol benzene/kgmol inert oil, and vapor feed composition $y_6 = 0.1$ kgmol benzene/kgmol air. Let's operate with units of minutes, so $L = 4/3$ kgmol inert oil/min and $V = 5/3$ kgmol air/min. Assume that the liquid molar holdup for each stage is $M = 20/3$ kgmol.

The numerical values yield the following matrices:

$$\mathbf{A} = \begin{bmatrix} -0.325 & 0.125 & 0 & 0 & 0 \\ 0.2 & -0.325 & 0.125 & 0 & 0 \\ 0 & 0.2 & -0.325 & 0.125 & 0 \\ 0 & 0 & 0.2 & -0.325 & 0.125 \\ 0 & 0 & 0 & 0.2 & -0.325 \end{bmatrix} \qquad (M6.13)$$

$$\mathbf{B} = \begin{bmatrix} 0.2 & 0 \\ 0 & 0 \\ 0 & 0 \\ 0 & 0 \\ 0 & 0.25 \end{bmatrix} \qquad (M6.14)$$

The steady-state input is:

$$\mathbf{u}_s = \begin{bmatrix} x_{fs} \\ y_{6s} \end{bmatrix} = \begin{bmatrix} 0.0 \\ 0.1 \end{bmatrix} \qquad (M6.15)$$

Solving for x at steady-state, we find $\mathbf{x}_s = -\mathbf{A}^{-1}\mathbf{B}\,\mathbf{u}_s$

$$\mathbf{x}_s = \begin{bmatrix} 0.0076 \\ 0.0198 \\ 0.0392 \\ 0.0704 \\ 0.1202 \end{bmatrix} \qquad (M6.16)$$

which is simply a vector of the liquid phase compositions. The liquid product composition (leaving stage 5) is $x_5 = 0.1202$.

The vapor product composition (leaving stage 1) is obtained from the linear equilibrium relationship $y_1 = 0.5 \, x_1 = 0.5 \, (0.0076) = 0.0038$ lbmol benzene/lbmol air.

M6.4 STEP RESPONSES

Here we will use the MATLAB function `step` to find the responses to a step change in feed compostion. The step function requires a linear state space model in deviation variable form.

$$\Delta \, \dot{x} = A \, \Delta x + B \, \Delta u$$

M6.4.1 Step Change in Vapor Feed Composition

At time $t = 5$ minutes, the composition of the vapor stream entering the column changes from $y_6 = 0.1$ kgmol benzene/kgmol air to $y_6 = 0.15$ kgmol benzene/kgmol air. Use the MATLAB `step` function to find how the compositions of the liquid and vapor streams (x_6 and y_1) leaving the column change with time. How do the transient responses for all of the stage liquid compositions (x_1 through x_5) differ?

`step` is used to solve the following equations

$$\Delta \, \dot{x} = A \, \Delta x + B \, \Delta u$$

and

$$\Delta y = C \, \Delta x + D \, \Delta u$$

where Δy is the output vector (deviation form). For this problem, we consider x_5 (composition of liquid leaving the bottom of the column) and y_1 (composition of the vapor stream leaving the top of the column) as the outputs.

$$\Delta y = \begin{bmatrix} \Delta x_5 \\ \Delta y_1 \end{bmatrix}$$

Therefore, the C and D matrices are:

$$C = \begin{bmatrix} 0 & 0 & 0 & 0 & 1 \\ 0.5 & 0 & 0 & 0 & 0 \end{bmatrix}$$

$$D = \begin{bmatrix} 0 & 0 \\ 0 & 0 \end{bmatrix}$$

```
[y,x,t] = step(A,B,C,D,u,2)
```

where the 2 in the final column indicates that the second input is being step changed.

A comparison of Figures M6.3 and M6.4 shows that the bottom composition (x_6) responds more quickly to the vapor feed change than the top vapor composition (y_1). This

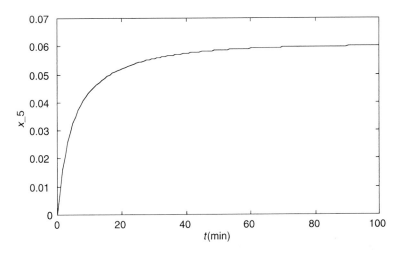

FIGURE M6.3 Response of bottom composition to a step increase in vapor feed composition of 0.05, at $t = 0$.

makes physical sense, because the disturbance must propagate through six stages (from the bottom to the top of the column) to affect the top composition.

Figure M6.5 shows that the magnitude of the change in stage composition is greatest on the bottom stage. Deviation variables $(x_i(t) - x_i(0))$ have been used for ease of comparison. The relative "speed of response" is faster the closer a stage is to the bottom of

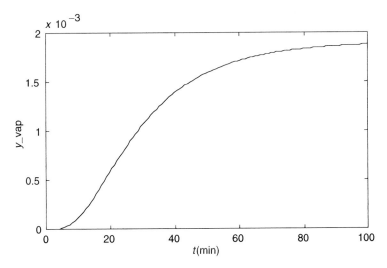

FIGURE M6.4 Response of top vapor composition to a step increase in vapor feed composition of 0.05, at $t = 0$.

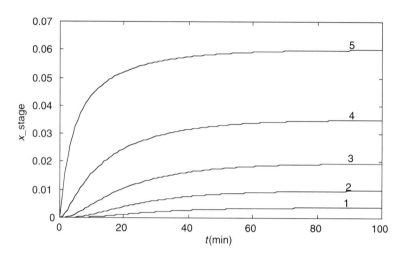

FIGURE M6.5 Response of stage compositions to a step increase in vapor feed of 0.05, at $t = 0$.

the column, as shown in Figure M6.6. Scaled deviation variables $(x_i(t) - x_i(0))/x_i(t = 100) - x_i(0))$ have been used for ease of comparison.

M6.4.2 Step Change in Liquid Feedstream Composition

At time $t = 0$, the composition of the liquid stream entering the column changes from $x_f = 0.0$ kgmol benzene/kgmol inert oil to $x_f = 0.025$ kgmol benzene/kgmol inert oil. Find how the compositions of the liquid and vapor streams (x_5 and y_1) leaving the column

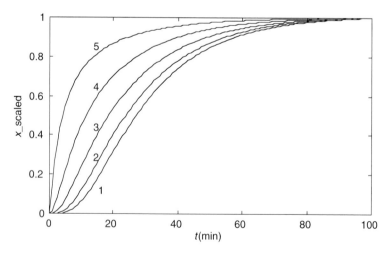

FIGURE M6.6 Response of stage compositions to a step increase in vapor feed composition of 0.05, at $t = 0$. Normalized deviation variables are plotted.

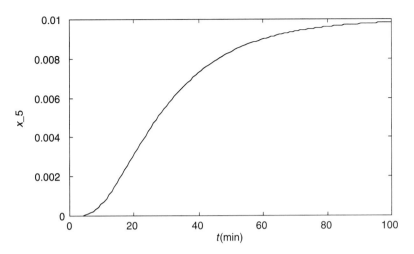

FIGURE M6.7 Response of bottom composition to a step increase in liquid feed composition of 0.025 at $t = 0$.

change with time. How do the transient responses for all of the stage liquid compositions (x_1 through x_6) differ?

A comparison of Figures M6.7 and M6.8 shows that the bottom composition (x_5) responds more slowly to the liquid feed change than the top vapor composition (y_1). This makes physical sense, because the disturbance must propagate through six stages (from the top to the bottom of the column) to affect the bottom composition.

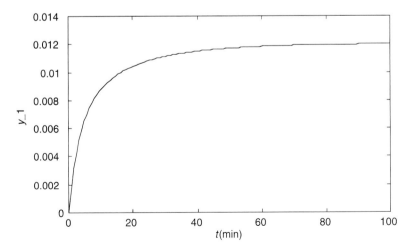

FIGURE M6.8 Response of top vapor composition to a step increase in liquid feed composition of 0.025 at $t = 0$.

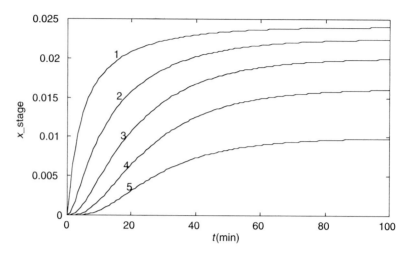

FIGURE M6.9   Response of stage compositions to a step increase in liquid feed composition of 0.025 at $t = 0$.

Figure M6.9 shows that the magnitude of the change in stage composition is greatest on the top stage. Deviation variables $(x_i(t) - x_i(0))$ have been used for ease of comparison. The relative "speed of response" is faster the closer a stage is to the top of the column, as shown in Figure M6.10. Scaled deviation variables $(x_i(t) - x_i(0))/(x_i(t = 150) - x_i(0))$ have been used for ease of comparison.

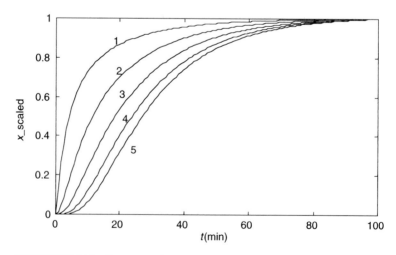

FIGURE M6.10   Response of stage compositions to a step increase in liquid feed composition of 0.025 at $t = 0$ min. Normalized deviation variables are plotted.

M6.4.3   Transfer Function Analysis

The reader should use the MATLAB function `ss2tf` to find the transfer functions relating each input to each state (student exercise 1). Show that the order of the numerator polynomial gets smaller the farther that the state is from the input variable. Also, find the zeros and poles for each transfer function and explain why the speed of response is faster the closer a state is to the input, in terms of the transfer functions.

M6.5   UNFORCED BEHAVIOR

Eigenvalue/eigenvector analysis can be used to understand the effect of perturbations in the initial conditions on the relative speeds of response. The eigenvalues span an order of magnitude, from -0.05 min^{-1} to -0.6 min^{-1}, as shown by the MATLAB command:

```
» [v,d]=eig(amat)

d =

   -0.59886127875258
   -0.05113872124742
   -0.32500000000000
   -0.16688611699158
   -0.48311388300842
```

The first eigenvalue is the fastest, while the second is the slowest. The first and second eigenvectors are associated with the fast and slow directions, respectively.

```
v =

Columns 1 through 4

  -0.16931748198570   -0.16931748198570   -0.31444675225915    0.27472527472527
   0.37095601705416   -0.37095601705416    0.00000000000000    0.34750303957894
  -0.54181594235423   -0.54181594235423    0.50311480361464   -0.00000000000000
   0.59352962728666   -0.59352962728666   -0.00000000000000   -0.55600486332631
  -0.43345275388338   -0.43345275388338   -0.80498368578343   -0.70329670329670

Column 5

  -0.27472527472528
   0.34750303957894
  -0.00000000000000
  -0.55600486332631
   0.70329670329670
```

The reader should show (see student exercise 2) that the following perturbation in initial condition:

$$\mathbf{x_0} = \begin{bmatrix} -0.1693 \\ 0.3710 \\ -0.5418 \\ 0.5935 \\ -0.4335 \end{bmatrix} *0.025$$

yields a response that is over ten times faster than the following perturbation in initial condition:

$$\mathbf{x_0} = \begin{bmatrix} -0.1693 \\ -0.3710 \\ -0.5418 \\ -0.5935 \\ -0.4335 \end{bmatrix} *0.025$$

SUMMARY

In this module we developed a linear model for an absorption column and used the MATLAB step function to study the effect of input composition changes. The magnitude of composition change was greatest on the stages closest to the step input change, and the "speed" of composition change was faster on the stages closest to the step disturbance. For example, for a step change in the vapor feed composition (which enters the bottom of the column) the effect was greatest (magnitude largest and speed fastest) on the compositions of the stages close to the bottom of the column.

FURTHER READING

The gas absorption example is a modification of an example presented by:

Luyben, W.L., & L.A. Wenzel. (1988). *Chemical Process Analysis: Mass and Energy Balances*. Englewood Cliffs, NJ: Prentice Hall.

EXERCISES

1. Use the MATLAB function `ss2tf` to find the transfer functions relating each input to each state. Show that the order of the numerator polynomial gets smaller the farther that the state is from the input variable. Find the zeros and poles for each transfer function. Explain why the speed of response is faster the closer a state is to the input, in terms of the transfer functions and the pole-zero values.

2. Use the MATLAB initial function to show the effect of the "direction" of a perturbation in initial condition. Use plots to compare the "fast" and "slow" directions.

APPENDIX: MATLAB `absorp_ex.m`

```
% absorp_ex.m
% absorption example - Process Dynamics
% orginal version - 29 June 93
% (c) B. Wayne Bequette
% revised - 7 June 1997
%
% aeq = gas/liquid equil coeff
% ns = number of stages
% liq = liquid molar flowrate
% vap = vapor molar flowrate
% mhold = liquid molar holdup per stage
% xfeed = liquid feed composition
% yfeed = vapor feed composition
%
% set parameters
%
   ns = 5;
   liq = 80/60;
   vap = 100/60;
   mhold = 20/3;
   xfeed = 0;
   yfeed = 0.1;
   aeq = 0.5;
%
   amat = zeros(ns,ns);
   bmat = zeros(ns,2);
%
%   calculate ratios
%
   lvratio = -(liq+vap*aeq)/mhold;
   lratio  = liq/mhold;
   vratio  = vap*aeq/mhold;
%
   amat(1,1) = lvratio;
   amat(1,2) = vratio;
%
   for i = 2:ns-1;
     amat(i,i-1) = lratio;
     amat(i,i)   = lvratio;
     amat(i,i+1) = vratio;
   end
%
   amat(ns,ns-1) = lratio;
   amat(ns,ns)   = lvratio;
```

```
%
   bmat(1,1) = lratio;
   bmat(ns,2) = vap/mhold;
%
%  calculate the steady-state
%
   xs = -inv(amat)*bmat*[xfeed;yfeed];
%
%  perform step tests using step
%
%  step change the vapor feed composition at t = 0
%
   cmat = zeros(2,ns);
   cmat(1,ns) = 1;
   cmat(2,1)  = aeq;
   dmat = zeros(2,2);
%
  [y,x,t]=step(amat,bmat,cmat,dmat,2);
%
% actual step magnitude was 0.05
%
   y = y*0.05;
   x = x*0.05;
%
%  plot the liquid composition leaving the
%  absorber (deviation variables)
%
   plot(t,y(:,1),'w')
   xlabel('t(min)')
   ylabel('x_liq')
%
   pause
%
%  plot the vapor composition leaving the absorber
%
   plot(t,y(:,2),'w')
   xlabel('t(min)')
   ylabel('y_vap')
%
   pause
%
%  plot each stage liquid composition (deviation)
   plot(t,x,'w')
   xlabel('t(min)')
   ylabel('x_stage')
   pause
%
%  plot each stage, normalized by the final change
%
   nt = length(t);
%
```

```
   for i = 1:ns;
     xscale(:,i) = x(:,i)/x(nt,i);
   end
%
   plot(t,xscale,'w')
   xlabel('t(min)')
   ylabel('x_scaled')
   pause
%  now, do a step change in liquid feed comp
%
   [y1,x1,t1]=step(amat,bmat,cmat,dmat,1);
%
%  actual input change was 0.05
   y1 = y1*0.025;
   x1 = x1*0.025;
%
%  plot the liquid composition leaving the
%  absorber (deviation variables)
%
   plot(t1,y1(:,1),'w')
   xlabel('t(min)')
   ylabel('x_5')
%
   pause
%
%  plot the vapor composition leaving the absorber
%
   plot(t1,y1(:,2))
%
   pause
%  plot each stage liquid composition (deviation)
   plot(t1,x1,'w')
   xlabel('t(min)')
   ylabel('x_stage')
   pause
%
%  plot scaled liquid compositions
%
   nt1 = length(t1);
%
   for i = 1:ns;
     xscale1(:,i) = x1(:,i)/x1(nt1,i);
   end
%
   plot(t1,xscale1,'w')
   xlabel('t(min)')
   ylabel('x_scaled')
```

ISOTHERMAL CONTINUOUS STIRRED TANK CHEMICAL REACTORS

After studying this module, the student should be able to:

- Determine the steady-state behavior of an isothermal CSTR (number of steady-states)
- Determine the stability of a particular steady-state
- Find the state-space and transfer function form of a linearized CSTR model at a particular steady-state
- Show that steady-state gains calculated from the steady-state equations are consistent with the transfer function gains

The major sections of the module are:

M7.1 INTRODUCTION

Chemical reactors are generally the most important unit operations in a chemical plant. Chemical reactors come in many forms, but two of the most common idealizations are the continuous stirred tank reactor (CSTR) and the plug flow reactor (PFR). These two types serve as limiting bounds for the behavior of many operating reactors. The CSTR is often

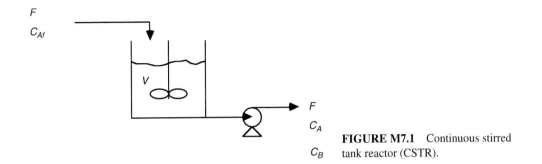

FIGURE M7.1   Continuous stirred tank reactor (CSTR).

used in dynamic modeling studies, because it can be modeled as a lumped parameter system. A dynamic model of a plug flow reactor consists of partial differential equations (also known as a distributed parameter system).

Consider the continuous stirred tank reactor shown in Figure M7.1. In this module we will assume that the reactor is operating at a constant temperature (it is isothermal), so we do not need an energy balance (and can also assume that the reaction rate parameters are constant). In addition, we will assume that the volume is constant. The standard single irreversible reaction system is presented in Section M7.2. The Van de Vusse reactor, a reaction scheme with a number of interesting characteristics is presented in Section M7.3.

M7.2 FIRST-ORDER IRREVERSIBLE REACTION

Recall that a model of a CSTR for a first-order irreversible reaction was developed in Chapter 2. Here we provide a quick review of the modeling equations, before presenting a detailed analysis.

Consider a single irreversible reaction $A \to B$. Assume that the rate of reaction per unit volume is first-order with respect to the concentration of A:

$$\text{molar rate of reaction of } A \text{ per unit volume} = r_A = kC_A$$

Each mole of A reacted creates a mole of B, so we have the following relationship for the rate of formation of B:

$$\text{molar rate of formation of } B \text{ per unit volume} = r_B = kC_A$$

M7.2.1 Component Material Balance on A

Our first step is to write the dynamic modeling equations for the reacting component, A:

$$\begin{array}{cccc} \text{rate of} & \text{(rate)} & \text{(rate)} & \text{(rate)} \\ \text{accumulation} = & \text{in by flow} - & \text{out by flow} - & \text{out by reaction} \end{array}$$

$$\frac{dVC_A}{dt} = F\,C_{Af} - F\,C_A - V\,k\,C_A$$

where k is the reaction rate constant. Since V is constant:

$$\frac{dC_A}{dt} = \frac{F}{V} C_{Af} - \frac{F}{V} C_A - k\, C_A$$

which we can write as:

$$\frac{dC_A}{dt} = \frac{F}{V} C_{Af} - \left(\frac{F}{V} + k\right) C_A \qquad (M7.1)$$

M7.2.2   Component Material Balance on *B*

It is natural to assume that there is no B in the feedstream, which yields the following modeling equation:

$$\frac{dVC_B}{dt} = -FC_B + V k\, C_A$$

where again, k is the reaction rate constant. Since V is constant:

$$\frac{dC_B}{dt} = -\frac{F}{V} C_B + k\, C_A \qquad (M7.2)$$

The two dynamic modeling equations are (M7.1) and (M7.2). Notice that the concentration of B does not play a role in equation (M7.1), so equation (M7.1) can be solved independently of (M7.2).

Before analyzing the dynamics of this system, it is important to understand the steady-state behavior.

M7.2.3   Steady-State Behavior

Assume that F/V is the input variable of interest. In the reaction engineering literature F/V is known as the "space velocity." Similarly, V/F is known as the "residence time" or "space time," that is, the amount of time that it takes for the reactor volume to be "swept out" by the flow. Equation (M7.1) in steady-state form can be arranged to solve for the steady-state concentration (here we use the subscript s to indicate steady-state value):

$$C_{As} = \frac{\dfrac{F_s}{V} C_{Afs}}{\dfrac{F_s}{V} + k} \qquad (M7.3)$$

Notice in (M7.3) that C_{As} is a monotonic function of F_s/V. As F_s/V gets large ($\to \infty$), then C_{As} approaches C_{Afs}. That is, the fluid is flowing so fast through the reactor that there is no time for any conversion of A to B. As F_s/V gets very small ($\to 0$), then C_{As} approaches 0, indicating complete conversion to B. This makes sense because $F_s/V = 0$ corresponds to a batch reactor, and at steady-state we would expect complete conversion since we have an irreversible reaction.

The steady-state gain between the input F_s/V and the output (C_{As}) is simply the derivative of the output with respect to the input. From (M7.3) we find:

$$\frac{\partial C_{As}}{\partial \dfrac{F_s}{V}} = \frac{kC_{Afs}}{\left(\dfrac{kV}{F_s} + k\right)^2} \qquad (M7.4)$$

We can also find the steady-state gain between the input (C_{Afs}) and the output (C_{As}) from (M7.3) by:

$$\frac{\partial C_{As}}{\partial C_{Afs}} = \frac{\dfrac{kV}{F_s}}{\dfrac{kV}{F_s} + k} \qquad (M7.5)$$

Notice that we can make a general plot of the solution of (M7.3) by dividing by C_{Afs} to find

$$\frac{C_{As}}{C_{Afs}} = \frac{1}{1 + \dfrac{kV}{F_s}} = \frac{\dfrac{F}{kV}}{\dfrac{F}{kV} + 1} \qquad (M7.6)$$

C_{As}/C_{Afs} can be considered the dimensionless concentration of A. The relationship in (M7.6) is shown plotted in Figure M7.2.

We can also plot the relationship in equation (M7.6) as shown in Figure M7.3. Notice that, if we consider F to be the manipulated variable and C_A to be the output variable, then we see that the steady-state gain (change in output / change in input) decreases as the flowrate increases. That is, the gain is high and low flowrates and low at high flowrates.

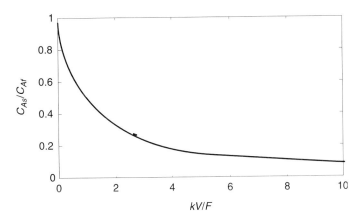

FIGURE M7.2   Dimensionless concentration as a function of kV/F.

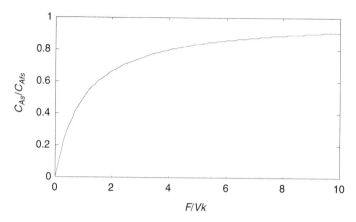

FIGURE M7.3 Dimensionless concentration (equation (M7.4)) as a function of F/kV.

This result is consistent with (M7.4). We will find the same results when using the Laplace transfer function analysis.

Notice from Figure M7.3 that to react most of A we need a large volume/feed ratio (residence time), which would require large vessels for a given flowrate. This indicates that economics must be used to guide the reactor design, and a trade-off must be made between the capital cost of the reactor versus the operating cost (which includes the difference in values of the reactants and products). The residence time is also related to the process time constant, which ultimately effects the quality of control that is possible.

Solving (M7.2) for the steady-state value of B we find:

$$C_{Bs} = \frac{k\,C_{As}}{\dfrac{F_s}{V}} \tag{M7.7}$$

Substituting (M7.5) into (M7.7):

$$C_{Bs} = \frac{k}{\left(\dfrac{F_s}{V}\right)} \frac{\dfrac{F_s}{V}\,C_{Afs}}{\dfrac{F_s}{V}+k} = \frac{k\,C_{Afs}}{\dfrac{F_s}{V}+k} \tag{M7.8}$$

which can be written:

$$\frac{C_{Bs}}{C_{Afs}} = \frac{1}{\dfrac{F_s}{k\,V}+1} = 1 - \frac{C_{As}}{C_{Afs}}$$

Notice in (M7.8) that the dimensionless concentration of B is also a monotonic function of the space velocity. As F_s/V gets large ($\to\infty$) then C_{Bs} approaches 0. That is, the fluid is

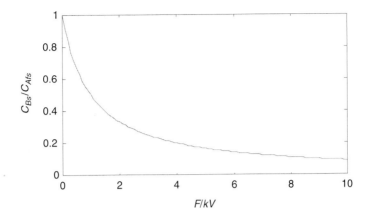

FIGURE M7.4 Dimensionless concentration of B as a function of reactor flowrate.

flowing so fast through the reactor that there is no time for any conversion of A to B. As F/V gets very small ($\rightarrow 0$) then C_{Bs} approaches C_{Afs}, indicating complete conversion of A to B (see Figure M7.4). Notice that this figure is the mirror image of Figure M7.3.

M7.2.4 Dynamic Behavior

Let the state variables in deviation form be represented by:

$$x_1 = C_A - C_{As} \tag{M7.9a}$$
$$x_2 = C_B - C_{Bs} \tag{M7.9b}$$

Also, let the input variables in deviation form be represented by:

$$u_1 = F - F_s \tag{M7.9c}$$
$$u_2 = C_{Af} - C_{Afs} \tag{M7.9d}$$

The state space model (by linearizing (M7.1) and (M7.2)) is then:

$$\frac{dx_1}{dt} = -\left(\frac{F_s}{V} + k\right) x_1 + (C_{Afs} - C_{As}) \, u_1 + \frac{F_s}{V} u_2 \tag{M7.10a}$$

$$\frac{dx_2}{dt} = k \, x_1 + \left(-\frac{F_s}{V}\right) x_2 - C_{Bs} \, u_1 \tag{M7.10b}$$

or,

$$\frac{dx_1}{dt} = a_{11} x_1 + b_{11} u_1 + b_{12} u_2 \tag{M7.10a}$$

$$\frac{dx_2}{dt} = a_{21} x_1 + a_{22} x_2 + b_{21} u_1 \tag{M7.10b}$$

Writing in state-space matrix notation:

$$
\begin{bmatrix} \dfrac{dx_1}{dt} \\ \dfrac{dx_2}{dt} \end{bmatrix} = \begin{bmatrix} a_{11} & 0 \\ a_{21} & a_{22} \end{bmatrix}\begin{bmatrix} x_1 \\ x_2 \end{bmatrix} + \begin{bmatrix} b_{11} & b_{12} \\ b_{21} & 0 \end{bmatrix}\begin{bmatrix} u_1 \\ u_2 \end{bmatrix}
$$

$$
\begin{bmatrix} \dfrac{dx_1}{dt} \\ \dfrac{dx_2}{dt} \end{bmatrix} = \begin{bmatrix} -\left(\dfrac{F_s}{V}+k\right) & 0 \\ k & -\dfrac{F_s}{V} \end{bmatrix}\begin{bmatrix} x_1 \\ x_2 \end{bmatrix} + \begin{bmatrix} C_{Afs}-C_{As} & \dfrac{F_s}{V} \\ -C_{Bs} & 0 \end{bmatrix}\begin{bmatrix} u_1 \\ u_2 \end{bmatrix}
$$

we can see that the eigenvalues of A are simply a_{11} and a_{22} (the reader should show this). Since the eigenvalues are negative, the system is stable.

Since dx_1/dt is not a function of x_2, we find (from the Laplace transform of (M7.10a)):

$$
x_1(s) = \frac{b_{11}}{s-a_{11}}u_1(s) + \frac{b_{12}}{s-a_{11}}u_2(s) \tag{M7.11a}
$$

which can be written in gain-time constant form as:

$$
x_1(s) = \frac{k_{11}}{\tau_1 s + 1}u_1(s) + \frac{k_{12}}{\tau_1 s + 1}u_2(s) \tag{M7.11b}
$$

where

$$
k_{11} = -\frac{b_{11}}{a_{11}} = \left[\frac{C_{Afs}-C_{As}}{\dfrac{F_s}{V}+k}\right] = \frac{k\,C_{Afs}}{\left(\dfrac{F_s}{V}+k\right)^2} \tag{M7.12}
$$

Notice that k_{11} is consistent with (M7.4).

$$
\tau_1 = -\frac{1}{a_{11}} = \left[\frac{1}{\dfrac{F_s}{V}+k}\right] = \left(\frac{V}{F_s}\right)\left(\frac{C_{As}}{C_{Afs}}\right) \tag{M7.13}
$$

$$
k_{12} = -\frac{b_{12}}{a_{11}} = \left[\frac{\dfrac{F_s}{V}}{\dfrac{F_s}{V}+k}\right] = \left(\frac{C_{As}}{C_{Afs}}\right) \tag{M7.14}
$$

Notice that k_{12} is consistent with (M7.5). From (M7.11) we see that the process gain for output 1 decreases as the flowrate increases. We see that τ_1 is always less than the reactor residence time from (M7.12). Also notice that decreases as the reactor flowrate increases. From (M7.14) we see that k_{12} is always less than 1.

Taking the Laplace transform of (M7.10a) and (M7.10b), we find:

$$
x_2(s) = \frac{b_{21}s + (b_{11}a_{21}-b_{21}a_{11})}{(s-a_{11})(s-a_{22})}u_1(s) + \frac{b_{12}a_{21}}{(s-a_{11})(s-a_{22})}u_2(s)
$$

which can be written in the form of:

$$x_2(s) = \frac{k_{21}(\tau_n s + 1)}{(\tau_1 s + 1)(\tau_2 s + 1)} u_1(s) + \frac{k_{22}}{(\tau_1 s + 1)(\tau_2 s + 1)} u_2(s)$$

where:

$$k_{21} = -\frac{k\, C_{Afs}}{\left(\dfrac{F_s}{V} + k\right)^2} = -k_{11} \tag{M7.15}$$

$$\tau_2 = -\frac{1}{a_{22}} = \left[\frac{1}{\dfrac{F_s}{V}}\right] = \frac{V}{F_s} \tag{M7.16}$$

$$\tau_n = \left[\frac{1}{\dfrac{F_s}{V}}\right] = \frac{V}{F_s} \tag{M7.17}$$

$$k_{22} = \frac{b_{12}a_{21}}{a_{11}a_{11}} = \left[\frac{k}{\dfrac{F_s}{V} + k}\right] = \tag{M7.18}$$

Notice that $\tau_n = \tau_2$, so there is pole-zero cancellation in $g_{21}(s)$ and the transfer function becomes first-order. Also, since the eigenvalues are always negative (the time constants are always positive), the system is always stable.

M7.2.5   Numerical Example

Consider a system with the following parameter and steady-state values:

$$\frac{F_s}{V} = 0.2 \text{ min}^{-1}$$

$$k = 0.2 \text{ min}^{-1}$$

$$C_{Afs} = 1.0 \frac{\text{gmol}}{\text{liter}}$$

which yields the following state-space model:

$$\begin{bmatrix} \dfrac{dx_1}{dt} \\ \dfrac{dx_2}{dt} \end{bmatrix} = \begin{bmatrix} -0.4 & 0 \\ 0.2 & -0.2 \end{bmatrix} \begin{bmatrix} x_1 \\ x_2 \end{bmatrix} + \begin{bmatrix} 0.5 & 0.2 \\ -0.5 & 0 \end{bmatrix} \begin{bmatrix} u_1 \\ u_2 \end{bmatrix}$$

The following figures compare responses of the nonlinear and linear systems

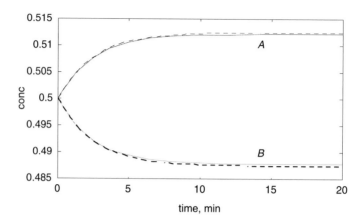

FIGURE M7.5   Small (0.01) step change in *F/V*. Solid = nonlinear, dotted = linear.

SMALL STEP CHANGE IN *F/V*

In Figure M7.5 we compare the responses of the linear and nonlinear models for a small (0.01) step change in *F/V*. Notice that the models yield virtually identical results.

LARGE POSITIVE STEP CHANGE IN *F/V*

In Figure M7.6 we compare the responses of the linear and nonlinear models for a large positive (0.2) step change in *F/V*. The nonlinear model has a lower "gain" than the linear model for both states.

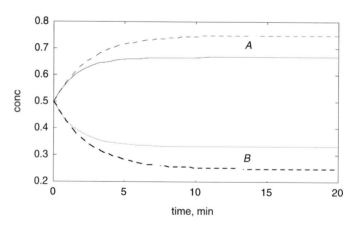

FIGURE M7.6   Large (0.2) step change in *F/V*. Solid = nonlinear, dotted = linear.

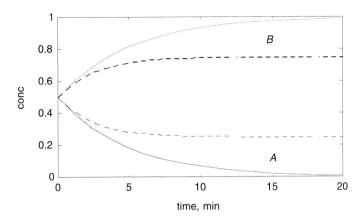

FIGURE M7.7   Large negative (−0.2) step change in F/V. Solid = nonlinear, dotted = linear.

LARGE NEGATIVE STEP CHANGE IN F/V

In Figure M7.7 we compare the responses of the linear and nonlinear models for a large negative (−0.2) step change in F/V. The nonlinear model has a higher "gain" than the linear model for both states.

M7.2.6   Phase Plane Analysis

To determine the important directions for this system, we find the eigenvalues and eigenvectors of the A matrix: The slow eigenvalue is −0.2, which has the associated eigenvector $\xi_1 = \begin{bmatrix} 0 \\ 1 \end{bmatrix}$. The fast eigenvalue is −0.4 with an eigenvector $\xi_2 = \begin{bmatrix} 0.7071 \\ -0.7071 \end{bmatrix}$. These results indicate that an initial perturbation in the second state variable (concentration of B) will respond slowly, while an equal magnitude pertubation in both state variables will respond twice as fast.

M7.2.7   Summary of Behavior of the First-Order Irreversible Chemical Reaction

We can make some interesting observations about the behavior of this simple system. We notice that both reactor concentrations have first-order transfer functions with respect to the dilution rate, F/V. The concentration of A is also first-order with respect to C_{Af}, but a second-order transfer function relates C_{Af} to the concentration of B. The response of B to a change in the feed composition of A is slower than the response of A for the same change.

M7.3   VAN DE VUSSE REACTION

Often reaction schemes will exhibit a maximum in the concentration of product versus flowrate curve. Here we will consider such a reaction. It turns out this this type of system can have significantly different input/output characteristics, depending on the operating condition chosen. Consider the reaction scheme consisting of the following irreversible reactions:

$$A \xrightarrow{k_1} B \xrightarrow{k_2} C$$

$$2A \xrightarrow{k_3} D$$

This scheme was presented by Van de Vusse (1964). Engell and Klatt (1993) note that the production of cyclopentinol from cyclopentadiene is based on such a reaction scheme, where:

A = cyclopentaddiene

B = cyclopentenol

C = cyclopentanediol

D = dicyclopentadiene

In the following development, we assume that the feedstream contains only component A.

M7.3.1   Modeling Equations

OVERALL MATERIAL BALANCE

Assuming constant density and constant volume:

$$\frac{dV}{dt} = 0 \quad \text{and} \quad F = F_i$$

COMPONENT A BALANCE

The balance on A is:

accumulation = in − out by flow − out by reaction 1 − out by reaction 2

$$\frac{d(VC_A)}{dt} = F(C_{Af} - C_A) - Vk_1C_A - Vk_3C_A^2$$

Since V is constant:

$$\frac{d(C_A)}{dt} = \frac{F}{V}(C_{Af} - C_A) - k_1C_A - k_3C_A^2 \tag{M7.19}$$

COMPONENT B BALANCE

Similarly, we can write:

$$\frac{d(C_B)}{dt} = -\frac{F}{V}C_B + k_1C_A - k_2C_B \qquad \text{(M7.20)}$$

COMPONENT C BALANCE

Also, for component C:

$$\frac{d(C_C)}{dt} = -\frac{F}{V}C_C + k_2C_B \qquad \text{(M7.21)}$$

COMPONENT D BALANCE

And component D:

$$\frac{d(C_D)}{dt} = -\frac{F}{V}C_D + \frac{1}{2}k_3C_A^2 \qquad \text{(M7.22)}$$

Look at equations (M7.19) through (M7.22). Notice that (M7.19) and (M7.20) do not depend on C_C or C_D. If we are only concerned about C_A and C_B, we only need to solve equations (M7.19) and (M7.20).

$$\frac{d(C_A)}{dt} = \frac{F}{V}(C_{Af} - C_A) - k_1C_A - k_3C_A^2 \qquad \text{(M7.19)}$$

$$\frac{d(C_B)}{dt} = -\frac{F}{V}C_B + k_1C_A - k_2C_B \qquad \text{(M7.20)}$$

Solving for the steady-state for (M7.19), we find a quadratic equation in C_{As}:

$$-k_3C_A^2 + \left(-k_1 - \frac{F_s}{V}\right)C_{As} + \frac{F_s}{V}C_{Afs} = 0$$

Solving this quadratic (using the positive root), we find:

$$C_{As} = \frac{-\left(k_1 + \dfrac{F_s}{V}\right)}{2k_3} + \frac{\sqrt{\left(k_1 + \dfrac{F_s}{V}\right)^2 + 4k_3\left(\dfrac{F_s}{V}\right)}}{2k_3}$$

and solving for C_{Bs}:

$$C_{Bs} = \frac{k_1C_{As}}{\dfrac{F_s}{V} + k_2}$$

These results lead to the input/output curves presented in Figure M7.8.

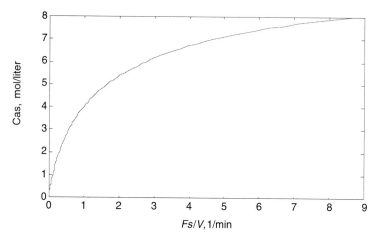

a. Concentration of *A* as function of the space velocity.

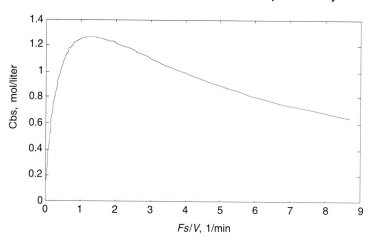

b. Concentration of *B* as a function of the space velocity.

FIGURE M7.8   Steady-state relationships for the Van de Vusse reactor.

Notice that there is a maximum value of the concentration of *B*. If the objective of the process is to maximize the production of *B*, then there exists an optimum residence time (F_s/V) of 1.292 min^{-1} (a function routine to find this is shown in the appendix).

STATE-SPACE MODEL

Here we linearize the nonlinear modeling equations to find the state-space form:

$$\dot{\mathbf{x}} = \mathbf{A}\,\mathbf{x} + \mathbf{B}\,\mathbf{u}$$
$$\mathbf{y} = \mathbf{C}\,\mathbf{x}$$

where the state, input, and output vectors are, in deviation form:

$$\mathbf{x} = \begin{bmatrix} C_A - C_{As} \\ C_B - C_{Bs} \end{bmatrix} = \text{state variables}$$

$$\mathbf{u} = \begin{bmatrix} \dfrac{F}{V} - \dfrac{F_s}{V} \end{bmatrix} = \text{input variable}$$

$$\mathbf{y} = \begin{bmatrix} C_A - C_{As} \\ C_B - C_{Bs} \end{bmatrix} = \text{output variables}$$

Recall that our two dynamic functional equations are:

$$\frac{dC_A}{dt} = f_1\left(C_A, C_B, \frac{F}{V}\right) = \frac{F}{V}(C_{Af} - C_A) - k_1 C_A - k_3 C_A^2$$

$$\frac{dC_B}{dt} = f_2\left(C_A, C_B, \frac{F}{V}\right) = -\frac{F}{V} C_B + k_1 C_A - k_2 C_B$$

The elements of the state-space A matrix are found by: $A_{ij} = \partial f_i / \partial x_j$

$$A_{11} = \frac{\partial f_1}{\partial x_1} = \frac{\partial f_1}{\partial (C_A - C_{As})} = \frac{\partial f_1}{\partial C_A} = -\frac{F_s}{V} - k_1 - 2 k_3 C_{As}$$

$$A_{12} = \frac{\partial f_1}{\partial x_2} = \frac{\partial f_1}{\partial (C_B - C_{Bs})} = \frac{\partial f_1}{\partial C_B} = 0$$

$$A_{21} = \frac{\partial f_2}{\partial x_1} = \frac{\partial f_2}{\partial (C_A - C_{As})} = \frac{\partial f_2}{\partial C_A} = k_1$$

$$A_{22} = \frac{\partial f_2}{\partial x_2} = \frac{\partial f_2}{\partial (C_B - C_{Bs})} = \frac{\partial f_2}{\partial C_B} = -\frac{F_s}{V} - k_2$$

$$B_{11} = \frac{\partial f_1}{\partial u_1} = \frac{\partial f_1}{\partial \left(\dfrac{F}{V} - \dfrac{F_s}{V}\right)} = \frac{\partial f_1}{\partial \dfrac{F}{V}} = C_{Afs} - C_{As}$$

$$B_{21} = \frac{\partial f_2}{\partial u_1} = \frac{\partial f_2}{\partial \left(\dfrac{F}{V} - \dfrac{F_s}{V}\right)} = \frac{\partial f_2}{\partial \dfrac{F}{V}} = -C_{Bs}$$

and the state-space model is:

$$A = \begin{bmatrix} -\dfrac{F_s}{V} - k_1 - 2k_3 C_{As} & 0 \\ k_1 & -\dfrac{F_s}{V} - k_2 \end{bmatrix}$$

$$B = \begin{bmatrix} C_{Afs} - C_{As} \\ -C_{Bs} \end{bmatrix}$$

Here we will consider several different steady-state operating points. Case 1 illustrates inverse response behavior, while Case 2 does not. Case 3 will illustrate operation at the maximum production point for compound B.

CASE 1

Consider a dilution rate of $F_s/V = 4/7$ min^{-1}, which is on the left side of the peak shown in Figure M7.8b. The steady-state concentrations are:

$$C_{As} = 3\,\frac{\text{mol}}{\text{liter}}$$

$$C_{Bs} = \frac{105}{94} = 1.117\,\frac{\text{mol}}{\text{liter}}$$

which yields the following state-space model:

$$A = \begin{bmatrix} -2.4048 & 0 \\ 0.8333 & -2.2381 \end{bmatrix}$$

$$B = \begin{bmatrix} 7.0000 \\ -1.1170 \end{bmatrix}$$

The transfer function relating input 1 to output 2 is:

$$g_{21}(s) = \frac{-1.1170s + 3.1472}{s^2 + 4.6429s + 5.3821}$$

which has a zero in the right-half plane:

$$\text{zero} = \frac{3.1472}{1.1170} = 2.8175$$

The process transfer function can also be written in the form:

$$g_{21}(s) = \frac{0.5848(-0.3549\,s + 1)}{0.1858\,s^2 + 0.8627\,s + 1}$$

CASE 2

Consider now the following dilution rate $F_s/V = 2.8744$ min^{-1}, which is on the right side of the peak in Figure M7.8b. This yields the following operating conditions:

$$C_{As} = 6.0870\,\frac{\text{mol}}{\text{liter}}$$

$$C_{Bs} = \frac{105}{94} = 1.117\,\frac{\text{mol}}{\text{liter}}$$

And the state-space model:

$$A = \begin{bmatrix} -5.7367 & 0 \\ 0.8333 & -4.5411 \end{bmatrix}$$

$$B = \begin{bmatrix} 3.9130 \\ -1.1170 \end{bmatrix}$$

The transfer function relating input 1 to output 2 is:

$$g_{21}(s) = \frac{-1.1170\,s - 3.1472}{s^2 + 10.2778\,s + 26.0508}$$

which has a zero in the left-half plane:

$$\text{zero} = -\frac{3.1472}{1.1170} = -2.8175$$

Notice that cases 1 and 2 are based on the same concentration of component B. Also note that case 1 has a right-half plane zero, while case 2 has a left-half plane zero of the same magnitude. This means that case 1 exhibits inverse response, while case 2 does not. The different types of behavior are shown in the following figures. Figures M7.9 and M7.10 are characteristic of a system with a positive gain and inverse response (right-half plane zero). Figure M7.9 shows that, for a small step change, the linear model is a good approximation to the nonlinear model. Figure M7.10 shows that the gain of the nonlinear model is significantly lower than the linear model, when a large step input is made.

Figures M7.11 and M7.12 are characteristic of a system with a negative gain. Figure M7.11 shows that, for a small step change, the linear model is a good approximation

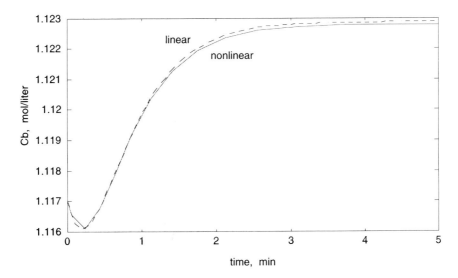

FIGURE M7.9   Small step change (0.01) in space velocity, Case 1 conditions.

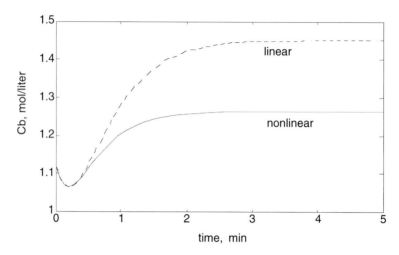

FIGURE M7.10   Response to a large step change (doubling, from 4/7 to 8/7 min^{-1}) in space velocity, Case 1 conditions.

to the nonlinear model. Figure M7.12 shows that the gain of the nonlinear model is lower than the linear model, when a large step input is made.

It is interesting to note that all operating points to the left of the peak in the input/output curve (Figure M7.8b) have right-half-plane zeros (inverse response) while all operating points to the right of the peak do not.

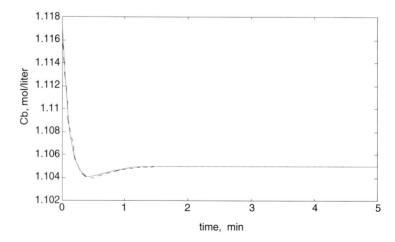

FIGURE M7.11   Response to a small step change in space velocity (0.1), case 2 conditions.

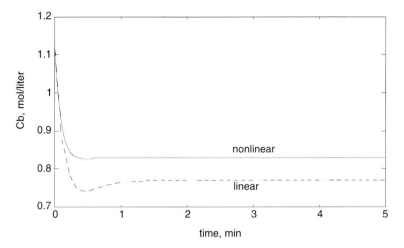

FIGURE M7.12   Response to a large step change (doubling of space velocity), case 2 conditions.

CASE 3   OPERATING AT THE PEAK

The optimum operating point for the production of B is at the following dilution rate:

$$\frac{F_s}{V} = 1.2921 \text{ min}^{-1}$$

which yields the composition:

$$C_{As} = 4.4949 \frac{\text{mol}}{\text{liter}} \quad C_{Bs} = 1.2660 \frac{\text{mol}}{\text{liter}}$$

and the state-space model:

$$A = \begin{bmatrix} -3.6237 & 0 \\ 0.8333 & -2.9588 \end{bmatrix}$$

$$B = \begin{bmatrix} 5.5051 \\ -1.2660 \end{bmatrix}$$

The transfer function relating input 1 and output 2 is:

$$g_{21}(s) = \frac{-1.2660 \, s}{s^2 + 6.5825 \, s + 10.7217}$$

which has a zero at the origin.

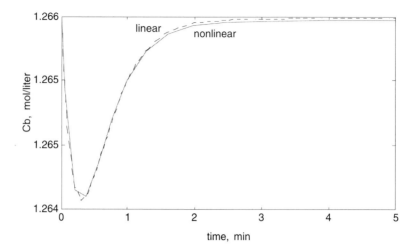

FIGURE M7.13   Response to a small step change, case 3 (peak operating point) conditions.

Notice that there is no steady-state change in the concentration of B for a small step change in space velocity, as shown in Figure M7.13; this behavior is consistent with the zero gain in the g_{21} transfer function.

SUMMARY

Two example isothermal reactor systems were presented in this chapter. We noted that V/F (residence time; or equivalently, F/V, space velocity) is a common design parameter for chemical reactors. We noted that a first-order reaction system has a monotonic relationship (no input or output multiplicity) between output concentration and reactor flowrate. A lower reactor flowrate leads to a higher gain with respect to the concentration of A.

The Van de Vusse reactor had the interesting characteristics of input multiplicity and right-half plane zeros (inverse response). At the maximum in the concentration versus flowrate curve the gain was zero.

REFERENCES AND FURTHER READING

The following reference presents a more detailed study of an exothermic reactor that has the same reaction scheme as the Van de Vusse reactor.

Engell, S., & K.-U. Klatt. (1993). Nonlinear control of a non-minimum-phase CSTR. *Proceedings of the 1993 American Control Conference*, 2041–2045.

There are a number of good reaction engineering texts. Most of the emphasis of these texts is on the steady-state behavior.

Fogler, H.S. (1992). *Elements of Chemical Reaction Engineering*, 2nd ed. Englewood Cliffs, NJ: Prentice-Hall.

Froment, G.F., & K.B. Bischoff. (1990). *Chemical Reactor Analysis and Design*, 2nd ed. New York: Wiley.

Levenspiel, O. (1972). *Chemical Reaction Engineering*, 2nd ed., New York: Wiley.

Smith, J.M. (1981). *Chemical Engineering Kinetics*, 3rd ed. New York: McGraw-Hill.

STUDENT EXERCISES

1. Consider the $A \rightarrow B$ example. Extend this to a reversible reaction $A \leftrightarrow B$ in a CSTR, where both the forward and reverse reactions are first-order, and where there is no B in the feed stream.

   a. Derive the state-space model.

   b. For the case where the reverse reaction rate constant is 1/2 the forward reaction rate constant (which is 0.1 min^{-1}), find the value of F/V where the conversion of A is 50%.

   c. For a step increase in the feed concentration of A of 10%, find the change in the A and B concentrations in the reactor.

2. Consider the Van de Vusse reactor. Include C and D as state variables and write the state-space model. Find the transfer functions relating F/V and C_C and C_D.

APPENDIXES

A1 Generate Steady-State Input/Output Curves for the Van de Vusse Reaction

```
function [fsov,cbs,cas] = vandss(k1,k2,k3,cafs);
%
%   25 july 94
%   (c) b.w. bequette
%
%   finds the steady-state input/outputcurves for
%   the vand de vusse reactor
%
%   fsov = steady-state space velocity, fs/v
%   cas  = steady-state concentration of a
%   cbs  = steady-state concentration of b
%   cafs = feed concentratioon of a
%
    fsov = 0:0.1:10;
```

```
%
   cas1 = -(k1 + fsov)/(2*k3);
   cas2 = sqrt((k1+fsov).^2 + 4*k3*fsov.*cafs)/(2*k3);
   cas  = cas1 + cas2;
%
%  cas = 0.01*cafs:cafs/100:0.80*cafs;
%
%  fsov = (k3.*cas.*cas + k1*cas)./(cafs - cas);
%
   cbs = k1*cas./(fsov + k2);
```

A2   Van de Vusse Function File (`vanmax.m`) for Finding the "Peak" in the Input/Output Curve

Finding the maximum in the input/output curve for the Van de Vusse reaction:

```
Use fmin

        fmin('vanmax',0,2)

  function cbsmin = vanmax(x);
%
%  25 july 94
%  (c) b.w. bequette
%
%  finds the maximum of cbs with respect to fsov for
%  the van de vusse reaction
%
%  use matlab function fmin which finds the minimum of a
          function
%    take -cbs as the function to be minimized,
%    which is equivalent to maximizing cbs
%
%  fsov = steady-state space velocity, fs/v
%  cas  = steady-state concentration of a
%  cbs  = steady-state concentration of b
%  cafs = feed concentratioon of a
%
   k1 = 5/6;
   k2 = 5/3;
   k3 = 1/6;
   cafs = 10;
%
   fsov = x;
%
   cas1 = -(k1 + fsov)/(2*k3);
   cas2 = sqrt((k1+fsov)^2 + 4*k3*fsov*cafs)/(2*k3);
   cas  = cas1 + cas2;
%
   cbs    =  k1*cas/(fsov + k2);
   cbsmin = - cbs;
```

A3   Van de Vusse Function File for Integration Using ode45, vanvusse.m

```
function xdot = vanvusse(t,x);
%
%  (c) b.w. bequette
%  25 July 94
%  revised 20 July 96
%
%  dynamic model for the van de vusse reaction
%
%  to use this function file with ode45:
%
%   [t,x] = ode45('vanvusse',t0,tf,x0]
%
%  where t0 is the initial time (usually 0)
%         tf is the final time
%         x0 is the initial condition for the states
%
%  x    = states
%  x(1) = concentration of A
%  x(2) = concentration of B
%  t    = time
%
%  fov  = dilution rate
%  ca   = concentration of a
%  cb   = concentration of b
%  caf  = feed concentration of a
%
%  the reaction scheme is
%
%  A --> B --> C
%    (k1)    (k2)
% 2A --> D
%    (k3)
%
   k1 = 5/6; % rate constant for A -> B (min^-1)
   k2 = 5/3; % rate constant for B -> C (min^-1)
   k3 = 1/6; % rate constant for A + A -> D (mol/(liter min)
   caf = 10; % feed concentration of A, mol/liter
%
%  for step changes in fov, use the following
%   fov  = 4/7 + delf;
%
%  use familiar notation for the states
%
   ca   = x(1);
   cb   = x(2);
%
%  modeling equations
%
   dcadt = fov*(caf - ca) -k1*ca - k3*ca*ca;
   dcbdt = -fov*cb + k1*ca - k2*cb;
%
```

```
%  derivatives placed in vector notation
%
   xdot(1) = dcadt;
   xdot(2) = dcbdt;
```

A4   Various Cases for the Van de Vusse Reactor

```
a1 =
   -2.4048          0
    0.8333    -2.2381

»[v1,d1] = eig(a1)

v1 =
          0     0.1962
     1.0000    -0.9806

d1 =
   -2.2381          0
         0    -2.4048

»a2 = [-5.7367 0;.8333 -4.5411]

a2 =
   -5.7367          0
    0.8333    -4.5411

»[v2,d2]= eig(a2)

v2 =
          0     0.8204
     1.0000    -0.5718

d2 =
   -4.5411          0
         0    -5.7367

»a3 = [-3.6237 0;.8333 -2.9588]

a3 =
   -3.6237          0
    0.8333    -2.9588

»[v3,d3] = eig(a3)

v3 =
          0     0.6237
     1.0000    -0.7817

d3 =
   -2.9588          0
         0    -3.6237
```

BIOCHEMICAL REACTORS

MODULE
8

After studying this module, the student should be able to

- Develop the dynamic modeling equations for a two-state biochemical reactor
- Understand the concept of "washout"
- Understand the different types of steady-state and dynamic behavior exhibited by the Monod and Substrate Inhibition models
- Find the number of steady-state solutions and to determine the stability of each steady-state

The major sections of this module are:

M8.1   BACKGROUND

Biochemical reactors are used to produce a large number of intermediate and final products, including pharmaceuticals, food, and beverages. Biochemical reactor models are similar to chemical reactor models, since the same type of material balances are per-

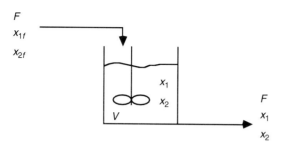

FIGURE M8.1   Biochemical reactor.

formed. In the simplest reactor we consider two components: biomass and substrate. The biomass consists of cells that consume the substrate. One example would be a wastewater treatment system, where the biomass is used to "eat" waste chemicals (substrate). Another example is fermentation, where cells consume sugar and produce alcohol.

Consider the schematic of a biochemical reactor shown in Figure M8.1.

In this module we assume that the reactor is perfectly mixed and that the volume is constant. We use the following notation:

$x_1 =$ biomass concentration $= \dfrac{\text{mass of cells}}{\text{volume}}$

$x_2 =$ substrate concentration $= \dfrac{\text{mass of substrate}}{\text{volume}}$

$r_1 =$ rate of cell generation $= \dfrac{\text{mass of cells generated}}{\text{volume} \cdot \text{time}}$

$r_2 =$ rate of substrate consumption $= \dfrac{\text{mass of substrate consumed}}{\text{volume} \cdot \text{time}}$

$F =$ volumetric flowrate $=$ volume/time

Now we can write the material balances to describe the behavior of this system.

M8.2   MODELING EQUATIONS

The dynamic model is developed by writing material balances on the biomass (cells) and the substrate (feed source for the cells). Biomass grows by feeding on the substrate.

M8.2.1   Biomass Material Balance

We write the biomass material balance as:

rate of accumulation = in by flow − out by flow + generation

$$\frac{dVx_1}{dt} = Fx_{1f} - Fx_1 + Vr_1 \qquad\qquad \text{(M8.1)}$$

where x_{1f} is the concentration of biomass in the feed stream and F is the volumetric flowrate.

M8.2.2 Substrate Material Balance

The substrate material balance is written:

rate of accumulation = in by flow − out by flow − consumption

$$\frac{dVx_2}{dt} = Fx_{2f} - Fx_2 - Vr_2 \tag{M8.2}$$

where x_{2f} is the concentration of substrate in the feed stream.

M8.2.3 Specific Growth Rate

The reaction rate (mass of cells generated/volume time) is normally written in the following form:

$$r_1 = \mu x_1 \tag{M8.3}$$

where μ is the specific growth rate coefficient. We can think of μ as being similar to a first-order reaction rate constant; however, μ is not constant—it is a function of the substrate concentration as shown in Section M8.2.6. The units of μ are time^{-1}.

M8.2.4 Yield

There is a relationship between the rate of generation of biomass and the rate of consumption of substrate. Define Y as the yield, that is, the mass of cells produced per mass of substrate consumed:

$$Y = \frac{\text{mass of cells produced}}{\text{mass of substrate consumed}} = \frac{r_1}{r_2} \tag{M8.4}$$

From (M8.4) we can write:

$$r_2 = \frac{r_1}{Y} \tag{M8.5}$$

and substituting (M8.3) into (M8.5), we find:

$$r_2 = \frac{\mu x_1}{Y} \tag{M8.6}$$

We assume in the subsequent analysis that Y is a constant.

M8.2.5 Dilution Rate

Assuming a constant volume reactor, we can write (M8.1) and (M8.2) as:

$$\frac{dx_1}{dt} = \frac{F}{V} x_{1f} - \frac{F}{V} x_1 + r_1 \tag{M8.7}$$

$$\frac{dx_2}{dt} = \frac{F}{V} x_{2f} - \frac{F}{V} x_2 - r_2 \tag{M8.8}$$

Defining F/V as D, the *dilution rate*, and using the rate expressions in (M8.3) and (M8.6), we find:

$$\frac{dx_1}{dt} = D\, x_{1f} - D\, x_1 + \mu x_1 \tag{M8.9}$$

$$\frac{dx_2}{dt} = D\, x_{2f} - D\, x_2 - \frac{\mu x_1}{Y} \tag{M8.10}$$

Generally, it is assumed that there is no biomass in the feed stream, so $x_{1f} = 0$. The bioreactor modeling equations are then normally written in the following form:

$$\frac{dx_1}{dt} = (\mu - D)\, x_1 \tag{M8.11}$$

$$\frac{dx_2}{dt} = D\, (x_{2f} - x_2) - \frac{\mu x_1}{Y} \tag{M8.12}$$

The dilution rate (D) is the same as the space velocity in the chemical reaction engineering literature. It is also the inverse of the reactor residence time and has units of time^{-1}.

The expressions for μ (specific growth rate) are developed in the following section.

M8.2.6   Growth Rate Expressions

The growth rate coefficient is usually not constant. A number of functional relationships between the growth rate coefficient and substrate concentration have been developed. The most common are (i) *Monod* and (ii) *Substrate inhibition*.

MONOD

The growth rate coefficient often varies in a hyperbolic fashion. The following form was proposed by Monod in 1942. Notice that μ is first-order at low x_2 and zero order at high x_2.

$$\mu = \frac{\mu_{max}\, x_2}{k_m + x_2} \tag{M8.13}$$

Notice that μ is first-order at low x_2 and zero order at high x_2. That is, when x_2 is low:

$$\mu \approx \frac{\mu_{max}}{k_m} x_2$$

and when x_2 is high:

$$\mu \approx \mu_{max}$$

Since the reaction rate is:

$$r_1 = \mu x_1$$

this means that the Monod description is similar to a second-order (bimolecular) reaction when x_2 is low, since

$$r_1 \approx \frac{\mu_{max}}{k_m} x_2 x_1$$

and to a first-order reaction when x_2 is high, since

$$r_1 \approx \mu_{max} x_1$$

Equation (M8.13) is the same form as the Langmuir adsorption isotherm and the standard rate equation for enzyme-catalyzed reactions with a single substrate (Michaelis-Menten kinetics).

SUBSTRATE INHIBITION

Sometimes the growth rate coefficient increases at low substrate concentration, but decreases at high substrate concentration. The physical reason may be that the substrate has a toxic effect on the biomass cells at a higher concentration. This effect is called substrate inhibition and is represented by the following equation:

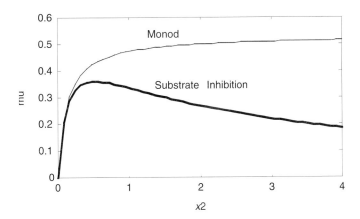

FIGURE M8.2 Comparison of Monod and substrate inhibition models for growth rate.

TABLE M8.1   Parameters for Monod and Substrate Inhibition Models

Monod		Substrate Inhibition	
μ_{max}	$= 0.53 \text{ hr}^{-1}$	μ_{max}	$= 0.53 \text{ hr}^{-1}$
k_m	$= 0.12 \text{ g/liter}$	k_m	$= 0.12 \text{ g/liter}$
		k_1	$= 0.4545 \text{ liter/g}$
Y	$= 0.4$	Y	$= 0.4$
x_{2fs}	$= 4.0 \text{ g/liter}$	x_{2fs}	$= 4.0 \text{ g/liter}$

$$\mu = \frac{\mu_{max}\, x_2}{k_m + x_2 + k_1\, x_2^2} \tag{M8.14}$$

Notice that the Monod equation is a special case of (M8.14), with $k_1 = 0$.

SPECIFIC GROWTH RATE RELATIONSHIPS

The characteristic relationships between substrate (x_2) and specific growth rate (μ) are quite different for Monod and substrate inhibition. The curves for μ as a function of x_2 for both models are compared in Figure M8.2. Notice that the substrate inhibition model exhibits a maximum in the growth rate curve, while Monod becomes zero-order at high substrate concentrations.

M8.3   STEADY-STATE SOLUTION

In this section, the MATLAB function `fsolve` will be used to solve for the steady-state values of the biomass and substrate concentrations. The numerical values used in our simulations are shown in Table M8.1.

We will study the following cases:

Case 1. Medium Dilution Rate, $D_s = 0.3 \text{ hr}^{-1}$
Case 2. Low Dilution Rate, $D_s = 0.15 \text{ hr}^{-1}$
Case 3. High Dilution Rate, $D_s = 0.45 \text{ hr}^{-1}$

EXAMPLE M8.1   Case 1 Results (D = 0.3)

The function file `bio_ss.m` (Appendix 1) is set for Case 1 (D = 0.3) and the substrate inhibition model (k1 = 0.4545). The MATLAB function fsolve is used to solve for the steady-state values by entering the following in the command window (with an initial guess of x(1) = 1 and x(2) = 1):

```
» x = fsolve('bio_ss',[1;1])
```

The steady-state solution obtained is:

```
x =
    0.9951
    1.5122
```

Different initial guesses result in two other solutions for the substrate inhibition model. Also, the Monod model has two steady-state solutions. The reader should find the following results using `fsolve` and `bio_ss.m`, by entering different initial guesses.

Monod (2 steady-state solutions)

Equilibrium 1	$x_{1s} = 0$	$x_{2s} = 4.0$
Equilibrium 2	$x_{1s} = 1.5374$	$x_{2s} = 0.1565$

Substrate Inhibition (3 steady-state solutions)

Equilibrium 1	$x_{1s} = 0$	$x_{2s} = 4.0$
Equilibrium 2	$x_{1s} = 0.9951$	$x_{2s} = 1.5123$
Equilibrium 3	$x_{1s} = 1.5302$	$x_{2s} = 0.1745$

Notice that Equilibrium 3 on the SI model is almost identical to Equilibrium 2 for the Monod model. In this section we have discussed case 1 results ($D = 0.3$) only. Cases 2 and 3 will be discussed in Section M8.7.

In the next section we will analyze the dynamic behavior of this system, and in Section M8.7 we will show how multiple steady-state solutions arise.

M8.4   DYNAMIC BEHAVIOR

In the previous section we found that the Monod and substrate inhibition models had two and three steady-state solutions, respectively, for the Case 1 parameter values. In this section we perform simulations of the dynamic behavior of this system. A function file named `bio.m` is shown in Appendix 2.

M8.4.1   Case 1 (D = 0.3), Substrate Inhibition Model

The initial simulation is with the substrate inhibition parameters under Case 1 conditions ($D = 0.3$). The simulations for two different initial conditions are shown in Figure M8.3.

```
» [t1,x1] = ode45('bio',0,30,[1;1]);

» [t2,x2] = ode45('bio',0,30,[0.75;2]);
```

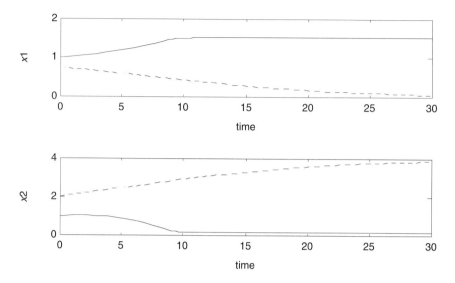

FIGURE M8.3   Substrate inhibition, Case 1. $x0 = [1,1]$ (solid), $x0 = [0.75,2]$ (dashed).

Although both initial conditions are reasonably close to the Equilibrium 2 solution found in section 3, one simulation converges to Equilibrium 1 (dashed line) while the other converges to Equilibrium 3 (solid line). We find in the next section that Equilibrium 2 is unstable. Further simulations will be performed and analyzed in the phase-plane (section 6).

M8.5   LINEARIZATION

In this section we find the linear state-space and transfer function models. So that there is no confusion in notation, we will use the following form:

$$\dot{\mathbf{z}} = \mathbf{A}\,\mathbf{z} + \mathbf{B}\,\mathbf{u}$$
$$\mathbf{y} = \mathbf{C}\,\mathbf{z}$$

where:

$$z_1 = x_1 - x_{1s}$$
$$z_2 = x_2 - x_{2s}$$

$$u_1 = D - D_s$$
$$u_2 = x_{2f} - x_{2fs}$$

The state-space matrices are:

$$\mathbf{A} = \begin{bmatrix} \mu_s - D_s & x_{1s}\mu_s' \\ -\dfrac{\mu_s}{Y} & -D_s - \dfrac{\mu_s' x_{1s}}{Y} \end{bmatrix}$$

$$\mathbf{B} = \begin{bmatrix} -x_{1s} & 0 \\ x_{2f} - x_{2fs} & D_s \end{bmatrix}$$

$$\mathbf{C} = \begin{bmatrix} 1 & 0 \\ 0 & 1 \end{bmatrix}$$

where it is assumed that both states are outputs. The notation μ_s' is used in the $\mathbf{A}$ matrix to represent the derivative of growth rate with respect to substrate concentration, evaluated at steady-state:

$$\mu_s' = \frac{\partial \mu_s}{\partial x_{2s}}$$

For the *Monod* model:

$$\mu_s' = \frac{\partial \mu_s}{\partial x_{2s}} = \frac{\mu_{max} k_m}{(k_m + x_{2s})^2} \tag{M8.15}$$

and for the *substrate inhibition* model:

$$\mu_s' = \frac{\partial \mu_s}{\partial x_{2s}} = \frac{\mu_{max}}{k_m + x_{2s} + k_1 x_{2s}^2} - \frac{\mu_{max} x_{2s}(1 + 2k_1 x_{2s})}{(k_m + x_{2s} + k_1 x_{2s}^2)^2}$$

$$\mu_s' = \frac{\mu_{max}(k_m + x_{2s} + k_1 x_{2s}^2) - \mu_{max} x_{2s}(1 + 2k_1 x_{2s})}{(k_m + x_{2s} + k_1 x_{2s}^2)^2} \tag{M8.16}$$

$$\mu_s' = \frac{\mu_{max}(k_m - k_1 x_{2s}^2)}{(k_m + x_{2s} + k_1 x_{2s}^2)^2}$$

M8.5.1   Substrate Inhibition Model

Here we analyze the substrate inhibition model under Case 1 conditions. A MATLAB m-file, `bio_jac.m` (Appendix 1), is used to generate the $\mathbf{A}$ matrix and the eigenvectors and eigenvalues.

EQUILIBRIUM POINT 1

The steady-state value (section 3) is $(x_{1s}, x_{2s}) = (0,4)$.
    The following command is entered:

```
»[jac,evec,lambda] = bio_jac([0;4])
```

where `jac` is the Jacobian (A matrix), `evec` is the eigenvector matrix and `lambda` are the eigenvalues.

```
jac =
    -0.1139            0
    -0.4652       -0.3000

evec =
         0        0.3714
    1.0000       -0.9285

lambda =
    -0.3000            0
         0       -0.1139
```

so,

$$\mathbf{A} = \begin{bmatrix} -0.1139 & 0 \\ -0.4652 & -0.300 \end{bmatrix}$$

$$\lambda_1 = -0.3 \qquad \xi_1 = \begin{bmatrix} 0 \\ 1 \end{bmatrix}$$

$$\lambda_2 = -0.1139 \qquad \xi_2 = \begin{bmatrix} 0.3714 \\ -0.9285 \end{bmatrix}$$

Since both eigenvalues are negative, the system is *stable* at equilibrium point 1, verifying the simulation results shown in Section M8.4.

EQUILIBRIUM POINT 2

The steady-state value is $(x_{1s},x_{2s}) = (0.9951,1.5122)$.

```
»[jac,evec,lambda]  =  bio_jac([0.9951;1.5122])

jac =
     0.0000       -0.0679
    -0.7500       -0.1302

evec =
     0.3714        0.2209
    -0.9285        0.9753

lambda =
     0.1698            0
         0       -0.3000
```

The positive eigenvalue (0.1698) indicates that equilibrium 2 is *unstable*.

EQUILIBRIUM POINT 3

The steady-state is $(x_{1s}, x_{2s}) = (1.5302, 0.1746)$.

```
»[jac,evec,lambda] = bio_jac([1.5302;0.1746])

jac =
    0.0000    0.9048
   -0.7500   -2.5619

evec =
    0.9492   -0.3714
   -0.3147    0.9285

lambda =
   -0.3000         0
         0   -2.2619
```

Both eigenvalues are negative, indicating that equilibrium point 3 is *stable*.

M8.6   PHASE-PLANE ANALYSIS

The m-file `bio_phas_gen.m` (Appendix 2) was used to generate the following phase-plane plot for the *substrate inhibition* model under Case 1 conditions (see Figure M8.4). Notice that all initial conditions converge to either the washout steady-state (trivial solu-

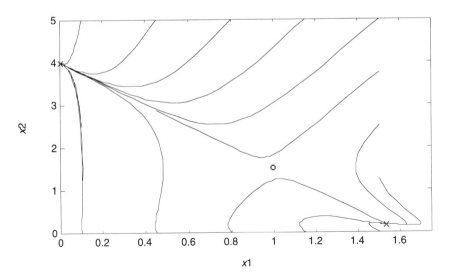

FIGURE M8.4 Phase-plane plot for *substrate inhibition* model, Case 1 conditions (x = stable steady-state, o = unstable steady-state).

tion, equilibrium 1) or equilibrium 3; while equilibrium 2 is a saddle point (unstable). These results are consistent with the stability analysis of Section M8.5.

A phase-plane plot for the Monod model was shown in Chapter 13.

M8.7   UNDERSTANDING MULTIPLE STEADY-STATES*

In this section we find analytically the steady-state solutions for the bioreactor model and determine their stability.

The steady-state solutions ($d_{x1}/dt = d_{x2}/dt = 0$) of (M8.11) and (M8.12) are:

$$0 = (\mu_s - D_s)\, x_{1s} \qquad\qquad\qquad (M8.17)$$

$$0 = D_s\,(x_{2fs} - x_{2s}) - \frac{\mu x_{1s}}{Y} \qquad\qquad (M8.18)$$

where the subscript s indicates steady-state.

There are two different types of solutions to (M8.17) and (M8.18). One is known as the trivial or "washout" solution. The other type is the nontrivial solution.

M8.7.1   Washout Condition

From (M8.17) and (M8.18) we can immediately see one solution, usually called the trivial solution.

$$x_{1s} = 0$$
$$x_{2s} = x_{2fs} \qquad\qquad\qquad\qquad (M8.19)$$

This is also known as the washout condition, since the reactor concentrations are equal to the feed concentrations; that is, there is no "reaction." Since there is no biomass in the feed stream, then there is no biomass in the reactor under these conditions; all of the cells have been "washed out" of the reactor.

M8.7.2   Nontrivial Solutions

From (M8.17), assuming that $x_{1s} \neq 0$, then:

$$\mu_s = D_s \qquad\qquad\qquad\qquad (M8.20)$$

which indicates that the specific growth rate is equal to the dilution rate, at steady-state.

From (M8.18) we find that:

$$D_s\,(x_{2fs} - x_{2s}) = \frac{\mu_s x_{1s}}{Y} \qquad\qquad (M8.21)$$

and from (M8.20) and (M8.21):

$$x_{1s} = Y\,(x_{2fs} - x_{2s}) \qquad\qquad\qquad (M8.22)$$

*This section contains a detailed analysis which the reader may wish to skip on a first reading.

We can solve for x_{2s}, by using the relationship for μ_s as a function of x_{2s} (either Monod or substrate inhibition), since we know that $\mu_s = D_s$ (from (M8.20)). Let $\mu_s(x_{2s})$ represent this general functionality. Then, we must solve:

$$\mu_s(x_{2s}) = D_s \tag{M8.23}$$

for x_{2s}, then substitute this value into (M8.22) to solve for x_{1s}. The specific cases of Monod and substrate inhibition are shown in the subsections below.

MONOD

From (M8.13), the dilution rate at steady-state is

$$\mu_s = \frac{\mu_{max} x_{2s}}{k_m + x_{2s}} \tag{M8.24}$$

Solving (M8.24) for x_{2s}, we find:

$$x_{2s} = \frac{k_m \mu_s}{\mu_{max} - \mu_s} \tag{M8.25}$$

and since $\mu_s = D_s$

$$x_{2s} = \frac{k_m D_s}{\mu_{max} - D_s} \tag{M8.26}$$

For (M8.26) to be feasible, we note that $D_s < \mu_{max}$. Actually, there is a more rigid requirement than that. From (M8.22) we note that the highest value that x_{2s} can be is x_{2fs}, otherwise x_{1s} will be less than zero. The maximum D_s in reality is then $\mu_s(x_{2fs})$, or (from (M8.22), letting $x_{2s} = x_{2fs}$):

$$D_s < \frac{\mu_{max} x_{2fs}}{k_m + x_{2fs}} \qquad (Monod) \tag{M8.27}$$

We also see from (M8.26) that there is a single solution for x_{2s} as a function of D_s. This means that there is a total of *two* steady-state solutions for the Monod model, since there is also the washout (trivial) steady-state.

SUBSTRATE INHIBITION

We found in the previous subsection that there are two possible steady-states for the Monod model, for a given dilution rate. In this subsection we find the number of possible steady-states for the substrate inhibition model.

From (M8.14) at steady-state:

$$\mu_s = \frac{\mu_{max} x_{2s}}{k_m + x_{2s} + k_1 x_{2s}^2} \tag{M8.28}$$

From (M8.28), we find that:

$$k_1 x_{2s}^2 + \left(1 - \frac{\mu_{max}}{\mu_s}\right) x_{2s} + k_m = 0 \qquad (M8.29)$$

Since $\mu_s = D_s$ (M8.20), we substitute into (M8.29) to find:

$$k_1 x_{2s}^2 + \left(1 - \frac{\mu_{max}}{D_s}\right) x_{2s} + k_m = 0 \qquad (M8.30)$$

Since (M8.30) is a quadratic equation, there will be two solutions for x_{2s}. This means that there are *three* steady-state solutions for substrate inhibition, since there is also the washout (trivial) steady-state.

We see from (M8.30) that for positive values of x_{2s} the coefficient multiplying x_{2s} must be negative. The implication is that μ_{max} must be greater than D_s (the same result as the Monod equation). This implication can be seen more clearly from the solution of the quadratic formula for (M8.30):

$$x_{2s} = \frac{-\left(1 - \frac{\mu_{max}}{D_s}\right) \pm \sqrt{\left(1 - \frac{\mu_{max}}{D_s}\right)^2 - 4k_1 k_m}}{2k_1} \qquad (M8.31)$$

So, for solutions with physical significance:

$$\left(1 - \frac{\mu_{max}}{D_s}\right)^2 > 4k_1 k_m \qquad (M8.32)$$

and

$$\mu_{max} > D_s \qquad (M8.33)$$

Because of (M8.33), we know that the term inside the brackets in (M8.32) is negative. For (M8.32) to be satisfied, then we know:

$$\left(1 - \frac{\mu_{max}}{D_s}\right) < -\sqrt{4k_1 k_m} \qquad (M8.34)$$

which implies that:

$$D_s < \frac{\mu_{max}}{1 + 2\sqrt{k_1 k_m}} \qquad \text{(\textit{substrate inhibition})} \qquad (M8.35)$$

We could have found the same result from viewing Figure M8.2. Notice that there is a peak in the μ_s curve, and again recall that $D_s = \mu_s$. The steady-state dilution rate, D_s cannot be above the peak in the x_{2s} versus μ_s curve. We can find the peak by finding $\partial\mu_s/\partial x_{2s} = 0$. From (M8.16):

$$\frac{\partial\mu_s}{\partial x_{2s}} = \frac{\mu_{max}(k_m - k_1 x_{2s}^2)}{(k_m + x_{2s} + k_1 x_{2s}^2)^2} = 0 \qquad (M8.36)$$

We see from (M8.36) that $\partial\mu_s/\partial x_{2s} = 0$ if:

$$x_{2s} = \sqrt{\frac{k_m}{k_1}} \tag{M8.37}$$

We can substitute this result into (M8.28) to find:

$$\mu_s = \frac{\mu_{max}\sqrt{\dfrac{k_m}{k_1}}}{k_m + \sqrt{\dfrac{k_m}{k_1}} + k_1\dfrac{k_m}{k_1}} = \frac{\mu_{max}}{k_m\sqrt{\dfrac{k_1}{k_m}} + 1 + k_m\sqrt{\dfrac{k_1}{k_m}}} \tag{M8.38}$$

$$= \frac{\mu_{max}}{1 + 2k_m\sqrt{\dfrac{k_1}{k_m}}}$$

$$\mu_s = \frac{\mu_{max}}{1 + 2\sqrt{k_1 k_m}} \tag{M8.39}$$

so the maximum dilution rate (for the nontrivial steady-state) is:

$$D_s = \frac{\mu_{max}}{1 + 2\sqrt{k_1 k_m}}$$

which is the same result as (M8.35).

M8.7.3   Summary of Steady-State—Monod and Substrate Inhibition

WASHOUT (BOTH MONOD AND SUBSTRATE INHIBITION)

Both Monod and substrate inhibition models have a washout (trivial) steady-state:

$$x_{1s} = 0 \qquad x_{2s} = x_{2fs} \tag{M8.19}$$

NONTRIVIAL STEADY-STATE FOR MONOD

The nontrivial steady-state solutions for substrate and biomass are:

$$x_{2s} = \frac{k_m D_s}{\mu_{max} - D_s} \tag{M8.26}$$

$$x_{1s} = Y(x_{2fs} - x_{2s}) \tag{M8.22}$$

with the requirement that $D_s < \mu_{max}\, x_{2fs}/k_m + x_{2fs}$ (that is $D_s < \mu_s(x_{2fs})$)

NONTRIVIAL STEADY-STATES FOR SUBSTRATE INHIBITION

The two nontrivial steady-state solutions for substrate are:

$$x_{2s} = \frac{-\left(1 - \dfrac{\mu_{max}}{D_s}\right) + \sqrt{\left(1 - \dfrac{\mu_{max}}{D_s}\right)^2 - 4k_1 k_m}}{2k_1}$$

(M8.31)

and the associated biomass concentration is:

$$x_{1s} = Y(x_{2fs} - x_{2s})$$

(M8.25)

with the requirement for dilution rate:

$$D_s < \frac{\mu_{max}}{1 + 2\sqrt{k_1 k_m}}$$

(M8.32)

M8.7.4   Stability of the Steady-States

The stability of each steady-state solution is determined from the eigenvalues of the Jacobian matrix (matrix A in the state-space form). For a two-state system we know that the eigenvalues are found by:

$$\det(\lambda I - A) = \lambda^2 - \operatorname{tr}(A)\lambda + \det(A) = 0$$

(M8.40)

From Chapter 13 we know that the following conditions must be satisfied for stability of a second-order system:

$$\operatorname{tr}(A) < 0$$

(M8.41)

$$\det(A) > 0$$

(M8.42)

That is, the eigenvalues (λ) will be negative if conditions (M8.41) and (M8.42) are satisfied. The Jacobian of the bioreactor modeling equations (M8.11 and M8.12) is:

$$\mathbf{A} = \begin{bmatrix} \mu_s - D_s & x_{1s}\mu_s' \\ -\dfrac{\mu_s}{Y} & -D_s - \dfrac{\mu_s' x_{1s}}{Y} \end{bmatrix}$$

(M8.43)

where we have used the notation μ_s' to represent the derivative of growth rate with respect to substrate concentration, evaluated at steady-state:

$$\mu_s' = \frac{\partial \mu_s}{\partial x_{2s}}$$

(M8.44)

The trace and determinant of A are:

$$\operatorname{tr}(A) = (\mu_s - D_s) - D_s - \frac{\mu_s' x_{1s}}{Y}$$

(M8.45)

$$\det(A) = -(\mu_s - D_s)\left(D_s + \frac{\mu_s' x_{1s}}{Y}\right) + \frac{x_{1s}\mu_s' \mu_s}{Y}$$

(M8.46)

We will use (M8.45), (M8.46), and the conditions shown in (M8.41) and (M8.42) to determine the stability of each steady-state.

STABILITY OF WASHOUT STEADY-STATE

Under washout conditions: $x_{2s} = x_{2fs}$ and $x_{1s} = 0$.

For stability of the washout steady-state, the following criteria then must be met. First, from the requirement that $tr(A) < 0$:

$$\mu_s - D_s - D_s < 0 \tag{M8.47}$$

and from the requirement that $det(A) > 0$:

$$-(\mu_s - D_s)(D_s) > 0 \tag{M8.48}$$

From (M8.47) we see that the requirement for stability is then:

$$D_s > \frac{\mu_s}{2} \tag{M8.49}$$

while from (M8.48) the requirement for stability is:

$$D_s > \mu_s \tag{M8.50}$$

Notice that μ_s is evaluated at the substrate feed concentration for the washout condition. Perhaps the expression $\mu_s(x_{2fs})$ should be used to designate this relationship. Comparing (M8.49) and (M8.50), we see that (M8.50) is the more rigorous requirement for stability of the washout steady-state.

The growth rate expression for Monod kinetics is:

$$\mu_s = \mu_s(x_{2fs}) = \frac{\mu_{max} x_{2fs}}{k_m + x_{2fs}} \tag{M8.51}$$

while for substrate inhibition kinetics:

$$\mu_s = \mu_s(x_{2fs}) = \frac{\mu_{max} x_{2fs}}{k_m + x_{2fs} + k_1 x_{2fs}^2} \tag{M8.52}$$

Notice that $\mu_s(x_{2fs})$ is simply a shorthand expression for the specific growth rate evaluated at the substrate feed concentration. We must use (M8.50) along with either (M8.51) or (M8.52) to determine the stability of the washout steady-state. Notice that the washout steady-state will only be stable if D_s is high enough. We can think of D_s as a dynamic bifurcation parameter, because the stability of the washout steady-state will depend on the value of the dilution rate.

Stability of Washout Steady-State for Monod.   From (M8.50) and (M8.51), the washout steady-state will be stable if:

$$D_s > \frac{\mu_{max} x_{2fs}}{k_m + x_{2fs}} \tag{M8.53}$$

and unstable if:

$$D_s < \frac{\mu_{max} x_{2fs}}{k_m + x_{2fs}} \tag{M8.54}$$

Stability of Washout Steady-State for Substrate Inhibition. From (M8.50) and (M8.52), the washout steady-state will be stable if:

$$D_s > \frac{\mu_{max} x_{2fs}}{k_m + x_{2fs} + k_1 x_{2fs}^2} \tag{M8.55}$$

and unstable if:

$$D_s < \frac{\mu_{max} x_{2fs}}{k_m + x_{2fs} + k_1 x_{2fs}^2} \tag{M8.56}$$

NONTRIVIAL STEADY-STATES

For the nontrivial steady-states, $D_s = \mu_s$. The stability requirements for the nontrivial steady-states are then:

$$-D_s - \frac{\mu'_s x_{1s}}{Y} < 0 \tag{M8.57}$$

from the tr(A) specification, and

$$\frac{x_{1s}\mu'_s \mu_s}{Y} > 0 \tag{M8.58}$$

from the det(A) specification. Since D_s (and therefore μ_s), x_{1s}, and Y are positive, (M8.57) and (M8.58) reduce to the requirement that:

$$\mu'_s > 0 \tag{M8.59}$$

for stability.

Stability of Monod at the Nontrivial Steady-State. From (M8.24):

$$\mu'_s = \frac{\mu_{max} k_m}{(k_m + x_{2s})^2} \tag{M8.60}$$

We see immediately that μ'_s is always positive for the Monod model at the nontrivial steady-state; therefore, the nontrivial steady-state is always stable. Recall that $D_s < \mu_s(x_{2fs})$ for a nontrivial steady-state solution.

Stability of Substrate Inhibition at the Nontrivial Steady-States. We can tell from the substrate inhibition curve in Figure M8.2 that a steady-state that is on the left side of the peak will be stable (since $\mu'_s > 0$), while a steady-state on the right side will be unstable (since $\mu'_s < 0$).

Numerically, from (M8.36):

$$\mu'_s = \frac{\mu_{max}(k_m - k_1 x_{2s}^2)}{(k_m + x_{2s} + k_1 x_{2s}^2)^2} \tag{M8.61}$$

The x_{2s} that is on the left side of the peak, and is therefore *stable*, is (from (M8.31)):

$$x_{2s} = \frac{\left(\dfrac{\mu_{max}}{D_s} - 1\right) - \sqrt{\left(1 - \dfrac{\mu_{max}}{D_s}\right)^2 - 4k_1 k_m}}{2k_1} \tag{M8.62}$$

The x_{2s} that is on the right side of the peak, and is therefore *unstable*, is (from (M8.31)):

$$x_{2s} = \frac{\left(\dfrac{\mu_{max}}{D_s} - 1\right) + \sqrt{\left(1 - \dfrac{\mu_{max}}{D_s}\right)^2 - 4k_1 k_m}}{2k_1} \tag{M8.63}$$

Also, recall that $D_s < \mu_{max}/1 + 2\sqrt{k_1 k_m}$ (which is equivalent to requiring a real nontrivial solution).

M8.7.5   Case 1 ($D_s = 0.3$)

The reader should find the following results:

Monod

Equilibrium 1—washout	$x_{1s} = 0$	$x_{2s} = 4.0$	unstable
Equilibrium 2—nontrivial	$x_{1s} = 1.5374$	$x_{2s} = 0.1565$	stable

Substrate Inhibition

Equilibrium 1—washout	$x_{1s} = 0$	$x_{2s} = 4.0$	stable
Equilibrium 2—nontrivial	$x_{1s} = 0.9951$	$x_{2s} = 1.5123$	unstable (saddle point)
Equilibrium 3—nontrivial	$x_{1s} = 1.5302$	$x_{2s} = 0.1745$	stable

M8.7.6   Case 2

For a steady-state dilution rate of $D_s = 0.15$, the reader should find the following results:

Monod

Equilibrium 1—washout	$x_{1s} = 0$	$x_{2s} = 4.0$	unstable
Equilibrium 2—nontrivial	$x_{1s} = 1.5811$	$x_{2s} = 0.0474$	stable

Substrate Inhibition

Equilibrium 1—washout	$x_{1s} = 0$	$x_{2s} = 4.0$	unstable
Equilibrium 2—nontrivial	$x_{1s} = -0.6104$	$x_{2s} = 5.5261$	not feasible
Equilibrium 3—nontrivial	$x_{1s} = 1.5809$	$x_{2s} = 0.0478$	stable

Although there is a mathematical solution for equilibrium 2, it is not physically feasible, since it corresponds to a negative biomass concentration.

M8.7.7   Case 3

For a steady-state dilution rate of $D_s = 0.6$, the student should find the following results:

Monod

Equilibrium 1—washout	$x_{1s} = 0$	$x_{2s} = 4.0$	stable
Equilibrium 2—nontrivial	$x_{1s} = 2.0114$	$x_{2s} = -1.0286$	not feasible

Steady-state 2 is not feasible, because it corresponds to a negative substrate concentration.

Substrate Inhibition

Equilibrium 1—washout	$x_{1s} = 0$	$x_{2s} = 4.0$	stable
Equilibrium 2—nontrivial	$x_{1s} = 4.13 - 0.20j$	$x_{2s} = -0.13 + 0.50j$	not feasible
Equilibrium 3—nontrivial	$x_{1s} = 4.13 + 0.20j$	$x_{2s} = -0.13 - 0.50j$	not feasible

The second and third steady-states are not feasible because the concentrations for both the biomass and the substrate are complex.

There are some very interesting changes in the dynamic behavior of these models as we vary the dilution rate (again, we can think of dilution rate as a bifurcation parameter). Let us discuss this in order of the lowest dilution rate to the highest dilution rate.

LOW DILUTION RATE

Case 2 had the lowest dilution rate ($D_s = 0.15$). The Monod model has two steady-states—the washout steady-state is unstable and the other (nontrivial) steady-state is stable. This means that any set of initial conditions will eventually converge to the nontrivial steady-state, for the Monod model. The substrate inhibition model has only two feasible steady-states—the washout steady-state is unstable and the high conversion steady-state is stable.

The interesting result is that at low dilution rates, the substrate inhibition model behaves like the Monod model.

MEDIUM DILUTION RATE

Case 1 had the next highest dilution rate ($D_s = 0.30$). The Monod model has two steady-states—the washout steady-state is unstable and the other (nontrivial) steady-state is stable. This means that any set of initial conditions will eventually converge to the nontrivial steady-state, for the Monod model. The substrate inhibition model has three feasible steady-states. The washout (no conversion) steady-state is stable, the medium conversion steady-state is unstable and the high conversion (low x_{2s}) steady-state is stable. This means that any set of initial conditions will converge to one of the two stable steady-states. A phase-plane must be drawn to determine if a particular set of initial conditions will lead to washout.

HIGH DILUTION RATE

Case 3 had the highest dilution rate ($D_s = 0.60$). Both models had only one feasible steady-state, the washout steady-state, and it was stable. The student should be able to "sketch" phase planes for all three conditions for each of the models (Monod and substrate inhibition).

M8.8   BIFURCATION BEHAVIOR

The conditions for stability developed in Section M8.7 can be used to develop steady-state input-output diagrams for the numerical example presented in the previous sections.

M8.8.1   Diagram for the Monod Model

The diagram for the Monod model is shown in Figure M8.5. As calculated, the Monod model has two steady-states for dilution rates that are less than the specific growth rate under the feed conditions, $D_s < \mu_s(x_{2fs})$. The nontrivial steady-state is stable under those conditions, while the washout steady-state is unstable. For $D_s > \mu_s(x_{2fs})$ there is a single steady-state, the washout steady-state, and it is stable.

M8.8.2   Diagram for the Substrate Inhibition Model

The diagram for the substrate inhibition model is shown in Figure M8.6. At low dilution rates, where $D_s < \mu_s(x_{2fs})$, there are two steady-states (like the Monod model). The nontrivial steady-state is stable under those conditions, while the washout steady-state is unstable. For the intermediate dilution rate range, $\mu_s(x_{2fs}) < D_s < \mu_{max}/1 + 2\sqrt{k_1 k_m}$,

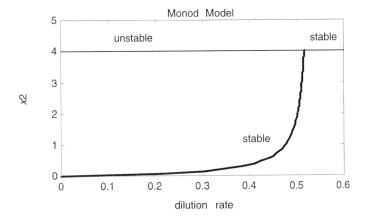

FIGURE M8.5   Input-output diagram for the Monod model.

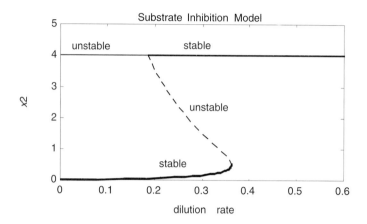

FIGURE M8.6   Input-output diagram for the substrate inhibition model.

there are three steady-states. Two of these are stable, while one is unstable. The stable steady-state that is attained will depend upon the initial conditions of the concentrations, or on the way that the process is started up. When the dilution rate meets the condition that $D_s > \mu_s(x_{2fs})$, there is a single steady-state, the washout steady-state, and it is stable.

M8.8.3   Hysteresis Behavior for the Substrate Inhibition Model

It is interesting to note that the way that the bioreactor is started up will determine the steady-state concentrations that the reactor achieves. Look at Figure M8.6. Notice that if we start at a very low dilution rate we will have only one stable steady-state, so the reactor must operate at that condition. If we slowly increase the dilution rate, we remain on the lower curve of Figure M8.6. When $D_s > \mu_{max}/1 + 2\sqrt{k_1 k_m}$ ($D_s = 0.36126$ for this example), the stable solution suddenly "leaps" to the upper stable steady-state (washout conditions). As we increase D_s further, we remain on the washout curve.

Now, assume that we are starting out at a high dilution rate along the upper curve, the washout conditions. As we slowly decrease the dilution rate, we remain on the washout curve until $D_s = \mu_s(x_{2fs})$, which is $D_s = 0.1861$ for this example. The stable steady-state then "jumps" down to the lower curve. As we continue to decrease the dilution rate further, we remain on the lower curve.

The type of behavior shown in Figure M8.6 is known as hysteresis and is exhibited by a number of processes, including exothermic chemical reactors and valves that "stick." The chemical reactor example is discussed further in Module 9.

The student should be able to show how the phase-plane behavior changes as a function of dilution rate, for the example shown in Figure M8.6.

SUMMARY

The modeling equations for a biochemical reactor were developed for Monod and substrate inhibition kinetics. We found that the Monod model normally has two steady-state solutions, while the substrate inhibition model normally has three steady-state solutions. At low dilution rates the substrate inhibition model behaves similarly to the Monod model, with a single stable steady-state. Washout will not be a problem at the low dilution rates.

At medium dilution rates the substrate inhibition model behaves quite differently from the Monod model. Depending on the initial conditions, the reactor will either converge to a high conversion or to washout conditions for the substrate inhibition model. It has not been discussed thus far, but if we wish to operate at an intermediate (unstable) conversion level, then feedback control must be used. Notice that the Monod model still has only one stable point, and there is no danger of wash-out.

At high dilution rates, both reactor models have only one feasible solution—washout. The flow is simply too high (residence time too low) for any cell growth.

FURTHER READING

An excellent source for an introduction to biochemical engineering is:

> Bailey, J.E., & D.F. Ollis. (1986). *Biochemical Engineering Fundamentals*, 2nd ed. New York: McGraw-Hill.

STUDENT EXERCISES

1. In this module we developed the modeling equations assuming that no biomass is fed to the reactor. Analyze the system studied for the case where the biomass feed concentration is 2.5% of the substrate feed concentration (so $x_{1f} = 0.1$ for the numerical values used in this module).

   Is there still the possibility of a washout steady-state?

2. Modify `bio_phas_gen.m` and `bio.m` to perform a phase-plane analysis for cases 2 and 3 with the substrate inhibition model.

3. Data for specific growth rate coefficient as a function substrate concentration for a biochemical reactor are shown below:

x_2, g/liter	μ, hr^{-1}
0	0
0.1	0.38
0.25	0.54

0.5	0.63
0.75	0.66
1	0.68
1.5	0.70
3	0.73
5	0.74

a. Estimate the parameter values for a Monod model (k_m, μ_{max})
b. The production rate of cells (biomass) is $D*x1$. Find the steady-state value of the dilution rate that maximizes the production rate of cells. The substrate feed concentration is 5 g/liter.
c. Find the steady-state concentration of biomass and substrate at this dilution rate.
d. Find the linear state-space model at this dilution rate, with dilution rate and substrate feed concentration as the input variables. Also find the transfer function relating dilution rate to biomass concentration.
e. Simulate the responses (using the nonlinear dynamic model) of the concentrations of biomass and substrate to step increases and decreases of 10% in the dilution rate (changes are from the dilution rate found in b.). Compare these results with those of the linear system (remember to convert deviation variables back to physical variables).

4. In this module we have analyzed how the biomass and substrate concentrations change depending on the dilution rate. If the purpose of a particular biochemical reactor is to produce cells, then we are more concerned with the production rate of cells. The production rate is mass of cells produced per unit time:

$$\text{steady-state production rate of cells} = D_s * x_{1s}$$

For both the Monod and substrate inhibition models presented in this module, find the dilution rate that maximizes the production rate of cells. Analyze the stability of the reactor under this condition.

5. Consider a biochemical reactor where the consumption of substrate (x_2) promotes the growth of biomass (x_1) and formation of product (x_3). The three modeling equations are:

$$\frac{dx_1}{dt} = (\mu - D)x_1$$

$$\frac{dx_2}{dt} = D(x_{2f} - x_2) - \frac{\mu x_1}{Y}$$

$$\frac{dx_3}{dt} = -Dx_3 + [\alpha\mu + \beta]x_1$$

where the specific growth rate is a function of both the biomass concentration and the product concentration:

$$\mu = \frac{\mu_{max}\,(1 - P/P_m)\,x_2}{k_m + x_2 + k_1 x_2^2}$$

with the following parameter values:[1]

Variable	Value	Variable	Value
Y	0.4 g/g	α	2.2 g/g
β	0.2 hr^{-1}	μ_{max}	0.48 hr^{-1}
P_m	50 g/liter	k_m	1.2 g/liter
k_1	0.04545 liter/g	x_{2f}	20 g/liter
D	0.202 hr^{-1}	x_1	6 g/liter
x_2	5 g/liter	x_3	19.14 g/liter

a. Compare and contrast this model with that of the two-state model with substrate inhibition kinetics presented in this module.
b. Verify that the steady-state values for x_1, x_2, and x_3 presented in the table above are correct. For a steady-state input of $D = 0.202$ (and all of the other parameters constant), are there any additional solutions for the states (for example, the trivial solution?). Analyze the stability of all steady-state solutions obtained.
c. Perform dynamic simulations of the nonlinear model, with step changes of ± 10% in the dilution rate. Discuss the results of your step changes (i.e., does an increase or decrease in D have a greater effect on the biomass concentration?). Compare your results with linear simulations.

APPENDIXES

1 Steady-State Biochemical Reactor Model, `bio_ss.m`

```
function f = bio_ss(x)
%
% b.w. bequette
% (c) 16 Nov 92
% revised 18 July 96
%
% find steady-states of bioreactor, using fsolve:
%
%   x = fsolve('bio',x0)
%
```

[1]This model is from Chapter 4 of the following monograph: Henson, M.A., & D.E. Seborg (ed.). (1997). *Nonlinear Process Control.* Upper Saddle River, NJ: Prentice-Hall.

```
% where x0 is a vector of initial guesses
%
% x(1) = biomass
% x(2) = substrate
%
% biomass ("bugs") consumes the substrate
%
% D     = dilution rate (F/V, time^-1)
% Y     = yield biomass/substrate
% mu    = specific growth rate
% mumax = parameter (both Monod and Substrate Inhibition)
% km    = parameter (both Monod and Substrate Inhibition)
% k1    = parameter (Substrate Inhibition only, k1 = 0 for Monod)
% sf    = substrate feed concentration
%
% the function vector consists of 2 equations
%
  f = zeros(2,1);
%
% parameter values
%
  D = 0.3;
  mumax = 0.53;
  Y = 0.4;
  km = 0.12;
  sf = 4.0;
  k1 = 0.4545;
%
%  Substrate Inhibition expression for specific growth rate
%
  mu = mumax*x(2)/(km+x(2)+k1*x(2)*x(2));
%
%  steady-state equations
%  fsolve varies x(1) and x(2) to drive f(1) and f(2) to zero
%
  f(1) = (mu - D)*x(1);
  f(2) = (sf - x(2))*D - mu*x(1)/Y;
```

2 `bio.m`, Function File for Dynamic Simulation Using `ode45`

```
  function xdot = bio(t,x)
%
% b.w. bequette
% (c) 18 July 96
%
% dynamic equations for bioreactor, integrated using ode45,
% using the following command
%
% [t,x] = ode45('bio',t0,tf,x0)
%
```

```
% where t0 is the initial time (usually 0), tf is
% the final time and x0 is the initial condition vector
% x0(1) = biomass initial condition
% x0(2) = substrate initial condition
%
% biomass ("bugs") consumes the substrate
%
% state variables
%
% x(1) = biomass
% x(2) = substrate
%
% D      = dilution rate (F/V, time^-1)
% Y      = yield biomass/substrate
% mu     = specific growth rate
% mumax  = parameter (both Monod and Substrate Inhibition)
% km     = parameter (both Monod and Substrate Inhibition)
% k1     = parameter (Substrate Inhibition only, k1 = 0 for Monod)
% sf     = substrate feed concentration
%
% the function vector consists of 2 equations
%
   f = zeros(2,1);
%
% parameter values
%
   D = 0.3;
   mumax = 0.53;
   Y = 0.4;
   km = 0.12;
   sf = 4.0;
   k1 = 0.4545;
%
% Substrate Inhibition expression for specific growth rate
%
   mu = mumax*x(2)/(km+x(2)+k1*x(2)*x(2));
%
% dynamic equations
%
   xdot(1) = (mu - D)*x(1);
   xdot(2) = (sf - x(2))*D - mu*x(1)/Y;
```

3 Phase-Plane Plot for Biochemical Reactor, `bio_phas_gen.m`

```
% bio_phas_gen.m
%
% b.w. bequette
% (c) 19 July 96
%
% generates phase-plane plots for the bioreactor
```

```
%
% set-up the axis limits
%
  axis([0 1.75 0 5]);
%
% stable and unstable points for substrate
% inhibition model
%
  x1u = [0.9951];
  x2u = [1.5122];
%
  x1s = [0;1.5302];
  x2s = [4;0.1746];
%
% place an 'x' on stable points
% place a 'o' on unstable points
%
  plot(x1u,x2u,'wo',x1s,x2s,'wx')
%
  hold on
%
% select different initial conditions
% x1 ranges from 0.1 to 1.5 (every 0.35)
% x2 ranges from 0 to 5 (every 1.25)
% total of 16 initial conditions
%
  x1init = [0.1 0.45 0.8 1.15 1.5 0.1 0.45 0.8 1.15 1.5];
  x2init = [ 0    0    0    0    0   5    5    5    5     5];
  x1inita = [1.5  1.5 1.5  0.1  0.1  0.1];
  x2inita = [1.25 2.5 3.75 1.25 2.5 3.75];
%
  x0 = [x1init x1inita;x2init x2inita];
%
% ncol = number of initial conditions
%
  [mrow,ncol] = size(x0);
%
% run simulations for each initial condition
%
  for i = 1:ncol;
%
        [t,x] = ode45('bio',0,30,[x0(:,i)]);
   %
        plot(x(:,1),x(:,2),'w')
   %
     end
%
  xlabel('x1')
  ylabel('x2')
  hold off
```

4   Direct Calculation of the Eigenvalues

Here we calculate the eigenvalues of the nontrivial solution:

$$\det(\lambda I - A) = \lambda^2 - \operatorname{tr}(A)\lambda + \det(A) = 0 \tag{M8.40}$$

where the trace and determinant are

$$\operatorname{tr}(A) = \mu_s - D_s - D_s - \frac{\mu_s' x_{1s}}{Y} \tag{M8.45}$$

$$\det(A) = -(\mu_s - D_s)\left(D_s + \frac{\mu_s' x_{1s}}{Y}\right) + \frac{x_{1s}\mu_s'\mu_s}{Y} \tag{M8.46}$$

For the nontrivial solution, $\mu_s = D_s$:

$$\operatorname{tr}(A) = -D_s - \frac{\mu_s' x_{1s}}{Y} \tag{M8.A1}$$

$$\det(A) = \frac{x_{1s}\mu_s'\mu_s}{Y} \tag{M8.A2}$$

The roots of (M8.40) are:

$$\lambda = \frac{\operatorname{tr}(A) \pm \sqrt{(\operatorname{tr} A)^2 - 4\det A}}{2} \tag{M8.A3}$$

and since $D_s = \mu_s$:

$$\lambda = \frac{-\mu_s - \frac{x_s\mu_s'}{Y} \pm \sqrt{\left(-\mu_s - \frac{x_s\mu_s'}{Y}\right)^2 - \frac{4x_s\mu_s'\mu_s}{Y}}}{2} \tag{M8.A4}$$

$$\lambda = \frac{-\mu_s - \frac{x_s\mu_s'}{Y} \pm \sqrt{\mu_s^2 + \frac{x_s^2\mu_s'^2}{Y^2} + \frac{2x_s\mu_s'\mu_s}{Y} - \frac{4x_s\mu_s'\mu_s}{Y}}}{2} \tag{M8.A5}$$

$$\lambda = \frac{-\mu_s - \frac{x_s\mu_s'}{Y} \pm \sqrt{\mu_s^2 + \frac{x_s^2\mu_s'^2}{Y^2} - \frac{2x_s\mu_s'\mu_s}{Y}}}{2} \tag{M8.A6}$$

$$\lambda = \frac{-\mu_s - \frac{x_s\mu_s'}{Y} \pm \sqrt{\left(-\mu_s - \frac{x_s\mu_s'}{Y}\right)^2}}{2} \tag{M8.A7}$$

$$\lambda = \frac{-\mu_s - \frac{x_s\mu_s'}{Y} \pm \left(\mu_s - \frac{x_s\mu_s'}{Y}\right)}{2} \tag{M8.A8}$$

and our roots are:

$$\lambda_1 = -\mu_s \quad \text{and} \quad \lambda_2 = -\frac{x_s \mu_s'}{Y} \tag{M8.A9}$$

and since $\mu_s = D_s$, we are assured that one pole will always be negative. The second root will only be positive if μ_s' is negative. Since μ_s' is positive for the Monod model, the nontrivial solution is stable as long as the solution is feasible ($D_s < \mu_s(x_{2fs})$). The μ_s' can be either positive or negative for the substrate inhibition model, so a nontrivial steady-state may either be stable or unstable.

DIABATIC CONTINUOUS STIRRED TANK REACTORS

The purpose of this module is to understand the steady-state and dynamic behavior of jacketed continuous stirred tank reactors (CSTRs), also referred to as diabatic CSTRs. After reviewing this module the student should be able to

- Understand the assumptions made in developing the classic CSTR model
- Understand the possible steady-state behavior based on an analysis of the heat generated by reaction and removed through the cooling jacket
- Understand what is meant by multiple steady-states. Develop the *hysteresis* (ignition/extinction behavior) input-output diagrams based on analysis of the heat generation and removal curves. Relate these to saddle-node and hysteresis bifurcations from Chapter 15
- Understand the dynamic behavior by linearization of nonlinear dynamic equations followed by eigenvalue analysis
- Use phase-plane analysis to understand which steady-state a given initial condition will converge to
- Realize the possibility of limit cycle behavior (Hopf bifurcations)

The major sections in this module are:

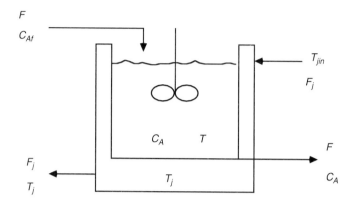

FIGURE M9.1   Continuous stirred tank reactor with cooling jacket.

M9.7   Understanding Multiple Steady-State Behavior
M9.8   Further Complexities
M9.9   Dimensionless Model

M9.1   BACKGROUND

The most important unit operation in a chemical process is generally a chemical reactor. Chemical reactions are either exothermic (release energy) or endothermic (require energy input) and therefore require that energy either be removed or added to the reactor for a constant temperature to be maintained. Exothermic reactions are the most interesting systems to study because of potential safety problems (rapid increases in temperature, sometimes called "ignition" behavior) and the possibility of exotic behavior such as multiple steady-states (for the same value of the input variable there may be several possible values of the output variable).

In this module we consider a perfectly mixed, continuously stirred tank reactor (CSTR),[1] shown in Figure M9.1. The case of a single, first-order exothermic irreversible reaction, $A \rightarrow B$, will be studied. We will find that very interesting behavior can arise in such a simple system.

In Figure M9.1 we see that a fluid stream is continuously fed to the reactor and another fluid stream is continuously removed from the reactor. Since the reactor is perfectly mixed, the exit stream has the same concentration and temperature as the reactor fluid. Notice that a jacket surrounding the reactor also has feed and exit streams. The jacket is assumed to be perfectly mixed and at a lower temperature than the reactor. Energy then passes through the reactor walls into the jacket, removing the heat generated by reaction.

---

[1]Sometimes this type of reactor is called "non-adiabatic." The term adiabatic means "no heat loss," so the term *non-adiabatic* constitutes a double negative (This point has been discussed by Barduhn, *Chem. Eng. Education*, 19, 171 (1985)). We prefer to use *diabatic* to describe this reactor.

There are many examples of reactors in industry similar to this one. Examples include various types of polymerization reactors, which produce polymers that are used in plastic products such as polystyrene coolers or plastic bottles. The industrial reactors typically have more complicated kinetics than we study in this module, but the characteristic behavior is similar.

M9.2   THE MODELING EQUATIONS

For simplicity we assume that the cooling jacket temperature can be directly manipulated, so that an energy balance around the jacket is not required. We also make the following assumptions:

- Perfect mixing (product stream values are the same as the bulk reactor fluid)
- Constant volume
- Constant parameter values

The constant volume and parameter value assumptions can easily be relaxed by the reader for further study.

M9.2.1   Parameters and Variables

The parameters and variables that will appear in the modeling equations are listed below for convenience.

A	Area for heat exchange
C_A	Concentration of A in reactor
C_{Af}	Concentration of A in feed stream
c_p	Heat capacity (energy/mass*temperature)
F	Volumetric flowrate (volume/time)
k_0	Pre-exponential factor (time^{-1})
R	Ideal gas constant (energy/mol*temperature)
r	Rate of reaction per unit volume (mol/volume*time)
t	Time
T	Reactor temperature
T_f	Feed temperature
T_j	Jacket temperature
T_{ref}	Reference temperature
U	Overall heat transfer coefficient (energy/(time*area*temperature))
V	Reactor volume
ΔE	Activation energy (energy/mol)
$(-\Delta H)$	Heat of reaction (energy/mol)
ρ	Density (mass/volume)

M9.2.2   Overall Material Balance

The rate of accumulation of material in the reactor is equal to the rate of material in by flow – the material out by flow.

$$\frac{dV\rho}{dt} = F_{in}\rho_{in} - F_{out}\,\rho$$

Assuming a constant amount of material in the reactor ($dV\rho/dt = 0$), we find that:

$$F_{out}\,\rho = F_{in}\,\rho_{in}$$

If we also assume that the density remains constant,[2] then:

$$F_{out} = F_{in} = F \quad \text{and} \quad dV/dt = 0$$

M9.2.3   Balance on Component A

The balance on component A, assuming a constant volume reactor, is:

$$V\frac{dC_A}{dt} = FC_{Af} - FC_A - rV \tag{M9.1}$$

where r is the rate of reaction per unit volume.

M9.2.4   Energy Balance

The energy balance, assuming constant volume, heat capacity and density, is:

$$V\rho c_p\frac{dT}{dt} = F\rho c_p(T_f - T) + (-\Delta H)Vr - UA(T - T_j) \tag{M9.2}$$

where $(-\Delta H)Vr$ is the rate of energy contributed by the exothermic reaction.

M9.2.5   State Variable Form of Dynamic Equations

We can write (M9.1) and (M9.2) in the following state variable form:

$$f_1(C_A,T) = \frac{dC_A}{dt} = \frac{F}{V}(C_{Af} - C_A) - r \tag{M9.1a}$$

$$f_2(C_A,T) = \frac{dT}{dt} = \frac{F}{V}(T_f - T) + \left(\frac{-\Delta H}{\rho c_p}\right)r - \frac{UA}{V\rho c_p}(T - T_j) \tag{M9.2a}$$

---

[2]It should be noted that the density of all streams does not need to remain constant for the modeling equations to hold. For example, Denn (1986) shows that as long as the density is a linear function of concentration, the final modeling equations are correct.

The reaction rate per unit volume (Arrhenius expression) is:

$$r = k_o \exp\left(\frac{-\Delta E}{RT}\right) C_A \tag{M9.3}$$

where we have assumed that the reaction is first-order.

M9.3 STEADY-STATE SOLUTION

The steady-state solution is obtained when $dC_A/dt = 0$ and $dT/dt = 0$, that is:

$$f_1(C_A,T) = 0 = \frac{F}{V}(C_{Af} - C_A) - k_o \exp\left(\frac{-\Delta E}{RT}\right) C_A \tag{M9.1s}$$

$$f_2(C_A,T) = 0 = \frac{F}{V}(T_f - T) + \left(\frac{-\Delta H}{\rho c_p}\right) k_o \exp\left(\frac{-\Delta E}{RT}\right) C_A - \frac{UA}{V\rho c_p}(T - T_j) \tag{M9.2s}$$

To solve these two equations, all parameters and variables except for two (C_A and T) must be specified. Given numerical values for all of the parameters and variables we can use Newton's method (Chapter 3) to solve for the steady-state values of C_A and T. For convenience, we use an "s" subscript to denote a steady-state value (so we solve for C_{As} and T_s).

In this module we will study the set of parameters shown for Case 2 conditions in Table M9.1. Cases 1 and 3 are left as exercises for the reader.

M9.3.1 Solution for Case 2 Parameters Using `fsolve` and `cstr_ss.m`

The function m-file for the steady-state equations is `cstr_ss.m` and is shown in Appendix 1. The command to run this file is

```
x = fsolve('cstr_ss',x0);
```

where x0 is a vector of the initial guesses and x is the solution. Before issuing this command the reactor parameters must be entered in the global parameter vector CSTR_PAR.

TABLE M9.1   Reactor Parameters

Parameter	Case 1	Case 2	Case 3
F/V, hr^{-1}	1	1	1
k_0, hr^{-1}	14,825*3600	9,703*3600	18,194*3600
$(-\Delta H)$, kcal/kgmol	5215	5960	8195
E, kcal/kgmol	11,843	11,843	11,843
ρc_p, kcal/(m^3°C)	500	500	500
T_f, °C	25	25	25
C_{Af}, kgmol/m^3	10	10	10
UA/V, kcal/(m^3°C hr)	250	150	750
T_j, °C	25	25	25

We find that different initial guesses for the concentration and temperature lead to different solutions.

When choosing initial guesses for a numerical algorithm, it is important to use physical insight about the possible range of solutions. For example, since the feed concentration of A is 10 kgmol/m^3 and the only reaction consumes A, the possible range for the concentration of A is $0 < C_A < 10$. Also, it is easy to show that a lower bound for temperature is 298 K, which would occur if there was no reaction at all, since the feed and jacket temperatures are 298 K. Notice also that there should be a correlation between concentration and temperature. If the concentration of A is high, this means that not much reaction has occurred so little energy has been released by reaction and therefore the temperature will not be much different than the feed and jacket temperatures.

GUESS 1

High concentration (low conversion), low temperature. Here we consider an initial guess of $C_A = 9$ and $T = 300$ K.

```
x = fsolve('cstr_ss',[9;300]);

x =
    8.5636
  311.1710
```

so the steady-state solution for guess 1 is $\begin{bmatrix} C_{As} \\ T_s \end{bmatrix} = \begin{bmatrix} 8.5636 \\ 311.2 \end{bmatrix}$, that is, high concentration (low conversion) and low temperature.

GUESS 2

Intermediate concentration and temperature.

```
x = fsolve('cstr_ss',[5;350])

x =
    5.5179
  339.0971
```

so the steady-state solution for guess 2 is $\begin{bmatrix} C_{As} \\ T_s \end{bmatrix} = \begin{bmatrix} 5.518 \\ 339.1 \end{bmatrix}$.

GUESS 3

Low concentration and high temperature.

```
x = fsolve('cstr_ss',[1;450])

x =
    2.3589
  368.0629
```

TABLE M9.2   Guesses and Solutions Using `fsolve`

Guess and Solution	Guess 1	Guess 2	Guess 3
x0(1), C_A guessed	9	5	1
x0(2), T guessed	300	350	450
x(1), C_A solution	8.564	5.518	2.359
x(2), T solution	311.2	339.1	368.1

so the steady-state solution for guess 3 is $\begin{bmatrix} C_{As} \\ T_s \end{bmatrix} = \begin{bmatrix} 2.359 \\ 368.1 \end{bmatrix}$, that is, low concentration (high conversion) and high temperature.

The results are summarized in Table M9.2.

Other initial guesses do not lead to any other solutions, so we see that there are three possible solutions for this set of parameters. In Section M9.7 we show how to use physical insight to determine the number of steady-state solutions for this problem.

M9.4   DYNAMIC BEHAVIOR

We noted in the previous section that were three different steady-state solutions to the Case 2 parameter set. Here we wish to study the dynamic behavior under this same parameter set. Recall that numerical integration techniques were presented in Chapter 4.

The m-file to integrate the modeling equations is `cstr_dyn.m`, shown in Appendix 2. The command to integrate the equations is

```
[t,x] = ode45('cstr_dyn',t0,tf,x0);
```

where `t0` is the initial time (usually 0), `tf` is the final time, `x0` is the initial condition vector, `t` is the time vector, and `x` is the state variable solution vector. Before performing the integration it is necessary to define the global parameter vector `CSTR_PAR`. To plot only concentration or temperature as a function of time, use `plot(t,x(:,1))` and `plot(t,x(:,2))`, respectively.

M9.4.1   Initial Condition 1

Here we use initial conditions that are close to the low temperature steady-state. The initial condition vector is [conc, temp] = [9,300]. The curves plotted in Figure M9.2 show that the state variables converge to the low temperature steady-state.

M9.4.2   Initial Condition 2

Here we use initial conditions that are close to the intermediate temperature steady-state. The initial condition vector for the solid curve in Figure M9.3 is [conc, temp] = [5,350], which converges to the high temperature steady-state. The initial condition vector for the

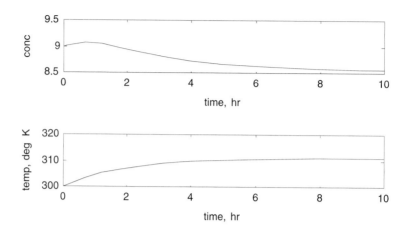

FIGURE M9.2   State variable responses with initial condition $x0 = [9;300]$.

dotted curve in Figure M9.3 is [conc, temp] = [5,325], which converges to the low temperature steady-state.

If we perform many simulations with initial conditions close to the intermediate temperature steady-state, we find that the temperature always converges to either the low temperature or high temperature steady-states, but not the intermediate temperature steady-state. This indicates to us that the intermediate temperature steady-state is *unstable*. This will be shown clearly by the stability analysis in Section M9.5.

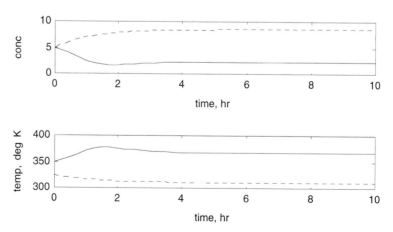

FIGURE M9.3   State variable responses with initial condition $x0 = [5;350]$ (solid) and $x0 = [5;325]$ (dashed).

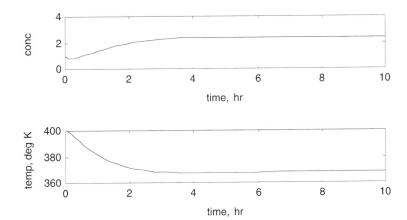

FIGURE M9.4　State variable responses with initial condition $x0 = [1;400]$.

M9.4.3　Initial Condition 3

Here we use initial conditions that are close to the high temperature steady-state. The initial condition vector is [conc, temp] = [1,400]. The curves plotted in Figure M9.4 show that the state variables converge to the high temperature steady-state.

　　In this section we have performed several simulations and presented several plots. In Section M9.6 we will show how these solutions can be compared on the same "phase-plane" plot.

M9.5　LINEARIZATION OF DYNAMIC EQUATIONS

The stability of the nonlinear equations can be determined by finding the following state-space form:

$$\dot{\mathbf{x}} = \mathbf{A}\,\mathbf{x} + \mathbf{B}\,\mathbf{u} \tag{M9.11}$$

and determining the eigenvalues of the $\mathbf{A}$ (state-space) matrix.

　　The nonlinear dynamic state equations (M9.1a) and (M9.2a) are:

$$f_1(C_A,T) = \frac{dC_A}{dt} = -\frac{F}{V}\,C_A - k\,C_A + \frac{F}{V}\,C_{Af} \tag{M9.1a}$$

$$f_2(C_A,T) = \frac{dT}{dt} = \left(\frac{-\Delta H}{\rho c_p}\right) k\,C_A - \frac{F}{V}\,T - \frac{UA}{V\rho C_p}\,T + \frac{UA}{V\rho C_p}\,T_j + \frac{F}{V}\,T_f \tag{M9.2a}$$

Let the state and input variables be defined in deviation variable form:

$$\mathbf{x} = \begin{bmatrix} C_A - C_{As} \\ T - T_s \end{bmatrix}$$

$$\mathbf{u} = \begin{bmatrix} T_j - T_{js} \\ C_{Af} - C_{Afs} \\ T_f - T_{fs} \end{bmatrix}$$

M9.5.1 Stability Analysis

Performing the linearization, we obtain the following elements for $\mathbf{A}$:

$$A_{11} = \frac{\partial f_1}{\partial x_1} = \frac{\partial f_1}{\partial C_A} = -\frac{F}{V} - k_s$$

$$A_{12} = \frac{\partial f_1}{\partial x_2} = \frac{\partial f_1}{\partial T} = -C_{As}k_s'$$

$$A_{21} = \frac{\partial f_2}{\partial x_1} = \frac{\partial f_2}{\partial C_A} = \frac{(-\Delta H)}{\rho C_p} k_s$$

$$A_{22} = \frac{\partial f_2}{\partial x_2} = \frac{\partial f_2}{\partial T} = -\frac{F}{V} - \frac{UA}{V\rho C_p} + \frac{(-\Delta H)}{\rho C_p} C_{As}k_s'$$

where we define the following parameters for more compact representation:

$$k_s = k_o \exp\left(\frac{-\Delta E}{RT_s}\right)$$

$$k_s' = \frac{\partial k_s}{\partial T} = k_o \exp\left(\frac{-\Delta E}{RT_s}\right)\left(\frac{\Delta E}{RT_s^2}\right)$$

or,

$$k_s' = k_s \left(\frac{\Delta E}{RT_s^2}\right)$$

From the analysis presented above, the state-space $\mathbf{A}$ matrix is:

$$\mathbf{A} = \begin{bmatrix} -\dfrac{F}{V} - k_s & -C_{As}k_s' \\ \dfrac{(-\Delta H)}{\rho C_p} k_s & -\dfrac{F}{V} - \dfrac{UA}{V\rho C_p} + \dfrac{(-\Delta H)}{\rho C_p} C_{As}k_s' \end{bmatrix} \qquad (M9.12)$$

The stability characteristics are determined by the eigenvalues of $\mathbf{A}$, which are obtained by solving $\det(\lambda\,\mathbf{I} - \mathbf{A}) = 0$.

$$\lambda \mathbf{I} - \mathbf{A} = \begin{bmatrix} \lambda - A_{11} & -A_{12} \\ -A_{21} & \lambda - A_{22} \end{bmatrix}$$

$$\det(\lambda \mathbf{I} - \mathbf{A}) = (\lambda - A_{11})(\lambda - A_{22}) - A_{12}A_{21}$$

$$= \lambda^2 - (A_{11} + A_{22})\lambda + A_{11}A_{22} - A_{12}A_{21}$$

$$= \lambda^2 - (\operatorname{tr} \mathbf{A})\lambda + \det(\mathbf{A})$$

The eigenvalues are the solution to the second-order polynomial:

$$\lambda^2 - (\operatorname{tr} \mathbf{A})\lambda + \det(\mathbf{A}) = 0 \qquad\qquad \text{(M9.13)}$$

The stability of a particular operating point is determined by finding the $\mathbf{A}$ matrix for that particular operating point, and finding the eigenvalues of the A matrix.

Here we show the eigenvalues for each of the three case 2 steady-state operating points. We use the function routine cstr_amat.m shown in appendix 3 to find the A matrices.

OPERATING POINT 1

The concentration and temperature are 8.564 kgmol/m^3 and 311.2 K, respectively.

```
»[amat,lambda]  = cstr_amat(8.564,311.2)

amat =
    -1.1680      -0.0886
     2.0030      -0.2443

lambda =
    -0.8957
    -0.5166
```

Both of the eigenvalues are negative, indicating that the point is stable, which is consistent with the results of Figure M9.2.

OPERATING POINT 2

The concentration and temperature are 5.518 and 339.1, respectively.

```
»[amat,lambda]  = cstr_amat(5.518,339.1)

amat =
    -1.8124      -0.2324
     9.6837       1.4697

lambda =
    -0.8369
     0.4942
```

One of the eigenvalues is positive, indicating that the point is *unstable*. This is consistent with the responses presented in Figure M9.3.

OPERATING POINT 3

```
»[amat,lambda] = cstr_amat(2.359,368.1)

amat =
    -4.2445    -0.3367
    38.6748     2.7132

lambda =
   -0.7657 + 0.9584i
   -0.7657 - 0.9584i
```

The real portion of each eigenvalue is negative, indicating that the point is stable; again, this is consistent with the responses in Figure M9.4.

M9.5.2   Input/Output Transfer Function Analysis

The input-output transfer functions can be found from:

$$\mathbf{G}(s) = \mathbf{C}(s\mathbf{I} - \mathbf{A})^{-1}\mathbf{B} \tag{M9.14}$$

where the elements of the **B** matrix corresponding to the first input ($u_1 = T_j - T_{js}$) are:

$$B_{11} = \frac{\partial f_1}{\partial u_1} = \frac{\partial f_1}{\partial T_j} = 0$$

$$B_{21} = \frac{\partial f_2}{\partial u_1} = \frac{\partial f_2}{\partial T_j} = \frac{UA}{V \rho c_p}$$

The reader should find the elements of the B matrix that correspond to the second and third input variables (see student exercise 8).

Here we show only the transfer functions for the low temperature steady-state for Case 2. The input/output transfer function relating jacket temperature to reactor concentration (state 1) is:

$$g_{11}(s) = \frac{-0.0266}{s^2 + 1.4123s + 0.4627} = \frac{-0.0575}{(1.1165s + 1)(1.9357s + 1)}$$

and the input/output transfer function relating jacket temperature to reactor temperature (state 2) is

$$g_{21}(s) = \frac{0.3s + 0.3504}{s^2 + 1.4123s + 0.4627} = \frac{0.7573(0.856s + 1)}{(1.1165s + 1)(1.9357s + 1)}$$

Notice that the transfer function for concentration is a pure second-order system (no numerator polynomial) while the transfer function for temperature has a first-order numerator and second-order denominator. This indicates that there is a greater "lag" between jacket temperature and concentration than between jacket temperature and reactor temperature. This makes physical sense, because a change in jacket temperature must first affect the reactor temperature before affecting the reactor concentration.

M9.6 PHASE-PLANE ANALYSIS

In Section M9.4 we provided the results of a few dynamic simulations, noting that different initial conditions caused the system to converge to different steady-state operating points. In this section we construct a phase-plane plot by performing simulations for a large number of initial conditions.

The phase-plane plot shown in Figure M9.5 was generated using `cstr_run.m` and `cstr.m` from the appendix. Three steady-state values are clearly shown; two are stable (the high and low temperature steady-states, shown as 'o'), while one is unstable (the intermediate temperature steady-state, shown as '+'). Notice that initial conditions of low concentration (0.5 kgmol/m^3) and relatively low-to-intermediate temperatures (300 to 365 K) all converge to the low-temperature steady-state. When the initial temperature is increased above 365 K, convergence to the high-temperature steady-state is achieved.

Now, consider initial conditions with a high concentration (9.5 kgmol/m^3) and low temperature (300 to 325 K); these converge to the low-temperature steady-state. Once the initial temperature is increased to above 325 K, convergence to the high-temperature steady-state is achieved. Also notice that, once the initial temperature is increased to around 340 K, a very high overshoot to above 425 K occurs, before the system settles down to the high-temperature steady-state. Although not shown on this phase-plane plot, higher initial temperatures can have overshoot to over 500 K before settling to the high-temperature steady-state. This could cause potential safety problems if, for example, secondary decomposition reactions occur at high temperatures. The phase-plane analysis then, is able to point out problem initial conditions.

Also notice that no initial conditions have converged to the intermediate temperature steady-state, since it is unstable. The reader should perform an eigenvalue/eigenvector analysis for the **A** matrix at each steady-state (low, intermediate, and high temperature) (see student exercise 3). You will find that the low, intermediate, and high temperature steady-states have stable node, saddle point (unstable), and stable focus behavior (see Chapter 13), respectively.

It should be noted that feedback control can be used to operate at the unstable intermediate temperature steady-state. The feedback controller would measure the reactor temperature and manipulate the cooling jacket temperature (or flowrate) to maintain the intermediate temperature steady-state. Also, a feedback controller could be used to make certain that the large overshoot to high temperatures does not occur from certain initial conditions.

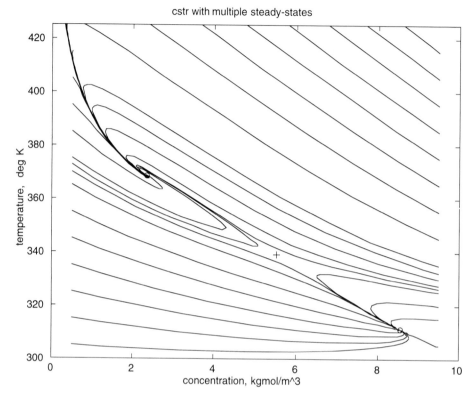

FIGURE M9.5   Phase-plane plot for Case 2. Stable points (conditions 1 and 3) are marked with 'o'. The unstable point (condition 2) is marked with '+'.

M9.7   UNDERSTANDING MULTIPLE STEADY-STATE BEHAVIOR

In previous sections we found that there were three steady-state solutions for Case 2 parameters. The objective of this section is to determine how multiple steady-states might arise. Also, we show how to generate steady-state input-output curves that show, for example, how the steady-state reactor temperature varies as a function of the steady-state jacket temperature.

M9.7.1   Heat Generation and Heat Removal Curves

In Section M9.3 we used numerical methods to solve for the steady-states, by solving two equations with two unknowns. In this section we show that it is easy to reduce the two equations in two unknowns to a single equation with one unknown. This will give us physical insight about the possible occurrence of multiple steady-states.

SOLVING FOR CONCENTRATION OF A AS A FUNCTION OF TEMPERATURE

The steady-state concentration solution ($dC_A/dt = 0$) for concentration is:

$$\frac{F}{V}(C_{Afs} - C_{As}) = k_o \exp\left(\frac{-\Delta E}{RT_s}\right) C_{As} \tag{M9.15}$$

We can rearrange this equation to find the steady-state concentration for any given steady-state reactor temperature, T_s:

$$C_{As} = \frac{\dfrac{F}{V} C_{Afs}}{\dfrac{F}{V} + k_o \exp\left(\dfrac{-\Delta E}{RT_s}\right)} \tag{M9.16}$$

SOLVING FOR TEMPERATURE

The steady-state temperature solution ($dT/dt = 0$) is:

$$\frac{F}{V}(T_s - T_{fs}) + \frac{UA}{V\rho C_p}(T_s - T_{cs}) = \left(\frac{-\Delta H}{\rho C_p}\right) k_o \exp\left(\frac{-\Delta E}{RT_s}\right) C_{As} \tag{M9.17}$$

The terms in (M9.17) are related to the energy removed and generated. If we multiply (M9.17) by $V\rho Cp$, we find that:

$$F_\rho C_p(T_s - T_{fs}) + UA(T_s - T_{js}) = -\Delta H \, V \, k_o \exp\left(\frac{-\Delta E}{RT_s}\right) C_{As} \tag{M9.18}$$

$$Q_{rem} \qquad\qquad = \qquad\qquad Q_{gen}$$

Energy removed by flow and heat exchange   Heat generated by reaction

Note the form of Q_{rem}

$$Q_{rem} = [-UA \, T_{js} - F\rho C_p T_{fs}] + [UA + F\rho C_p] \, T_s \tag{M9.19}$$

Notice that this is an equation for a line, where the independent variable is reactor temperature (T_s). The slope of the line is $[UA + F\rho C_p]$ and the intercept is $[-UA \, T_{js} - F\rho C_p T_{fs}]$. Changes in jacket or feed temperature shift the intercept, but not the slope. Changes in UA or F affect both the slope and intercept.

Now, consider the Q_{gen} term:

$$Q_{gen} = (-\Delta H) \, V \, k_o \exp\left(\frac{-\Delta E}{RT_s}\right) C_{As} \tag{M9.20}$$

Substituting (M9.16) into (M9.20), we find that:

$$Q_{gen} = -\Delta H\,V\,\frac{k_o \exp\!\left(\dfrac{-\Delta E}{RT_s}\right)\left(\dfrac{F}{V}C_{Afs}\right)}{\dfrac{F}{V} + k_o \exp\!\left(\dfrac{-\Delta E}{RT_s}\right)} \qquad \text{(M9.21)}$$

Equation (M9.21) has a characteristic "S" shape for Q_{gen} as a function of reactor temperature.

From equation (M9.18) we see that a steady-state solution exists when there is an intersection of the Q_{rem} and Q_{gen} curves.

M9.7.2   Effect of Design Parameters

In Figure M9.6 we show different possible intersections of the heat removal and heat generation curves. If the slope of the heat removal curve is greater than the maximum slope of the heat generation curve, there is only one possible intersection (see Figure M9.6a). As the jacket or feed temperature is changed, the heat removal lines shifts to the left or right, so the intersection can be at a high or low temperature depending on the value of jacket or feed temperature.

Notice that as long as the slope of the heat removal curve is less than the maximum slope of the heat generation curve, there will always be the possibility of three intersections (see Figure M9.6b) with proper adjustment of the jacket or feed temperature (intercept). If the jacket or feed temperature is changed, the removal line shifts to the right or left, where only one intersection occurs (either low or high temperature). This case is analyzed in more detail in Section M9.7.3.

M9.7.3   Multiple Steady-State Behavior

In Figure M9.7 we superimpose several possible linear heat removal curves with the S-shaped heat generation curve. Curve A intersects the heat generation curve at a low temperature; curve B intersects at a low temperature and is tangent at a high temperature;

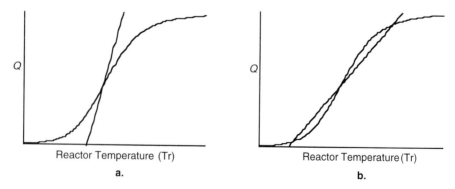

FIGURE M9.6   Possible intersections of heat generation and heat removal curves.

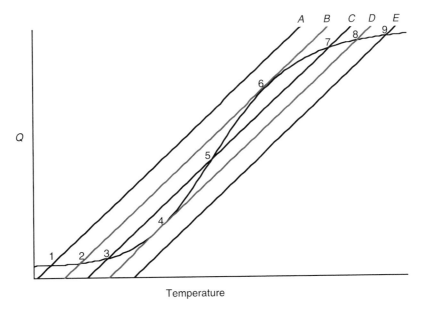

FIGURE M9.7 CSTR energy generated and energy removed as a function of reactor temperature.

curve C intersects at low, intermediate, and high temperatures; curve D is tangent to a low temperature and intersects at a high temperature; curve E has only a high temperature intersection. Curves A, B, C, D, and E are all based on the same system parameters, except that the jacket temperature increases as we move from curve A to E (from equation (M9.7) we see that changing the jacket temperature changes the intercept but not the slope of the heat removal curve). We can use Figure M9.7 to construct the steady-state input-output diagram shown in Figure M9.8, where jacket temperature is the input and reactor temperature is the output. Note that Figure M9.8 exhibits hysteresis behavior, which was first discussed in Chapter 15.

The term hysteresis is used to indicate that the behavior is different depending on the "direction" that the inputs are moved. For example, if we start at a low jacket temperature the reactor operates at a low temperature (point 1). As the jacket temperature is increased, the reactor temperature increases (points 2 and 3) until the low temperature "limit point"[3] (point 4) is reached. If the jacket temperature is slightly increased further, the reactor temperature jumps (*ignites*) to a high temperature (point 8); further jacket temperature increases result in slight reactor temperature increases.

Contrast the input-output behavior discussed in the previous paragraph (starting at a low jacket temperature) with that of the case of starting at a high jacket temperature. If

---

[3]Also known as a saddle-node bifurcation point. See Chapter 15.

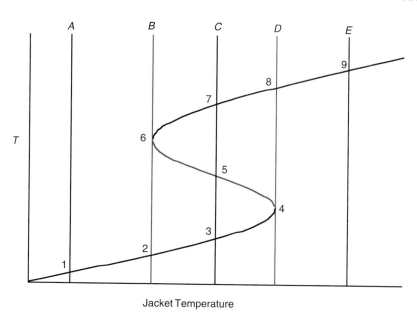

Jacket Temperature

FIGURE M9.8   Reactor temperature as a function of jacket temperature.

one starts at a high jacket temperature (point 9) there is a single high reactor temperature, which decreases as the jacket temperature is decreased (points 8 and 7). As we move slightly lower than the high temperature limit point (point 6), the reactor temperature drops (also known as *extinction*) to a low temperature (point 2). Further decreases in jacket temperature lead to small decreases in reactor temperature.

The hysteresis behavior discussed above is also known as *ignition-extinction* behavior, for obvious reasons. Notice that region between points 4 and 6 appears to be unstable, because the reactor does not appear to operate in this region (at least in a steady-state sense). Physical reasoning for stability is discussed in the following section.

M9.7.4   Multiple Steady-States: Stability Considerations (Steady-State Analysis)

Consider the system described by the energy removal curve C in Figure M9.7. Notice that the steady-state energy balance is satisfied for the operating points 3, 5, and 7, that is, there are three possible steady-states. What can we observe about the stability of each of the possible operating points, solely from physical reasoning?

LOWER STEADY-STATE (OPERATING POINT 3)

Consider a perturbation from operating point that is colder than point 3, say $T_3 - \delta T$. At this point we are generating more heat than can be removed by the reactor, so the temperature begins to rise—eventually moving to T_3. If we start at $T_3 + \delta T$, more energy is being removed than we are generating, so the temperature begins to decrease—eventually moving to T_3.

The lower temperature intersection, T_3, *may* be a *stable* operating point.

The open-loop stability cannot be assured until an eigenvalue analysis is performed (see Section M9.5).

MIDDLE STEADY-STATE (OPERATING POINT 5)

If we start at $T_5 - \delta T$, we are generating less energy than is being removed, causing T to decrease, eventually causing the temperature to go to T_3. If we start at $T_5 + \delta T$, more energy is being generated than is being removed, causing the temperature to rise, increasing until T_7 is reached.

The middle temperature intersection, T_5, *is* an *unstable* operating point.

Here the steady-state analysis is enough to determine that point 5 is an unstable operating point.

UPPER STEADY-STATE (OPERATING POINT 7)

Using the same arguments as T_3, we find:

The high temperature intersection, T_7, *may* be a *stable* operating point.

Again, the stability of the upper steady-state can only be determined by eigenvalue analysis (Section M9.5).

We have used physical reasoning to determine the open-loop characteristics solely from steady-state analysis. Recall that when we studied the Case 2 conditions in Section M9.5, we used eigenvalue analysis to find that the lower and upper temperature steady-states were stable, while the intermediate steady-state was unstable.

M9.7.5   Generating Steady-State Input-Output Curves

There are two different ways to find how the steady-state reactor temperature varies with the jacket temperature. One way is to solve the nonlinear algebraic (steady-state) equations for a large number of different jacket temperatures and construct the curve relating jacket to reactor temperature. It turns out that there is another, simpler, way for this particular problem.

Here we can easily solve for concentration as a function of reactor temperature. We can also directly solve for the jacket temperature required for a particular reactor temperature. Then we simply plot reactor temperature versus jacket temperature. We show the calculations in detail below.

As in Section M9.7.1, we can find the steady-state concentration for any steady-state reactor temperature, T_s (from the material balance equation for reactant A):

$$C_{As} = \frac{\frac{F}{V} C_{Afs}}{\frac{F}{V} + k_o \exp\left(\frac{-\Delta E}{RT_s}\right)}$$

From the steady-state temperature solution ($dT/dt = 0$) we can solve for the steady-state jacket temperature, T_{cs}:

$$T_{cs} = T_s + \frac{\left[F\rho c_p(T_s - T_{fs}) - (-\Delta H)Vk_o \exp\left(\frac{-\Delta E}{RT_s}\right) C_{As}\right]}{UA}$$

Similarly, if we assume that the jacket temperature remains constant, we can solve for the required feed temperature:

$$T_{fs} = T_s + \frac{\left[UA(T_s - T_{cs}) - (-\Delta H)Vk_o \exp\left(\frac{-\Delta E}{RT_s}\right) C_{As}\right]}{F\rho c_p}$$

The function file `cstr_io.m` shown in Appendix 4 is used to generate the input-output curves for the Case 2 parameters shown in Figures M9.9 and M9.10.

```
»Tempvec = 300:2.5:380;
»[Tjs,Tfs,conc] = cstr_io(Tempvec);
»plot(Tjs,Tempvec)
»plot(Tfs,Tempvec)
```

M9.7.6   Cusp "Catastrophe"

In Chapter 15 we presented a simple example of a cusp catastrophe that occurs when the input-output behavior moves from monotonic through a hysteresis bifurcation to multiple steady-state behavior. This can be shown clearly in Figure M9.11 where we find that, as we increase F/V, we move from monotonic to multiple steady-state behavior.

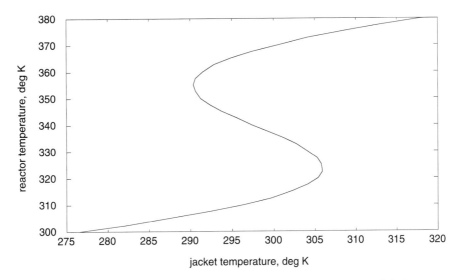

FIGURE M9.9 Steady-state reactor temperature as a function of jacket temperature, Case 2 parameters.

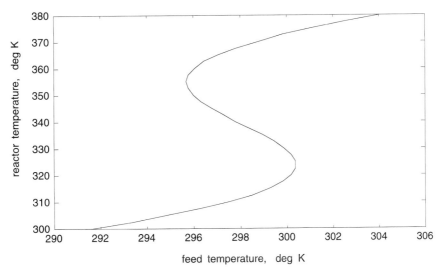

FIGURE M9.10 Steady-state reactor temperature as a function of feed temperature, Case 2 parameters.

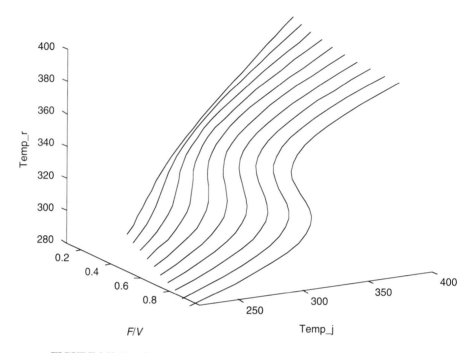

FIGURE M9.11   Cusp catastrophe for the diabatic CSTR, Case 2 conditions. The input/output (jacket temperature/reactor temperature) behavior changes as a function of the space velocity (F/V).

M9.8   FURTHER COMPLEXITIES

In this module we have only been able to cover only a small part of the interesting behavior that can occur in CSTRs due to nonlinear dynamics. In this section we provide a concise overview of other interesting topics in CSTR analysis. First (Section M9.8.1), we will draw a connection between our physical-based approach (heat removal and generation curves) for analyzing multiplicity behavior and the more mathematical bifurcation theory. Then (Section M9.8.2), we will discuss the effect of cooling jacket flowrate on the behavior of a CSTR. Finally (Section M9.8.3), we discuss limit cycle behavior and Hopf bifurcations.

M9.8.1   Connecting Bifurcation Theory with Physical Reasoning

In Chapter 15 we presented mathematical conditions for bifurcation behavior; we have focused on physical arguments in this module. The goal of this section is to draw a connection between the mathematical conditions and the physical conditions for multiple steady-state behavior.

First, recall the mathematical conditions for a bifurcation. Let x represent the state variable, μ a bifurcation parameter, and $g(x,\mu)$ a single nonlinear algebraic equation that must be satisfied at steady-state. If the following bifurcation conditions are satisfied:

$$g(x,\mu) = \frac{\partial g}{\partial x} = \frac{\partial^2 g}{\partial x^2} = \ldots = \frac{\partial^{k-1} g}{\partial x^{k-1}} = 0$$

and

$$\frac{\partial^k g}{\partial x^k} \neq 0$$

then there exists k steady-state solutions in the vicinity of the bifurcation point.

In our case, we were able to write a single nonlinear algebraic equation that must be satisfied for a steady-state solution:

$$[F\rho C_p(T_s - T_{fs}) + UA(T_s - T_{js})] - \left[(-\Delta H)Vk_o \exp\left(\frac{-\Delta E}{RT_s}\right)\frac{\frac{F}{V}C_{Afs}}{\frac{F}{V} + k_o \exp\left(\frac{-\Delta E}{RT_s}\right)}\right] = 0$$

For simplicity, we write this equation as:

$$g(T_s,\mu) = Q_{\text{rem}}(T_s,\mu) - Q_{\text{gen}}(T_s,\mu) = 0$$

where T_s is the state variable (steady-state reactor temperature) and μ is used to represent the vector of physical parameters that can possibly be varied.

The steady-state solution is obtained by solving:

$$g(T_s,\mu) = 0$$

or,

$$Q_{\text{rem}}(T_s,\mu) = Q_{\text{gen}}(T_s,\mu)$$

which is simply the intersection of the heat removal and heat generation curves. The first derivative condition:

$$\frac{\partial g(T_s,\mu)}{\partial T_s} = 0$$

is simply

$$\frac{\partial Q_{\text{rem}}(T_s,\mu)}{\partial T_s} = \frac{\partial Q_{\text{gen}}(T_s,\mu)}{\partial T_s}$$

These conditions are satisfied by curves B and D in Figure M9.12 below. The slope conditions are satisfied at a high temperature for curve B and a low temperature for curve D, and we directly see how a "limit point" on the input-output curve (Figure M9.12b) corre-

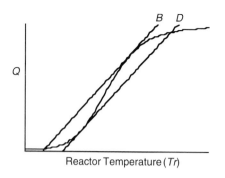

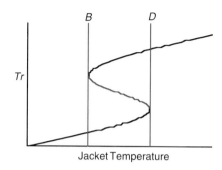

a. Heat Removed and Generated   **b. Input/Output**

FIGURE M9.12   Illustration of saddle-node (limit point) bifurcation.

sponds to the heat generation and removal curves (Figure M9.12a). We also note that the following condition also holds:

$$\frac{\partial^2 g(T_s,\mu)}{\partial T_s^2} \neq 0$$

Since, although the second derivative of the heat removal line is always zero,

$$\frac{\partial^2 Q_{rem}(T_s,\mu)}{\partial T_s^2} = 0$$

the heat generation curve is not at an inflection point:

$$\frac{\partial^2 Q_{gen}(T_s,\mu)}{\partial T_s^2} \neq 0$$

and we see that the conditions for saddle-node behavior are satisfied.

Now we illustrate the conditions for a hysteresis bifurcation, which are:

$$g(T_s,\mu) = \frac{\partial g(T_s,\mu)}{\partial T_s} = \frac{\partial^2 g(T_s,\mu)}{\partial T_s^2} = 0$$

All of these conditions are satisfied in Figure M9.13b on the next page. We show Figures M9.13a and M9.13c to illustrate the progression to satisfying the bifurcation conditions.

The progression shown in Figure M9.13 involved a decrease in the heat transfer capability from a to c. This is also shown in the cusp diagram of Figure M9.14, where there is no hysteresis at high UA values, the onset of hysteresis (bifurcation point) at an intermediate value of UA, and complete hysteresis behavior at low values of UA.

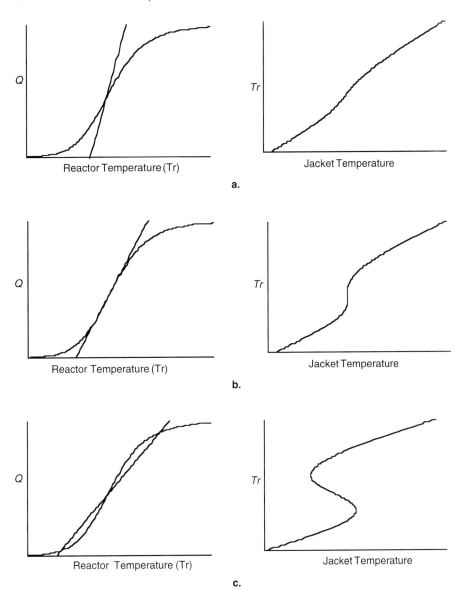

FIGURE M9.13 Illustration of a hysteresis bifurcation. Bifurcation conditions are satisfied in **b**.

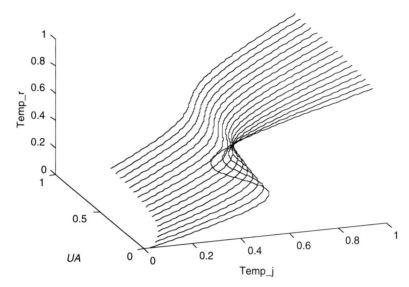

FIGURE M9.14   Cusp diagram.

M9.9   DIMENSIONLESS MODEL

The CSTR modeling equations have a large number of parameters. It is clear that these parameters are not independent. For example, the behavior is dependent on the ratio F/V, which can be considered a single parameter. Most of the analysis of CSTR behavior has been performed on dimensionless equations, with a minimum number of dimensionless parameters involved. One possible set of dimensionless parameters is shown in student exercise 6. For a more complete discussion of the development and use of dimensionless equations, see Aris (1994).

SUMMARY

Diabatic CSTRs are perhaps the most studied unit operation in the chemical process dynamics literature. They can exhibit exotic and interesting steady-state and dynamic behavior, including: steady-state multiplicity, Hopf bifurcations, periodic behavior, and chaos. In this module we have provided an elementary introduction to the steady-state and dynamic behavior of these systems.

The reader is encouraged to use MATLAB to "play with" the model (particularly for the parameter sets for Cases 1 and 3) presented in this chapter.

REFERENCES AND FURTHER READING

A number of undergraduate textbooks on chemical reaction engineering present the ideas of multiplicity behavior in a diabatic CSTR. These include the following books by Aris and Fogler:

Aris, R. (1989). *Elementary Chemical Reactor Analysis*. Boston: Butterworths. This is a reprint of the book originally published by McGraw Hill (1969).

Fogler, H.S. (1992). *Elements of Chemical Reaction Engineering*, 2nd ed. Englewood Cliffs, NJ: Prentice Hall.

The following book does an excellent job of developing the modeling equations for a diabatic CSTR:

Denn, M.M. (1986). *Process Modeling*, New York: Longman.

Diabatic CSTRs have been analyzed for complex behavior (multiple steady-states, Hopf bifurcations, quasi-periodic, chaos) more than any other chemical process. An excellent review of the literature on these topics is presented in the following paper:

Razon, L.F., & R.A. Schmitz. (1987). Multiplicities and instabilities in chemically reacting systems—a review. *Chem. Eng. Sci.*, 42(5): 1005–1047.

The following book provides an excellent discussion of the use of dimensionless models, with a diabatic CSTR as a major example.

Aris, R. (1994). *Mathematical Modelling Techniques*. New York: Dover. This is a reprint of the work originally published by Pitman, London (1978).

STUDENT EXERCISES

1. Consider the Case 1 parameters shown in Table M9.1.
   a. Find the steady-state using `fsolve`.
   b. Determine the stability of the steady-state using linearization and eigenvalue analysis.
   c. Simulate the dynamic behavior by perturbing the steady-state concentration and temperature values by ± 10 %. Use `ode45` for the simulations.
2. Consider the Case 3 parameters shown in Table M9.1.
   a. Find the steady-state using `fsolve`.
   b. Determine the stability of the steady-state using linearization and eigenvalue analysis.

c. Simulate the dynamic behavior by perturbing the steady-state concentration and temperature values by ± 10 %. Use `ode45` for the simulations.

3. a. For the Case 2 conditions, sketch the phase-plane plots for the process linearized (A matrices) at each of the steady-states (operating points 1, 2 and 3) in Section M9.5.1. Use the eigenvalues and eigenvectors of the A matrices to sketch your plots by hand.

b. Generate the phase-plane plot (Figure M9.5) for the nonlinear system of Section M9.5.1. Use `cstr_run.m` and `cstr.m` from the Appendix.

4. a. Modify the parameters in the file `cstr_amat.m` (from the Appendix) to generate the state-space A matrix for the steady-state operating point for Case 1. Use the eigenvector/eigenvalue analysis of the A matrix to sketch the phase-plane.

b. Modify `cstr_run.m` (change initial conditions) and `cstr.m` (change cstr parameters) to perform the phase-plane calculation for the nonlinear system for Case 1. Discuss the phase-plane results.

5. a. Modify the parameters in the file `cstr_amat.m` (from the Appendix) to generate the state space A matrix for the steady-state operating point for Case 3. Use the eigenvector/eigenvalue analysis of the A matrix to sketch the phase-plane.

b. Modify `cstr_run.m` (change initial conditions) and `cstr.m` (change cstr parameters) to perform the phase-plane calculation for the nonlinear system for Case 3. Discuss the phase-plane results.

6. The CSTR modeling equations have a large number of parameters. It is clear that these parameters are not independent. For example, the behavior is dependent on the ratio F/V, which can be considered a single parameter. Define the dimensionless variables as:

$$\beta = \frac{(-\Delta H)C_{Af}}{\rho C_p T_{f0}} \qquad \tau = \frac{F_0}{V}t \qquad \gamma = \frac{E}{RT_{f0}}$$

$$\delta = \frac{UA}{\rho C_p F_0} \qquad \phi = \frac{V}{F_0}k_0 e^{-\gamma} \qquad q = \frac{F}{F_0}$$

$$x_1 = \frac{C_A}{C_{Af0}} \qquad x_2 = \frac{T - T_{f0}}{T_{f0}}t\gamma \qquad x_3 = \frac{T_c - T_{f0}}{T_{f0}}\gamma$$

$$x_{1f} = \frac{C_{Af}}{C_{Af0}} \qquad x_{2f} = \frac{T_f - T_{f0}}{T_{f0}}\gamma$$

where F_0, C_{Af0}, and T_{f0} are the nominal steady-state values.

a. Show that the dimensionless modeling equations are:

$$\frac{dx_1}{d\tau} = -\phi x_1 \kappa(x_2) + q(x_{1f} - x_1)$$

$$\frac{dx_2}{d\tau} = \beta\phi x_1 \kappa(x_2) - (q + \delta)x_2 + \delta x_3 + qx_{2f}$$

where $\qquad \kappa(x_2) = \exp\left(\dfrac{x_2}{1 + x_2/\gamma}\right)$

and if the inputs are at their steady-state values, the modeling equations are:

$$\frac{dx_1}{d\tau} = 1 - (1 + \phi\kappa(x_2))\, x_1$$

$$\frac{dx_2}{d\tau} = \beta\phi x_1\, \kappa(x_2) - (1 + \delta)x_2 + \delta x_3$$

7. Consider a system with the following parameters. Show that steady-state temperature is 350 K, and the corresponding steady-state concentration is 0.5 gmol/liter. Determine the number of steady-state solutions and their stability characteristics. Generate a phase-plane diagram.

   Consider the response to ± 5 K changes in the jacket temperature. For one direction you should find oscillatory behavior, while the other direction is stable. This indicates that a Hopf bifurcation (Chapter 16) occurs for some value of T_j between 295 K and 305 K. Find the value of T_j where this occurs (recall that a Hopf bifurcation occur when the complex eigenvalues have zero real portion.

F	100 liters/min
V	100 liters
T_j	300 K
k_0, min^{-1}	7.2×10^{10}
$(-\Delta H)$, J/gmol	50000
E/R, K	8750
ρc_p, J/liter K	239
T_f, K	350
C_{Af}, kgmol/m^3	10
UA, J/min K	50000

8. Consider the input-output transfer function matrix for a CSTR, $G(s)$, from Section M9.5.2. Find the elements of the B matrix corresponding to inputs 2 and 3, and the resulting transfer functions for each of the Case 2 operating conditions. For the stable steady-states (low and high temperature) compare the step responses of the transfer functions with those of the nonlinear system. Use input changes of ± 0.1 mol/m^3 and ± 1 deg K in inputs 2 (feed concentration) and 3 (feed temperature), respectively. Remember that for comparison on the same plots you must convert from deviation to physical variables for the transfer function responses.

9. Often concentration measurements are not available in chemical reactors. Show how a steady-state energy balance around the reactor can be used to infer the rate of reaction, and therefore the "heat flow" of the reaction, $q_r = rV(-\Delta H)$. Think of two ways: (a) assuming UA is known and the jacket outlet temperature is available, and (b) assuming UA is unknown, but the jacket flowrate and jacket inlet and outlet temperatures are available.

APPENDIXES

1   Function m-file for Steady-State CSTR Solution

```
function f = cstr_ss(x)
%
% 2-state diabatic CSTR model
% Use fsolve to find steady-state concentration and
%    temperature
%
% The call statement from the command window is:
%
%    x = fsolve('cstr_ss',x0);
%
% where x is the solution vector (x = [conc;temp])
% and x0 is an initial guess vector
%
% define the following global parameter so that values
% can be entered in the command window
%       global CSTR_PAR
% parameters (case 2)
% CSTR_PAR(1):  frequency factor (9703*3600)
% CSTR_PAR(2):  heat of reaction (5960)
% CSTR_PAR(3):  activation energy (11843)
% CSTR_PAR(4):  density*heat cap. (500)
% CSTR_PAR(5):  heat trans coeff * area (150)
% CSTR_PAR(6):  ideal gas constant (1.987)
% CSTR_PAR(7):  reactor volume (1)
% CSTR_PAR(8):  feed flowrate (1)
% CSTR_PAR(9):  feed concentration (10)
% CSTR_PAR(10): feed temperature (298)
% CSTR_PAR(11): jacket temperature (298)
%
% states 1 and 2
% ca     = concentration of A, kgmol/m^3
% Temp   = reactor temperature, deg K
%
% 16 July 96 - (c) B.W. Bequette
% revised 2 January 1997 - added CSTR_PAR
%
   f = zeros(2,1);
%
   global CSTR_PAR
%
% below we use common notation for the states to
% improve code readibility
%
   ca     = x(1);
   Temp   = x(2);
%
% parameter values, case 2 shown in parentheses:
%
```

```
  k0    = CSTR_PAR(1);   % frequency factor (9703*3600)
  H_rxn = CSTR_PAR(2);   % heat of reaction (5960)
  E_act = CSTR_PAR(3);   % activation energy (11843)
  rhocp = CSTR_PAR(4);   % density*heat cap. (500)
  UA    = CSTR_PAR(5);   % ht trans coeff * area (150)
  R     = CSTR_PAR(6);   % gas constant (1.987)
  V     = CSTR_PAR(7);   % reactor volume (1)
  F     = CSTR_PAR(8);   % feed flowrate (1)
  caf   = CSTR_PAR(9);   % feed concentration (10)
  Tf    = CSTR_PAR(10);  % feed temperature (298)
  Tj    = CSTR_PAR(11);  % jacket temperature (298)
%
% ratios used in the equations
%
  fov   = F/V;
  UAoV  = UA/V;
%
% modeling equations:
%
  rate = k0*exp(-E_act/(R*Temp))*ca;
%
  dcadt = fov*(caf - ca) - rate;
  dTdt  = fov*(Tf - Temp) + (H_rxn/rhocp)*rate -
    (UAoV/rhocp)*(Temp - Tj);
%
% below we convert back to function notation.  fsolve
% seeks values of x(1) and x(2) to drive f(1) and f(2) to
    zero.
%
  f(1) = dcadt;
  f(2) = dTdt;
```

2   Function `m-file` for Integration of Dynamic Equations

```
  function xdot = cstr_dyn(t,x)
%
% 2-state diabatic CSTR model
% Use ode45 to find concentration and temperature
%
% The call statement from the command window is:
%
%   [t,x] = ode45('cstr_dyn',t0,tf,x0);
%
% where t is the time vector
%       x is the solution vector (x = [conc;temp])
%       t0 is the initial time
%       tf is the final time
%       x0 is an initial condition vector
%
% define the following global parameter so that values
% can be entered in the command window
%     global CSTR_PAR
```

```
% parameters (case 2)
% CSTR_PAR(1):  frequency factor (9703*3600)
% CSTR_PAR(2):  heat of reaction (5960)
% CSTR_PAR(3):  activation energy (11843)
% CSTR_PAR(4):  density*heat cap. (500)
% CSTR_PAR(5):  heat trans coeff * area (150)
% CSTR_PAR(6):  ideal gas constant (1.987)
% CSTR_PAR(7):  reactor volume (1)
% CSTR_PAR(8):  feed flowrate (1)
% CSTR_PAR(9):  feed concentration (10)
% CSTR_PAR(10): feed temperature (298)
% CSTR_PAR(11): jacket temperature (298)
%
% states 1 and 2
% ca     = concentration of A, kgmol/m^3
% Temp   = reactor temperature, deg K
%
% 16 July 96 - (c) B.W. Bequette
% revised 2 January 1997 - added CSTR_PAR
%
%

  global CSTR_PAR
%
% below we use common notation for the states to
% improve code readibility
%
  ca     = x(1);
  Temp   = x(2);
%
% parameter values, case 2 shown in parentheses:
%
  k0    = CSTR_PAR(1);  % frequency factor (9703*3600)
  H_rxn = CSTR_PAR(2);  % heat of reaction (5960)
  E_act = CSTR_PAR(3);  % activation energy (11843)
  rhocp = CSTR_PAR(4);  % density*heat cap. (500)
  UA    = CSTR_PAR(5);  % ht trans coeff * area (150)
  R     = CSTR_PAR(6);  % gas constant (1.987)
  V     = CSTR_PAR(7);  % reactor volume (1)
  F     = CSTR_PAR(8);  % feed flowrate (1)
  caf   = CSTR_PAR(9);  % feed concentration (10)
  Tf    = CSTR_PAR(10); % feed temperature (298)
  Tj    = CSTR_PAR(11); % jacket temperature (298)
%
% ratios used in the equations
%
  fov   = F/V;
  UAoV  = UA/V;
%
% modeling equations:
%
  rate = k0*exp(-E_act/(R*Temp))*ca;
%
```

```
  dcadt = fov*(caf - ca) - rate;
  dTdt  = fov*(Tf - Temp) + (H_rxn/rhocp)*rate -
    (UAoV/rhocp)*(Temp - Tj);
%
% below we convert back to ode45 function notation.
%
  xdot(1) = dcadt;
  xdot(2) = dTdt;
```

3   cstr_amat.m

```
  function [amat,lambda] = cstr_amat(ca,Temp)
%
% 16 July 96
% (c) B.W. Bequette
% find the A matrix for the 2-state CSTR to determine the
% stability by eigenvalue analysis:
%
% amat  = A matrix
% lamba = vector of eigenvalues
%
% parameters
% k0     = frequency factor, hr^-1
% fov    = F/V, hr^-1
% H_rxn  = heat of rxn, kcal/kgmol
% E_act  = activation energy, kcal/kgmol
% rhocp  = density*heat capacity, kcal/m^3 deg K
% Tf     = feed temperature, deg K
% caf    = feed concentration of A,
% UAoV   = heat trans. coeff.*area/volume, kcal/m^3 deg K hr
%
% manipulated input
% Tj     = jacket temperature, deg K
%
% states
% ca     = concentration of A, kgmol/m^3
% Temp   = reactor temperature, deg K
%
% parameter values
% case 2:
  fov = 1;
  k0  = 9703*3600;
  H_rxn = 5960;
  E_act = 11843;
  rhocp = 500;
  Tf    = 298;
  caf   = 10;
  UAoV  = 150;
  R     = 1.987;
  Tj    = 298;
%
  ks = k0*exp(-E_act/(R*Temp));
```

```
    ksprime = ks*(E_act/(R*Temp*Temp));
%
  amat(1,1) = -fov - ks;
  amat(1,2) = -ca*ksprime;
  amat(2,1) = ks*H_rxn/rhocp;
  amat(2,2) = -fov - UAoV/rhocp + (H_rxn/rhocp)*ca*ksprime;
%
  lambda = eig(amat);
```

4 *cstr_run.m* (Script File to Generate Phase-Plane Plot)

```
% cstr_run.m
% b.w. bequette
% 2 Dec 96
% revised - 4 Dec 96 - more documentation
%
% generate phase-plane plot for nonlinear cstr (cstr.m)
%
% set of initial conditions
%
  x01a= [0.5;305];
  x01 = [0.5;315];
  x02 = [0.5;325];
  x03 = [0.5;335];
  x04 = [0.5;345];
  x05 = [0.5;355];
  x06 = [0.5;365];
  x07 = [0.5;375];
  x08 = [0.5;385];
  x09 = [0.5;395];
  x010= [0.5;405];
  x011= [0.5;415];
  x012a=[9.5;305];
  x012= [9.5;315];
  x013= [9.5;325];
  x014 = [9.5;335];
  x015 = [9.5;345];
  x016 = [9.5;355];
  x017 = [9.5;365];
  x018 = [9.5;375];
  x019 = [9.5;385];
  x020 = [9.5;395];
  x021 = [9.5;405];
  x022 = [9.5;415];
%
% use ode45 for numerical integration
%
  [t1a,x1a] = ode45('cstr',0,10,x01a);
  [t1,x1] = ode45('cstr',0,10,x01);
  [t2,x2] = ode45('cstr',0,10,x02);
```

```
[t3,x3]   = ode45('cstr',0,10,x03);
[t4,x4]   = ode45('cstr',0,10,x04);
[t5,x5]   = ode45('cstr',0,10,x05);
[t6,x6]   = ode45('cstr',0,10,x06);
[t7,x7]   = ode45('cstr',0,10,x07);
[t8,x8]   = ode45('cstr',0,10,x08);
[t9,x9]   = ode45('cstr',0,10,x09);
[t10,x10] = ode45('cstr',0,10,x010);
[t11,x11] = ode45('cstr',0,10,x011);
[t12,x12] = ode45('cstr',0,10,x012);
[t12a,x12a] = ode45('cstr',0,10,x012a);
[t13,x13] = ode45('cstr',0,10,x013);
[t14,x14] = ode45('cstr',0,10,x014);
[t15,x15] = ode45('cstr',0,10,x015);
[t16,x16] = ode45('cstr',0,10,x016);
[t17,x17] = ode45('cstr',0,10,x017);
[t18,x18] = ode45('cstr',0,10,x018);
[t19,x19] = ode45('cstr',0,10,x019);
[t20,x20] = ode45('cstr',0,10,x020);
[t21,x21] = ode45('cstr',0,10,x021);
[t22,x22] = ode45('cstr',0,10,x022);
%
 x040 = [0.5;370];
 x041 = [0.5;372.5];
 x049 = [9.5;320];
 x050 = [9.5;326.25];
 x051 = [9.5;327.5];
 x052 = [9.5;330];
 x053 = [9.5;332.5];
[t40,x40] = ode45('cstr',0,10,x040);
[t41,x41] = ode45('cstr',0,10,x041);
[t49,x49] = ode45('cstr',0,10,x049);
[t50,x50] = ode45('cstr',0,10,x050);
[t51,x51] = ode45('cstr',0,10,x051);
[t52,x52] = ode45('cstr',0,10,x052);
[t53,x53] = ode45('cstr',0,10,x053);
%
% phase-plane plot
%
 plot(x1(:,1),x1(:,2),'w')
 hold on
 plot(x1a(:,1),x1a(:,2),'w')
 plot(x2(:,1),x2(:,2),'w')
 plot(x3(:,1),x3(:,2),'w')
 plot(x4(:,1),x4(:,2),'w')
 plot(x5(:,1),x5(:,2),'w')
 plot(x6(:,1),x6(:,2),'w')
 plot(x7(:,1),x7(:,2),'w')
 plot(x8(:,1),x8(:,2),'w')
 plot(x9(:,1),x9(:,2),'w')
 plot(x10(:,1),x10(:,2),'w')
```

```
    plot(x11(:,1),x11(:,2),'w')
    plot(x12(:,1),x12(:,2),'w')
    plot(x12a(:,1),x12a(:,2),'w')
    plot(x13(:,1),x13(:,2),'w')
    plot(x14(:,1),x14(:,2),'w')
    plot(x15(:,1),x15(:,2),'w')
    plot(x16(:,1),x16(:,2),'w')
    plot(x17(:,1),x17(:,2),'w')
    plot(x18(:,1),x18(:,2),'w')
    plot(x19(:,1),x19(:,2),'w')
    plot(x20(:,1),x20(:,2),'w')
    plot(x21(:,1),x21(:,2),'w')
    plot(x22(:,1),x22(:,2),'w')
%
% mark the stable and unstable steady-states (o = stable, +
%     = unstable)
%
    plot([8.564],[311.2],'wo',[5.518],[339.1],'w+',[2.359],
        [368.1],'wo')
%
%   some initial conditions not used
%   x030 = [5.768;339.1];
%   x031 = [5.368;339.1];
%   x032 = [5.518;338.5];
%   x033 = [5.518;339.7];
%   [t30,x30] = ode45('cstr',0,10,x030);
%   [t31,x31] = ode45('cstr',0,10,x031);
%   [t32,x32] = ode45('cstr',0,10,x032);
%   [t33,x33] = ode45('cstr',0,10,x033);
%
    plot(x40(:,1),x40(:,2),'w');
    plot(x41(:,1),x41(:,2),'w');
    plot(x49(:,1),x49(:,2),'w');
    plot(x50(:,1),x50(:,2),'w');
    plot(x51(:,1),x51(:,2),'w');
    plot(x52(:,1),x52(:,2),'w');
    plot(x53(:,1),x53(:,2),'w');
%
% plot(x30(:,1),x30(:,2),'w');
% plot(x31(:,1),x31(:,2),'w');
% plot(x32(:,1),x32(:,2),'w');
% plot(x33(:,1),x33(:,2),'w');
%
% set limits on plot, and label axes and provide plot
%     title
%
    axis([0 10 300 425])
    xlabel('concentration, kgmol/m^3')
    ylabel('temperature, deg K')
    title('cstr with multiple steady-states')
%
```

5 `cstr_io.m`

```
function [Tjs,Tfs,concs] = cstr_io(Tempvec)
%
% Generate the steady-state i/o curve for the diabatic CSTR.
% Tempvec is a vector of reactor temperatures generated
% before this routine is called.
%
% This routine solves for vector of jacket temperatures
% that would yield the required reactor temperature vector
% (assuming a constant feed temperature).
%
% In addition, this routine solves for the vector of
% feed temperatures that would yield the required reactor
% temperature vector (assuming constant jacket temperature).
%
% For consistency with cstr_ss.m and cstr_dyn.m, we use
% a global parameter vector. The elements of this vector
% should be set in the command window before this routine
% is run.
%    global CSTR_PAR
% parameters (case 2)
% CSTR_PAR(1):   frequency factor (9703*3600)
% CSTR_PAR(2):   heat of reaction (5960)
% CSTR_PAR(3):   activation energy (11843)
% CSTR_PAR(4):   density*heat cap. (500)
% CSTR_PAR(5):   heat trans coeff * area (150)
% CSTR_PAR(6):   ideal gas constant (1.987)
% CSTR_PAR(7):   reactor volume (1)
% CSTR_PAR(8):   feed flowrate (1)
% CSTR_PAR(9):   feed concentration (10)
% CSTR_PAR(10): feed temperature (298)
% CSTR_PAR(11): jacket temperature (298)
%
% 16 July 96 - (c) B.W. Bequette
% revised 2 January 1997 - added CSTR_PAR
%
   global CSTR_PAR
%
% parameter values, case 2 shown in parentheses:
%
   k0    = CSTR_PAR(1);   % frequency factor (9703*3600)
   H_rxn = CSTR_PAR(2);   % heat of reaction (5960)
   E_act = CSTR_PAR(3);   % activation energy (11843)
   rhocp = CSTR_PAR(4);   % density*heat cap. (500)
   UA    = CSTR_PAR(5);   % ht trans coeff * area (150)
   R     = CSTR_PAR(6);   % gas constant (1.987)
   V     = CSTR_PAR(7);   % reactor volume (1)
   F     = CSTR_PAR(8);   % feed flowrate (1)
   caf   = CSTR_PAR(9);   % feed concentration (10)
   Tf    = CSTR_PAR(10); % feed temperature (298)
```

```
  Tj     = CSTR_PAR(11); % jacket temperature (298)
%
% ratios used in the equations
%
  fov    = F/V;
  UAoV   = UA/V;
%
for i = 1:length(Tempvec);
  Temp    = Tempvec(i);
  k       = k0*exp(-E_act/(R*Temp));
  ca      = fov*caf/(fov + k);
  rate    = k0*exp(-E_act/(R*Temp))*ca;
  stuff1  = fov*rhocp*(Temp - Tf);
  stuff2  = H_rxn*rate;
  Tjs(i)  = Tempvec(i) + (stuff1 - stuff2)/UAoV;
  concs(i) = ca;
  stuff3  = UAoV*(Temp - Tj);
  stuff4  = H_rxn*rate;
  Tfs(i)  = Tempvec(i) + (stuff3 - stuff4)/(fov*rhocp);
end;
```

6 `cstr_cusp.m`—Cusp Catastrophe

```
% cstr_cusp.m
% generates the cusp diagram, based on case 2 parameters
%
  case2;
%
clear Tempr Tjs Tfs concs;
  Tempr = 280:5:400;
% vary UA or F
  npts = length(Tempr);
  CSTR_PAR(8) = 0.1;
  F_plt = ones(npts)*CSTR_PAR(8);
  [Tjs,Tfs,concs] = cstr_io(Tempr);
  plot3(F_plt,Tjs,Tempr,'w-')
  hold on
 for kindex = 1:9;
  CSTR_PAR(8) =  0.1+ 0.1*kindex;
  F_plt = ones(npts)*CSTR_PAR(8);
  [Tjs,Tfs,concs] = cstr_io(Tempr);
  plot3(F_plt,Tjs,Tempr,'w-')
 end
  hold off
%   axis([0.25 1.375 260 360 300 380])
  axis([0.1 1.0 220 400 280 400])
  view(60,20)
  xlabel('F/V')
  ylabel('Temp_j')
  zlabel('Temp_r')
```

IDEAL BINARY DISTILLATION

MODULE
10

After studying this module, the student should be able to

- Develop the dynamic modeling equations for ideal binary distillation
- Solve for the steady-state
- Linearize and find the state space model
- Understand the dynamic behavior of distillation columns
- Use MATLAB for steady-state and dynamic simulation

The major sections of this module are:

M10.1   BACKGROUND

Distillation is a common separation technique for liquid streams containing two or more components and is one of the more important unit operations in chemical manufacturing processes. Design and control of distillation is important in order to produce product streams of required purity, either for sale or for use in other chemical processes.

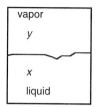

FIGURE M10.1   Closed system with liquid and vapor in equilibrium.

Distillation is based on the separation of components of a liquid mixture by virtue of the differences in boiling points of the components. For illustration purposes, we will base our discussion on the separation of liquid streams containing two components (binary mixture). We will refer to the pure component that boils at a lower temperature as the *light* component and the pure component that boils at a higher temperature as the *heavy* component. For example, in a mixture of benzene and toluene, benzene is the light component and toluene is the heavy component.

A saturated liquid mixture of two components at a given concentration is in equilibrium with a vapor phase that has a higher concentration of the light component than the liquid phase. Let x represent the mole fraction of the light component in the liquid phase and y represent the mole fraction of the light component in the vapor phase. Consider Figure M10.1 as a conceptual representation of phase (vapor/liquid) equilibrium. The saturated liquid is in equilibrium with a saturated vapor. The concentration of the light component will be larger in the vapor phase than the liquid phase.

Figure M10.2 is an example of an equilibrium diagram that represents the relationship between the liquid and vapor phase compositions (mole fraction). For example, if the liquid composition is 0.5 mole fraction of the light component, we find from Figure M10.2 that the vapor composition is 0.7.

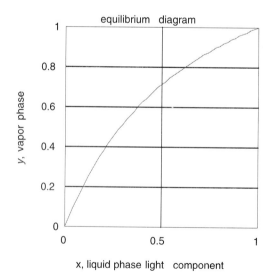

FIGURE M10.2   Vapor/liquid equilibrium diagram.

For ideal mixtures, it is common to model the phase equilibrium relationship based on constant relative volatility

$$y = \frac{\alpha x}{1 + (\alpha - 1)x} \tag{M10.1}$$

where α is known as the relative volatility. Figure M10.2 was generated based on equation (M10.1) with $\alpha = 2.5$.

M10.2   CONCEPTUAL DESCRIPTION OF DISTILLATION

The following is a conceptual description of the operation of a binary (two-component) distillation column. The feed typically enters close to the middle of the column (above the feed stage), as shown in Figure M10.3. Vapor flows from stage to stage up the column, while liquid flows from stage to stage down the column. The vapor from the top tray is condensed to liquid in the overhead condenser and a portion of that liquid is returned as reflux. The rest of that vapor is withdrawn as the overhead product stream; this overhead product stream contains a concentrated amount of the light component. A portion of the liquid at the bottom of the column is withdrawn as a bottoms product (containing a con-

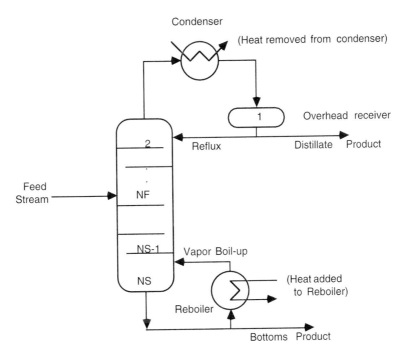

FIGURE M10.3   Schematic diagram for a distillation column.

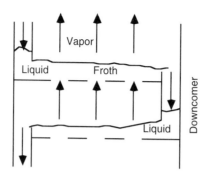

FIGURE M10.4   Schematic diagram for a distillation column tray.

centrated amount of the heavy component), while the rest is vaporized in the reboiler and returned to the column.

The liquid from one tray goes over a weir and cascades down to the next tray through a downcomer. As the liquid moves across a tray, it comes in contact with the vapor from the tray below. The schematic diagram for a sieve tray is shown in Figure M10.4.

Generally, as the vapor from the tray below comes in contact with the liquid, turbulent mixing is promoted. Assuming that the mixing is perfect, allows one to model the stage as a lumped parameter system, as shown in Figure M10.5. Notice that the vapor from stage i is modeled as a single stream with molar flowrate V_i and light component vapor composition (mole fraction) y_i. The liquid leaving stage i through the downcomer is modeled as a single stream with molar flowrate L_i and light component liquid composition (mole fraction) x_i.

The conceptual diagram for the feed stage is shown in Figure M10.6. It differs from Figure M10.5 in that an additional input to the stage is from the feed to the column.

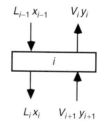

FIGURE M10.5   Conceptual material balance diagram for a typical stage.

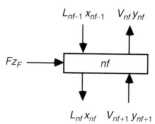

FIGURE M10.6   Conceptual material balance diagram for the feed stage.

M10.3 DYNAMIC MATERIAL BALANCES

M10.3.1 All Stages Except Feed, Condenser, and Reboiler

The component balance for the liquid phase of a typical stage as shown in Figure M10.5 is:

$$\underset{\substack{\text{accumulation}}}{\frac{dM_i x_i}{dt}} = \underset{\substack{\text{liquid from}\\\text{tray above}}}{L_{i-1} x_{i-1}} + \underset{\substack{\text{vapor from}\\\text{tray below}}}{V_{i+1} y_{i+1}} - \underset{\substack{\text{liquid}\\\text{leaving}}}{L_i x_i} - \underset{\substack{\text{vapor}\\\text{leaving}}}{V_i y_i} \qquad \text{(M10.2)}$$

where M_i is the liquid molar holdup on stage i.

For this simple binary distillation model, we will make the common assumption of equimolal overflow (King, 1980). For any stage except the feed stage, we assume that the vapor flowrate from one stage is equal to the vapor molar flowrate of the stage below:

$$V_i = V_{i+1} \qquad \text{(M10.3)}$$

and that the liquid leaving the stage is equal to the liquid flowing from one stage above:

$$L_i = L_{i-1} \qquad \text{(M10.4)}$$

M10.3.2 Feed Stage

Let q_F represent the *quality* of the feedstream. If the feed is a saturated liquid, then $q_F = 1$, while $q_F = 0$ for a saturated vapor. The vapor molar flowrate leaving the feed stage is (where NF = number of the feed stage)

$$V_{NF} = V_{NF+1} + F(1 - q_F) \qquad \text{(M10.5)}$$

Similarly, the liquid molar flowrate of the stream leaving the feed stage is:

$$L_{NF} = L_{NF-1} + F q_F \qquad \text{(M10.6)}$$

M10.3.3 Condenser

A total condenser removes energy from the overhead vapor, resulting in a saturated liquid. Assuming a constant molar holdup in the distillate receiver, the total liquid flowrate from the distillate receiver (reflux + distillate flows) is equal to the flowrate of the vapor from the top tray:

$$L_D + D = V_2 \qquad \text{(M10.7)}$$

where L_D and D represent the reflux and distillate molar flowrates, respectively.

M10.3.4   Reboiler

A total material balance around the reboiler yields:

$$B = L_{NS-1} - V_{\text{reboiler}} \qquad (M10.8)$$

where V_{reboiler} is the reboiler molar flowrate and B is the bottoms product molar flowrate.

M10.3.5   Summary of the Modeling Equations

The rectifying section (top section of column, above the feed stage) liquid molar flowrates are:

$$L_R = L_D \qquad (M10.9)$$

The stripping section (bottom section of column, below the feed stage) liquid molar flowrates are

$$L_S = L_R + Fq_F \qquad (M10.10)$$

The stripping section vapor molar flowrates are:

$$V_S = V_{\text{reboiler}} \qquad (M10.11)$$

The rectifying section vapor molar flowrates are:

$$V_R = V_S + F(1 - q_F) \qquad (M10.12)$$

In the following we assume a constant liquid phase molar holdup $(dM_i/dt) = 0$.
    The overhead receiver component balance is:

$$\frac{dx_1}{dt} = \frac{1}{M_D}[V_R(y_2 - x_1)] \qquad (M10.13)$$

The rectifying section component balance is (from $i = 2$ to $NF-1$):

$$\frac{dx_i}{dt} = \frac{1}{M_T}[L_R x_{i-1} + V_R y_{i+1} - L_R x_i - V_R y_i] \qquad (M10.14)$$

The feed stage balance is:

$$\frac{dx_{NF}}{dt} = \frac{1}{M_T}[L_R x_{NF-1} + V_S y_{NF+1} + Fz_F - L_S x_{NF} - V_R y_{NF}] \qquad (M10.15)$$

The rectifying section component balance is (from $i = NF+1$ to $NS-1$):

$$\frac{dx_i}{dt} = \frac{1}{M_T}[L_S x_{i-1} + V_S y_{i+1} - L_S x_i - V_S y_i] \qquad (M10.16)$$

And the reboiler component balance is:

$$\frac{dx_{NS}}{dt} = \frac{1}{M_B}[L_S x_{NS-1} - Bx_{NS} - V_S y_{NS}] \qquad (M10.17)$$

EQUILIBRIUM RELATIONSHIP

It is assumed that the vapor leaving a stage is in equilibrium with the liquid on the stage. The relationship between the liquid and vapor phase concentrations on a particular stage can be calculated using the constant relative volatility expression:

$$y_i = \frac{\alpha x_i}{1 + (\alpha - 1)x_i} \tag{M10.18}$$

M10.4 SOLVING THE STEADY-STATE EQUATIONS

To obtain the steady-state concentrations we must solve the system of equations, $\mathbf{f}(x) = 0$.
From the overhead receiver component balance:

$$f_1 = y_2 - x_1 = 0 \tag{M10.19}$$

From the rectifying section component balance ($i = 2$ to $NF-1$):

$$f_i = L_R x_{i-1} + V_R y_{i+1} - L_R x_i - V_R y_i = 0 \tag{M10.20}$$

From the feed stage balance:

$$f_{NF} = L_R x_{NF-1} + V_S y_{NF+1} + F z_F - L_S x_{NF} - V_R y_{NF} = 0 \tag{M10.21}$$

From the stripping section component balance ($i = NF+1$ to $NS-1$):

$$f_i = L_S x_{i-1} + V_S y_{i+1} - L_S x_i - V_S y_i = 0 \tag{M10.22}$$

And from the reboiler component balance:

$$f_{NS} = L_S x_{NS-1} - B x_{NS} - V_S y_{NS} = 0 \tag{M10.23}$$

where $B = L_S - V_S$.

We must realize that (M10.19)–(M10.23) constitute a set of nonlinear algebraic equations, since the relative volatility relationship (M10.18) is nonlinear in the state variable. Equations (M10.19)–(M10.23) are NS equations in NS unknowns. A Newton-based technique will be used to solve the equations.

EXAMPLE M10.1   Steady-State Operation of a 41-Stage Column

Consider a 41-stage column with the overhead condenser as stage 1, the feed tray as stage 21 and the reboiler as stage 41. The following parameters and inputs apply

α = 1.5
F = 1 mol/min
z_F = 0.5 mole fraction of light component
R = 2.706 mol/min
D = 0.5 mol/min
q_F = 1 (sat'd liquid feed)

From an overall material balance, the bottoms product flowrate is:

$$B = F - D = 1 - 0.5 \, \text{mol/min}$$

the stripping section flowrate is:

$$L_S = R + Fq_F = 2.706 + 1 = 3.706 \, \text{mol/min}$$

and a balance around the reboiler yields:

$$V_S = L_S - B = 3.706 - 0.5 = 3.206 \, \text{mol/min}$$

The m-file `dist_ss.m` (shown in the Appendix) is used to solve for the steady-state compositions.

```
x  =  fsolve('dist_ss',x0)
```

The resulting compositions are shown in Figure M10.7. Notice the strong sensitivity to reflux
flowrate.

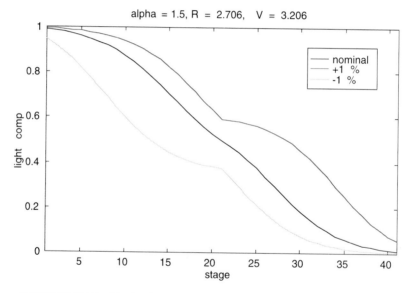

FIGURE M10.7   Liquid phase composition (mol fraction of light component)
as a function of stage number. Solid = nominal reflux, dashed = +1% reflux,
dotted = −1% reflux.

The overhead composition (stage 1) is 0.99 and the bottoms composition (stage 41) is 0.01 for
the nominal reflux rate (2.706 mol/min).

Steady-State Input-Output Relationships

The sensitvity to reflux rate is also shown by the plot in Figure M10.8. The steady-state gain
(change in output/change in input) for distillate composition is large when reflux is less than 2.7,

but small when the reflux is greater than 2.71 mol/min. The opposite relationship holds for bottoms composition, where the gain is small when reflux is less than 2.7 mol/min, but large for reflux greater than 2.71 mol/min. This sensitivity has important ramifications for control system design.

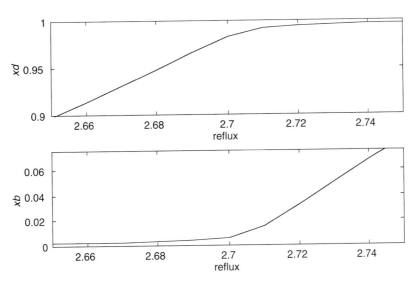

FIGURE M10.8   Steady-state input (reflux)—output (distillate or bottoms composition) relationship.

M10.5 SOLVING THE NONLINEAR DYNAMIC EQUATIONS

Equations (M10.13)–(M10.17) are a set of initial value ordinary equations, which can be solved using numerical integration techniques. The next example uses the variable step size MATLAB routine `ode45` to perform the integration.

EXAMPLE M10.2   Dynamic Response

Consider now the previous problem, with the initial conditions of the stage compositions equal to the steady-state solution of Example M10.1. The additional parameters needed for the dynamic simulation are the molar holdups on each stage. Here we use the following parameters:

$$M_1 = M_D \quad = \quad \text{overhead receiver molar holdup} \quad = 5 \text{ mol}$$
$$M_2 = \quad \quad \quad \text{feed tray molar holdup} \quad = 0.5 \text{ mol}$$
$$M_3 = 5 \quad = \quad \text{bottoms (reboiler) molar holdup} \quad = 5 \text{ mol}$$

To illustrate the nonlinear behavior we compare the results of ± 1% step changes in the reflux rate at time t = 5 minutes.

```
[t,x]=ode45('dist_dyn',0,400,x0)
```

Note that the current version of `ode45` does not allow model parameters to be passed through the argument list, so global parameters are defined in the m-file `dist_dyn.m` shown in the Appendix.

The following results are shown in Figure M10.9. A positive 1% step change in the reflux rate yields a small increase in the distillate composition; this makes since because the maximum possible increase in distillate purity is 0.01 (the composition cannot be greater than 1 mole fraction) while it can decrease much more than that. A negative 1% step change in reflux causes a larger change in the distillate purity. The opposite effects are observed for bottoms composition, where a positive reflux change yields a large bottoms composition change. A negative reflux change yields a small bottoms composition change.

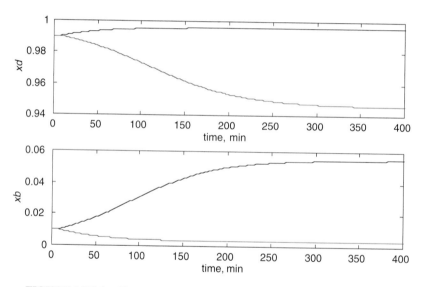

FIGURE M10.9   Illustration in nonlinear response of distillate and bottoms compositions to step changes in reflux. Solid line = +1%, Dashed line = −1%.

M10.6   STATE-SPACE LINEAR DISTILLATION MODELS

Linear state space models are useful for stability analysis and control system design. Here we develop models of the form:

$$\dot{\mathbf{x}} = \mathbf{A}\,\mathbf{x}' + \mathbf{B}\,\mathbf{u}' \tag{M10.24}$$

$$\mathbf{y}' = \mathbf{C}\,\mathbf{x}' \tag{M10.25}$$

where $'$ is used to represent the deviation variables, $\mathbf{x}' = \mathbf{x} - \mathbf{x}_s$, $\mathbf{u}' = \mathbf{u} - \mathbf{u}_s$ (the subscript s indicates the steady-state values). Defining

$$K_i = \frac{\partial y_i}{\partial x_i} = \frac{\alpha}{(1 + (\alpha - 1)x_i)^2} \tag{M10.26}$$

and linearizing the dynamic equations (M10.13)–(M10.17)

$$A_{1,1} = \frac{\partial f_1}{\partial x_1} = -\frac{V_R}{M_D} \tag{M10.27}$$

$$A_{1,2} = \frac{\partial f_1}{\partial x_2} = \frac{V_R K_2}{M_D} \tag{M10.28}$$

For $i = 2$ to $NF-1$:

$$A_{i,i-1} = \frac{\partial f_i}{\partial x_{i-1}} = \frac{L_R}{M_T} \tag{M10.29}$$

$$A_{i,i} = \frac{\partial f_i}{\partial x_i} = -\left(\frac{L_R + V_R K_i}{M_T}\right) \tag{M10.30}$$

$$A_{i,i+1} = \frac{\partial f_i}{\partial x_{i+1}} = \frac{V_R K_{i+1}}{M_T} \tag{M10.31}$$

For the feed stage:

$$A_{NF,NF-1} = \frac{\partial f_{NF}}{\partial x_{NF-1}} = \frac{L_R}{M_T} \tag{M10.32}$$

$$A_{NF,NF} = \frac{\partial f_{NF}}{\partial x_{NF}} = -\left(\frac{L_S + V_R K_i}{M_T}\right) \tag{M10.33}$$

$$A_{NF,NF+1} = \frac{\partial f_{NF}}{\partial x_{NF+1}} = \frac{V_S K_{NF+1}}{M_T} \tag{M10.34}$$

For $i = NF+1$ to $NS-1$:

$$A_{i,i-1} = \frac{\partial f_i}{\partial x_{i-1}} = \frac{L_S}{M_T} \tag{M10.35}$$

$$A_{i,i} = \frac{\partial f_i}{\partial x_i} = -\left(\frac{L_S + V_S K_i}{M_T}\right) \tag{M10.36}$$

$$A_{i,i+1} = \frac{\partial f_i}{\partial x_{i+1}} = \frac{V_S K_{i+1}}{M_T} \tag{M10.37}$$

and for the reboiler (stage NS)

$$A_{NS,NS-1} = \frac{\partial f_i}{\partial x_{i-1}} = \frac{L_S}{M_B} \tag{M10.38}$$

$$A_{NS,NS} = \frac{\partial f_i}{\partial x_i} = -\left(\frac{B + V_S K_{NS}}{M_B}\right) \tag{M10.39}$$

Now, for the derivatives with respect to the inputs; $u_1 = L_R = L_1$ and $u_2 = V_S = V_{\text{reboiler}}$:

$$B_{1,1} = \frac{\partial f_1}{\partial u_1} = 0 \qquad\qquad B_{1,2} = \frac{\partial f_1}{\partial u_2} = 0 \tag{M10.40}$$

For $i = 1$ to $NS-1$:

$$B_{i,1} = \frac{\partial f_1}{\partial u_1} = \frac{x_{i-1} - x_i}{M_T} \qquad\qquad B_{1,2} = \frac{\partial f_1}{\partial u_2} = \frac{y_{i+1} - y_i}{M_T} \tag{M10.41}$$

and for the bottom stage:

$$B_{NS,1} = \frac{\partial f_1}{\partial u_1} = \frac{x_{i-1} - x_i}{M_{NS}} \qquad\qquad B_{NS,2} = \frac{\partial f_1}{\partial u_2} = \frac{x_{NS} - y_{NS}}{M_{NS}} \tag{M10.42}$$

If the output variables are the overhead and bottoms compositions, then:

$$C_{1,1} = 1, \text{ while } C_{1,i} = 0 \text{ for } i \neq 1 \tag{M10.43}$$

$$C_{2,NS} = 1, \text{ while } C_{1,i} = 0 \text{ for } i \neq 1 \tag{M10.44}$$

M10.6.1   Transforming the State Space Linear Models to Transfer Function Form

The matrix transfer function is:

$$G(s) = C(sI - A)^{-1}B \tag{M10.45}$$

It is easy to generate MATLAB m-files to calculate each of the state-space matrices (A,B,C) for a particular set of parameters (and steady-state compositions). For a column of reasonable size (say the 41-stage example) the denominator polynomial in (M10.45) would be quite large (say 41st order). What is often more useful is to be able to directly calculate the steady-state gain matrix, as shown below.

The steady-state gain matrix is:

$$G = -CA^{-1}B \tag{M10.46}$$

where, again, it is easy to generate a MATLAB m-file to perform this calculation.

M10.7   MULTIPLICITY BEHAVIOR

Even simple ideal binary distillation columns have been shown recently to have interesting steady-state and dynamic behavior, including multiple steady-states. Nice examples are shown by Jacobsen and Skogestad (1991, 1994). The key assumption that must be made for this behavior to occur is that mass flows, rather than molar flows, are manipulated. The reader is encouraged to read these papers and modify the MATLAB m-files presented in this chapter to illustrate the behavior shown by Jacobsen and Skogestad.

SUMMARY

In this chapter we have developed modeling equations to describe the steady-state and dynamic behavior of ideal, binary distillation columns. The 41-stage column example shows that steady-state distillate and bottom compositions are a nonlinear function of the manipulated inputs (distillate and vapor boil-up flows). Also, the dynamic responses of these compositions depends on the magnitude and direction of changes in the manipulated inputs.

REFERENCES AND FURTHER READING

The following undergraduate chemical engineering texts develop the steady-state modeling equations for ideal binary distillation:

King, C.J. (1980). *Separations Processes*. 2nd ed. New York: McGraw-Hill.

McCabe, W.L., & J.C. Smith. (1976). *Unit Operations of Chemical Engineering*, 3rd ed. New York: McGraw-Hill.

The dynamic modeling equations for distillation are presented by:

Luyben, W.L. (1990). *Process Modeling, Simulation and Control for Chemical Engineers*, 2nd ed. New York: McGraw-Hill.

More advanced treatments of steady-state and dynamic distillation models are presented by:

Holland, C.D. (1981). *Fundamentals of Multicomponent Distillation*. New York: McGraw-Hill.

Holland, C.D., & A.I. Liapis. (1983). *Computer Methods for Solving Dynamic Separation Problems*. New York: McGraw-Hill.

The parameters for Example M10.1 are presented in the following two references:

Skogestad, S., & M. Morari. (1988). Understanding the dynamic behavior of distillation columns. *Ind. Eng. Chem. Res.*, 27(10): 1848–1862.

Morari, M., & E. Zafiriou. (1988). *Robust Process Control*. Englewood Cliffs, NJ: Prentice-Hall.

The possibility of multiple steady-state behavior in ideal binary distillation is presented by:

Jacobsen, E.W., & S. Skogestad. (1991). Multiple steady-states in ideal two-product distillation. *AIChE J.*, 37(4): 499–511.

Jacobsen, E.W., & S. Skogestad. (1994). Instability of distillation columns. *AIChE J.*, 40(9): 1466–1478.

STUDENT EXERCISES

1. Consider a simple 1 tray (3 stage) column with the overhead condenser as stage 1, the feed tray as stage 2 and the reboiler as stage 3. Use the following parameters and inputs:

$$\alpha = 5$$
$$R = 3 \text{ mol/min}$$
$$q_F = 1$$
$$F = 1 \text{ mol/min}$$
$$D = 0.5 \text{ mol/min}$$
$$z_F = 0.5 \text{ mole fraction of light component}$$

Find the bottoms product flowrate, the stripping section flowrate and the vapor boil-up rate (stripping section vapor flowrate). Use `fsolve` and `dist_ss.m` to find the resulting compositions:

$$\mathbf{x} = \begin{bmatrix} 0.703 \\ 0.486 \\ 0.297 \end{bmatrix} = \begin{bmatrix} \text{distillate composition} \\ \text{composition of stage 2 (the feed tray)} \\ \text{bottoms product composition} \end{bmatrix}$$

Consider now the dynamic behavior, with the initial conditions of the stage compositions equal to the steady-state solution. The additional parameters needed for the dynamic simulation are the molar holdups on each stage. Use:

$$M_1 = M_D = \text{overhead receiver molar holdup} = 5 \text{ mol}$$
$$M_2 = \text{feed tray molar holdup} = 0.5 \text{ mol.}$$
$$M_3 = 5 = \text{bottoms (reboiler) molar holdup} = 5 \text{ mol}$$

At time zero, the reflux is changed from 3.0 mol/min to 3.2 mol/min. Use `ode45` and `dist_dyn.m` to simulate the dynamic behavior shown in the figures below.

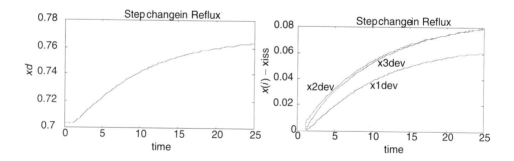

The Reflux is step changed from 3.0 to 3.2 at $t = 1$ minute.

APPENDIX

dist_ss.m

```
function f = dist_ss(x);
%
% solve for the steady-state stage compositons in an ideal
% binary distillation column using fsolve.
%
% (c) 1993 B. Wayne Bequette - 21 june 93
% revised 31 Dec 96
%
% All flowrates are molar quantities.  Stages are numbered
% from the top down.  A total condenser is assumed.
% The overhead receiver is stage 1.  The partial reboiler
% is stage ns (the number of equilibrium "trays" is then
% ns-1). The column parameters should be specified in the
% DIST_PAR array.
%
% to use this function, enter the following in the command
% window, or from a script file (after defining parameters
% in the DIST_PAR array:
%
% x = fsolve('dist_ss',x0)
%
% where x0 is a vector of initial guesses for the liquid
% phase stage compositions (length(x0) = ns)
%
  global DIST_PAR
%
% DIST_PAR is a vector of distillation column parameters
%          used by both dist_ss.m and dist_dyn.m
%
  if length(DIST_PAR) < 8;
    disp('not enough parameters given in DIST_PAR')
      disp(' ')
      disp('check to see that global DIST_PAR has been defined')
      return
  end
%
  alpha   = DIST_PAR(1);   % relative volatility (2.5)
  ns      = DIST_PAR(2);   % total number of stages (3)
  nf      = DIST_PAR(3);   % feed stage (2)
  feed    = DIST_PAR(4);   % feed flowrate (1)
  zfeed   = DIST_PAR(5);   % feed composition, light comp (0.5)
  qf      = DIST_PAR(6);   % feed quality (1 = sat'd liqd,
%                                    0 = sat'd vapor) (1)
  reflux  = DIST_PAR(7);   % reflux flowrate (3)
  vapor   = DIST_PAR(8);   % reboiler vapor flowrate (3.5)
%
% DIST_PAR(9:19) used by dist_dyn.m (distillation dynamics)
%   dist     = distillate product flowrate
```

```
%   f(i)       = ith comp mat bal equation
%   lbot       = bottoms product flowrate
%   lr         = liquid flow in rectifying section (top)
%   ls         = liquid flow in stripping section (bottom)
%   vr         = vapor flow - rectifying sec (= vapor + feed*(1-qf)
%   vs         = vapor flow - stripping section (= vapor)
%   x(i)       = mole frac light component on stage i, liq
%   y(i)       = mole frac light component on stage i, vap
%
% rectifying and stripping section liquid flowrates
%
      lr    =   reflux;
      ls    =   reflux + feed*qf;
%
% rectifying and stripping section vapor flowrates
%
      vs    =   vapor;
      vr    =   vs +  feed*(1-qf);
%
% distillate and bottoms rates
%
      dist  = vr - reflux;
      lbot  = ls - vs;
%
      if dist < 0
         disp('error in specifications, distillate flow < 0')
         return
        end
        if lbot < 0
         disp('error in specifications, stripping section ')
         disp(' ')
         disp('liquid flowrate is negative')
         return
        end
%
% zero the function vector
%
      f = zeros(ns,1);
%
% calculate the equilibrium vapor compositions
%
      for i=1:ns;
        y(i)=(alpha*x(i))/(1.+(alpha-1.)*x(i));
      end
%
% material balances
%
% overhead receiver
%
      f(1)=(vr*y(2)-(dist+reflux)*x(1));
%
% rectifying (top) section
```

```
%
      for i=2:nf-1;
        f(i)=lr*x(i-1)+vr*y(i+1)-lr*x(i)-vr*y(i);
      end
%
% feed stage
%
      f(nf)   =   lr*x(nf-1)+vs*y(nf+1)-ls*x(nf)-
                  vr*y(nf)+feed*zfeed;
%
% stripping (bottom) section
%
      for i=nf+1:ns-1;
        f(i)=ls*x(i-1)+vs*y(i+1)-ls*x(i)-vs*y(i);
      end
%
% reboiler
%
      f(ns)=(ls*x(ns-1)-lbot*x(ns)-vs*y(ns));
```

dist_dyn.m

```
  function xdot = dist_dyn(t,x);
%
% solve for the transient stage compositions in an ideal
% binary distillation column using ode45.
%
% (c) 1997 B. Wayne Bequette - 24 Jan 1997
% revised 31 Dec 96
%
% All flowrates are molar quantities.  Stages are numbered
% from the top down.  A total condenser is assumed.
% The overhead receiver is stage 1.  The partial reboiler
% is stage ns (the number of equilibrium "trays" is then
% ns-1). The column parameters should be specified in the
% DIST_PAR array.
%
% to use this function, enter the following in the command
% window, or from a script file (after defining parameters
% in the DIST_PAR array:
%
% [t,x] = ode45('dist_dyn',t0,tf,x0)
%
% where x0 is a vector of initial values for the liquid
% phase stage compositions (length(x0) = ns)
%
  global DIST_PAR
%
% DIST_PAR is a vector of distillation column parameters
%          used by both dist_ss.m and dist_dyn.m
```

```
%
   if length(DIST_PAR) < 11;
     disp('not enough parameters given in DIST_PAR')
       disp(' ')
       disp('check to see that global DIST_PAR has been defined')
       return
   end
%
   alpha    = DIST_PAR(1);  % relative volatility (1.5)
   ns       = DIST_PAR(2);  % total number of stages (41)
   nf       = DIST_PAR(3);  % feed stage (21)
   feedi    = DIST_PAR(4);  % initial feed flowrate (1)
   zfeedi   = DIST_PAR(5);  % initial feed composition, light comp
                                  (0.5)
   qf       = DIST_PAR(6);  % feed quality (1 = sat'd liqd,
%                                 0 = sat'd vapor) (1)
   refluxi  = DIST_PAR(7);  % initial reflux flowrate (2.706)
   vapori   = DIST_PAR(8);  % initial reboiler vapor flowrate
                                  (3.206)
   md       = DIST_PAR(9);  % distillate molar hold-up (5)
   mb       = DIST_PAR(10); % bottoms molar hold-up (5)
   mt       = DIST_PAR(11); % stage molar hold-up (0.5)
%
   if length(DIST_PAR) == 19;
    stepr   = DIST_PAR(12); % magnitude step in reflux (0)
    tstepr  = DIST_PAR(13); % time of reflux step change (0)
    stepv   = DIST_PAR(14); % magnitude step in vapor (0)
    tstepv  = DIST_PAR(15); % time of vapor step change (0)
    stepzf  = DIST_PAR(16); % magnitude of feed comp change (0)
    tstepzf = DIST_PAR(17); % time of feed comp change (0)
    stepf   = DIST_PAR(18); % magnitude of feed flow change (0)
    tstepf  = DIST_PAR(19); % time of feed flow change (0)
   else
    stepr = 0; tstepr = 0; stepv = 0; tstepv = 0;
    stepzf = 0; tstepzf = 0; stepf = 0; tstepf = 0;
   end
%
% DIST_PAR(9:19) used by dist_dyn.m (distillation dynamics)
%  dist      = distillate product flowrate
%  lbot      = bottoms product flowrate
%  lr        = liquid flow in rectifying section (top)
%  ls        = liquid flow in stripping section (bottom)
%  vr        = vapor flow - rectifying sec (= vapor + feed*(1-qf)
%  vs        = vapor flow - stripping section (= vapor)
%  x(i)      = mole frac light component on stage i, liq
%  xdot(i)   = light component ith stage mat bal equation
%  y(i)      = mole frac light component on stage i, vap
%
%
%  check disturbances in reflux, vapor boil-up, feed composition
%        and feed flowrate
%
   if t < tstepr;
```

```
    reflux = refluxi;
  else
    reflux = refluxi + stepr;
  end
%
  if t < tstepv;
    vapor = vapori;
  else
    vapor = vapori + stepv;
  end
%
  if t < tstepzf;
    zfeed = zfeedi;
  else
    zfeed = zfeedi + stepzf;
  end
%
  if t < tstepf;
    feed = feedi;
  else
    feed = feedi + stepf;
  end
%
% rectifying and stripping section liquid flowrates
      lr   =   reflux;
      ls   =   reflux + feed*qf;
%
% rectifying and stripping section vapor flowrates
%
      vs   =   vapor;
      vr   =   vs +  feed*(1-qf);
%
% distillate and bottoms rates
%
      dist  = vr - reflux;
      lbot  = ls - vs;
%
      if dist < 0
         disp('error in specifications, distillate flow < 0')
         return
       end
       if lbot < 0
         disp('error in specifications, stripping section ')
         disp(' ')
         disp('liquid flowrate is negative')
         return
       end
%
% zero the function vector
%
      xdot = zeros(ns,1);
%
```

```
% calculate the equilibrium vapor compositions
%
        for i=1:ns;
          y(i)=(alpha*x(i))/(1.+(alpha-1.)*x(i));
        end
%
% material balances
%
% overhead receiver
%
        xdot(1)=(1/md)*(vr*y(2)-(dist+reflux)*x(1));
%
% rectifying (top) section
%
        for i=2:nf-1;
          xdot(i)=(1/mt)*(lr*x(i-1)+vr*y(i+1)-lr*x(i)-vr*y(i));
        end
%
% feed stage
%
        xdot(nf)   = (1/mt)*(lr*x(nf-1)+vs*y(nf+1)-ls*x(nf)-
          r*y(nf)+feed*zfeed);
%
% stripping (bottom) section
%
        for i=nf+1:ns-1;
          xdot(i)=(1/mt)*(ls*x(i-1)+vs*y(i+1)-ls*x(i)-vs*y(i));
        end
%
% reboiler
%
        xdot(ns)=(1/mb)*(ls*x(ns-1)-lbot*x(ns)-vs*y(ns));
```

INDEX